Edible & Medicinal Wild Plants

of Minnesota & Wisconsin

Matthew Alfs

OTBH

2001

DISCLAIMER -- PLEASE READ CAREFULLY

The identification, selection, and processing of any weed or wild plant for ingestion or other personal use requires reasonable care and attention to details. Certain parts of some plants are wholly unsuitable for use and, in some instances, are even toxic. In this book, every effort has been made to describe each plant with utmost fidelity; nevertheless, some variations in their actual appearance may be encountered in the field as a result of seasonal and/or geographic factors. Because attempts to use any weed or wild plant for food or other personal use depends upon various factors controllable only to the individual, neither the author, publisher, printer, or distributors of this book assume any responsibility whatsoever for adverse health effects of such failures as might be encountered in the individual case and should not be held liable to any person or entity with respect to any loss, damage, or injury caused, or alleged to be caused, directly or indirectly, by the information contained in this book.

Be fully aware that while this book examines how individual plants have been used by native Americans, herbalists, and other healers as health aids, such information is not meant, and should not be construed by the individual, as incentive for substituting such information for professional medical care. Both the author and the publisher urge readers to consult their professional health-care provider with respect to any personal illness or injury.

EDIBLE & MEDICINAL WILD PLANTS OF MINNESOTA & WISCONSIN, by Matthew Alfs

Copyright © 2001 by Matthew Alfs

All Rights Reserved.

Alfs, Matthew.
   Edible & medicinal wild plants of Minnesota & Wisconsin / Matthew Alfs.
   -- 1st ed.
   p. cm.
   Edible and medicinal wild plants of Minnesota and Wisconsin
   Includes bibliographical references and index.
   LCCN: 2001130556
   ISBN: 0-9612964-3-7

   1. Wild plants, Edible--Minnesota. 2. Wild plants, Edible--Wisconsin. 3. Medicinal plants--Minnesota. 4. Medicinal plants--Wisconsin.   I. Title. II. Title: Edible and medicinal wild plants of Minnesota and Wisconsin

QK98.5.U6A54 2001   581.6'3'0977
                  QBI01-200748

FIRST EDITION

Published by

OLD THEOLOGY BOOK HOUSE
P O BOX 120342
NEW BRIGHTON MN 55112 USA
E-mail: Viskichi@aol.com

MADE IN THE U.S.A.

ISBN: **0-9612964-3-7**

Library of Congress Catalog Card Number: **2001 130556**

All photos by the author.
Photo on the cover:
Chicory (*Cichorium intybus*)

Acknowledgments

The author wishes to express his thanks and appreciation to the following persons for their assistance and/or encouragement in the production of this book: JAMES ALFS and CAROLYNE BUTLER. Pre-publication reviews provided by FRANCOIS COUPLAN, EDELENE WOOD, and JIM DUKE are also hereby gratefully acknowledged.

CONTENTS

*

100 PLANT MONOGRAPHS & WERE TO FIND THEM

Preface

*

The region encompassing our great states of Minnesota and Wisconsin has a varied landscape—coniferous forests in the north, prairie to the west and south, and deciduous woodlands and fields in the central portion. In view of this variety, one can surmise that a tremendous assortment of wild plants must flourish throughout this bio-region. In fact, this is the case: Even looking at just the state of Minnesota, over 2000 species of vascular plants have recently been verified as carpeting its wild lands.*(Ownbey & Morley 1991:vii)* Astoundingly, that is about 1/10 of all the known vascular species covering the entirety of North America (e.g., 21,641)!*(Moerman 1996:1-22)*

A large number of our area's plants were utilized as food and medicine by our native tribes. For Wisconsin, these have primarily consisted of the Menominee (also spelled "Menomini"), Hoshungras[Winnebago], Potawatomi, Oneida, Mesquaki[Fox] (largely associated with Iowa today), and Ojibwe[Chippewa] peoples. Minnesota has been home to the latter tribe (whose name in the Ojibwe language is actually Anishinabe, a term that some modern tribesmen prefer to Ojibwe), as well as to the Mesquaki, plus the Santee Dakota[Sioux]. In the present day, in addition to traditional-lists among the remnants of these tribes on state reservations, over two dozen wild-foods teachers and/or practitioners of Western herbology—mostly centered in Minneapolis-Saint Paul, Duluth-Superior, Milwaukee, Madison, and Green Bay—enthusiastically perpetuate the hallowed tradition of a symbiotic interaction with the native flora in a culinary and/or healing context.

In view of our area's rich flora and the protracted and colorful history of its utilization by the inhabitants of our two states, it seems almost reprehensible that no significant study has yet been published on the edible and medicinal uses of either Minnesota's or Wisconsin's wild plants. Owing to the sorry lack as just described, the present study is offered as a serious attempt to remedy this deficiency. In doing so, it culls information from a variety of sources: phytochemical and therapeutic studies from the scholarly literature, historical uses by native American tribes, the concerted uses of the plants so carefully outlined by America's Eclectic physicians of the 1800s and early 1900s, clinical uses by herbalists past and present, and finally my own foraging and utilization of our area's plants.

I should state here that the present work is an initial volume in a projected series designed to cover the full spectrum of utilizable plants in the Minnesota-Wisconsin region. This initial volume discusses 100 plants, all of which are herbaceous. Succeeding volumes are projected to cover additional herbaceous plants as well as the *vines, shrubs,* and *trees* in our region.

The intent of the present volume is to chart a course that not only attempts to shed new light on the commoner plants, but also to focus on those plants which, while rich in medicinal and nutritional content, have thus far been largely neglected. Thus, it has attempted to "open new doors" for those engaged in wild-plant studies by providing fresh and invigorating information. I leave it to the reader to determine whether I have reached such a goal.

Matthew Alfs, M.H.
St. Paul, MN
Spring 2001

General Introduction
The Art of Wild-Plant Foraging

*

So many changes we have seen in the last century! My grandfather used to marvel when considering that when he was born, people were still riding around in horse-drawn carriages and yet when he had reached his senior years, a rocket ship had landed a man on the moon!

Yet, in serious contemplation, how many of the changes that we have witnessed have *truly* been for our benefit, as well as for that of the other creatures on this planet—not to mention the planet itself? Certainly we "baby boomers" have, for most of our lives, dwelled in relative comfort—just a hop-skip-and-jump away from the local supermarket, pharmacy, and department store where our food, medicine, and clothing have all been easily obtained. Yet, might we still be missing something on a deeper level that previous generations have richly experienced? Here I remember that it was quite a revelation for me, as a lad, to learn that people of only a couple of generations ago—especially the American Indians and early settlers/pioneers—had subsisted largely off of nature's bounty for their food, medicine, and clothing, which they did through a symbiosis with the land that preserved both it and them. Even with this realization, I wondered if we had lost something—something, at least, of the spirit of adventure.

"Progress" powerfully changed the previous, self-sustaining way of life, I came to learn. Big cities, industry, and technology divorced us from the land as our provider so that we thereby lost the feeling of kinship with it that we once had possessed. By the time of my childhood years (late 1950s/early 1960s), the wilderness had become a terrifying monstrosity to most, with the general motto being: "Tame that godforsaken 'wasteland'! Civilize it! Pave it over!"

Nevertheless, with the "back-to-the-land" movement initiated by Rachel Carson, Euell Gibbons, Bradford Angier, and others in the 1960s and 1970s, a sizable minority of the population began to adjust its attitude back to that possessed by its forebears. The wilderness, these people discovered, was not an enemy, but the most intimate of friends. I think now that it was inevitable that, as I grew to become a teen, I would stumble across the books of the aforementioned naturalists. And here I quickly found myself in harmony with the thoughts that they so clearly and logically presented. It was then that I began to realize something else: that there was more than just *adventure* to this way of life, but that it vibrated with a sublime *spirituality* as well.

1

Here, though, my own sense of integrity compelled me to practice what I was now preaching. Not only that, but Gibbons' books, especially, imbued in me a burning desire to live an existence whereby nature's bounty would provide at least a portion of my food and medicine. So, again almost inevitably, I, too, became a naturalist and eventually an herbalist as well, spending long hours in the bush identifying, harvesting, and later preparing wild foods and medicines from hundreds of different plants.

In the present work, dear reader, I gladly pass on this accumulated knowledge and experience to you, admonishing you and entrusting you to use it wisely and care-fully, as well as closely observing guidelines of ecology and safety. Here my solemn wish is that you, too, can experience the oneness with nature that has so delighted my own heart and soul over these past three decades!

But the path to a forager's nirvana requires some training in the art. And make no mistake about it, foraging for wild plants *is* an acquired art. Not only must you learn where to look, but you need to know something about botany, conservation, harvesting, storage techniques, cooking, drying. . . and the list goes on and on! I hope to make clear some of the specifics below, but this way of life is really a matter of continuous education—where first-hand experience becomes the chief of many crucial factors. But let's see what we can do to get you prepared for your first forays into foraging. . . .

Where to Look

Before I suggest where you might forage for wild plants, I want to stress a very important point: *Do not ever collect plants from roadsides, no matter how tempting that may be.* Exhaust from autos, despite increasing antipollution measures, contains many harmful substances (including the toxic metal cadmium—also found in cigarette smoke—which collects in the body causing great harm), and it settles on nearby vegetation like a noxious fog. Unfortunately, to further compound the problem, the structure of many wild plants—such as their possession of fine hairs or exterior resins or other "inviting" characteristics—makes them natural magnets for such pollutants. (A study of the effects of pollution on roadside cattails, for instance, found them heavily contaminated with toxic heavy metals, including cadmium.-*Erickson&Lindzey 1983:550-55*) So, then, take the above-stated warning to heart. Your health is too precious to risk for a few scrumptious meals!

Having said the above, we are left with the question: Where *does* one look for edible/medicinal wild plants? Interestingly, the best foraging places are often the most overlooked. For example, the strip of land bordering each side of railroad tracks is often quite productive (as long as these areas haven't been sprayed to kill vegetation—many aren't, but some still are). Here you may stumble across plants not commonly found in the immediate vicinity; out-of-state plants may have become rooted in the area from wind-blown seeds that hitched a ride onto railroad cars and were later blown off.

A variety of wonderful wild plants can be spotted on river banks and on the shores of lakes, which means that an activity such as canoeing or fishing could prove to be doubly rewarding. (After all, fish and greens make a terrific combo!)

How about that vacant lot you've noticed a few blocks down the road? Such spots are frequently quite rewarding for the forager in that they usually contain "the plants that follow the white man," as our Indian friends used to describe the flora that seemed to sprout in the footsteps of the settlers. This highly nutritious and often easily recognized category of urban weeds includes such characters as: yellow wood sorrel, purslane, lamb's quarters, dandelion, chicory, and plantain (which can even commonly be found between sidewalk slabs in downtown areas!).

Meadows are prime foraging locations, as are the borders of swamps and marshes. Open woods yield some interesting plants not found elsewhere, but woods in general are not as prime a location as the novice may imagine. On the other hand, the perimeters of woods can sometimes prove to be excellent natural pantries.

Wasteland is often quite ripe for foraging, but you should be sure to get permission to pick from wasteland that may be privately owned. The same is true of the vacant lots mentioned above. For a small fee, you can usually check ownership of land in a plat book at the government office for the county in question.

Even where public land is concerned, there are regulations that you should consider. Numerous endangered species are protected by law; this is understandable and should be respected. Then, too, parks on all levels—national, state, regional, county, and city—usually prohibit the uprooting of plants (though some of them allow berry picking). Sometimes you can obtain a foraging permit from the park system authorities to take leaf samples, etc., in the interests of wildlife education and exploration. At any rate, should you be restricted in what you can collect, you can still use the parks for identifying plants and for studying them in their natural habitat. A good camera with a macro lens for close-up photography becomes an invaluable tool here.

It is a sad-but-true fact that so-called conservationists are pressing for firmer "look-but-don't-touch" regulations on other public lands as well. These individuals, whom I call "prima-donna conservationists" (like the dolled-up wife who tells her husband: "Look, but don't touch!") lack an education in, and an appreciation for, true wilderness ecology. As well-known herbalist Michael Tierra has so ably phrased it: "Pseudo-Ecologists. . .make futile attempts to maintain natural environments as aesthetic monuments with no functional purpose, leaving signs saying 'do not touch,' 'do not pick the plants,' etc. The herbalist, along with the American Indian, appreciates nature not only for its beauty but also for the valuable resource of wild foods and medicines that grow in these all-giving bowers. Thus the herbalist views nature as a positive force, and as a provider and teacher." *(Tierra 1990:xix-xx)*

What the prima-donna conservationists, though well-meaning, fail to realize is that man, after all, is *part* of nature, not a bystander off to the sidelines. Far from putting nature on a pedestal, then, humans must *live in relationship with it* in order to truly appreciate it and genuinely wish to preserve it. We have a classic example of this, as both Tierra and myself have noted above, in the American Indians who dwelled in our land for centuries before the white man arrived, *using* (not merely "observing") its many resources, but treating them with wisdom and respect so as to keep them *renewable*. Such a true ecology (see more on this below—page 8) was a natural consequence of their interaction with nature, for in continually experiencing the wilderness to be their provider on a day-to-day basis, there grew in these native Americans a deep reverence for it in direct derivation of such intimacy. Sadly, however, if the prima-donna conservationists have their way, there will be tougher and more far-reaching restrictions for foraging on public land than there are now, even including prohibitions against utilizing the commonest of weeds (the most important plants for food and medicine!) instead of simply designating protection for rare species of plants, as certainly should be done. Such overzealous edicts would effectively rob many people of a simpler, nomadic way of life that they so cherish and to which they are Constitutionally entitled in the free expression of religious belief (Nature as Provider) and in their pursuit of life, liberty, and happiness.

Even worse than the prima-donna conservationists, however, are the greedy land-developers who are draining swamps, bulldozing woods, and spraying everything in sight with herbicides. Euell Gibbons said of this situation: "[A] real menace is the huge array of herbicides sprayed on roadside, in fields, and over lawns. These are poisons and badly upset the balance of nature. They kill earthworms and wash into streams where they kill aquatic life. Another enemy is the engineering mentality that wants to bulldoze all the hills, fill all the marshes, pave over all open areas—never even knowing the names and natures of the billions of life forms they are destroying." *(Gibbons 1973:148)*

Gibbons penned those words almost thirty years ago. Yet, the situation is now considerably worse; foraging and other natural resources are disappearing like the pro-verbial Cheshire cat. Sadly, unlike that fairy-tale feline, these natural wonders—once vanished—will never reappear.

Searching for Particular Plants

As you study the field guide following this introductory material, you will see that certain plants favor particular locales, such as wasteland, meadows, moist woods, dry woods, hills, deserts, mountains, swamps, marshes, ponds, or other bodies of water. Study these habitats and you will be better prepared to scout out the particular species that may be appealing to you. For instance, if you come across boneset in a swampy region, you can—armed with knowledge of habitat—anticipate finding blue vervain or Joe-Pye weed growing along with it. You can also expect that wood sorrel and lamb's-quarters might be growing at the edges of clearings or paths, not to mention shepherd's purse, plantain, or peppergrass. Yes, you can become skilled enough so that if you need a particular wild plant for a recipe or herbal formula, you will know just where to look for it out in the wilds. That is not only a time-saver, but it can also be quite rewarding on a personal level.

Remember, too, that certain plants bloom only at particular times of the day (examples being spiderwort, chicory, and evening primrose) or in particular weather (the blossoms of chickweed and purslane open only on sunny days). Knowing this will help you to plan your foraging so that you can hunt for these plants when they are in bloom and thus easier to identify or locate.

Personal Safety During Foraging

It might seem strange to think that you could possibly encounter hazards while foraging, yet there definitely can be dangers. To begin with, the forager needs to be careful about where he/she is walking. One shouldn't be so intent on weed watching that one saunters into a hole, trips on a rock, or marches into an overhanging branch that pokes one's eyes! These things may seem laughable, but all of them have happened to foragers I have known. I myself was once knocked off my pins by a fast-racing bicyclist while crossing a paved park path to get a look at some attractive, pond dwelling plants

A second caution has to do with insect pests. Mosquitoes and deer flies can ruin a foray into the wild if you are not prepared for their onslaught. Commercial sprays may contain substances that are damaging to the body's systems or organs. Several studies have indicted "DEET" (N,N-diethyl-meta-toluamide—the ingredient common to many sprays that, according to its sponsors and some independent studies, best repels mosquitoes) as having the potential to cause serious problems. Since up to 15 percent of the chemical can be absorbed into the skin, allergic or toxic reactions may arise, and seizures have even been known to occur in children that have been doused with it.*(Silverstein 1990:86)* Showing sensitivity to these concerns, New York state, in the early 1990s, restricted sales of repellents containing more than 30 percent DEET.*(Hodgson 1992:1)* Other states have followed suit since that time.

Several commercial sprays coming on the market now reflect the increased awareness of the potential hazards posed by large concentrations of DEET. Some of these sprays, though, still contain artificial chemicals that may provoke reactions in sensitive persons. Various herbal and other natural measures, however, may provide some protection without damaging side effects. Herbal infusions made from chamomile, pineapple-weed, yarrow, or garlic—kept in a spray bottle in a pocket for repeated application—have proven helpful to some people (though first-time users should be cautious with use of these herbs; some people are sensitive to the sesquiterpene lactones occurring in the first three, which can provoke contact dermatitis in such ones; others are senstive to the sulfur compounds and other chemicals in garlic). Commercially-available essential oils of patchouli, eucalyptus, and/or pennyroyal have been employed with success by many. Because these are highly concentrated, a few drops are mixed with an innocuous base-paste or oil or otherwise diluted and applied to pulse points (wrists, ankles, and neck). (Most essential oils should never be applied full-strength. Also, a bottle of pennyroyal essential oil should only be opened when outdoors—*never* in a vehicle, as in such a closed environment it has the potential to cause respiratory collapse. This is especially a hazard for infants or small children, but it can happen to adults, too!)

Some sources have suggested that ingestion of vitamin B_1(thiamine) prior to traversing into the wilds might cut down on mosquito attacks, with the idea being that it causes the skin to emit an unpleasant odor. Another interesting theory posits that mosquitoes prefer people whose carbohydrate-to-protein ingestion ratio is higher than normal. In accord with this theory, the suggestion is that one avoid sweets, and instead consume only high-protein foods (especially seeds, nuts, grains, or fish) just prior to going outside. Still another line of research suggests that sulfur compounds on one's tongue—the same compounds that cause "bad breath"—attract mosquitoes, so that brushing or scraping one's tongue before wilderness travel might offer some protection.

A more serious threat is that posed by ticks, especially the deer tick (*Ixodes dammini*), which can carry the crippling Lyme disease. This tick, which used to be confined to only a few areas of theUnited States, is now spreading rapidly. In its early life as a larva, it often latches on to the white-footed mouse, picking up from it the spirochete that can cause Lyme disease in humans. It chooses a large mammalian host (occasionally a human, but more often a deer) during its later nymphal and adult stages, sometimes transmitting to this host the infectious spirochete. Experts have found that, contrary to popular belief, the tick usually transmits spirochetes to humans during its nymphal stage (late spring), as opposed to its adult (tick) stage (though considerable transmission at this later stage is suspected as well). Because both the adult tick and the nymph are very small and difficult to spot on one's body, the problem is frightfully compounded.

How, then, can you protect yourself from ticks and similar parasites? While there are no surefire procedures, the following tips may prove helpful:

- Wear light-colored clothing.

- Wear long-sleeved shirts and long pants, tucking the pants' legs into socks.

- Ticks hate garlic. In view of this, some choose to make a garlic infusion and spray the contents onto their neck, wrists, or ankles (the exterior pulse points). Or they use an oil preparation in which the garlic has been heavily diluted by olive oil. (*Never* apply full-strength garlic to skin, though, as it may burn it! The skin of some people may even be too sensitive for dilutions.) Some persons, if coming across the related wild onion while foraging, elect to rub some of the crushed leaves or bulb *on the outer layer of their pant legs* (again, not directly on the skin, as it can burn it).

- After returning home, take off your clothes and put them in the dryer. Set on "high," then let them tumble for at least 20 minutes.

- Visually inspect yourself in a full-length mirror for any ticks.

- Shower. This last tip may not be as helpful as the others. I have observed clinging ticks while under the shower, and they simply "hunker down" and bear with it! (The shower may knock off any wandering ticks, however.)

In addition to these tips, there is this *very* important one: *All foragers with long hair should bunch up their hair underneath a cap or hat.*

Seven years ago, *The New England Journal of Medicine*(Nocton et al 1994:229-34) announced a new, more accurate method of Lyme diagnosis employing polymerase chain reaction to detect the tick-borne spirochete. This new diagnostic test, replacing the previous—quite unsatisfactory—procedures, was hailed as a breakthrough. Current *treatment* for Lyme disease, however, calls for at least a month-long course of massive oral antibiotics or a two-week course of intravenous antibiotics. While necessary to kill the spirochetes, which left untreated could eventually cause heart failure or permanent damage to joints, such an antibiotic regimen can seriously upset important bodily functions. Antibiotics often induce the proliferation of yeast cells and perhaps even initiate the transformation of these cells into an invasive fungal form, thus precipitating chronic symptoms of a nagging, or even harmful, nature. Thus, it is far better to do all that you can to prevent the disease in the first place. Whenever antibiotics are advisable, however, the supplementation of beneficial intestinal bacteria (such as contained in capsule form from the health-food store or in recently-produced, unsweetened yogurt labelled to possess live *Lactobacillus* and *Bifidus* cultures) should be considered, as such organisms discourage yeast overgrowth. (It is usually advisable to maintain this regimen for at least several weeks after the termination of the antibiotic regimen, as it is *after* the antibiotics have been discontinued that intestinal flora can be most effectively restored.)

Not long ago, too, the FDA approved a vaccine for Lyme Disease, said to be 78% effective if used in three doses over a period of a year.(anon 1998:1A) Called Lymerix, and made by SmithKline Beecham Pharmaceuticals, it was approved for persons 15 years of age and older. It should be emphasized that it is promoted as being effective only for the common form of Lyme disease, not for human granulocytic ehrlichiosis disease, a sometimes fatal bacterial infection also spread by the deer tick and which was discovered only a few years ago in the Duluth area.

There are also some important cautions related to the particular act of harvesting wild plants. These will be discussed immediately below.

How to Harvest

Foraging for plants is something to be taken seriously. One must take care, not only to identify the plant carefully (cross-referencing in *several* field guides is strongly recommended), but also to pick so that neighboring plants are not accidentally pulled with the desired one(s). One must also exercise caution so as to utilize only the particular parts of a plant that may be edible and to prepare these parts in a fashion that renders them safe and palatable. Also, individual allergies vary, so even wild plants that have been positively identified as edible must be sampled with caution. Some wild plants may even be related to domestic foods to which a person may be allergic.

In other words, this is not an activity that should be engaged in by some "space cadet" from a fast-food restaurant who can't even get a "hold-the-onions" order straight! We are dealing with a serious activity, requiring an organized mind. Mistakes can be hazardous, even deadly. Contrariwise, with care for the art, one can reap great rewards.

The serious forager, if anything, is well equipped. He/she is never without certain implements, all of which can be carried in one of those strap-on waist pouches designed for carrying small objects. These include the following:

- Collecting bags (use cloth or paper instead of plastic, because plastic induces specimens to mold quickly).

- A small garden spade or a solid "digging stick" (employed by the American Indians for foraging).

- Clean, round-edged scissors (children's scissors).

- Small plastic containers for fragile specimens such as berries.

When gathering herbs to be dried for future use as a tea, collect only on dry, sunny days, because the volatile oils are at their high point in this sort of weather. Then, too, leaves collected while wet tend to mold more easily when dried at home than do those collected on dry, sunny days.

Plant parts should be collected when they are at their greatest point of utilization in the growth cycle. The American Indians developed the following methodology in this regard, which modern herbalists have also adopted: *Roots* of annuals were collected in spring, prior to flowering; those of biennials were dug in the autumn of the first year; and those of perennials were gathered either in the fall or in the spring before the leaves sprouted. *Leaves* were usually collected only up to the time when the plant would blossom. *Flowers* were collected shortly after they began to appear. *Bark* was usually collected in the early spring, when it was most easily removed, with few exceptions.

When harvesting, be conservation-minded. Do not take lone plants, but look for colonies. When these are located, take only a few plants, at most, bearing in mind that wild animals often depend on such colonies for food. The American Indians cultivated this sort of fine attitude and even felt that if a lone plant was chanced upon, it could be supplicated to lead them to its plant brothers growing in a colony.

Consider, as well, what parts of a given plant you will really need before harvesting. Taking one or two leaves from a plant that has a half dozen or more will not hurt it, but bear in mind that taking more than that amount may kill the plant, because leaves are needed so that it can produce enough food to survive through the winter. Harvesting a root will kill a plant for sure. If you feel a real need to take a root, break or cut off part of it, leaving the section that is attached to the stem, and then replant the herb. It may survive. I have done this with certain species, and it sometimes works.

Again, the American Indians provided the proper example in wild-plant conservation. Typical of the praiseworthy attitude of contemporary Indian tribes, those of the Missouri River region related these cherished instructions, inherited from their forefathers, to ethnobotanist Melvin Gilmore: "Do not needlessly destroy the flowers on the prairies or in the woods. If the flowers are plucked there will be no flower babies (seeds); and if there be no flower babies then in time there will be no people of the flower nations. . . . Then the earth will be sad. . . . The world would be incomplete and imperfect without them." *(Gilmore 1977:97–98)*

How to Clean Harvested Foods

Nature's processes normally clean wild plants quite effectively, but if you have serious doubts about possible recent contamination of harvested plants by dog urine, human spittle, or other pollution, you may elect to disinfect these plants before consumption. Boiling kills most harmful microorganisms, of course. But what if you wish to consume the plant raw, as a trail nibble or in a salad? Foraging guides of past years would often recommend soaking the plants or plant parts in a solution of a teaspoon or tablespoon of bleach or hydrogen peroxide to a gallon of water for at least ten minutes, afterwards rinsing the plants thoroughly before consuming. Fortunately, we now have available a natural substance to do the trick: grapefruit-seed extract, available at health-food stores. This amazing substance has been shown to possess activity against a wide variety of bacteria and other harmful microorganisms. The recommended dose is usually only ten drops per two quarts of water, as it is very strong stuff! Of course, even if choosing to apply this natural disinfecting agent, plant parts should be thoroughly rinsed afterwards with clean water before they are ingested.

Any plants gathered from fresh waters (as opposed to seawater)—such as cattail roots, watercress, etc.—*must* be cleaned if you intend to eat them raw, as such plants may harbor harmful water parasites. Cattail is fine when cooked (which usually kills any clinging parasites), but because watercress is best raw, you would be wise to exercise special care in cleaning and disinfecting it should you choose to consume it in this state. (I recommend bypassing it and securing this herb from commercial sources, where it is grown in protected beds.)

How to Store Harvested Foods

If intending to dry and use all the aerial parts of an herb for a tea, it is best to hang the whole plant upside down in a cool, dry place out of direct sunlight. But if only particular parts of a plant are needed, the method of choice involves the use of a drying rack. You can make a simple but efficient rack out of cheesecloth (not wire screen, which may impart a metallic taste) and some wood for frames. The important thing is to allow for air to circulate to all parts of the plant, as otherwise mold may set in. Not only is mold undesirable from the standpoint of taste or appearance, but certain molds can be quite dangerous (e.g., those appearing in sweet clovers or in brambles.).

When fully dried, break up, but do not fully crumble, the contents in your hand and place them in a dark-colored, glass bottle or jar with a screw-on lid. Store in a cupboard or other dry, dark place until ready to use. Check stock after a day or two (and periodically thereafter) for condensation or signs of molding. If molded, throw the leaves away. But if you find no mold and see only condensation (indicating that the herb was not fully dried), the contents can often be removed, redried, and rebottled. When finally desirous of making an infusion, simply crumble the amount needed in your hand.

Herbs do deteriorate. Some are even hygroscopic, that is, they absorb moisture from the air, which reactivates their enzymes, precipitating deterioration. Normally your stock should be stored for only a year (aromatic herbs) to a year-and-a-half (other herbs). Some herbs (such as shepherd's purse) should not be kept any longer than six months, as such can become worthless—or even somewhat toxic—if kept beyond that point.

Roots slated to be dried for decoctions should first be cleaned of all clinging dirt. If the roots are thick, slit lengthwise one or two times before drying.

Fresh leaves, stems, flowers, berries, and roots can be stored in a refrigerator. Some keep for a long time, others for not so long. Trial and error will teach you which is which. Generally, roots and tubers will keep best, but some—such as Jerusalem artichokes—spoil in about a week and a half.

Berries, larger fruits, and vegetable shoots can often be canned for winter storage. Be sure to consult a *post-1989* professional manual on this, because improper canning procedures (some of which were not known to be risky when the older canning manuals were published) can result in the production of the botulism toxin, which could prove fatal.

Roots can often be kept fresh in a leaf-insulated pit in the ground, with a board placed over the pit and then more leaves and branches heaped thereon. You can access the pit by turning over the board, thereby dumping the leaves and any snow.

Freezing is a fine form of preservation, but even here certain procedures and cautions should be understood. Fruits are easiest to freeze. Berries and smaller fruits (e.g., wild plums) can simply be washed and frozen as is, although you may want to cut up larger fruits. Store them all in sealed containers, with headspace for expansion.

Wild veggies, however, need a bit more work. Once washed and cut up, they should be blanched in boiling water, then cooled in ice water, then drained or patted dry, and finally, sealed in plastic bags or containers for freezing. If you use containers for freezing, leave headspace in them for expansion; if you use plastic bags, squeeze out the air in them. For the blanching, use a wire basket (preferably non-aluminum) to dip the veggies into the boiling water—usually for 2-3 minutes, but see variations enumerated in the field guide under individual plant entries. Then plunge the veggies into the cooling, ice-water bath. Bear in mind that wild veggies, like domestic ones, must be used immediately upon thawing and cannot be refrozen for later use.

How to Prepare and Cook Harvested Foods

As a health enthusiast, I believe that, when possible, fruits and vegetables should be consumed raw. This allows one to benefit from the cleansing and invigorating power of the plant enzymes, which would be destroyed should the vegetation be cooked. Then, too, I believe that the solar energy trapped in the plant's cells can be utilized most efficiently by the body if the plant is consumed raw. *(Kenton and Kenton 1984:64–65;Jensen 1977:55–56,60–61)*

Still, light steaming usually preserves, at least to some degree, some of these same benefits. Cooked food also requires less calories to digest, and this can be crucial in a survival situation. And, of course, certain plants are only edible *after* being cooked. Some of these must even be cooked in several changes of water or be dried before being cooked. (These variations will be noted in the text.) Here it should be stressed that plants being prepared as a potherb should almost never be immersed in cold water and then boiled. Rather, in most cases, they should be placed directly in water that is already boiling; this is crucial to insure optimal flavor. Also, most plants requiring short boiling periods should be put in very little water—generally, just enough to cover the plants, and sometimes less. The lesser volume of water you give the plant to disperse its natural flavors and oils into, the better your plant will taste upon emerging and the more bursting with nutrients it will be. You will, however, need a greater amount of water with plants requiring longer boiling periods (such as cleavers).

Although steaming is often preferable to boiling because it not only better preserves the taste but also the plant's beneficial nutrients, you should not steam any plants for which several changes of water are required. (These will be noted in the field guide.) Many of such plants are treated with this succession of water changes in the first place because they have poisonous or bitter principles that need to be boiled off into the earlier water(s). Steaming might not allow these principles to fully escape.

Most roots and tubers are best if baked. (I refer to a regular oven here, as I personally frown on the use of microwave ovens for cooking food.[*]) They usually do

---

[*] I once read of an experiment where kirlian photography was used to assess the quality of the auras that foods (as organic substances) possess after being cooked by various methods. The experiment found that microwave cooking completely dissipated the energy aura of foods, while other cooking procedures left it intact to varying degrees. Although I have never been fully convinced as to the accuracy or merits of kirlian photography, reading about this experiment disturbed me nevertheless, especially after considering it in the light of some comments I also read in a scholarly work on food, to wit: "When microwave heating was first introduced nobody knew for sure what chemical changes would take place in the food. Were they exactly the same as those produced by conventional heating, or did the microwave field produce particular breaks and recombinations in molecules different from those produced by traditional cooking? Although nobody knew the answer, microwaved food never caught the attention of regulatory agencies the way irradiated food did. It was *simply assumed* that no significant amounts of harmful products were present in microwaved food."*[italics mine].(Diehl 1990:147)*I don't know about you, but I am leery about entrusting my health to those who make mere 'assumptions' about the safety of my food when assaulted by modern technology. Then, too, recent scientific research*(anon 1999:9)* suggests that there may be some palpable evidence for a concern in this regard: In the early 1990s, a Swiss scientist, Hans Hertel, conducted a qualitative study on the effects of microwaved food on human volunteers and specifically on their blood and physiology, publishing his results jointly with a Lausanne University professor in 1991. This study found crucial changes in the blood of those volunteers, including a decrease in hemoglobin and cholesterol values and a significant change in their lymphocytes. Further evidence revealed that the *microwave energies themselves* were actually*passed into the physiologyof the human volunteers*! The conclusion of the study was that microwave cooking produced changes in the nutrients that may actually intitiate *a pre-cancerous change in the blood*! There is an interesting accumulation of this and other evidences for the deleterious effects of microwaved food on an Internet website(www.healthfree.com/paa/paa0001.htm), wherein the origin of microwave ovens is traced back to Nazi Germany, whose scientists were said to have developed it for field-ration use by German solders on the Russian front. After the war, according to this summary, both the Americans and the Russians retrieved the Nazi documents relative to these microwave ovens. The Russians are said to have subsequently done extensive research on the effects of microwave cooking and consequently banned it as dangerous, issuing an international warning. The summary of the Russian research—including details on the changes in food that microwaves are said to produce—appears on the website. Very interesting—and scary!—information.

not do well being steamed, but they can often be palatably boiled. If boiling is undertaken, however, one should bear in mind that roots often need more water than other plant segments because they must be cooked for a longer period of time in order to reach their appropriate tenderness.

Teas made from roots are often quite delicious, too, and are usually prepared by boiling (called a *decoction*), while those made from leaves or stems are normally prepared instead by infusion. (See these, and other, common plant-preparation techniques in the appendix section entitled "How herbal extracts may be made at home.")

Well, that's about it! I trust that you will find the above information helpful in maintaining your own joy of foraging. In conclusion, I leave you with this final wish: May you experience the wondrous way of weeds, not only during many daylight hours, but also in your nighttime dreams, and that for many years to come!

Auxiliary Introduction

Health, Medicine, and Weeds

*

Even as there are numerous benefits to health that can be gleaned from domestic vegetables, so this is true of wild plants, and sometimes even more so. Some of such advantages present themselves on the most basic of levels. Take, for instance, sight. As to why we humans so often yearn to "get out in nature," especially when under great stress, it may be that the very *color* of wild vegetation, green, holds the key. Dr. Bernard Jensen has put it this way: "It is no haphazard occurrence that the color green is so prevalent in Nature. Green acts on the nervous system as a sedative. We find it is helpful in insomnia, exhaustion and nervous irritability. The color green can lower the blood pressure, dilate the capillaries, produce a sensation of tranquillity. Neuralgia, headaches, and nervous conditions often respond to its use. . . . Green is a calming color, the restful color. It is suggestive of life and immortality, the emblem of youth and happiness and prosperity. It counteracts the brightness of the sun, excites the eye less than any other color, including black, and induces repose." *(Jensen 1977:39)*

Chlorophyll is the pigment that conveys this wondrous green tint to plants. This unique substance also appears to contribute substantially to human health when ingested, giving rise to an upswing in energy, improved overall health, and a heightened sense of well-being. Scientific studies have revealed, in addition, that chlorophyll can actually regenerate the blood, help heal some intestinal diseases, re-energize geriatric patients, and possibly even help prevent cancer. *(Wigmore 1985:120-21)* Popular health writers such as Bernard Jensen and Ann Wigmore have done much to acquaint the public with the wonders of this powerful green vitalizer, but knowledge of its benefits needs yet to be far more engrained.

Quality Nutrition

But aside from chlorophyll, there are numerous other health benefits that edible weeds can provide. Euell Gibbons did much to bring this knowledge to the public by means of the many plant analyses he did and reported in his books, yet today much of this information remains largely unknown. This is ironic, in that wild fruits and vegetables are often far superior to their domestic counterparts in vitamin and mineral content (not to mention taste!), and they lack harmful preservatives, dyes, waxes, etc.

Take, for instance, lamb's-quarters, a sort of wild spinach, which dwarfs domestic spinach in protein, vitamin C, and (pro-)vitamin A.*(Gibbons 1971:161)* That is quite an accomplishment, especially for vitamin A, because cultivated spinach contains more of this vitamin than any other marketed vegetable (approximately 8,100 units per 100 grams, compared to 10,000–15,000 for lamb's-quarters).*(Zennie & Ogzewalla 1977:77)* Many wild edibles, however, overflow with valuable nutrients in amounts that surpass most or all of our domestic veggies. Just staying with vitamin A for now, several other common edible weeds surpass spinach for the amount of this vitamin, including plantain (with 10,000 units per 100 grams)*(Zennie & Ogzewalla 1977:77)* violet (with an incredible 15,000–20,000 units per 100 grams), *(Zennie & Ogzewalla 1977:77)* and peppergrass (9,300 units per 100 grams).

Another example of the ultra potency of the wild is the stinging nettle, an edible food (when properly prepared) which contains an incredible amount of protein—one source finding 42 percent by dry weight!*(Tull 1987:16)* That may be more than is contained in the leafy green portion of any green plant, wild or domestic! But that is not all: nettle is also one of the single best sources of chlorophyll known in the plant world—so rich that it has been cultivated for commercial extraction of this substance. In fact, it is rich in a wide spectrum of nutrients—especially calcium, chromium, magnesium, zinc, and vitamin C. Its iron content, though not phenomenally high by measurement, has nevertheless proven to be particularly bioavailable in that the plant's high content of vitamin C serves to enhance its absorption.

Arrowhead tubers, which were a staple of many American Indian tribes (and the knowledge of which, gleaned from kindly Indians, saved the lives of the Lewis & Clark expedition, according to their own testimony) are an incredible source of energy, containing up to 400 calories per serving, and possessing a rich amount of potassium and phosphorous. (They are also very high in thiamine, which, according to some research, may help to minimize mosquito molestation.)

The familiar chickweed plant, a staple of yards and gardens as well as available to the forager in various wild forms (especially in the Minnesota-Wisconsin area as the delicious water chickweed, *Stellaria aquatica*) has one of the highest contents of iron known! It is also rich in magnesium, silicon, manganese, and zinc—all in all, a genuine nutritional powerhouse. Many foragers remark that more chickweed passes their lips than any other plant, and I confess to being among them!

Numerous other examples could be cited. But, suffice it to say that edible weeds can be incomparably nutritious. And proper nutrition is medicine of the best kind—preventive medicine. This was acknowledged by that ancient sage of medicine, Hippocrates (460–377 BC), who put the matter succinctly, yet appropriately, when he stated: "Let your food be your medicine."

Biologically- and Pharmacologically-Active Substances

Aside from nutritional benefits, numerous wild plants contain one or more biologically, even pharmacologically, active substances or properties that have made them popular in healing remedies. In some cases, this is because certain amounts or combinations which occur in particular plant species lend these plants for very specific uses in human health. Thus, an individual herb may possess a combination of constituents that gives it anti-inflammatory properties, or antiseptic properties, or antispasmodic properties, or sedative properties, or carminative (gas-expelling) properties, or some other.

Interestingly, many of these properties have been known and used by man for hundreds—sometimes thousands—of years, though without detailed scientific knowledge of the mechanisms of effectiveness. The wound-healing and blood-stemming (styptic) properties of yarrow, for instance, were employed by Achilles in tending his injured soldiers during the Trojan War—almost two and a half millennia ago! Hippocrates and other ancient physicians were acquainted with many of the botanical properties and discussed them in their extant works, though of course without a knowledge of the specifics as to how they worked. North America's Indian tribes implemented over 3,650 vascular plants for medicinal or culinary benefits.

The discovery of the therapeutic properties in herbs is shrouded in mystery, but may have been achieved through trial-and-error, by accident, by watching what other higher mammals (such as bears) ate when ill or lacking in some way, and/or probably by other means not known today. According to the tenth chapter of the ancient Jewish work known as the *Book of Jubilees* (composed 1st to 2nd century B.C.), God's faithful angels transmitted to Noah the knowledge of how to heal with the various plants and he in turn passed this information on to his descendants. But regardless of the means, the accumulated knowledge was passed on to succeeding cultures. In fact, many North American herbalists—even pharmacists—owe much of their plant-healing knowledge to the American Indians mentioned above, who kindly shared their carefully cultivated storehouse of herbal knowledge with early settlers who found the labor of taming a new land to be compounded by various ills. This knowledge was in turn recorded in the various herbals and pharmacopoeias that so many pharmacognosists and phytochemists have consulted over the years in initiating their own plant investigations.

In regard to the nuts-and-bolts specifics of the efficaciousness of herbal medicines, we are now living in a most exciting time. As Delena Tull has so well elaborated: "Today, for the first time in history, we have the tools for distinguishing myth from fact. Indeed, modern biochemical techniques have brought about a revolution in our understanding of the chemical constituents in plants and their values in medicine.... Through biochemical analysis, researchers have discovered thousands of plant alkaloids that we never knew existed before this century. In the 1940s, scientists were aware of only 1000 alkaloids, but by 1969, 3350 new ones had been isolated. In addition, many glycosides, saponins, flavonoids, and more than 2000 other organic plant substances are now known. Any or all of these may have value in medicine." *(Tull 1987:7)*

All of such discoveries, coupled with an increasing frustration by the public with orthodox medicine (including the many thousands of deaths each year which have been established to be iatrogenic—that is, physician-induced) have led to what in recent years has appropriately been named an "Herbal Renaissance." Thus, in 1996, many millions of Americans spent a total of 3.24 *billion* dollars on herbal medicines! Botanical remedies now abound—not only in health-food stores, but also, as they did at the turn-of-the-century and earlier, in pharmacies, with some 73% now carrying them (although a recent university study showed that only 2% of pharmacists assessed themselves as feeling competent in their use!), according to data published by the American Botanical Council's Herb Education & Research Center.

To provide some insight into the medicinal usage of weeds, however, let's take a cursory look at how some of the plant compounds enumerated by Tull above are currently being understood as to their impact on human health. . . .

Alkaloids

Alkaloids occur in about 10–15 percent of vascular plants, and are defined as generally toxic substances that affect the central nervous system of living creatures. They contain heterocyclic nitrogen and are produced in plants from amino acids and related substances. Although, as stated above, being generally toxic (they are often the substance that makes poisonous plants toxic—even deadly), a number of alkaloids have been put to use in small amounts, or in certain forms, as medicinal agents. In fact, alkaloids were among the first substances isolated from plants for medicinal purposes. Examples of commonly known alkaloids include quinine, ephedrine, morphine, codeine, nicotine, and caffeine (note the *-ine* endings). In the present work's field guide, you will note various alkaloids referenced in conjunction with their therapeutic functions.

Glycosides

Glycosides are abundant in the plant kingdom. They are defined as substances in which a sugar molecule is paired by chemical bonding with a non-sugar molecule (called an aglycone or genin). In their compound form they are often (though not always) inert, but when broken down into their basic components (aglycone + sugar)—such as can occur by human ingestion as well as by other means, including sometimes a mere picking or damaging of their plant source—the aglycone part is often rendered pharmacologically active. Unfortunately, the nature of glycosides as just explained has accounted for some of their unique and important antiviral properties being "masked" until recently when their aglycone components have been more rigorously investigated. *(Hudson 1990:119)*

Many references to various glycosides are scattered throughout the text. Note that certain of them are cardioactive, i.e., they affect the heart. References will be made as well to anthraquinone glycosides, which are laxative in small amounts and toxic in larger amounts. Various other types of glycosides will be referred to as well, such as . . .

A.) Coumarin Glycosides

Coumarin glycosides, widely sprinkled among the many plant species, are characterized by an oxidized, phenolic sort of aglycone known as a coumarin. They possess anticoagulant, antibacterial, anthelmintic, analgesic, estrogenic, and sedative abilities.*(Farnsworth 1966:165)* Psoralen, a type of coumarin known as a furocoumarin [furanocoumarin] (where a coumarin has fused with a furan ring) and occurring in the field herb yarrow, is currently proving useful in the treatment of psoriasis.*(Fuller & McClintock 1986:320-21)* Furocoumarins also have proven effective against many gram-positive bacteria, and it has been demonstrated that they possess the rare ability to bind DNA. *(Fuller & McClintock 1986:320-21)* A coumarin derivative known as scopoletin (technically known as a"hydroxy-coumarin") has been found to be a powerful anti-inflammatory and antispasmodic agent, useful in the treatment of allergies, menstrual cramps, and other troublesome conditions. This chemical is found in nettle and in some other plants.

B.) Iridoid Glycosides

These consist of a group of bitter-tasting chemicals technically known as monoterpenoid lactones. They include aucubin, a powerful antibiotic compound found in plantain, mullein, and several other wild plants and in appreciable enough of an amount to make those plants useful as topical antiseptics for minor wounds. Aucubin is also anticatarrhal and increases the excretion of uric acid by the kidneys.

C.) Flavonoids

Flavonoids occur in all vascular plants, sometimes in free form but often as the aglycone component of glycosides. Under intensive study since the 1940s, flavonoids are subdivided into several categories.

Flavonoids that have been known to be biologically active have historically been designated as "bioflavonoids." They are thought to play a vital role in restoring—if not also maintaining—human health.*(Middleton 1988:103-44)* It has long been known, for instance, that bioflavonoids serve an important function in the maintenance of the healthy state of capillary walls (so much so that a common indicator of bioflavonoid deficiency can be frequent nosebleeds). In fact, initial enthusiasm on the part of a number of researchers had even designated them (collectively or sometimes individually) as a vitamin—Vitamin P. This designation was later shown to be technically inappropriate. Still, evidence accumulated to demonstrate that bioflavonoids do work synergistically with vitamin C in the body (in an antioxidant-dependent, vitamin-C-sparing way).*(Middleton 1988:103-44)* Time has revealed their role in human health to be significant, contrary to what orthodox nutritionists had once so confidently asserted.

Although their role in maintaining human health has not yet been fully explicated, the evidence for their role as therapeutic agents in the alleviation of many health problems has been powerfully elucidated, stemming from research initiated way back in the 1940s. In fact, two bioflavonoids—rutin and hesperidin—have been employed in orthodox medicine for almost half a century. Quercetin (found in many edible weeds) has captured the attention of biologists and other researchers for about twenty-five years now, proving its worth toward the restoration of human health in some very important respects. One of these, of great interest to those afflicted with hayfever and allergies, is its explicated role as an inhibitor of Type-A (histamine-mediated) allergic reactions.*(Middleton 1988:103-44)* More recently investigated has been the potent antiviral effects that this flavonoid has been shown to possess against some eleven kinds of"tough guy" viruses.*(Middleton 1988:103-44; Selway et al 1986 :521-36)*

Genistein, a type of flavonoid known as an isoflavone, occurs in red clover and has been shown to be an important anticarcinogenic agent.*(see refs. cited in Mills & Bone 2000:55-56)* Its existence in red clover's chemical profile may partly elucidate that plant's frequent inclusion in the many herbal formulas that have historically been used to help heal people afflicted with cancer.

Then there is a flavonoid compound known as silymarin (composed of silybin, silydianin, and silychristin), obtained from the lowly milk thistle (*Silybum marianum*), which possesses a special antioxidant ability that has proven its mettle with respect to counteracting liver toxins such as phalloidon (the deadly substance conveyed via *Amanita* mushrooms) and carbon tetrachloride (once prescribed by orthodox physicians to kill hookworms and roundworms!)*(Middleton 1988:108–9; Hikino & Kiso 1988:40-45)*

Anthocyanins, given a lot of press in the last few years, are nitrogenous, water-soluble flavonoids responsible for the tints of flower petals and fruits in the red-violet-blue range and for certain colors occurring in autumnal leaves. Their contributions to human health have been unappreciated until very recently, but now are under intense study. For example, like some other flavonoids, anthocyanins have been shown to strengthen blood vessels, such as in the eyes, where they also regenerate the visual purple, thereby improving night vision and even myopia.*(Alfieri & Sole 1966:1590)* Research further suggests that they improve cardiac circulation, dilate the coronary arteries, and lower LDL cholesterol levels.*(Seranillos et al 1983:7; Mazzag & Miniti 1993; Millet et al 1984:439)* Then, because ingestion of these flavonoids in appreciable amounts (at least 300 mgs. a day) also helps the adrenal glands to retain their vitally needed vitamin C,*(Sturua et al 1971:21575)* consumption of wild foods containing them may also aid immunity and stress tolerance—two major functions of the adrenals—in the human organism. In addition, much recent research has focused on the potent antioxidant (free-radical-scavenging) qualities that anthocyanins have been shown to possess.*(Santrucek et al 1988:974-96)* All in all, the anthocyanins in wild fruits and flowers are thought to have made crucial contributions to the therapeutic applications of these plant segments that have been implemented by herbal healers of various traditions, including our native Americans.

Partaken in their natural form in edible foods (both domestic and wild), most—if not all—flavonoids would appear to be safe, with any potentially harmful effects negated by other substances in the foods or by reactions occurring in the digestive processes, while the positive benefits of these pigments shine through unhindered.

D.) Saponins

Then there are the highly interesting saponins. Chemically, they are glycosides, possessing a terpenoid aglycone (called a sapogenin). Though not as widespread in the plant kingdom as flavonoids, they still probably occur in most flowering plants (the current state of phytochemical knowledge has isolated them in about 600 plant species from almost 100 different plant families). Curiously, saponins are natural soap particles (the term "saponin" is from the Latin word *sapo*, meaning "soap"!), frothing greatly when mixed or shaken with water, though they are non-alkaline. One common wild plant containing them in abundance, bouncing bet, is also known as "soapwort," because its leaves and stems can be crushed and rubbed on one's hands to produce a soapy froth so as to clean them—a feature heartily made use of by Indians and settlers of times gone by.

Because of the nature and amount of saponins that saturate soapwort, however, this plant is potentially toxic if ingested (although, interestingly, several American Indian tribes have discovered various preparative and ingestive methodologies to offset this). Soapwort toxicity produces nausea, gastrointestinal irritation, and vomiting. And human poisonings from several other saponin-containing plants (e.g., bittersweet nightshade) are hypothesized to occur because the saponins in these other plants are thought to damage the intestinal mucosa so as to allow absorption of accompanying toxins.

In the form and concentration in which they can be found in certain other wild plants, however, a number of saponins are thought to convey some health-restorative benefits upon ingestion of the plants in which they occur as constituents. Thus, the saponin content in mullein helps to make an infusion of that plant a useful expectorant for those afflicted with colds, bronchitis, asthma, etc.

In their concentrations in certain other weeds, saponins appear to offer diuretic and laxative functions to the user. In yet others (e.g., chickweed), they yield a crude anti-inflammatory effect. (Certain saponins are even implemented by pharmaceutical chemists to synthesize anti-inflammatory steroid hormones.) Saponins have also been shown to be fine antifungal and antibacterial agents. *(Hiller 1987:175–84)* Finally, research conducted over the last fifteen years has found certain orally administered saponins to be immunomodulators, stimulating—in animal experiments—humoral and cell-mediated aspects of the immune system, including the enhancing of T-cell and B-cell activity. *(Hudson 1990:140–41)*

Sesquiterpene Lactones

Sesquiterpene lactones, occurring widely in particular plant families, are aromatic compounds, usually concentrated in the leaves and flowers of plants, and especially in their glandular hairs (trichomes). They are being investigated with fervor owing to their recently discovered anticarcinogenic properties. *(Hausen 1992:228)* But a number of them also possess antifungal, antibacterial, analgesic, and anti-inflammatory properties. Prominent in the latter category are the azulenes, occurring in the wild plants yarrow and pineapple-weed, no doubt elucidating the traditional use of these plants for sore throats and stomach inflammation. And certain sesquiterpene lactones occurring in the marsh weed boneset are thought to be partly responsible for that plant's long-renowned, and recently verified, antitumor and immunostimulating properties. Anthelmintic properties are also attributed to sesquiterpene lactones, so that plants possessing them in abundance (e.g., boneset, pineapple-weed, and wormwood) are among the herbal kingdom's best parasite fighters.

Tannins

Tannins are phenolic substances occurring in many weeds and especially in the bark of shrubs and trees. They occur in two forms: (1) Condensed tannins, including proanthocyanidins [procyanidins] (abbreviated PCOs or OPCs), which are related to flavonoids; (2) Hydrolysable tannins, derived from simple phenolic acids.

Tannins possess the unusual ability to precipitate proteins. This property makes them useful in treating certain diseases or conditions—if used judicially. That last phrase is significant, for in large amounts (such as can be obtained through excessive consumption of black tea—and even a number of herbal teas), tannins irritate intestinal mucosa. In small amounts, however, they serve the useful function of precipitating the protein contained in the mucosal cells, thus rendering these cells impermeable to any irritating substances which may be present. Not surprisingly, therefore, tannins are used to heal ulcers, diarrhea, and various afflictions of mucous membranes throughout the body. Their well-documented use in aiding the healing of burns is also attributed to the aforementioned property. Likewise, they can inhibit infectious agents by disrupting their proteins and/or by starving them of their protein food source   Even some viruses are not immune to the powerful effects of tannins.*(Hudson 1990:159–61; Beladi et al 1977:358-64)*

However, like a living paradox, tannins appear to contain both carcinogenic and anticarcinogenic properties. Of course, carcinogenic substances occur in many staple foods—corn, wheat, and celery being examples—as Dr. Bruce Ames demonstrated in his groundbreaking study published in *Science* magazine about two decades ago.*(Ames 1983:1256–64)* As with these foods, it may be that the anticarcinogenic substances in tannins nullify the alleged carcinogenic ones when it comes to human consumption. We now know conclusively, thanks to Japanese research, that tannins are powerful free-radical scavengers, and the accumulation in the body of excessive free radicals has long been understood to be a strongly predisposing factor in the development of cancer. But there is no proof that any anticarcinogenic effects of tannins outweigh any carcinogenic effects. This uncertainty, coupled with the strong suspicion on the part of many scientists that throat and stomach cancer may be related to excessive ingestion of tannins, would suggest that frequent consumption would be unwise. And irregardless of the cancer factor, such overconsumption can definitely hinder absorption of B vitamins (and thereby iron), as well as protein, vitamin A, and other important nutrients.

Essential Oils (Volatile Oils)

Essential (or volatile) oils are important constituents of many plants and are characterized by the strong aroma that they produce. They often display remarkable antibacterial and antifungal properties.*(Farnsworth 1966:268)* It is thought that one reason for their antiseptic action has to do with their ability to increase the flow of blood when coming in contact with bodily mucosa. Likewise, they may aid a distressed digestive system by increasing the flow of gastric juices or by acting as a carminative. Species in the mint family are renowned for their essential oils. Menthol, occurring more highly in wild mint than in any other form known (including even the cultivated peppermint), is well-known and appreciated for its decongestant and carminative properties, but has especially found use of late in the treatment of sports injuries, since one study found its implementation to surpass the more traditionally used salicylates for effectiveness. (See under "wild mint" in the text.) Thymol, well known to orthodox medicine as a powerful antiseptic, antifungal, and anthelmintic, occurs in the most concentrated form known in the lovely plants horsemint and bergamot—meadow-loving herbs marked by powerful, perfume-like scents.

Polysaccharides

"Many sugars" is the etymology of polysaccharides, and such they are, being composed of at least ten sugar units(monosaccharides) in a row. Starch, glycogen, pectin, and cellulose are the most familiar of these in the plant kingdom. Another interesting one is inulin, which occurs in Jerusalem artichoke, dandelion, wild onion (as well as in the cultivated species), and other plants discussed in the text. Because, unlike starch, it does not convert rapidly to glucose, foods containing this substance are often of aid in the restricted diets of those with blood-sugar problems. However, being difficult to digest, this polysaccharide can cause flatulence.

Mucilage is another polysaccharide occurring in a number of wild plants, such as the common weeds plantain and mullein. When ingested, mucilage reacts with water present to form plastic-like masses that protect mucous membranes from irritants. Thus, it has proven popular in remedies for the relief of conditions involving irritated mucous membranes, such as sore throat, ulcers, etc.

Another fascinating group of polysaccharides, heteroglycans, has been found to be immunostimulating in nature. These compounds are water soluble and possess a high molecular weight. Included here are the xyloglucurans from boneset, designated as 4-0-methylglucuronoxylans.

Organic Acids

Plants possess the unique ability to accumulate organic acids into their tissues. These are colorless, chemically stable, water soluble liquids, recognized by their low pH. Some of these acids, called tricarboxylic, are important in the Krebs cycle of living organisms, which process occurs in the mitochondria of cells and is responsible for the production of ATP, the all-important energy currency of the body. An example is malic acid, which occurs in cherries, apples, plums, grapes, and rhubarb. Citric acid is another tricarboxylic and occurs in strawberries (including the wild form), gooseberries, and black currants. Yet other organic acids include acetic acid, the precursor of fatty acids, and ascorbic acid, or vitamin C. Oxalic acid, which gives a lemony flavor to many favorite edible plants of wild-foods foragers (including dock, sheep sorrel, yellow wood-sorrel, lamb's-quarters, purslane, and others), is still another. However, this acid can be fatally toxic in excessive amounts, and thus appropriate cautions are mentioned in the text in the monographs for the abovenamed plants.

Pharmacognosy, Cartesian Reductionism, and Holism

The study of the physical and chemical properties of plant- and animal-based crude drugs, including the plant components we have isolated above, is defined by the term *pharmacognosy*. It is a wonderful science, explaining a great deal relative to the healing powers of wild plants. But it does not explain *everything*, though unfortunately some of its professionals seem to think that it does. As heirs of the Cartesian philosophy that attempts to define everything by the sum of its components, Western scientists have failed to grasp some important considerations that have not so readily eluded Chinese and other health researchers who have embraced a more holistic (wholistic) method of appraisal. Holism is a science that looks at things in their entirety, and does not judge the whole solely by the sum of its parts. It also does not downplay the testimony of history as to how herbs have affected human health for a period of hundreds—sometimes even thousands—of years.

On that latter point, let's get specific: Boneset was used by both the American Indians and early white settlers (as well as by black slaves and—after the Civil War—emancipated Negro communities) to deal with colds, influenza (including the deadly Spanish flu epidemic of 1918), and even diseases such as dengue and typhoid fever. Various health practitioners of the time (including many medical doctors) recorded in great detail the application and tremendous efficacy of this herb. However, with the advent of modern drugs, it fell out of use and has since been held by orthodox medicine to be practically worthless, devoid of significant biological or pharmacological activity. Since the 1970s, however, the herb has attracted the attention of several phytochemists and other scientists who, in a number of skilled analyses and studies published in scientific journals, have confirmed many of the herb's renowned uses, especially its immune-stimulating and fever-fighting ones (see under "Boneset" in the text). Another example, as we have already noted, is red clover, which has traditionally been used in herbal formulas to fight cancer. Such an application, especially highlighted in the heated controversy over one brew which has contained it, the "Hoxsey Formula" of the 1950s, had previously been fiercely criticized by orthodox scientists. Now, though, we know, as we have earlier seen, that the isoflavone genistein, occurring in this plant, has marked anticancer properties.

Here the question to be posed is this: How many years of potential benefit to the ill were lost because orthodox medical practitioners and investigators employed a strictly reductionistic, and not a holistic, approach to evaluating these herbs—dismissing them prior to the present time because they competed with drugs being hyped and because nobody cared enough to verify that they—despite their pedigree of therapeutic successes—contained any of the "right stuff"? The tragedy of waste, not to mention the uncalled-for slurs on natural healers who had been employing these herbs all along, is assuredly a black mark on the record of orthodox medicine. And waste abounds still in the pharmacological field where researchers choose to experiment with extending the properties of existing drugs in preference to investigating the properties of the wide variety of crude drugs already time-tested as to efficacy in natural medicine. How many untapped powerhouses are thus being neglected remains anybody's guess!

This is not to say that the current state of holistic thought has all the answers. In fact, holistic researchers and practitioners are often divided as to how medicinal plants work. Some of them believe that all herbs primarily accelerate the immune system in a general way so that the body can heal itself. Others hold more specifically to a theory known as "energetics." Basically, this school holds that respective herbs may be possessed of particular energies (i.e., being either heating or cooling, drying or dampening, etc.), while diseases or conditions are likewise possessed of specific energies, so that with the application of an herb possessing an energy opposite to that of the disease or condition, healing is precipitated. Still others—probably the majority—combine phytochemical evidence and theories with one of the above and/or yet another theory or theories. Hence, it certainly cannot be said that holistic health researchers have closed the book on the why's and wherefore's of the healing power of herbs.

In fact, holistic health researcher and herbalist Peter Holmes would say that what is needed now on the part of scientists and researchers in both the reductionist and holistic realms is "multi-paradigm thinking." As he puts it in more detail: "Modern quantum physics has educated us to let go of the idea of the naive belief in the correctness of a single point of view. Writers from Werner Heisenberg and David Bohm to Fritjof Capra have stressed that different models of reality can and should be applied to different situations, whether in daily life or in respect to a technical field. . . . A variety of views of reality need to be adopted if we are to do justice to its ever-changing, limitless and ungraspable nature. Clearly then, the fact that multiple models or theories of reality exist, many to be used equally and advantageously, points to the conclusion that here we are dealing with flexible particular emblems rather than fixed universal truths. . . . Now that the universal validity of a single approach to reality has been seen as a naive belief, it is of great importance to deal with reality models in a flexible way. We should now be prepared to exchange one theoretical model for another if and when reality requires it. Rather than blindly abide by a single truth, [we have] the opportunity to become *inclusive* in our ways rather than exclusive." *(Holmes 1989, 1:35)* [Emphasis his.]

Of course, the advances in phytochemistry and biology are very helpful and much to be lauded. Even without this information, however, the herbal knowledge accumulated through centuries of use has refined the various applications to an art.[*] Hence, skilled herbal practitioners such as can be found among herbalists, naturopaths, holistic chiropractors, and some other alternative-medicine practitioners have, in our modern day, often helped persons abandoned as hopeless by the orthodox medical realm.[#] This is not said as a blanket endorsement of these practices, for there are of course some charlatans and incompetents in these fields—even as there are in the medical field. But the statement is simply made in harmony with the idea that herbs do not need phytochemical "verification" of their efficacy to be successfully employed. Rather, in the hands of a skilled healer, the consistent *results* obtained—culture after culture, person after person, animal after animal—are the only testimony that herbs have ever truly needed.

In conclusion, then, as we have discussed above, the health-conscious person may benefit from the judicious use of weeds in a variety of ways. Such benefits are discussed in more detail in the individual plant entries appearing in the field guide. Of course, these herbs are not cure-alls, but the advantages they offer to man have, unfortunately, been all too neglected.

---

[*] An excellent example of this can be found in the recent and fascinating study by two scientists of the healing practice of country herbalist Tommie Bass, in which the researchers uncovered and collated much scientific evidence in support of Bass' time-tested "country remedies," who himself had been unacquainted with the scientific body of literature but only skilled in herbal culture and having had his skills sharpened by long years of clinical experience and observation. *(Crellin & Philpott, 1989, passim)*

[#] Chagrined by such successes, and unable to deny the plain evidence that many people got better after taking herbs, a number of orthodox medical practitioners of the 1950s, 1960s, and early 1970s endeavored to explain these cases by means of what has been called the "placebo" theory, which alleged that herbs possess no innate therapeutic properties but that they work because the user thinks they will, so that the mind heals the body. Of course, this contention has been excluded for several decades now by the escalating phytochemical attestation to the pharmacological properties possessed by herbs. But it was never advanced by careful thinkers in the first place, owing to the long-demonstrated evidence that herbal therapies worked, not only on countless past generations of humans who never possessed modern medicine, but on infants and animals as well! (In fact, many animal herbals have been published over the years.)

Introduction to the Field Guide

The field guide lists plants in alphabetical order according to their most-often-used common name. (If you prefer, however, any plant can be located by its scientific name by means of the index in the back of the book.) At the beginning of each monograph, a detailed physical *Description* of the plant is given, in which reference is made as well to one or more numbers corresponding to color photos that can be found immediately following the field guide. (These photos are matched to the text by means of reference numbers appearing in their captions.) This descriptive material is followed by information on the plant's *Range & Habitat* and thereafter on its known *Constituents.*

In most monographs, detailed information is given on how the particular plant may be consumed as *Food* (i.e., whether raw or cooked or both) and instructions are usually given for cooking. Sometimes, too, skeleton recipes are provided.

Since the late 1960s, there has been a great resurgence of interest in herbal medications. In accord with this, a large and detailed section *(Health/Medicine)* on the therapeutic benefits of the respective plants is provided. This serves, in most instances, both as an herbal **REPERTORY** and as a **MATERIA MEDICA**. Here the reader will note that physiological functions of the plants are listed in small letters and bold italics (thus, *astringent*), while pathologies or other health conditions are listed in capital letters and bold italics (thus, *PNEUMONIA*). Some readers, in consulting this information, may be surprised to learn that some of the most potent or multi-utilizable herbal medications are derived from the commonest of weeds. However, this simply underscores what Ralph Waldo Emerson said in his essay of 1878 entitled *Fortune of the Republic*: "What is a weed? A plant whose virtues have not yet been discovered." Nowadays, as noted in the previous Introduction, a plethora of new information on the pharmacological properties of these "lowly" weeds is being processed ever increasingly by scientists. Hence, in referring to therapeutic uses old and new, I have made a special effort to reference published studies conducted by researchers which support the herbal uses that I have listed.

Here, however, an important point needs to be stressed: While specifics for herbal use are sometimes related, it should be understood that such material is included by way of historical, scientific, and otherwise informational purposes, and is not intended by way of prescription for maladies. Moreover, self-treatment, as well as self-diagnosis, can be risky. Overuse of a self-prescribed and administered herbal formulation may prove harmful, just as a weak formula—or an infrequent enough consumption of a properly made formula—may prove ineffective. Thus, precious time could be wasted while other methods of healing might have been skillfully implemented by a professional.

Of course, many practitioners of natural therapy (herbalists, naturopaths, holistic chiropractors, etc.) consult with the public relative to the use of herbs. One who wishes to pursue an herbal regimen might choose to inquire into the services of one of these professionals and/or seek out a knowledgeable medical doctor who would be willing to consider one's inquiries as to an herbal regimen or at least one in conjunction with his or her prescribed drug therapy. Here there may be little or no conflict between the use of drugs with particular herbs, especially if they are taken apart from each other during the day. But, there are a number of glaring exceptions, and a person wishing to choose both regimens would no doubt be wise to carefully discuss this notion with his/her physician as well as a knowledgeable pharmacist and/or herbalist. Obviously, the potential for a dangerous situation would most likely arise if one were to take an herb that acted on a condition in either an *opposite*, or conversely, *identical* manner as did the drug. While, in the first instance, the drug's effects might be neutered, in the latter instance its effects could be overamplified, with conceivably negative potential in certain situations. (One can imagine how this might be the case with conditions like diabetes, heartbeat problems, and nervous-system disorders, where a delicate balance of the body's chemicals needs to exist) On the other hand, herbal therapy usually dovetails best with a drug therapy when the chosen herb acts as an adjunct to the drug—working on a peripheral area—and not as an amplification or negation of the drug.

As to other features of the field guide: **Cautions** are given at the end of each monograph, often including contraindications for pregnancy and lactation. One should pay special attention to these, as they are there to help offset potential problems. My only regret is that this material is not exhaustive. Our present state of knowledge simply does not allow for certainties here, and therefore the cautions listed must be viewed as preliminary only and not by any means as comprehensive.

Referencing in the text is by author, year, and page, to be matched up in the **REFERENCES** section at the rear of this volume. **APPENDICES** following the text include a detailed index to the plants based on physiological function (i.e.,*astringent, demulcent*, etc.) and a list of microorganisms inhibited by the various plants.

A **GLOSSARY** at the end of the field guide attempts to clarify the usage of certain technical terms that may bewilder the neophyte student of wild-plant ways. The study rounds out, as mentioned earlier, with both a bibliography (by way of the alphabetized *References* section) and a very lengthy—indeed, nearly comprehensive— **INDEX.**

I trust that conservationists will appreciate that I (a conservationist myself) have made a special effort to include in the following monographs only those plants which are commonly thought of as "weeds" or otherwise are common enough not to be threatened by judicious foraging. Thus, I have omitted numerous endangered species sometimes listed in other foraging manuals (e.g., ginseng, goldenseal, gentian, etc.)

IMPORTANT NOTE: Before foraging or even studying the field guide, you are advised to read both the disclaimer on the copyright page as well as the two Introductions preceding the present Introduction, as this material contains some important cautions that need to be observed.

ARROWHEAD
[WAPATO; DUCK POTATO]
(Sagittaria latifolia)

Description

[See photos #1 & 2 in rear of book.]

An erect, perennial, aquatic to semi-aquatic plant that grows 5-40 ins. tall.

Arrowhead is characterized by large, *broad,* dark-green, blatantly arrow-shaped leaves that are terminally situated on a basal leafstalk.

A leafless flower-stalk bears waxy, white, rounded, three-petaled flowers (each petal 1-2 cms.) whorled in groups of three at spots along the stalk. Flowers can be either male (with a fuzzy yellow center) or female (having green mounds).

In the autumn, the flowers are replaced by the fruits, which consist of fuzzy, greenish-brown, globular heads.

Long roots lie submerged beneath the mud, with the main one terminating (often at some distance from the main plant) in a tuber that is possessed of a milky juice.

Range & Habitat

Throughout our area, at the edges of ponds, slow streams, lakes, and marshes.

Seasonal Availability

Tubers ripe for collection in October-November.

Chief or Important Constituents

Diterpene ketones(trifoliones A, B, C, & D), diterpene glucosides(sagittariosides A & B), other diterpenes, and arabinothalictoside. Nutrients include exceptional amounts of phosphorous, potassium, and thiamine.

Food

In the autumn, arrowhead's long—often angling—roots terminate in a tuber about the size of an acorn. If a quantity of these can be gathered and cooked, they are a culinary delight. (Raw tubers, while passably edible as a survival food, should not normally be consumed as they contain a bitter, milky juice that can produce a sting in the throat. This principle is destroyed during cooking.)

Arrowhead tubers can be prepared exactly as one would prepare potatoes. The Potawatomi Indians relished a dish made from these "potatoes" mixed with deer meat and crystallized maple syrup.*(Smith 1933:95)* Our area's Ojibwe (Chippewa) Indians dried them and strung them up for use during winter when food was scarce.*(Black 1973:123)*

Obtaining the tubers, however, is no simple task. Reaching down with one's arm is often disappointing, as well as messy. However, the Indians used to dispatch their women into the water, barefooted, to dig up the tubers with their toes. Once thus released from the roots, the tubers would float to the water's surface, facilitating easy collection. This was usually done in the cold, autumn waters when the tubers were at maximum growth, which was usually when the seed stock was left alone of the plant above the surface of the water (or at the least, situated next to its shriveled remains).

Modern foraging manuals recommend using a rake to accomplish the same task. Yet, somehow this seems unfair to me. After all is said and done, I still think it best to employ the time-tested methods of the wilderness' original and most experienced students, the American Indians. Now, don't be timid, gals—the water's just fine!

Health/Medicine

Ojibwe Indians made a tea from the tubers for **INDIGESTION.***(Densmore 1974:342-42; Smith1932:353)* The Iroquois valued such for **CONSTIPATION.***(Herrick1995:239)* They also employed an infusion of the leaf for **RHEUMATISM.** *(Herrick 1995:239).*

The famous British herbal by Maud Grieve states that arrowhead is **antiscorbutic** and **diuretic.** *(Grieve 1971:1:57)* The latter attribute was especially appreciated by early American healers, with contemporary botanical scholar Constantine Rafinesque pointing out that the "roots" were viewed as especially "useful" when "applied to feet or yaws and dropsical legs." Rafinesque added that "the leaves [are also] applied to breast to dispel milk of nurses [as an **antigalactic**]." *(Rafinesque 1830:259)*

A recent investigation found **antihistamine** properties experimentally verified for the diterpene constituents of the tubers.*(Yoshikawa et al 1996:992-99)*

Cautions

* Don't eat raw tubers, which can leave a sting in the throat.

* Don't confuse arrowhead with the poisonous Arrow Arum *(Peltandra virginica)*, the tubers of which are toxic unless treated (including a thorough drying) in a long, involved fashion. Both plants grow at the edges of lakes and marshes and they have, at first glance, a very similar appearance to their leaves. However, arrow arum's leaf veins all run diagonally from the midrib, while those of arrowhead run parallel to it, having originated from a common point at the center of the leaf. Arrow arum's flower is also wrapped tightly in a spathe, unlike arrowhead's.

* Don't confuse arrowhead with the toxic Wild Calla [Water Arum] (*Calla palustris*), which grows in the same habitat, but which has heart-shaped leaves without conspicuous lobes. When flowering, it possesses a large, white flower situated in a spathe. (As mentioned above, arrowhead's flowers are not in a spathe.)

ASTER

(Aster spp.)

Description

[See photos #3 & 4 in rear of book.]

There are over 25 species of asters in our region. These members of the *Compositae* (or, *Asteraceae*) family have flower heads composed of numerous ray flowers surrounding a central disk flower that is almost always yellowish. Our area's major species include the following:

Aster cordifolius (heart-leaved aster) grows 1-5 feet high and is possessed of hairy, toothed, heart-shaped leaves that appear on very slender stalks. The middle stem leaves are most sharply toothed. The ray flowers are violet-blue to rose in color. Dark-green tips adorn the flower bracts.

A. lanceolatus (panicled aster) is tall and smooth-stemmed and has willowlike leaves that have little or no teeth. They are also conspicuously short-stalked.

A. macrophyllus (large-leaved aster; big-leaved aster) reaches a height of 1-4 feet and has large, basal leaves that are thick and rough to the touch. The stem leaves are small, ovate, and stalkless. The ray flowers are lavender colored, while the disk flower tends to have a reddish tinge. The flower stalks are rough, possessing tiny glands.

A. novae-angliae (large blue aster; New England aster) grows 2-7 feet high and displays bristly, hairy leaves that are lanceolate and which clasp the stalk. They are 1½-5 inches long. The ray flowers are a vibrant violet. The flower bracts are narrow, hairy, and sticky—as is the flower stalk.

A. puniceus (purple-stemmed aster) can range anywhere from 2-8 feet tall. Its stout stem is hairy and reddish in color. The leaves gradually taper down at their base, where they clasp the stalk. Here the ray flowers are also violet, but sometimes quite pale.

Other species common to our area include *A. borealis* (rush aster), *A. ciliolatus* (Lindley's aster), *A. ericoides* (heath aster), *A. laevis* (smooth aster), *A. lateriflorus*, *A. ontarionis* (Ontario aster), *A. oolentangiensis* (azure aster), *A. sericeus* (silky aster), *A. umbellatus* (flat-top aster), and *A. hesperius* (Gray root).

Range & Habitat

Throughout our area in open, dry woods; also in fields, thickets, clearings, damp meadows, and swamps.

Seasonal Availability

Asters become conspicuous in the autumn, when they flower. (I wouldn't doubt if the expression "late bloomer" even stems from the last-minute appearance of their blossoms!)

Chief or Important Constituents

Tannins.

Food

The heart-leaved aster (*A. cordifolius*), known in Maine as "tongue," has long been used in this state as a green. *(Perkins 1929)* Pillager Ojibwe Indians treasured the roots of large-leaved aster (*A. macrophyllus*), using them as stock for soups. *(Smith 1932:398)* Algonquin Indians of Quebec preferred the leaves of this species, as did our own area's Ojibwe (Chippewa) Indians, who boiled them with fish. *(Densmore 1974:307, 32; Black 1980:108;)* The Flambeau Ojibwe also consumed the young and tender leaves as food. *(Smith 1932:398)* Indeed, as the wild-foods authorities Fernald and Kinsey have pointed out, only the very young leaves are edible, because as the leaves mature they become tough and leathery. *(Fernald & Kinsey 1958:355)*

Health/Medicine

A. cordifolius was used in nineteenth-century North Carolina as "an aromatic nervine in the form of a decoction and as an infusion for rheumatism and by old women, as *partus accelerandum*." *(Jacobs & Burlage)* The early-nineteenth-century botanical scholar Constantine Rafinesque emphasized the nervine connection, suggesting that aster be tried in the place of valerian *(Valeriana* spp.) to deal with **SPASMS, EPILEPSY,** and **HYSTERICS.** *(Rafinesque 1830:198)*

A. lancoelatus was used by Zuni Indians as an inhalant to stop **NOSEBLEEDS**; the plants were crushed and sprinkled on live coals and the smoke inhaled. *(Stevenson 1915)* This tribe also utilized *A. hesperius* in the same way. In addition, they put it to work to heal **WOUNDS** from arrows or bullets, which they did by soaking a cloth in a tea made from the plant, twisting the cloth, and washing out the wound therewith. The wound was then washed a second time, with more care and precision, by means of a twig wrapped with raw cotton soaked in the tea. *(Youngken 1925:165)*

A. macrophyllus was used by the Flambeau Ojibwe as a **FEMALE REMEDY**. The young roots were also made into a tea to wash the head so as to relieve a **HEAD-ACHE**. *(Smith 1932:363)* The Iroquois Indians made a tea from this species and four other plants (including sweet cicely and bloodroot—*see under those headings*) to treat **VENEREAL DISEASE**. *(Herrick 1995:230)*

ASTER

The Cherokee Indians poulticed pulverized roots of *A. novae-angliae* on **PAIN-FUL AREAS OF THE BODY**.*(Hamel&Chiltoskey1975:24)* Rafinesque tells of how he learned from a Dr. Lawrence of New Lebanon that a decoction of the tea was indicated in "many eruptive diseases of the skin" and that applied externally to **POISON-IVY** and **POISON-SUMAC RASHES,** it could even succeed in 'removing the poisonous state of the skin.' *(Rafinesque 1830:198)*

The Cherokees also implemented a tea made from the roots of this species to treat **DIARRHEA** and **FEVER**.*(Hamel & Chiltoskey 1975:24)* The Iroquois likewise appreciated it for the latter complaint, putting the roots and leaves from two plants in six quarts of water, boiling this down to one quart, and then drinking this often when the fever was bad.*(Herrick 1995:230)*

A. puniceus was listed in the 21st ed of the *U.S. Dispensatory* as an **astringent** and was also said to have been appreciated as a "stimulating diaphoretic in rheumatic and catarrhal affections." *(Wood et al 1926:1213)* The Woods Cree Indians found, indeed, that a root decoction served as a **diaphoretic** to reduce a **FEVER**.*(Leighton 1985:31)* The Iroquois implemented it to the same end, infusing 3-4 roots in two quarts of water—taking it warm if the patient was chilly, but cooled if he/she was hot.*(Herrick 1995:230)* This species was also used by the Woods Cree to alleviate illness that accompanied **TEETHING,** as well as to help women who **FAILED TO MENSTRUATE.** Viewed also as a valuable **anodyne**, it was chewed and placed on an **ACHING TOOTH** to provide relief. It seems further to have been used in the treatment of **FACIAL PARALYSIS,** especially where a contortion or grimace was observed similar to what occurs with Bell's palsy.*(Leighton 1985:31)*

Caution

In soils rich in selenium, asters can accumulate large amounts of this mineral into their structure. In that selenium is toxic if excessively incorporated into the human diet, one should take care to limit consumption of this plant.

BLOODROOT
[RED PUCCOON]
(Sanguinaria canadensis)

Description

[See photo #5 in rear of book.]

An odd-looking, but most beautiful, plant, topping off at 6-12 inches. Its flowering stem and rhizome are filled with a reddish-orange juice that exudes when it is damaged.

The lone, large (4-to-7-inch) leaf is basal and palmate, with 5-9 lobes. It is wrapped around the single flower bud as it emerges and then expands as the flower blooms.

The short-lived blossom is about an inch and a half in diameter and opens only on sunny days. It has 8-12 oblong petals and appears at the end of the leafless stem. Numerous golden stamens combine with the snow-white petals to give the flower a most lovely and striking appearance.

Bloodroot flowers early in the spring, before many other plants open their blossoms. In that it tends to grow in colonies, this makes for quite a sight on an other-wise flowerless terrain during the muddy days of late April to early May.

The fruit consists of a capsule with two parts. It is pointed at the ends.

Range & Habitat

Bloodroot thrives in colonies in rich woods throughout our area, but it is getting harder and harder to find.

Chief or Important Constituents

Isoquinoline alkaloids(incl. sanguinarine, sanguidimerine, berberine, protopine, chelerythrine, chelidonine), organic acids(citric, malic), resin

Health/Medicine

This is a plant with significant, even blatant, physiological activity, having been listed in the *U. S. Pharmacopoeia* from 1820 to 1926 and in the *National Formulary* up until 1965. It possesses *drying* and *cooling* energies when used in small amounts and more *heating* energies when used in larger amounts.

Bloodroot has been a respected *vulnerary* in various cultures, especially among the American Indians. Modern herbalists still use a bloodroot paste or vinegar extract of the fresh roots as a topical treatment for **RINGWORM, SCABIES, ATHLETE'S FOOT, TINEA, WARTS**, and *ECZEMA,* always being careful to apply it only to the diseased part of the skin.*(Gunn 1859-61:756;Tierra 1992:385)* As to the latter complaint, a 1907 study published in the *Journal of the American Medical Association* described the successful use of a bloodroot fluid extract in the treatment of papular and squamous eczema.*(Graber 1907:705)*

Probably bloodroot's most famous use has been as a salve for *SURFACE CANCERS,* and that as a powerful *escharotic.* This interesting treatment has had a powerful lay following.*(Naiman 1999)* Previously, however, it was also appreciated by physicians,*(Lewis & Elvin-Lewis 1977:123-24)* having been given prominence in orthodox circles due to the investigations of J.W. Fell, a nineteenth-century physician. Fell, having heard about how Indians living on the shores of Lake Superior had successfully applied it to cancer, had his curiosity aroused and ruminated about trying it on some of his "hopeless" patients with surface tumors. He eventually decided to try mixing a bloodroot extract with flour, water, and zinc chloride, smear the resultant paste on cotton or cloth, and then cover their tumors with this. This he implemented and continued on a daily basis. Eventually, after some days, the tumors developed encrustation, and as they did so, Fell made incisions approximately one-half-inch apart in the hardened mass and then inserted the bloodroot paste into the cuts. In a mere 2-4 weeks, the cancer was reported to have been destroyed, with the mass falling out 10-14 days later, leaving a flat sore that soon healed.

Fell's technique was next tested in a controlled setting at Middlesex Hospital in London, where twenty-five cases (mainly involving breast cancer) were treated by his methods. Fascinatingly, remission occurred with each and every case! Three out of ten persons did eventually relapse and return for further treatment two years later, but this favorably compared to the eight out of ten also returning who had been treated by the standard medical procedures.*(Fell 1857)* Judged in this light, the treatment appears to have been most successful.

Fell's procedures were abandoned not long after his death, but were revived by several physicians in the 1960s, led by J. T. Phelan, who experienced dramatic results with a bloodroot/zinc-chloride paste for carcinomas of the nose and ear. These physicians reported that most of their patients were totally healed and that only a few experienced reoccurrences. Their results were widely published in respected surgical journals.*(Phelan et al 1962:25-30; Phelan et al 1963:310-14;Phelan &Juardo 1963:224-46)* In 1972, the *American Journal of Surgery* even carried an article detailing the successful use of a similarly structured paste in the topical treatment of ulcerated breast lesions (grade III adenocarcinoma), which proved to be quite successful and concerning which a scrupulous follow-up found the subjects cancer-free some eighteen months later.*(Sonneland 1972:391-93)* Detailed analysis since has revealed that bloodroot's alkaloids are the active *antineoplastic* agents, having been shown to produce therapeutic results on carcinomas and sarcomas in mice.*(Stickl 1929:801-67; Shear et al 1960:19-24)*

33

Although bloodroot is most famous for its use in cancer, most herbalists no doubt have utilized the plant's root more often for afflictions of the respiratory tract, for which region it seems to possess a special affinity. Thus, the Pillager Ojibwe implemented the root juice for **SORE THROAT**,*(Smith 1932:377)* while the Forest Potawatomi squeezed out this juice onto maple sugar as a throat lozenge for same. They also used an infusion of the root for **DIPHTHERIA,** which they recognized as a throat disease.*(Smith 1933:68)* Bloodroot has, however, been most predominantly viewed (e.g., in *the U.S. Pharmacopoeia* and *U.S. Dispensatory*) and implemented as a powerful **expectorant**, either by way of a tincture or as a decoction (the latter finding a teaspoon of the root boiled in a pint of water and then a teaspoon of this tea taken three to six times a day). This effect seems to be primarily due to the sanguinarine content, which comprises some 50% of bloodroot's alkaloidal constituents.*(Karlowsky 1991)* The *U.S. Dispensatory* has recognized its use for "subacute and chronic bronchitis" through many editions. As to the indications for such a use, Eclectic physician John Scudder wrote in the 1870s that bloodroot was "valuable in bronchitis with increased secretion" and that "in minute doses [1/2 to 5 drops of the tincture, max] we employ it in cases of cough with dryness of the throat and air passages, feeling of constriction in the chest, difficult and asthmatic breathing, with sensation of pressure."*(Scudder 1870:208; cf. Ellingwood 1983:242)* Herbalist Peter Holmes, explicating the energetic principles involved in its indication, says that bloodroot is best used with what Traditional Chinese Medicine (TCM) would call "lung phlegm cold" (manifesting a productive cough with sternum pain) or "lung phlegm dryness" (manifesting an unproductive and harsh cough with a throat tickle).*(Holmes 1997:1:232)*

As for the "asthmatic breathing" mentioned by Scudder, southwest-American herbalist Michael Moore finds bloodroot useful with **ASTHMA** that is largely extrinsic and which is characterized by a dry, spastic cough.*(Moore 1994:4.2)* California-based herbalist Michael Tierra finds it most helpful for asthma when there is cold and thick phlegm. Tierra also finds it invaluable for **PNEUMONIA,** especially when dosed at one or two drops, repeated often during the day.*(Tierra 1992:384)* Early American physician Asahel Clapp wrote that he had used it successfully for over thirty years for both spasmodic asthma and pneumonia, as well as for **VESICULAR EMPHYSEMA** and **TUBERCU-LOSIS**.*(Clapp 1852:735)* The Micmac Indians had likewise used it for tuberculosis.*(Lacey 1993:6)* Of especial interest here is that an aqueous extract of the root was shown in 1954 to exert a powerful influence against a virulent human strain of TB, even in as high a dilution as 1:80!*(Fitzpatrick et al 1954:520)*

Cherokee Indians implemented small amounts of the root for **LUNG INFLAMMATION**.*(Hamel & Chiltoskey 1975:26)* Dr. Clapp had also found, over his thirty-year span of implementing the herb, that it "frequently cures or relieves pneumonic inflammation."*(Clapp 1852:735)* Then, too, the leaves of this plant were implemented by the Micmac Indians to treat **RHEUMATISM**,*(Lacey 1993:6)* and an infusion of the root was used for the same purpose by several other tribes.*(Lewis & Elvin-Lewis 1977:166; Vogel 1970:355; Chamberlain 1888:156))* Such applications should not be surprising in view of a 1976 study which found that a sample of bloodroot's aerial portions (the roots were not tested) possessed **anti-inflammatory** properties.*(Benoit et al 1976:167)*

Aside from the usage for tuberculosis, referred to above, various Amerindian tribes cherished a spectrum of uses for bloodroot that we today would also classify under the heading of "antimicrobial." For example, the Malecite Indians would steep bloodroot, then boil it down some more, then soak a rag with it, and finally compress this onto **INFECTED CUTS**, applying a fresh rag as soon as the previous one had dried.*(Mechling 1959:246)* Both Indians and southern whites utilized bloodroot for **INTER-MITTENT FEVERS** (malaria, dengue, etc.).*(Vogel 1970:355)* Not surprising, then, is that bloodroot's chief alkaloid, sanguinarine, has been demonstrated to possess significant and wideranging **antimicrobial** activity, including against both gram-positive and gram-negative bacterial strains, fungal pathogens such as *Candida*, and protozoa such as *Trichomonas* (a sometimes cause of vaginitis in women).*(Godowski 1989:96-101)*

Another alkaloid in the root, chelidonine (also found in celandine [*Chelidonium majus*]), has demonstrated anti-HIV activity in test-tube studies*(Kakiuchi et al 1987:22-27)* Taking the clue from this and other research, certain scientists adminstered freeze-dried bloodroot and celandine to thirteen patients with **HIV INFECTION**, with the result that marked improvements were noted in their CD8 values as well as in their persistent lymphadenopathy.*(D'Adamo 1992:31-34.)* A year later, a drug based on chelidonine named Ukrain was available to be injected into AIDS patients, which was found to markedly reduce their Kaposi's sarcoma lesions and to improve their CD4+ cells.*(Martinez et al 1993:401)*

Our area's Ojibwe (Chippewa) Indians implemented a tea of *Sanguinaria*'s root for **ABDOMINAL CRAMPS**.*(Densmore 1974:344)* Dr. Scudder, quoted above, used it similarly, finding that it was "valuable in...atonic conditions of stomach and bowels with increased secretions of mucus."*(Scudder 1870:208)* Bloodroot has also traditionally been used for **AMENNORHEA** and **DYSMENORRHEA**. Eclectic physician Finley Ellingwood wrote that "the tincture in full doses is an emmenagogue, restoring the menses when suppressed from cold" but that "it is not to be given if menstrual deficiency is due to anemia, although it is tonic and stimulant in its influence upon the reproductive organs."*(Ellingwood 1983:243)*

Bloodroot also yields **emetic***(Lewis&Elvin-Lewis1977:278;Speck1917:318)* and **cardiotonic***(Newell et al 1996:42)* properties. The latter would seem to be at least partly due to the protopine content, which has an antiarrhythmic action.*(Lewis & Memory-Lewis 1977:191)*

Renewed interest in the plant by scientists occurred in the 1980s and 1990s, when numerous published studies demonstrated that sanguinarine helped to prevent and reduce **GINGIVITIS** and **DENTAL PLAQUE**, evidently by killing the bacteria involved and by blocking enzymes that degenerate gum tissue by destroying its collagen.*(Godowski 1989:96-101; Grossman et al 1989:435-40; Dzink & Socranksy 1985:663-65; Harper et al 1990:352-58; Mauriello & Bader 1988:238-43; Harper et al 1988:359-63)* As might have been expected, several mouthwashes and at least one brand of toothpaste came to add sanguinarine to their list of ingredients once these studies began to proliferate.

Cautions

* Bloodroot is **CONTRAINDICATED** for use in *PREGNANCY* and *LACTATION* owing to the potentially toxic nature of its alkaloids and their uterine-stimulating properties. It is contraindicated as well with *GLAUCOMA*. (Persons afflicted with such a condition should probably also avoid using toothpastes or mouthwashes containing sanguinarine.)

* *This is not a plant for the layman to harvest and prepare, owing to its toxic potential!* In addition, bloodroot is becoming somewhat scarce in our region, making it further incumbent for one to use only the commercial preparations grown from cultivated plants. Even when using commercial preparations, however, one needs to be very careful not to exceed the recommended dosages. Overdosing would be very irritating to the system and could cause violent vomiting and possibly even death. (Not only that, but high doses simply *do not work* for the conditions delineated in the text above; one to several drops of the tincture, or a teaspoon of the tea repeated several times a day, are the effective ranges.) Also, when taking bloodroot tincture internally, one needs to take care to *dilute* it *thoroughly*, as it would be most irritating otherwise. Bloodroot is also caustic when used externally, so it should likewise be diluted in these cases (unless such an effect is actually being sought via the topical application—as for example, with warts).

BLUEBEAD LILY
[CORN LILY; CLINTONIA; YELLOW CLINTONIA]
(Clintonia borealis)

Description

[See photos # 6 & 7 in rear of book.]

Named after the famous governor of New York, DeWitt Clinton (1768-1829), who was also a naturalist, this gorgeous plant adorns the northern woodlands.

Growing 6-18 ins. tall, clintonia has 2-5 (usually 3) broad, shiny, succulent, parallel-veined, basal leaves. These grow 5-12 inches long.

Inflourescence consists of 3-10 drooping, yellow-green, bell-like flowers on a leafless stalk that ranges from 6-16 ins tall. The blossoms possess six stamens

The fruit consists of steel-blue, oval-shaped berries, each about 8 mms. thick.

Range & Habitat

Clintonia flourishes east of the Mississippi River, from Newfoundland and Labrador south to Tennessee and Georgia. In our area, it thrives in the upper half except for Minnesota's western counties. The preferred habitat is cool, rich, open woods with acid soil.

Chief or Important Constituents

Little chemical work has been done on this plant. The fruits are known to contain sugars(levulose, dextrose) and acids(citric, acetic, tartaric). The steroidal saponin diosgenin occurs in the roots. There is a strong anti-inflammatory component present in the leaves (as there is in clintonia's close cousin, false Solomon's seal, for which *see entry on pp. 126-28*), which to my knowledge has not yet been identified, but which may consist partly of saponins.

Food

The young (still curled or newly unfurled) leaves are edible and delicious in a salad, tasting a lot like cucumber! They are, without a doubt, one of my favorite spring greens. As they age, however, they become tough and bitter and leave an unpleasant aftertaste.

Clintonia can be eaten as a potherb, too. The wonderful foraging manual by Fernald and Kinsey informs us in this regard: "The very young leaves of this plant are extensively used as a potherb by country people in parts of Maine under the name of Cow Tongue." *(Fernald & Kinsey 1958:134)* Here the leaves are boiled for only about 10 minutes.

Health/Medicine

Although neglected today, this herb was widely utilized by American Indians. First and foremost in this regard was its use as a *vulnerary*. Thus, Algonquin Indians employed the leaves as a poultice for *BRUISES* and *SORES*, laying them on after wetting or bruising them.*(Rafinesque 1830:211)* They also poulticed them on *INFECTIONS* and *OPEN WOUNDS*.*(Black 1980:138)* Bella Coola Indians toasted a lone leaf to apply to *SORE EYES* and *WOUNDS*.*(Smith 1929:53)* Other Indian tribes in the Quebec region found bluebead lily useful for the suppuration of *TUMORS*.*(Erichsen-Brown1989:346)* Our own area's Ojibwe (Chippewa) Indians used the plant topically for various kinds of skin injuries, including *BURNS* and *SCROFULOUS SORES*.*(Densmore1974:354-55)* Washington state's Cowlitz Indians used the juice from a smashed plant of the Western species, *C. uniflora*, for topical application to *CUTS*.*(Gunther 1973:25)*

The Iroquois Indians found that a compound decoction inclusive of bluebead lily was useful in managing *DIABETES*.*(Herrick 1977:283)* They also implemented a decoction of this plant to benefit *HEART PROBLEMS*. *(Herrick 1995:244; Herrick 1977:283)* In that the herb is closely related to false Solomon's seal (see entry under that heading), which contains the cardioactive glycoside convallarin, it is possible that this chemical may exist in this plant as well and thereby elucidate the cardiotonic usage.

The Micmac Indians appreciated this plant when *GRAVEL* was present, for which they drank the expressed juice from the roots.*(Speck 1917:317)* Among the Flambeau Ojibwe, a clintonia root infusion served an important role in aiding *CHILDBIRTH* *(Smith 1932:373)* Might it be significant in this regard that the roots contain diosgenin*(Marker 1942:1283-85)*, the precursor to progesterone? Or was the infusion instead simply used because of its regenerative effects upon tissue, in the manner of its usual implementation (as a vulnerary)?

One use for bluebead lily still occasionally employed in our area is the plant's insect-repellent properties, gleaned from the Tete-de-Boule Indians, who had found that rubbing the crushed leaves on the face and hands largely protected one from *MOSQUITO BITES*.*(Raymond 1945:126)* Otherwise, sadly, this herb is mostly neglected nowadays when "Pop Superstar Herbs" are all that most herbal researchers are inclined to investigate. Yet, in view of bluebead lily's many uses as enumerated above, as well as the inclusion of a steroidal saponin (the valuable diosgenin) in its chemical profile, I am inclined to agree with Jim Duke ("the Duke of Herbs") when he writes with a sense of urgency of regarding this plant: "Science should investigate!"*(Foster & Duke 1990:100)*

Cautions

Although certain early writers described the tempting blue fruits as "sweetish, edible,"*(Rafinesque1830:211)* it appears that they are actually inedible and perhaps even mildly toxic.*(Fernald &Kinsey1958:134-35)*

BLUE COHOSH
[PAPOOSE ROOT; SQUAW ROOT]
(Caulophyllum thalictroides)

Description

[See photos # 8 & 9 in rear of book.]

A native perennial growing 1-3 feet tall and possessed of a waxy, whitish bloom on its purplish stem until the leaves are fully opened. When that happens, the bloom disappears and the stem becomes greenish.

There are only two leaves, which are stalkless and compound—possessed of numerous wedge-shaped to egg-shaped leaflets. The upper leaf, which is the largest, is divided into 7 to 9 leaflets, while the lower and smaller leaf has 27 leaflets, each growing 1-3 inches long.

The pretty, greenish-yellow flowers are small—only about a half an inch in diameter. The corolla displays six, small, hoodshaped petals and six larger, pointed sepals. Several blossoms can be found in loosely branched, terminal clusters.

What look like blueberries (about their size and color) appear in place of the flowers in the summer, on small and inflated stalks. These, however, are not the plant's fruits, but actually its seeds!

Range & Habitat

This is a plant of moist, rich woods. It occurs in such environs throughout our area, but is not as plentiful as it used to be.

Chief or Important Constituents

Lupine alkaloids(n-methylcytisine[caulophylline], anagrine, baptifoline), iso-quinoline alkaloids(magnoflorine[thalictrine]), flavonoids(leontin), triterpene saponins (incl. caulosapo[gen]nin[hederagenin], caulophyllosaponin, cauloside D), phytosterols, phosphoric acid, resin, citrullol, tannin.

Health/Medicine

Blue cohosh is a powerful and important herb, possessed of *warming* and *drying* energies. It has had broad applications in herbal medicine, although today it has pretty much been pigeonhold into only one category in the mind of the public. However, as stated, its uses are legion....

First, Omaha Indians reverenced this plant as the best of all *febrifuges,* giving a decotion of the roots for *FEVERS.(Gilmore 1991:31)* Iroquois Indians found that it worked as a febrifuge as well, in fact for "any kind of fever."*(Herrick 1995:125)*

It was also treasured by native Americans as an *emetic.* Our area's Ojibwe (Chippewa) Indians utilized it in such a manner,*(Smith 1932:358)* whereas the Iroquois Indians implemented an infusion of the smashed roots to more specifically help persons with gallstones to vomit.*(Herrick 1977:333)* Blue cohosh is also listed as a *vermifuge,* and here the same principles may be at play.*(Wren 1988:84; PDR Hrbs)*

Caulophyllum is *antispasmodic(Smith 1999:16-17)* and was thus used by the Ojibwe for *STOMACH CRAMPS(Densmore 1974:344)* and for *INDIGESTION.(Densmore 1974:342)* Then, too, a study conducted in the mid-1970s found that it significantly *inhibited inflammation* in the carageenan-induced, rat-paw-edema model,*(Benoit et al 1976:163)* which effect may be due to the leontin contained in the plant.*(Willard 1991:41)* At any rate, the Cherokees found that they could rub the leaves on *POISON-OAK RASH* to achieve relief from that irksome inflammation.*(Hamel & Chiltoskey 1975:30)* It has also proven helpful for *ARTHRITIC PAINS*, especially when in the fingers or toes.*(Winston 1999:35)* Another classic use has been for *RHEUMATISM,(Wren 1988:84)*, with Indians tribes such as the Iroquois*(Herrick 1977:333)* and Cherokees*(Hamel & Chiltoskey 1975:30)* using it for such an affliction. Aside from its *anti-inflammatory* properties, a *diuretic* potential in the plant may also contribute to its successes with rheumatism. *(Wren 1988:84; PDR Hrbs)*

There appears to be some affinity for the respiratory system as well. The Ojibwe found it helpful for *LUNG TROUBLES*, making a decoction from two roots and one quart of water and dosing the tea at one swallow.*(Densmore 1974:340-41)* These Indians also took a small quantity of the finely scraped root and squeezed it through a white cloth in warm water to treat *LUNG HEMORRHAGES.(Densmore 1974:346-47)* Eclectic physician Finley Ellingwood noted that blue cohosh had been used with repeated success in treating *BRONCHITIS, PNEUMONITIS,* and *WHOOPING COUGH.(Ellingwood 1983:481)* Interesting here is that the triterpene glycosides of the Russian species, *C. robustum,* have been shown to be *antimicrobial.(Anisimov et al 1972:834-37)* Lab experiments with mice have revealed that they even *stimulate phagocytosis.(Di Carlo et al 1964:84)*

Blue cohosh seems to possess an affinity for the genito-urinary system, too. Our area's Mesquaki (Fox) Indians gave it to men with problems in this system,*(Smith 1928:205)* while Mohegan Indians used it for *KIDNEY DISORDERS.(Tantaquidgeon 1972:71, 128-29)* But it is best known as a "female remedy," implemented in a variety of conditions. Here the Cherokees valued it for an *INFLAMED WOMB,(Hamel & Chiltoskey 1975:30)* while Ellingwood recommended it for "constant ovarian irritation" as well as for "chronic disease of the uterus or ovaries or of the cervix."*(Ellingwood 1983:481)* Present-day herbalist Amanda McQuade-Crawford finds that it serves ably to shrink painful *UTERINE FIBROIDS.(McQuade-Crawford1996:136)* Ojibwe Indian women treasured it for *MENSTRUAL CRAMPS.(Smith1932:358)* Ellingwood agreed that "in painful menstruation it has an established reputation."*(Ellingwood 1983:481)* He found it invaluable for *AMENORRHEA* in young women, especially when used at the onset of menstruation.*(Ellingwood 1983:481)* Southwest-American herbalist Michael Moore says that it is helpful for women with **CONGESTIVE DYSMENORRHEA** who are experiencing cycles of more than thirty days.*(Moore 1994a:13.3; cf. Mills & Bone 2000:242)* Menominee and Forest Potawatomi Indians boiled the roots to obtain a tea that was drunk to diminish **PROFUSE MENSTRUATION**.*(Smith 1923:28; Smith 1933:43)*

But blue cohosh's best known use is as a *parturient.* It was listed in the *U.S. Pharmacopeia* from 1882 until 1905 for just such a function. The root is, indeed, a labor aid *par excellence,* specific for **STALLED LABOR,** and studies conducted in the 1950s suggest that the active agent in its oxytocic effects is its caulosaponin.*(Tyler 1993:47)* The Eclectics noticed that blue cohosh's action seems to be especially helpful with"delicate women" and where the labor has been slowed by pain or fatigue.*(Harrar & O'Donnell 1997:37; Willard 1991:42; Moore 1994a:14.2; McIntyre 1995:26)* Eclectic physician John Scudder well explained that it "exerts a very decided influence upon the parturient uterus, stimulating normal contraction, both before and after delivery. Its first use, in this case, is to relieve false labor pains; its second, to effect co-ordination of the muscular contractions; and third, to increase the power of these. The first and second are most marked, yet the third is quite certain. Still if any one expects the marked influence of Ergot, in violent and continued contractions, he will be disappointed."*(Scudder 1870:100)* Scudder said that blue cohosh is best used as a tincture of the recently dried root.*(Scudder 1870:99)* Susun Weed, discussing blue cohosh's parturient action, recommends taking 10-20 drops in a small glass of water, repeating hourly as needed to achieve delivery.*(Weed 1986:65)*

Blue cohosh's **oxytocic** properties have also been enlisted for *starting*, not merely assisting, labor. Even so, Weed stresses that its usage would only be effective *once the cervix has ripened.(Weed 1986:60-61)* In such a situation, she says, she has heard of good results with the application of a 200x homeopathic dose of blue cohosh, repeated every half-hour for two hours. She herself particularly suggests, however, the use of three to eight drops of the herbal tincture in a glass of warm water, repeated every half hour until contractions are regular. She says that if labor is not underway after four hours, this dose can be increased to a dropperful under the tongue every hour up to four more hours until the contractions become powerful and reliable.

Very small amounts of the root—always in a formula with herbs like black haw *(Virburnum prunifolium)* and several others—have traditionally been used to **prevent miscarriage** from the third to the ninth month of gestation. One example of such a formula is the the famed "Mother's Cordial" of the Eclectics. Ellingwood tells us that the *Caulophyllum* component of such a formula "prevents premature delivery by a superior tonicity, which it induces in all the reproductive organs. It has caused many cases to overrun their time a few days, and yet easy labors and excellent recoveries have followed."*(Ellingwood 1983:481)*

However, a popular notion among women who have a casual familiarity with herbs is that blue cohosh should be used—either alone or merely in conjunction with black cohosh (*Actaea* [formerly, *Cimicifuga*] *racemosa*)—for several weeks before the due date in order to tone the uterus and to keep the baby from running overterm. Here it is true that some modern herbalists—including Susun Weed—have discussed the use of blue cohosh in this regard. (Weed herself, however, sets a daily limit of two cups of tea or two divided doses of 20-30 drops of the tincture and advises use only when accompanied with black cohosh, quoting a midwife as saying that the use of blue cohosh alone can lead to "preciptious labor."-*Weed 1986:23*) The Eclectics, however, do not (with possibly the exception of Ellingwood-*Ellingwood 1983:481*) seem to have used blue cohosh by itself as a partus preparator, but only in small amounts as one ingredient in a mixture of five herbs composing the famed "Mother's Cordial," as noted above.

In view of the lack of any definable trend in historic use, then,*(see MH 12.1:12-15)* it is difficult to assess the reliability or safety of utilizing blue cohosh as a partus preparator, either alone or in conjunction with only black cohosh. There are, in fact, two modern cases which suggest that it may *not* be appropriate as a partus preparator: One woman who took three capsules of blue cohosh a day for three weeks before birth delivered a baby with heart problems.*(Jones & Lawson 1998:550-52; Edmunds 1999:34-35)* (The plant's saponins are definitely cardioactive; see below, under "Cautions.") Another who used both blue and black cohosh prior to birth delivered a baby who experienced seizures.*(Gunn & Wright 1996:410-11; Wright 1998:550-52; Baillie & Rasmussen 1996:410-11)* Until more is known about the margins of safety relative to the use of blue cohosh during pregnancy, its use as a partus preparator—especially by the layman—is strongly discouraged.

Cautions

   * Don't confuse this plant with the various meadow rues (*Thalictrum* spp.), which have similar leaves but quite different flowers and fruits.

   * Touching this plant can cause **DERMATITIS** in sensitive persons.

   * The blue "berries" (which, as we've seen above, are really the plant's seeds) are toxic and should not be consumed. Roasting them appears to eliminate, or reduce, the toxic elements,*(Spoerke 1990:42)* but there may be variables and intricacies of detail here that mandate preparation by a seasoned expert.

   * Blue cohosh should be avoided by persons with **HEART DISEASE** or **HIGH BLOOD PRESSURE** because of the cardioactive effects of its chemicals, methylcitisine (which increases blood pressure) and caulosaponine (which constricts coronary blood vessels).*(Tyler 1993:47; Spoerke 1990:42)*

   * *DO NOT USE BLUE COHOSH IN THE FIRST TRIMESTER OF PREGNANCY.* If desirous of using it in the second trimester for any miscarriage-preventative effects, *do not use it alone,* but if at all, only in a commercial preparation with other, time-tested, anti-abortive herbs (including black haw, *Viburnum prunifolium*), and be sure not to exceed the maximum recommended dose.

   * For parturitive effects during stalled labor, this herb is best used under the guidance of a skilled herbalist or midwife.

BLUE-EYED GRASS
[CHILL GRASS]
(Sisyrinchium spp.)

[See photo #10 in rear of book.]

Description

A perennial plant with a stiff, grasslike appearance and fibrous roots. The stem is wiry and two-edged. The lovely flowers have yellow centers and six, petal-like appendages (3 true petals, alternating with 3 sepals), distinguished by their pointed tips.

Several species can be found in our area, including...

Sisyrinchium campestre (prairie blue-eyed grass) has white or pale-blue flowers.

S. montanum (common blue-eyed grass) has blatantly flattened stems, deep blue or violet flowers on short stalks, and leaves that are 1/4-1/2 inch wide.

S. mucronatum (slender blue-eyed grass) has, as its common name implies, very slender stems and leaves (the latter only 1/12 an inch wide!). In this species, the leaves are shorter than the flowering stems.

Range & Habitat

This plant thrives in fields, meadows, and marshes throughout our area.

Health/Medicine

The Mesquaki Indians made a tea of *S. campestre* to treat **HAYFEVER**.*(Smith 1928:224)* Blue-eyed grass has also been used in Appalachian folk medicine for **FEVER** and **CHILLS,** according to country herbalist Tommie Bass.*(Crellin & Philpott 1989:105)* Early American practitioners of botanical medicine used various species of blue-eyed grass for **deobstruent, antiscorbutic, tonic,** and **laxative** purposes.*(Crellin & Philpott 1989:105)*

As to the latter, both the Cherokee and Iroquois Indians used the plant to "keep regular." Here the Iroquois imbibed a decoction of the root and stalks (½ cupful of the roots/stalks in 1½ quarts of water) just prior to breakfast to insure a morning bowel movement.*(Hamel & Chiltoskey 1975:26; Herrick 1977:288)*

However, the steeped root of at least one species, *S. angustifolium,* was used instead to quell **DIARRHEA,** especially that which occurred in children.*(Hamel & Chiltoskey 1975:26)* The Iroquois likewise used it for "summer complaint"—a nineteenth-century term for the epidemic diarrhea that especially affected children following hot, damp weather.*(Herrick 1995:247)* Interestingly, in the early 1950s, this species was found to possess antimicrobial activity, very powerfully inhibiting a virulent human strain of tuberculosis.*(Fitzpatrick 1954:529)* Might it also, then, inhibit the microbes that can cause diarrhea in children—so prevalent nowadays in day-care centers following hot, damp weather (analogous to the "summer complaint" of nineteenth-century medical literature)? There would seem to be, indeed, a marked effect upon the gastrointestinal tract, especially since **INTESTINAL WORMS** and **STOMACHACHE** were other complaints addressed by the *Sisyrinchium* genus, primarily by California's Mahuna Indians.*(Romero 1954:6)* I would suggest, then, that this genus would be a good candidate for rigorous phytochemical investigation.

BLUE FLAG
[WILD BLUE IRIS; LIVER LILY]
(Iris versicolor)

Description

[See photo #11 in rear of book.]

A divinely beautiful, native perennial that reaches a height of three feet. Its sturdy stem springs from a horizontal rhizome.

This iris' leaves are long (8-32 inches), flat, entire, toothless, and swordlike. They arise from a common basal cluster.

The gorgeous blossoms, reaching three inches in diameter, appear on branched stems. Each blossom displays what appears to be nine petals, in three whorls of three. However, only the middle whorl is composed of true petals—three, small, ascending, violet ones, blotched with yellow and white at their bases. The innermost whorl actually consists of three, pale-violet, crestlike branches of the flower pistil, which are flared at their tips. Most showy, however, is the outermost whorl of three, large sepals, which curve outward and then downward. The bases of these lovely structures are whitish with a blatant yellow blotch and bold purple veins running through this lighter coloration.

Long, erect, green pods eventually replace the breathtaking blossoms.

Range & Habitat

Blue flag thrives throughout our area in moist woods, wet meadows, marshes, ditches, and on lake shores and streambanks.

Chief or Important Constituents

Volatile oil(furfural), resins, iridin[irisin], a triterpenoid(iriversical), sterols, tannin, isophthalic acid, traces of other acids(salicylic, stearic, lauric, and palmitic)

Health/Medicine

Formerly official in the *U.S. Pharmacopoeia* (1820-95) and in *the National Formulary* (1916-42), the rhizome of blue flag is a classic *alterative*(Smith 1900:17; Mills & Bone 2000:254) and thus often helpful in chronic skin problems—especially where rough and greasy. This includes conditions such as *ECZEMA, PSORIASIS,* and other dysfunction of the sebaceous glands (e.g., *ACNE*).(Smith 1999:17; Felter & Lloyd)

It has often proven helpful for *NON-MALIGNANT ENLARGEMENTS* of bodily structures or tissues as well, especially of the *THYROID GLAND* and of the various *LYMPHATIC STRUCTURES*.(Scudder 1870:146; Felter & Lloyd; Clymer 1973:67-68) As to thyroid enlargement, it is the herb of choice when there is *FLUID RETENTION* coexisting with **DEPRESSED IMMUNITY,***(Moore1994a:7.5)* or in women when there are accompanying problems involving the reproductive organs.*(Smith 1999:17)* As to lymphatic dysfunction, it is primarily indicated with *soft, yielding lymph nodes(Smith 1999:17)* and it is specific for **LYMPHANGITIS** that is lingering and subacute.*(Moore 1994:21.6)* (Southwest-American herbalist Michael Moore says, however, that for lymphatic conditions, blue flag is best used in combination with other lymphatic herbs rather than by itself.*-Moore 1979:40)*

Because it has also *diuretic* properties, our iris can ably drain the lymph and blood of impurities,*(Ellingwood 1983:313; Scudder 1870:146)* so that it has often proven helpful in conditions such as **DROPSY** (at one time being widely used for such by physicians of Russia and early America*-Hutchens1991:58; Rafinesque 1830:232*), **RHEUMATISM,***(Tierra 1998:100)* (even as the Oklaloma Delaware Indians had discovered*-Tantaquidgeon 1972:118-19; Tantaquidgeon 1942:30*) **SYPHILIS,***(Scudder 1870:146; Hutchens 1991:57)*, **FLUID-FILLED CYSTS** that are very boggy,*(MH 9:2:12)* and sometimes even **OBESITY,***(Tierra 1998:100)* especially where this problem is connected with poor glandular function.

This plant is a powerful *sialagogue,* increasing not only the secretions of salivary glands but also those of other glands/organs, including the pancreas, spleen, intestines, and especially the liver and gallbladder. It is quite preeminently a *cholagogue*, and a very trustworthy one, indicated in poor hepatic management of bile as manifested by "biliousness," i.e., by the following symptom set: a"sick" headache, dizziness, indigestion and nausea (especially after eating fatty foods), and constipation with eventual clay-colored stools.*(Clymer1973:67-68; Moore1994:8.1)* Nineteenth-century medical botanist Laurence Johnson remarked that "in sick headache dependent upon indigestion, small doses, frequently repeated, often act most happily."*(Johnson 1884:271)* When used moderately, too, it is a mild *laxative* for the abovementioned**CONSTIPATION** attributable to poor liver function.*(Felter & Lloyd; Smith 1999:17;Millspaugh 1974:694)* In larger doses, however, it is *cathartic* as well as *emetic,(Wood et al 1926:597)* which various American Indian tribes had discovered and put to good use to attain such goals when necessary.*(Smith 1932:371; Bartram 1958:288)*

This plant was widely used by the American Indians, implemented most often as a *vulnerary*.*(Vogel 1970:283-84)* As to such, topical application of the rhizome was used for **SORES, BRUISES, SWELLINGS, and WOUNDS,** including by our own area's Ojibwe Indians.*(Densmore 1974:366; Gilmore 1991:20; Smith 1928:224; Speck 1917:315; Raymond 1945:129; Smith 1933:60; Gilmore 1933:126)* In the early 1700s, an American colonel with the last name of Lydius was healed of horrible sores on his legs by Indians who had poulticed the lightly boiled and crushed rhizome over his sores and had simultaneously washed his legs with the decoction.*(Kalm 1937:2:606)* These were probably **STAPH SORES**, for which blue flag has tradtionally been found helpful.*(Moore1979:40)* Indeed, the rhizome has been experimentally shown to manifest *antimicrobial* ability.*(Fitzpatrick 1954:530)* Further on this, Charles Millspaugh pointed out that the pulped, fresh root has been justly considered to be one of the very best topical applications for a painful**FELON** on the finger, another bacterial infection.*(Millspaugh 1974:694)* Then, too, the Omaha-Ponca Indians mixed the pulverized rootstock with saliva and droppered this mixture into the ear to alleviate **EARACHE**, which can also often involve bacteria.*(Gilmore 1991:20)*

Energetically, blue flag is held to be *cooling* and *drying* *(Holmes 1998:2:689)* and is appreciated as a plant for relieving stagnation. Looking at it from the standpoint of Traditonal Chinese Medicine (TCM), herbalist Michael Tierra apprises it as being especially applicable to excess, Yang conditions of a hypermetabolic nature. *(Tierra 1998:100)*

As to form and dosage, a liquid extract (tincture) of the fresh rhizome is sometimes used. This is the preparation discussed by Millspaugh. *(Millspaugh 1974:694)* The famous eclectic physician John Scudder also utilized the fresh-rhizome tincture, employing 76% alcohol, because as he asserted: "The dried root...possesses no more medicinal property than sawdust, and preparations from it...are an imposition." *(Scudder 1870:146)* Ed Smith, an herbalist of the western USA, is another who discusses using a tincture of the fresh rhizome, and he gives guidelines *(Smith 1999:17-18)* for using his own commercial tincture of such in one of two ways: (1) for regular, gentle use: mixing 40 to 60 drops into a half-cup of water and imbibing a teaspoon of the mixture every one or two hours; (2) for more intensive use: mixing 10 to 20 drops into a half-cup of water, adding a bit of carminative and/or demulcent herbs, *(cf.Tierra 1998:100)* and drinking immediately.

Other herbalists, however, would choose not to use preparations of the fresh rhizome. Revered American herbalist John Christopher stressed that the fresh rhizome is "somewhat acrid and purgative but the dried form retains its healing virtues, without the negative characteristics." *(Christopher 1976:76)* Christopher therefore recommended using a liquid extract consisting of 1/2 to 2 fluid drachms of the dried rhizome. *(Christopher 1976:76)* With reference to the closely allied species *I. missouriensi*, Moore encourages the use of only the dried rhizome internally and stresses that a tincture should never be made of the fresh rhizome. *(Moore 1979:40)* His recommended dose for a dried-rhizome tincture of this species (and of *I. versicolor*) is 5-20 drops, up to three times a day, but used with caution. *(Moore 1994:14)* Herbalist David Hoffmann describes the use of a decoction of the dried rhizome, which is accomplished by putting 1/2 to 1 teaspoon of the rhizome into a cup of water, bringing it to a boil, and simmering it for 10-15 minutes, thereafter using up to 3 cups a day. *(Hoffmann 1986:177)* The official preparations in U.S. pharmacology likewise preferred the dried rhizome.

Cautions

* Do *not* use when **PREGNANT** or **LACTATING**. *Do not give to small children.*

* The constituent iridin (irisin) allows for this plant to have toxic potential. Even moderate overdoses can cause severe GI blistering, facial neuralgia, nausea, vomiting, and diarrhea. Merely *touching* it can cause dermatitis in some people! Therefore, *extreme caution must be exercised in the use of blue flag.* This is not an herb that the layperson should wildcraft and prepare for himself/herself; but if desirous of implementing this plant, he/she should use only commerical extracts, and in so doing always dilute them liberally and never exceed specified doses. (Naturopath Francis Brinker points out that a mere three teaspoons of the tincture would be highly toxic, while herbalist Ed Smith cautions against using more than 60 drops in any 24-hour period or utilizing the extract for longer than a week. *--Brinker 2000:122; Smith 1999:17)*

BLUE VERVAIN
[WILD HYSSOP; AMERICAN VERVAIN]
(Verbena hastata)

Description

[See photos #12-14 in rear of book.]

Growing from 1½ to 5 feet tall, this very attractive wild plant has a stem that is square, erect, and grooved.

The leaves are opposite, narrow (lanceolate to oblong), and toothed (evenly to unevenly). Occasionally the lower leaves are three-lobed. The length of the leaves varies greatly, all the way from 1½ to 7 inches.

The beautifully colored flowers are bluish to bluish-pink to reddish-blue. They are clustered at the ends of the stalks in slender spikes growing to 8 inches long. A helpful identifier is that they often blossom haphazardly, or in a loose band, advancing from the bottom of the spike toward the tip and leaving behind ripening grain. The seeds are tiny nutlets (approximately 2-mm long).

Range & Habitat

Blue vervain grows in dry marshes, moist meadows and fields, pastures, prairies, and on the sides of streams. The plant can be found west of the Mississippi and in adjacent portions of Canada. It is scattered throughout about 70% of our area.

Chief or Important Constituents

Iridoid glycosides(verbenalin, verbenin[aucubin], hastatoside), phenylpropanoid glycosides(verbascoside, eukovoside), flavonoids(luteolin-7-diglucuronide), an unidentified alkaloid, volatile oil(incl. geraniol, verbenone, and limonene), stachyose, tannin, mucilage. Nutrients include vitamins(A, C, and E), calcium, and magnesium.

Food

American Indians used to roast the seeds and grind them into flour. The taste is somewhat bitter (though not unpleasantly so) and the work involved quite tedious; but, if blue vervain can be found growing in large patches, allowing the gathering of a quantity of the seeds in a short time, the activity should not be readily dismissed.

As an emergency trail-nibble, the husked seeds can be eaten raw. I have enjoyed these on numerous occasions, especially during winter foraging, when they are often still available.

Health/Medicine

Blue vervain is a mild *diaphoretic*, having been listed in the U.S. *National Formulary* with that function (1916-1926). It is also a verified *galactagogue* (owing to its aucubin content-*Oliver-Bever 1986:239*) and an *emmenagogue* (a function also experimentally confirmed-*Mabey1988:124*). Dr. John Gunn, in his famed household medical guide published in the 1860s, had said of the latter: "The root...is an excellent emmenagogue, one of the best and safest known, in all cases of suppressed menses; to be used freely in strong decoction,...half a teaspoonful or more, three or four times a day." *(Gunn 1859-61:875)*

This plant can also serve as a reliable *emetic(Spoerke 1990:176)* and was widely used in this regard during the Revolutionary War. Moreover, its iridoid content renders it a *mild laxative* as well. *(Inouye et al 1974:285-88)*

Blue vervain has *diuretic* and *lithotriptic* properties, too, attributed to its saponins. *(Grases et al 1994:507; Holmes 1989:1:133; Dobelis et al 1986:323)* Thus, our area's Menominee Indians found that they could rely on this plant to clear up urine when it was cloudy. *(Smith 1923:58)*

Styptic properties are in *Verbena's* bullpen as well: Ojibwe Indians from our region found that a snuff of the powdered blossoms was the remedy *par excellence* for halting a nosebleed. *(Densmore1974:294)* The remedy worked because verbenalin, the glycoside found in significant dosage in this plant, hastens coagulation of blood. *(Duke 1985:508)* Internally, too, this glycoside is quickly transmitted to the brain, bringing on a feeling of peace and order and thus making blue vervain useful both as a *nervine* and as an *antidepressant. (Mowrey1986:226;Wren1988:275)* Not surprisingly, then, one of the herb's most dramatic uses is as an agent for easing **STOMACHACHES** due to **STRESS,** *(Moore 1979:160)* an application much appreciated by the Teton Indians. *(Gilmore 1991:59)* **HEADACHES** (including some **MIGRAINES,** *Holmes 1989:1:133*) often respond to blue vervain's magic as well, especially those *triggered by stress or exhaustion*. I use this marvelous herb quite frequently as a nervine with clients and the red-letter indications I look to here are: (1) the predomination of a future-related anxiety, (2) tight/spastic muscles (esp. in the upper back and shoulders), and (3) stools that tend toward constipation as opposed to diarrhea.

In accord with its nervine functions, herbalist Michael Moore also suggests the tea for **FIDGETY CHILDREN**, especially when first afflicted with a **COLD** or the **FLU**. *(Moore 1979:160)* The plant has a reputation for fighting colds and flu *per se* and even the *National Formulary* (1906-1926) had held it to be a trusted *expectorant. Antipyretic* properties have also been consistently held for it; *(Rafinesque1830:274)* indeed, to such an end it was used by the Cherokee, Delaware, and Mahuna Indians. *(Moerman1986:1:507)* The herb's *antitussive* ability, recently confirmed in the related *V. officinalis* in a Chinese study *(Gui et al 1985:35),* is also appreciated when colds are present. Then, too, the Eclectic writers Felter and Lloyd lauded its *tonic* effects in this regard: "Taken cold, the infusion forms a good tonic in some cases of debility, anorexia, and during convalescence from acute diseases." *(Felter & Lloyd)*

A mid-1970s study found that *Verbena hastata* markedly inhibited inflammation in rats. *(Benoit et al 1976:169)* Research conducted in Japan has confirmed that

V. officinalis also possesses ***anti-inflammatory***—even ***analgesic***—properties.*(Sakai1963:6–17)* Both species are commonly used for ***GOUT****(Holmes 1986:1:133)* and both prove to be of aid relative to **INFLAMMATION** of the **GALLBLADDER** and **LIVER**, helping the body to battle **HEPATITIS, CIRRHOSIS** and **JAUNDICE,***(Tierra 1990:242;Wren 1988:275)* even as the Houma Indians had earlier discovered.*(Speck 1941:65)*

Any, or all, of blue vervain's nervine, emmenagogue, or anti-inflammatory properties may elucidate the plant's reputation in folk medicine as a treatment for **MENOPAUSE** and **PMS-A** (where *anxiety* is the dominant symptom). Another possible factor here is that a ***luteinizing action*** has been suspected of this plant by numerous clinicians. The likelihood of this may perhaps be gauged by the fact that *Verbena* is closely related to the popular herb vitex, or chaste-tree berry *(Vitex agnus-castus),* which has a widely studied and heralded luteinizing action that increases the ratio of progesterone to estrogen and thus takes the edge off of PMS-A. In fact, so comparable is the overall activity of the two plants that vitex, like vervain, is also a galactagogue.

Blue vervain's renowned ***antispasmodic*** properties come to the fore with reference to **MENSTRUAL CRAMPS**, although herbalist David Winston well points out that extracts made from the *fresh* plant are much more preferable than those from the dried.*(Winston1998)* The antispasmodic properties presently lack detailed scientific confirmation, although in the mid-1980s it was reported that verbascoside assisted the anti-tremor action of the Parkinson's drug, levodopa.*(Oliver-Bever 1986)* Long-standing clinical use, however, leaves no doubt, going back (in America) to use by the Mesquaki (Fox) Indians, who are recorded as having implemented an extract of the root to successfully deal with **FITS.***(Smith 1928:193)* Independently—and apparently unaware—of Mesquaki use, the noted American herbalist Joseph Meyer studied this application thoroughly owing to a clinical report brought to him regarding its use in a long-standing case of epilepsy: "Its effects in that particular case were so remarkable, that I was led to study it more carefully. I found, after close investigation and most elaborate experiment that, prepared in a certain way and compounded with [three other herbs], and the best of whiskey, it has no equal for the cure of fits, epilepsy, or anything of a spastic nature.... A more remarkable plant is not found within the whole range of the herbal pharmacopoeia."*(cited in Shook 1978:276)* The anticonvulsant activity is now listed as one of the "medicinal uses" of the plant in the newest edition of *Potter's Cyclopaedia of Botanical Drugs and Preparations.(Wren 1989:275)*

Cautions

  * Because this plant is a ***uterine stimulant*** (confirmed by animal studies) *(Farnsworth 1975:535-98),*, it should be strictly **AVOIDED DURING PREGNANCY.**

  * The glycoside verbenalin found in this plant is an ***emetic,*** so one needs to bear in mind that drinking too much of the tea—or drinking it too quickly—can cause **VOMITING.**

BONESET
[THOROUGHWORT; FEVERWORT; INDIAN SAGE; AGUEWEED]
(Eupatorium perfoliatum)

Description

[See photos #15 & 16 in rear of book.]

Boneset is a perennial plant indigenous to North America. It can grow as high as five feet, but usually tops off in the 2- to 4-foot range. The stem is erect, cylindrical, hairy, and branches at the top.

The wrinkled-looking, finely toothed leaves are usually about four inches long toward the top of the plant, but grow to around eight inches toward the bottom of the plant. The undersides of the leaves are downy looking.

Of curious interest is that the leaves are attached in an odd sort of fashion—a way that makes identification of boneset fairly simple: The species name, *perfoliatum*, gives us the clue here, indicating that each pair of leaves is united at their base, that is, they are joined *across* the stem, giving the appearance of but one long leaf *perforated* by the stem! (*See photos accompanying this entry.*) This is an important means of distinguishing this plant from toxic relatives (*see below*, under "Cautions.").

In midsummer, white florets, grouped in flat-topped clusters, appear at the top of the main stem and its branches. One can detect from these flowers only a weak odor.

The eventual fruit is a tufted achene.

Range & Habitat

Boneset sometimes grows in very moist woods, but its preferred habitat is swampy regions and the borders of marshes and ponds. Found throughout our area in such environs.

Chief or Important Constituents

Essential oil consisting of sesquiterpene lactones of the germacranolide (euperfolin and euperfolitin) and guaianolide(eufoliatin and eufoliatorin) variety, terpenes, diterpenes(hebenolide, hebeclinolide, dendroidinic acid), and triterpenes (alpha-amyrin, dotriacontane, 3-beta-hydroxy-ursa-20-ene), iso-humulene, chromenes, polysaccharides(4-0-methyl-glucuronoxylan, inulin), flavonoids(astragalin, kaempferol, kaempferol-3-rutinoside, quercetin, rutin, eupatorin), an uncharacterized alkaloid(?), sterols(sitosterol and stigmasterol), gum, resin, gallic acid, tannic acid.

51

Health/Medicine

   With the possible exceptions of plantain and mullein (*see under those entries*), my cupboards are stuffed with more boneset than any other medicinal plant. As I believe will become evident from the monograph below, there are numerous good reasons for "stocking up" on this odd-looking marsh plant. Indeed, as Alabama folk herbalist Tommie Bass once so well put it: "It is as good a one as the Lord put out there."*(Crellin & Philpott 1990b:106)*

   First of all, boneset was used with good success by early American settlers as a healing herb once the onset of a cold or flu was first suspected as developing. The famed Eclectic pharmacist John Uri Lloyd, who wrote a treatise on this herb in 1918, remarked how that boneset was so esteemed by early Americans for cold and flu that it could be "found in every well-regulated household" of these times.*(Lloyd 1921:137)* The author of *American Medicinal Plants,* Charles Millspaugh, further explained how that, by the late nineteenth century, boneset could be encountered hanging from the rafters of nearly every woodshed or attic in the country, ready for use at the sound of the slightest symptom of a cold or flu.*(Millspaugh 1974:312-14)* In this regard, Mrs. William Starr Dana offered her own poignant recollections as follows: "To one whose childhood was passed in the country some fifty years ago the name and sight of this plant is fraught with unpleasant memories. The attic or woodshed was hung with bunches of the dried herb, which served so many grewsome[sic] warnings against wet feet, or any over-exposure which might result in cold or malaria. A certain Nemesis, in the shape of a nauseous draught which was poured down the throat under the name of 'boneset tea,' attended such a catastrophe."*(Dana 1900)*

   Such therapeutic use of the plant seems to have been gleaned from the American Indians, for the available records clearly reveal that numerous tribes—including the Cherokees,*(Hamel&Chiltoskey1975:27)* Iroquois,*(Herrick1995:232)* Menominee,*(Smith1923:30)* and Mohegan,*(Tantaquidgeon 1928:265)*—also used the herb for treating **COLDS, INFLUENZA,** and **FEVERS**. Delaware Indians used an infusion of the root, and sometimes the leaves as well, for a fever with **CHILLS**.*(Tantaquidgeon1972:33,118-19;Tantaquidgeon1942:28)* Nanticoke Indians used it in combination with wild thyme for same.*(Tantaquidgeon 1972:98, 126-27)* The plant's implementation for colds and flu in such a wide variety of cultures strongly suggests efficaciousness for these illnesses, and I can certainly add my own testimony, having used it with great success in both regards over a period of some 15 years. Then, too, the lone clinical trial on record, published in 1981, even demonstrated that a mere homeopathic preparation of boneset (as opposed to the full-strength herbal tincture) was *just as effective as aspirin* in relieving the symptoms of a cold—both test-groups showing equal improvement based on lab tests, body temperature, and subjective complaints.*(Gassinger et al 1981:732-33)* Strangely, though, a series of good, clinical trials investigating boneset's effect upon colds and flu has not yet been undertaken.

   Boneset tea was a popular treatment in early America for various feverish, infectious diseases of a more serious nature, too. For example, it was the most popular treatment for **MALARIA** before the introduction of quinine.*(Bloyer 1901:837; Felter 1924:201; Hocking 1997:296)* Then, too, during the Civil War, when Peruvian bark, the source of quinine, became largely unavailable, it was recommended to Confederate troops for this illness by the prominent medical doctor Francis Porcher because, as he said, it was "thought by many physicians to be even superior to the dogwood, willow or poplar [the bark of these trees producing analgesia], as a substitute for quinine." *(Porcher1863:9)* One of such physicians was the noted John Sappington, who, in his classic work entitled *Theory and Treatment of Fevers,* published in 1844, identified boneset and dogwood as the best indigenous substitutes for quinine. *(Hall 1974:527; cf. Jackson 1876:234)*

Then, too, early American physicians implemented boneset for **YELLOW FEVER**. *(Barton 1817-18:2:135-37; Millspaugh 1974:314; Hocking 1997:296)* Dr. Benjamin Smith Barton, writing near the birth of the eighteenth century, reported how this herb was particularly instrumental in halting a yellow-fever epidemic that had occurred in Philadelphia in 1793. *(Barton 1810:1:28; 2:22-26, 55)* Then, too, Dr. William P. C. Barton related of the work of Samuel C. Hopkins, MD, of Woodbury, NJ, who had used *Eupatorium perfoliatum* most effectively as the sole treatment for **TYPHUS** and other serious fevers. *(Barton 1817-18:2:135)* Indeed, boneset was widely used for both typhus *(Rafinesque1828:177;Barton1817-18:2:135; Millspaugh1974:314)* and **TYPHOID FEVER**. *(Bloyer1901:837; Grieve1971:1:119; Hocking1997:296)*

But boneset's most famous historic role unfolded during a different and more dastardly Colonial plague. Constantine Rafinesque, noted compiler of a respected tome on medical botany published in the early 1800s, informs us that a medical authority he consulted, N. Chapman (author of several works on therapeutics and materia medica published from 1817–22), "relates that it cured the kind of influenza called Breakbone fever." *(Rafinesque 1828:176)* This so-called "breakbone fever" was a miserable viral contagion that produced joint and muscle pains so excruciating that its victims felt as if their bones were literally breaking (thus, the virus' name)! By checking this epidemic, *Eupatorium perfoliatum* earned the common name by which it is now most widely known, for in dealing with this virus, it was felt by the victims to have re-"set" their "bones"(thus, "boneset"). (This breakbone fever of Colonial times is now thought by historians of medicine to have been either a very bad strain of *INFLUENZA* or a form of the mosquito-transmitted virus known as *DENGUE*. *-Crellin&Philpott 2:107; Millspaugh 1974:314*)

As a result of its well-attested successes as enumerated above, boneset came to be widely implemented by nineteenth-century physicians. Rafinesque wrote of it in 1828: "It has been introduced extensively into practice all over the country. . .and inserted in all our medical works." *(Rafinesque 1828:176)* Dr. Asahel Clapp, in his 1852 report to the American Medical Association, emphasized that boneset "deservedly holds a high rank among our indigenous medical plants." *(Clapp 1852:791-92)* By the 1870s, John Jackson, writing in the *Canadian Pharmaceutical Journal,* acknowledged that the herb was being "used by many physicians." *(Jackson1876:234)* The beloved boneset even received official pharmaceutical recognition, appearing in the *U. S. Pharmacopoeia* (and that for over eighty years—from 1820–1916!) and in the *National Formulary* (for most of the first half of the twentieth century).

But how did boneset accomplish its task with breakbone fever—and indeed with those other, feverish, viral contagions so prevalent in early America? In many ways, it seems: We note first that boneset was used by the Mohegan Indians for general

DEBILITY(*Tantaquidgeon 1972:72*) and that it was listed in the USP(1820-1916) and NF (1926-50) as a *stimulant,*(*cf. Bloyer 1901:837; Gathercoal & Wirth 1936:721; Culbreth; Pedersen 1998:55*) so that it no doubt proved to be of great benefit for the severe lethargy that accompanied such feverish viral afflictions. (I might add here that I have witnessed it transform a person with a bad fever who was lying in bed and looking like a corpse into a human dynamo who, within one hour of use, proceeded to get up to do household chores and to go on to express a desire to go back to work!)

Its strong *diaphoretic* properties were undoubtedly of great use as well, helping victims to "sweat out" the fever. Botanist and physician Laurence Johnson, writing in the 1880s, observed in this regard: "Taken warm in large doses, the infusion or decoction produces copious diaphoresis, and is employed in the acute stages of catarrhal affections and in fevers, especially of the intermittent or remittent type."(*Johnson1884:173;cf PDRHrbs; Leung 1989:40; Wren 1988:40*) An interesting historical testimonial to its aid here survives from the pen of the British botanist Thomas Nuttall, who, when he was sick in 1819 with "intermittent fever," reported that because none of the standard medicines were at hand, he "took in the evening about a pint of a strong and very bitter decoction, of the *Eupatorium cuneifolium*, the *E. perfoliatum* or Bone-set, not being to be found in the neighborhood. This dose, though very nauseous, did not...operate as an emetic, but acted as a diaphoretic and gentle laxative, and prevented the proximate return of the disease."(*Nuttall 1905:244*)

Boneset is also a powerful *immunostimulant,*(*PDRHrbs*) and this was undoubtedly part of the reason for its success during the feverish contagions of colonial and pioneer times. Recent chemical analyses have revealed some fascinating information in this regard. For example, one of the plant's sesquiterpene lactones, eufoliatin, has experimentally shown particularly strong immunostimulating activity of a general sort.(*Woerdenbag 1993:180; Wagner et al 1985:139-44;Newall et al 1996:48; Wren1998:40*) Then there are the polysaccharides in boneset, which have been demonstrated to possess potent immunostimulating functions as well. Here several scientists employing three immunological test systems—carbon-clearance, granulocyte-, and chemiluminescence—found that the type of water-soluble polysaccharides occurring in the plant demonstrated a marked phagocytosis-enhancing effect, especially in the granulocyte test.(*Wagner et al 1984:659-61; Wagner et al 1985:139-44;Vollmar et al 1986:377-81*) This means that they stimulated certain of the body's white blood cells to perform their job of phagocytosis more effectively, which task involves an engulfing (through their cell membranes) of invading organisms so as to destroy them from the body. A similar study found this immune-enhancing effect to be equal to that possessed by the polysaccharides of the popular immune-enhancing herb echinacea *(Echinacea purpurea; E. angustifolia)* when both samples were in a concentration of .001 mg/ml, but *ten times stronger* when both were in a concentration of .01 mg/ml! (*Wagner et al 1985:139-44; cf. Wagner et al 1984:660*) In another study, a combination of boneset, arnica(*Arnica montana*), wild indigo(*Baptisia tinctoria*), and *Echinacea angustifolia* showed over 50% greater immune stimulation than did *E. angustifolia* alone.(*Wagner & Jurcic 1991:1072-76*) My own careful observations over the last 15 years have convinced me that boneset proves to be far superior as an immune-enhancer over echinacea for Type-O blood types, while echinacea proves to be more effective for

Type-A blood types.

Aside from aiding immunity in a general sense, however, the plant seems to possess direct **antibacterial** and **antiviral** activity, judging by both clinical and experimental evidence. Activity against several kinds of bacteria has been reported for *E. perfoliatum* in lab experiments on several occasions,*(Habtemarium & MacPherson 2000:575-77; Mockle 1955:89; Woerdenbag 1993:180)* as well as in other *Eupatorium* species.*(Croom 1983:60 & refs cited)* Moreover, two related species (*E. rotundifolium* and *E. squalidum)* have even recently been shown experimentally to have a direct effect against the malaria organism (plasmodium) in particular.*(Carvalho & Krettli 1991:181-84; Croom 1983:62 & ref. cited)* Modern American herbalists have also effectively used boneset (alone or in combination) against **HERPES INFECTIONS**, including even the genital herpes (Type II).*(MH 1995-96:10)* (My own favorite application here is a formula consisting of boneset, self-heal *[Prunella vulgaris]*, St. John's Wort [*Hypericum perforatum*], and lemon balm [*Melissa officinalis*].) In this regard, it is of interest that early American physicians also employed it also against what was thought to be a "herpes" of sorts, the so-called"James Valley Ringworm."*(Barton1:28, 2:22-26, 55)* A very recent study done in Madrid, Spain, has revealed some fascinating confirmatory evidence for these uses: Scientists working at one of the universities there have demonstrated that extracts of two species of *Eupatorium* (*E. articulatum and E. glutinosum*) are powerfully active against *Herpes simplex* Type 1 (the only type tested).*(Abad et al 1999:142-46)* In addition to this, a tincture of the related species, *E. odoratum*, which has traditionally been used in Guatemala for **VENEREAL DISEASES**, was recently shown in a test-tube study to be highly active against five strains of **GONORRHEA**.*(Caceres et al 1995:85-88)*

Boneset possesses as well some **anti-inflammatory** properties,*(PDR Hrbs; Benoit et al 1976:160;)* possibly owing to its flavonoids,*(Newall et al 1996:48)* so that it proves helpful where inflammation of the sinuses, ears, or bronchials are involved. It is also **anti-catarrhal**,*(Ellingwood 1983:269; Bloyer 1901:837; Santillo 1984:92; Grieve 1971:1:119)* and can work wonders with unyielding respiratory mucous, which it breaks up and expectorates most ably.*(Wren 1988:40; Pedersen 1988:55; Leung 1989:40)* (In fact, the **expectorant** constituents in the related Asian species, *E. fortunei,* used in Traditional Chinese Medicine, were identified as far back as the early 1980s.*-Cai 1983:30-31)* Thus, it is especially indicated where there is a *dry, irritable cough* corresponding to what TCM calls "lung wind-heat/dryness."*(Holmes 1989:1:131)* Country herbalist Tommie Bass particularly appreciated boneset for the type of **COUGH** associated with **CROUP**.*(Crellin & Phillpot 1990a:187)* America's Eclectic physicians, practicing in the late 1800s and early 1900s, found that it "relieves chest pain and cough," especially in "bronchial pneumonia," serving as well to allay the"irritable after-coughing" often associated with this condition.*(Felter 1922:201)* They also found it invaluable in the aged or debilitated "where there is an abundance of secretion but lack of power to expel it."*(Felter 1924:201)* Here the Eclectics had a wealth of information and experience to draw upon, as numerous articles had appeared in *The Eclectic Medical Journal* outlining its history and efficacies, including one in 1924 which had noted: "During the severe pandemic of 1918-19 [the Spanish Influenza] it was one of the safest and most successful remedies employed and contributed much to the successful management of the disease under Eclectic treatment."*(Felter1924:201; also Best 1928:93)*

But boneset has a plethora of useful applications aside from its legendary aid for those afflicted with infectious, febrile diseases. For one thing, the root was used by both the Mesquakie and Ojibwe (Chippewa) Indians to treat *RATTLESNAKE BITES.*(Smith 1928:214) Then, too, it has long been employed as a *vermifuge*, even against the stubborn *TAPEWORM.* (Woerdenberg 1993:179; Mockle 1955:89; Smith 1928:214; Locock 1990:229; Vogel 1970:226; Grieve 1971:1:119) One person who made this latter application somewhat famous in his own area of the country was a Mesquaki Indian herbalist named McIntosh, who used a tea of the leaves and flowers.(Smith1928:214;Shemluck1982:330) Early American physicians also often relied upon boneset as an *anthelmintic*.(Christianson 1987:73)

Owing to its very bitter sesquiterpene lactones, boneset also favorably *tonifies* and *stimulates* the digestive processes. Physician and botanist Laurence Johnson observed in this regard: "The infusion, taken cold in moderate doses, is tonic, and is employed in debility of the digestive organs and in convalescence."(Johnson 1884:173) Early American physician Jacob Bigelow had further pointed out: "Given in moderate quantities,...in cold infusion or decoction, it promotes digestion, strengthens the viscera, and restores tone to the system."(Bigelow 1817-20:36) It thus works well as an *appetite stimulant* and has been especially employed by herbalists to treat*LACK OF APPETITE IN THE AGED*.(PDRHrbs;Hocking1997:296;Spoerke1990:43;Grieve 1971:1:119) Then, too, as Rafinesque observed, it is "particularly useful in the indigestion of old people."(Rafinesque 1828:176)

In view of its bitter components, it is not surprising that boneset also has *hepatic* properties.(Grieve1971:1:119; Holmes1989:1:130) Thus, according to the great naturalist Pliny the Elder, king Mithridates Eupator of Pontus (134-63 B.C.E.) had implemented a *Eupatorium* species for liver problems. (Eupator was, in fact, so, fond of this genus that it wound up being named after him!) The physiomedicalist W.H. Cook would later point out that boneset is 'relaxing to the hepatobiliary apparatus, promoting secretion of bile and also expulsion from the gallbladder into the digestive tract.' (Cook 1985) Herbalist Peter Holmes notes that it is primarily indicated for*stagnant* (with right-sided pain) and *hot* (with constipation) hepatic conditions (according to the TCM model).(Holmes 1989:1:131) Boneset's favorable effect upon the biliary system may further elucidate its classic and reliable implementation as a mild *laxative*,(Wood et al 1937:446; Leung 1989:40; Felter 1922; Culbreth; Wren 1988:40) in that the free and proper flow of bile softens and moves stool through the colon. Despite this fine testimony as to boneset's liver-friendly effects, however, many popular writers on herbs have claimed that our herb contains pyrrolizidine alkaloids which are *toxic* to the liver, so that it should therefore never be used! What about this bold claim?

Well, first of all, we need to understand that pyrrolizidine alkaloids (hereafter abbreviated "PA's") are of two types: unsaturated PA's, which are potentially toxic to the liver, and saturated PA's, which are *not* hepatotoxic. Many species in the *Eupatorium* genus contain the toxic (unsaturated) PA's (including the widely-used Eur-Asian species *E. cannabinum* and the two likewise-often-used Asian species*E. fortunei* and *E. chilense*), but some (e.g.,*E. album*) contain no PA's whatsover,(Woerdenbag 1993:173, 175) and others (e.g., *E. semialatum*) contain only *non*-toxic (saturated) PA's.(Lang et al 2001:143-47) What is especially significant, however, is that *no* pyrrolizidine alkaloids—of *any* kind—have ever been isolated from *E. perfoliatum!(Newall 1996:48; Fetrow & Avila 1999:96)* A respected European phytochemist puts it succinctly: "E. perfoliatum: toxic PA? No."*(Woerdenbag 1993:173)* This plant scholar cites the investigative report of pyrrolizidine alkaloids found in the German journal *Pharmaceutische Zeitung,(anon. 1990:2532-33; 2623-24),* which explains that the German government's BGA (Bundesgesundheitsamt), which has strict regulations against plants with toxic PAs being used internally and has carefully investigated the phytochemistry of marketed herbs, does *not* restrict *E. perfoliatum.(anon. 1990:2532-33; 2623-24)* (A lone investigation claimed to have found alkaloids in our species. These, however, were not indentified as pyrrolizidines, but rather were said to be uncharacterized.*-Locock 1965*).

The plain truth, then, is that the popular species known as boneset,*Eupatorium perfoliatum*, is pefectly safe to use, as long as the proper dosage and cautionary guidelines (see below, re: pregnancy) are observed. As a point aside, however, even in those *Eupatorium* species containing toxic PA's (see, in the present work, under "Joe-pye Weed"), a toxic reaction with the liver has not usually been elicited: It has in one species, *E. adenophorum.(Kaushal et al 2001:615-19; Katoch et al 2000:309-14)* However, a rat study employing three other species (*E. fortunei, E. japonicum, and E. chilense*) failed to show hepatotoxicity as against controls, even after prolonged administration.*(Zhao et al 1987:59-67)* And another species, *E. cannabinum*, although possessed of *several* toxic PA's, has actually been demonstrated to be hepato*protective* against carbon-tetrachloride-induced heptatoxicity!*(Lexa et al 1989:127-32)*

As to yet other uses for boneset, our marsh-loving plant has had good success in treating **RHEUMATISM** of a **NEURALGIC** or **MUSCULAR** sort.*(Hocking 1997:296; Bloyer 1901:837; Rafinesque 1828:177; Barton 1817-18:2:138)* Boneset's anti-inflammatory property, as noted above, is undoubtedly of some aid here. However, the plant seems to be of assistance as a connective-tissue aid in other respects as well. For many years, for instance, the Iroquois Indians had used a cold, compound infusion inclusive of the herb's leaves as a poultice for broken bones.*(Herrick 1995:232)* Concordantly, in modern times, herbalist and naturalist Tom Brown, Jr., who was mentored under the revered Apache herbalist Stalking Wolf, has offered his lifetime of personal experience for boneset as a connective-tissue aid, finding it most effective when a small amount of the herb is steeped fifteen to thirty minutes in hot water.*(Brown 1985:67)*

Finally, a tincture of boneset leaves has been shown to possess*cytotoxic* activity comparable to chlorambucil, a cytotoxic agent in regular use.*(Habtemariam&MacPherson 2000:575)* Cytotoxic agents are utilized against cancer cells. In fact, boneset's flavonoid eupatorin has, by itself, been shown to be cytotoxic in laboratory tests.*(Midge & Rao 1975:541; Kupchan et al 1965:929-30; Vollmar et al 1986:377; Woerdenbag 1993:179; Herz et al 1977:2264-71; Heywood & Harborne 1977:421)* The sesquiterpene lactones present in the herb have also been shown to produce similar effects against cancer cells,*(Herz et al 1977:2264-71; Cassidy et al 1969:522)* as have those of *E. cannabinum*(Woerdenbag et al 1989:68-75; idem 1986:245-51) and *E. cuneifolium.(Kupchan et al 1973:2189)*

As to manner of implementation, generally boneset is used today as a tincture, although some prefer to make the fully dried leaves into a tea, which, as we've seen, was the most common form of the herb taken in early America. The tea is drunk hot, warm, or cold—cold as a tonic, hot or warm as a cold-and-flu combatant and/or as a diaphoretic. It should also be noted that boneset tea taken in large doses during a short period of time serves as an **emetic** (induces vomiting).   This is probably largely due to the chemical eupatorin.*(Spoerke 1990:43)* Emetics can be useful at times, and it is of interest that the 1968 edition of the *Merck Index* (not to be confused with the *Merck Manual*) noted that eupatorin has been used in emergency medicine as an emetic in situations of poisoning where gastric lavage was unavailable or inadvisable.

Cautions

   * Boneset should be *USED WITH CAUTION DURING PREGNANCY*, due to a tendency to absorb nitrates from the soil,*(Stephens 1980:152;Woerdenbag 1993:174-75,181;Duke 1985:188)* and since it has been reported that cattle eating large amounts of the herb in areas where the nitrate concentration was especially high had suffered abortions.*(Sund & Wright 1957:278-79)*

   * If gathering boneset from the wild, be very careful not to confuse it with *its very deadly cousin,* white snakeroot (*Eupatorium rugosum*), *to which it bears a close resemblance.* (The toxic chemical in white snakeroot is tremetol, a term denoting a mixture of benzofuranoid compounds—i.e., tremetone, dehydrotremetone, hydroxytremetone, and toxol.*-Heywood & Harbourne 1977:312-13, 489*). Both have nearly identical flowers and a similar overall "look" to them (after all, they're closely related!). However, white snakeroot's leaves, while opposite like those of boneset, do not have the through-the-stem, "uni-leaf" look that characterizes boneset, but rather are stalked. As a point aside, it should be noted here that, contrary to what is stated in some pop herbal works, neither tremetol nor its components has been listed in any of the reputable chemical profiles available for boneset (*E. perfoliatum*), including those by Duke, Moore, etc. The heavy concentation of that chemical in white snakeroot, however, is responsible for "the trembles," or "milk sickness," because it is soluble in the milk of cows who have consumed it. This horrible and fatal illness has been traced to bovine consumption of white snakeroot on small, family farms (whereas in large commerical farms, any tremetol present winds up being so diluted by conglomeration of the milk of the many cows that it becomes innocuous), but *never* to consumption of boneset. Unfortunately, *E. perfoliatum* and *E. rugosum* are often confused, such as in the entry for boneset in the first two editions of the popular, health-reference book *Prescription for Nutritional Healing* (although the recently-released 3rd ed. has it correct). However, they are not only quite varied species of the same genus, but even grow in distinctly different locales (white snakeroot in woodsy areas and boneset in marshy areas, though where these areas intersect the two may be found growing together).

BOUNCING BET
[SOAPWORT; LATHERWORT; FULLER'S-HERB; SWEET BETTY]
(Saponaria officinalis)

Description

[See photo #17 in rear of book.]

A European perennial that escaped from gardens during pioneer times. It grows 1-1½ feet tall and thrives in colonies. Its stems are smooth, but swollen at the nodes.

Bouncing bet's leaves are oval to lanceolate, 2-3 inches long, and prominently veined. They are entire, smooth-edged, and grow in pairs.

The plant's lovely flowers, often also growing in pairs, are pink or white and possessed of five, notched petals that are reflexed downwards. They are arranged in dense, terminal cymes.

Range & Habitat

Grows in waste places and alongside roads and RR tracks throughout our area.

Chief or Important Constituents

Saponin(to 5%), saporins, flavonoids(incl. vitexin), resin, mucilage, gum

Health/Medicine

Bouncing bet is perhaps best known as an *expectorant*,*(Weiss1988:204)* due to its large concentration of saponins which irritate the stomach, causing a reflex action to the lungs whereby mucus is expelled. Recent research by Russian scientists has shed much light on the *immunostimulating* properties of these saponins.*(Bogoiavlenskii et al 1999:229-32)* Yet another line of research has confirmed that the similarly-sounding saporins contained in the plant are deadly to human *LYMPHOMA, LEUKEMIA, MELANOMA* and *BREAST-CANCER CELLS in vitro*, although clinical studies are currently lacking in verification.*(Gasperi-Campani et al 1991:1007-11;Tecce et al 1991:115-23;Siena et al 1989:3328-32)* There is some concern, however, in that the saporins, by themselves, have been shown to be hepatotoxic to some degree.*(Stripe et al 1987:259-71)* However, bouncing bet has classically been used to *heal* "liver affections" (including jaundice),*(Chervallier1996:264)* not cause them, so perhaps balancing agents in the plant serve to offset the toxic effect of the saporins to such an extent that the herb actually winds up being slightly hepatic. Thus, America's Eclectic physicians of the late-nineteenth and early-twentieth centuries revered bouncing bet as "a remedy in the treatment of...jaundice, liver affections," as well as"alterative... [for] syphilitic, scrofulous and cutaneous diseases" They also found it to be a fairly good *emmenagogue*.*(Felter & Lloyd)*

Bouncing bet has been taken by decoction (*never* infusion!) of the dried and finely shaved root*(Potterton 1983:175)*, to the tune of one teaspoon per cup of water.*(GermCommE)* Historically, this tea has been imbibed only one teaspoon at a time, with a total consumption of not more than 2-4 fl. ozs a day, owing to a toxic potential. *If allowed to macerate, the tea has the potential to cause paralysis and tremor.* Thus, boiling the root for only a few minutes and then straining right away has been considered the safest procedure for preparing a tea that is to be taken internally.*(Potterton 1983:175)*

External applications have also abounded. For example, we are informed that the Cherokee Indians would poultice **BOILS** with bouncing bet.*(Hamel & Chiltoskey 1975:26)* The herb has also been classically used in European and American herbalism as a topical application for **ERYSIPELAS.***(de Bairacli Levy1974:135)* Since both erysipelas and boils are caused by bacteria (the latter usually by *Staph. aureus*), the aforementioned applications are interesting in light of a 1954 study which found bouncing bet to inhibit bacteria (in this particular study, *Mycobacterium tuberculosis*)*(Fitzpatrick1954:530)* Such varied antibacterial applications are understandable in light of the plant's high concentration of saponins, which are powerful antimicrobial agents.*(Hiller 1987:175-84)*

A topical wash made from a decoction of bouncing bet's roots remains a time-honored treatment for **POISON IVY.** It has been assumed that this works because of the skin-cleansing ability of the plant's saponins.*(Dobelis et al 1986:118)* But as the herb has been lauded as a treatment even *after* the rash has broken out, and not just prior to that time as a preventative, perhaps an **anti-inflammatory** aspect comes to the fore in soothing the rash, which property has been tapped in sundry other ways by various cultures. For example, California Indians poulticed the ground, pulpy leaves over the **SPLEEN** when inflamed, leaving them there for three to four hours.*(Romero1954:40)* Then, too, noted European herbalist Juliette de Bairacli Levy finds bouncing bet to be useful topically as a rub for **RHEUMATISM** and **ARTHRITIS.** *(de Bairacli Levy1974:135)*

Utilitarian

I frequently borrow a few leaves here and there from colonies of bouncing bet when I'm out enjoying my wilderness activities, thereafter adding some spittle (or water from my canteen) and then vigorously rubbing my hands with this combination in order to free them of dirt and grime. This natural soap works astonishingly well, owing to the proliferation of the plant's saponins. (Please note that is important *never to wash one's hands in a body of water with bouncing bet*, however, as saponins are extremely toxic to cold-blooded animals and just a little bit will poison a wide area and kill many of our aquatic brothers very quickly.)

Caution: Due to the large amount of saponins in this plant, and the consequent possibility of toxicity (manifested as gastrointestinal inflammation, diarrhea, or vomiting) from overuse or otherwise improper ingestion of it, any *internal uses of bouncing bet should be left to skilled hands only.* Furthermore, the herb's other major chemical constituent, saporin, appears to be hepatotoxic to some degree.*(Stripe et al 1987:259-71)*

BUGLEWEED
[WATER HOREHOUND]
(Lycopus virginicus; L. americanus)

Description

[See photos #18 & 19 in rear of book.]

A member of the mint family, with the consequent square stem and paired leaves. The flowers occur clustered on the stem at the leaf axils. Our species include:

Lycopus americanus (cut-leaved water horehound; Paul's betony) grows 6-24 inches tall and has deeply-lobed lower leaves and conspicuously-toothed upper leaves.

L. asper (rough bugleweed; western water horehound) has sharply toothed leaves with ascending teeth. It is often hairy, giving a rough feeling to the touch.

L. uniflorus (northern bugleweed; common water horehound) has hairless stems and fine-toothed leaves with short (1-4mm) stalks. The small, whitish flowers are sessile and they have flared lobes. They appear in clusters of 3-10 at the leaf axils.

L. virginicus (Virginia bugleweed) grows 6-24 inches tall and has dark-green—sometimes purple-tinged—leaves that narrow at both ends and are sharply toothed. Its calyx lobes are also noticeably blunt.

Range & Habitat

One or another species of bugleweed flourishes throughout our area in rich, moist soil—lakeshores, riverbanks, swamps, marshes, etc.

Chief or Important Constituents

Bugleweed is especially rich in tannins and pseudo-tannins(ellagic and gallic acids) and in other plant acids(including caffeic, chlorogenic, lithorspermic, and rosmarinic). The plant's essential oil contains pulegone and limonene. Flavonoids include luteolin-6-glucoside and apigenin. The glycoside lycopin is also present.

Food

L. uniflorus has crisp tubers, available from autumn until early spring, but none of our other species do. Our region's native Americans relished these tubers raw or cooked (boiled for 10-15 minutes), as do many modern foragers.

Health/Medicine

L. americanus was used as one herb in a compound mixture employed by the Mesquaki (Fox) Indians for **STOMACH CRAMPS**. *(Smith 1928:225)*

L. virginicus is one of several herbs (the others being motherwort, *Leonurus cardiaca—see under that heading—*and lemon balm, *Melissa officinalis*) classically used in the treatment of **HYPERTHYROIDISM**, including **GRAVE'S DISEASE**. It has been demonstrated to exhibit potent **antithyrotropic** effects, meaning specifically that it inhibits iodine metabolism and the peripheral deiodination—and thereby release—of thyroxine(T_4), *(PDRHrbs;Weiss1988:279)* the thyroid hormone that overproduces in hyperthyroid disorders. The plant accomplishes this partly by interfering with the ability of thyroid-stimulating hormone(TSH) to bind to receptor sites on the thyroid membrane. Since Grave's immunoglobulins resemble TSH closely in both structure and function, bugleweed may also prevent their attachment to the membrane. *(Auf M'kolk et al 1985:1687-93; Winterhoff et al 1988:101-06)*

Dosage for hyperthyroidism is dependent on individual situations, but one source lists 1 to 2 grams of bugleweed administed by infusion per day or a daily tincture dosage approximating 20 milligrams of the herb. *(PDRHrbs)*

Bugleweed also reduces the **RAPID HEARTBEAT** associated with this condition and with other conditions—a characteristic appreciated by America's Eclectic physicians of the nineteenth century *(Scudder 1870:156)* Of especial value here is the fact that, as the great botanical writer C. S. Rafinesque had pointed out, *(Rafinesque 1830:29)* bugleweed doesn't need to accumulate in the system to do so, unlike digitalis. *(Millspaugh 1974:459)* Then, too, because it can lower prolactin levels *(Brinker 1990:1-14)*, *L. virginicus* is also of use in **PMS-A,** where a lowered prolactin level—or at least a sensitivity to this hormone—often exists and consequent **TENSION** and **ANXIETY** arise. *(PDRHrbs)*

Bugleweed is a fine **hemostat,** useful in **MENORRHAGIA** and **PASSIVE INTERNAL HEMORRHAGE.** *(Millspaugh 1974:459)* The Natura physician R. S. Clymer found it to be of particular value in **HEMORRHAGE** of the **LUNGS**and **BLADDER,** as well as for **URINARY INCONTINENCE.** *(Clymer 1973:150)* Since the plant contains tannic acid and since herbal writers as diverse as Rafinesque and folk herbalist Tommie Bass have witnessed an oft-successful use for **DIARRHEA,** *(Rafinesque1830:2:29;Crellin&Philpott 1990:2:120)* we can assume that **astringent** properties are at least partly responsible for the hemostatic action.

Cautions

* Concurrent use with thyroid hormone preparations can be dangerous and should not be attempted unless under a physician's guidance. It should be borne in mind that using buglweed for Grave's disease will interfere with diagnosis via radioactive isotopes. *(PDRHrbs)*

• The use of this plant is not without side effects. Aside from inhibiting thyroid hormone, it at the same time inhibits hormones of the gonads, including HCG, LH and FSH. *(Winterhoff et al 1988:101-06;Brinker 1990:1-14)* Such effects may interfere with contraceptive or fertility medications and they also *clearly contraindicate the use of this herb during* **PREGNANCY** *and* **LACTATION.**

BULRUSH

(Scirpus acutus; S. validus, & related spp.)

[See photos #20 & 21 in rear of book.]

Description

Occasionally an annual, but usually a perennial, this interesting plant grows 3-10 feet high and thrives in large colonies. The stem is erect, circular or triangular, and fails to branch. Numerous small, drab flowers cluster on several pedicels just below the stem tip.

Several species thrive in our area, but predominating are *S. validus* (great bulrush, great American bulrush, or soft-stemmed bulrush) and *S. acutus* (Tule, Tule Bulrush, or hard-stem bulrush).

The former has a soft stem, lacks leaves entirely, and possesses small, reddish-brown flowers on several stalked, drooping clusters. It is also characterized by a reddish and scaly rhizome. The latter possesses a stiff stem and has long, toothless, grasslike, pointed leaves arising from the base only. It is also bulkier than *S. validus* and its flower clusters are cylinder shaped and grayish brown. The rhizome is also grayish brown.

Fruits are small, black, shiny nutlets.

Range & Habitat

Found throughout the USA and Canada (except within the Arctic Circle), bulrush dwells in wet ground or in shallow, stagnant waters.

Food

The Thompson Indians collected the pollen from the flower spike of a species of bulrush to use as flour, mixing it with other flours to make biscuits or cakes.*(Steedman 1928:484),* Many modern wild-foods foragers follow suit.

The young stems are a valuable survival food: they can be peeled and eaten raw, even as the Cheyenne Indians discovered.*(Hart 1981:8)* The tender white heart at the base of the mature stems is especially sweet and juicy and is one of the wild foods that I personally relish the most. The stems can also be added to soups, just like the Goisute Indians sometimes did in bygone days.*(Chamberlin 1911:381).*

Younger rhizomes can also be peeled and eaten raw, even as our area's Ojibwe Indians have traditionally done, pulling them in midsummer.*(Black 1973:125)* They can alternately be sliced and added to salads. Crushed and boiled, these young rhizomes are full of natural sugar and yield a sweet syrup if cooked for a considerable length of time.

Or they can instead simply be baked for about two to three hours and then eaten that way. The buds on the ends of the older rhizomes make a crispy, sweetish snack.

In the fall, the seeds can be collected, parched, ground, and prepared with heated water to make mush—a useful survival food. They can also be combined with flour secured from the rhizomes (accomplished by drying them, pounding them, and sifting out the fibers) and then pressed into cakes and baked. These are relatively sweet and sustaining.

I might add that Karl Knutsen, in his delightful foraging manual of the mid-1970s, provided a mouth-watering recipe for "Creamed Bulrush & Shrimp" that simply can't be ignored! *(Knutsen 1975:46)*

Health/Medicine

The Ramah Navajo and Cherokee Indians coaxed *emetic* properties out of a decoction of this plant when needed. *(Vestal 1952:19; Hamel & Chiltosky 1975:27)* On the other hand, *styptic* properties were utilized by the Woods Cree Indians, who applied bulrush stem-pith under a dressing to halt bleeding from *WOUNDS*. *(Leighton 1985:59)* Likewise, burned and cooled ashes from the stalk were applied by the Thompson Indians to a newborn baby's navel to stop *BLEEDING* there. *(Turner et al 1990:115)*

Bulrush's blood-stemming ability would seem to relate largely, if not entirely, to the plant's evident *astringency*, which property was proffered and exposited for such by early American physicians. *(Porcher 1849:851)* Such astringency also explains the use of bulrush by Maritime Indians as an agent for *SORE THROAT* *(Mechling 1959:247)* and elucidates Bradford Angier's observation that the plant's chopped stems have long been applied to *SORES* and *BURNS* as well as taken internally for *DIARRHEA*. *(Angier 1978:73)*

A mid-1970s' study found that three species of bulrush tested *(S. validus, S. americanus, and S. atrovirens)* possessed moderate *anti-inflammatory* properties. *(Benoit 1976:166)* This may explain why the Malecite and Micmac Indians were able to utilize the pounded root of a related species *(S. rubrotinctus)* —after mixing it with water to make a paste—as a topical application to an *ABSCESS,* so as to ease and reduce it. *(Mechling 1959:247)*

Herbalist Mary Ann P. Chai also finds bulrush to be *depurative, antipyretic*, and an *emmenagogue*. *(Chai 1978:92)*

BUNCHBERRY

[CANADIAN BUNCHBERRY; DWARF DOGWOOD; PIGEON BERRY]
(Cornus canadensis)

Description

[See photos #22 & 23 in rear of book]

A 2-to-9-inches-tall perennial (usually 4 to 6 inches tall), often growing in dense colonies due to the tendency of its underground rhizomes to spread. One would never know it, but this plant is so closely related to dogwood shrubs and trees that they all share the same genus (*Cornus*)!

Leaves appear about two-thirds of the length up the stem in a nearly sessile whorl of six (occasionally four, if the plant is a non-flowering one). They are pointed, egg-shaped, prominently lined, and 1½ - 3 inches long. Upon careful inspection, one can find beneath this whorl, and a bit further down the stalk, yet another pair—occasionally two pairs—of smaller leaves.

In spring, the tiny plant possesses what appears to be a solitary flower, ostensibly consisting of four white petals with a greenish-white center disk. However, the center section is actually a cluster of many, minuscule, *complete* flowers, while the surrounding white "petals" are really four, large, white, involucral bracts (= floral leaves).

Late summer finds the flowers replaced with pea-sized, reddish-orange "berries" (technically, drupes), usually tightly clustered, with calyx remains (just a bit, though) visible at the tips. These are soft when ripe.

Range & Habitat

As its alternate name, "Canadian bunchberry," suggests, this colorful herb is found pretty much throughout Canada (except for the northernmost strip). But it also flourishes in the northern strip of the US, dipping well into California in the West and into mountainous areas of West Virginia in the East. In our area, it is found in the upper two-thirds except for the westernmost counties of Minnesota. The preferred habitat is cool, moist, coniferous woods—the type of environment where one might find bluebead lily, wild strawberry, wild sarsaparilla, and the like.

Seasonal Availability

Fruits are available in August-September.

Chief or Important Constituents

Cornic acid, cornine, plant sterols(stigmasterol, beta-sitosterol, campesterol), a variety of flavonoids(including quercetin and hyperoside), phenylethylamine, alpha- and beta-amyrin, tannins. The leaves alone contain astragalin, hexahydroxydiphenic acid, procyanidin, and juglanin, while the fruit alone contains cyanin and pelargonin. The seeds contain a goodly amount of all three of the major essential fatty acids: linoleic, linolenic, and oleic.

Food

The berries are edible, although numerous foraging manuals describe them as "insipid" in taste. The popular naturalist Tom Brown, Jr., noting that this was a common assessment on the part of fellow foragers he has known, remarked in one of his books*(Brown 1985:73–75)* that the maligned little bunchberry really has a wonderful taste that comes through on a more subtle level if one gives it a chance. The problem, Brown intimates, is our MSG-laden society which demands strong seasonings on practically every food item, always clamoring for the ultra tantalizing. Applying ourselves to savoring the taste of the bunchberry, he explains, forces us to slow down and take a look at the finer things that life has to offer. The Indians demonstrated the proper attitude in this respect, relishing the beautiful little rubies as good food—our area's Potawatomi and Ojibwe (Chippewa) Indians, among other tribes, feasting with great appreciation on them.*(Erichsen-Brown 1989:304)* The naturalist P. H. Gosse, while traveling through the Quebec wilds in the first half of the nineteenth century, also reminisced fondly of the bunchberries: "We ate many; they are farinaceous and agreeable."*(Gosse 1971:299)*

Recipes abound for this berry, but one of the best is bunchberry pudding—a good and healthful recipe for which can be found in the excellent foraging cookbook by the Halls.*(Hall & Hall 1980:323)*

Health/Medicine

Data on the medicinal properties of the various *Cornus* species is sorely lacking. This is regrettable, in that some information exists to the effect that these species were held in respect by Amerindian healers. The Micmac Indians, for example, found bunchberry leaves, berries, and roots useful in treating *FITS,(Chandler 1979:56; Mechling 1959:256)* while the Montagnais implemented an infusion of the whole plant for *PARALYSIS.(Speck 1917:315)* Powdered, toasted leaves were sprinkled on *SORES* by the Thompson Indians.*(Steedman 1928:458)* The Micmac Indians applied them to *WOUNDS*, primarily to stop bleeding and to speed the healing process, for which application they were first chewed and moistened.*(Lacey 1993:80)* *KIDNEY AILMENTS* and *ENURESIS* in children were also treated with an infusion of the plant by the Micmacs.*(Lacey 1993:80)*

Boiled with wintergreen (*Gaultheria procumbens*), the bunchberry plant was held by the Tete de Boule Indians to be a treasured part of a most effective remedy for **COLDS**.*(Raymond1945:128)* Similarly, the Iroquois Indians employed the whole bunchberry plant as a decoction for **COUGH** and **FEVER**.*(Herrick 1977:402; Herrick 1995:177)* Michael Moore, a modern American herbalist who uses bunchberry with clients, also regards the herb as valuable for fevers, emphasizing that it works best with those that *tend to be chilling*—that is, with goose bumps and/or shivers.*(Moore1993:97)* Moore also finds the plant helpful for **HEADACHES**, and particularly those which are *clammy* in nature (i.e., accompanied by dampness of neck and back and with slight nausea).*(Moore1993:97)* Interestingly, the Delaware-Oklahoma Indians had also implemented bunchberry as an *analgesic*, finding it useful for **PAIN** most anywhere in the body.*(Tantaquidgeon 1942:26, 74)*

Known constituents shed some light on these uses. The cornic acid present, which has an analgesic effect similar to aspirin (though milder), undoubtedly at least partly elucidates usage of the berries or dried plant as an infusion for **FEVER** and **HEADACHE PAIN.** Pointing out that both the cornine and the flavonoids are anti-inflammatory, Moore suggests that this effect, in conjunction with the herb's astringent tannins, would also lend the bunchberry for use in treating **COLITIS, DIARRHEA, CHRONIC GASTRITIS**, and **FLATULENCE**.*(Moore 1993:97)* Interestingly, a gastrointestinal symptom profile like this was recognized earlier in history by several Amerindian tribes as calling for bunchberry: The Flambeau Ojibwe Indians, as one example, infused bunchberry roots to treat **INFANT COLIC**.*(Smith 1932:366-67)* The Micmac Indians found the plant useful for certain **STOMACH PROBLEMS.** *(Lacey 1993:80)* And understandably in view of this species' *anti-inflammatory* and *astringent* properties as elaborated above, the Southern Carrier Indians found an infusion of the plant (minus berries) to be an excellent eyewash for **SORE EYES**.*(Smith 1929:62)*

Another fascinating use for bunchberry is as an *appetite stimulant.* The Scots have long called bunchberry the "herb of gluttony." It is appropriately named, as I can personally testify: I recall how, on a trip to northern Minnesota, I once gorged myself on bunchberries and was immediately afterwards so consumed with appetite that I proceeded to open a large can of tuna, make three sandwiches, and wolf all of them down! Understandably, then, herbalists have long taken advantage of the plant's amazing appetite-stimulating properties to help those who have lost their desire for food, preeminently by steeping the whole plant to produce a tea.*(Schofield 1989:83)*

Lewis and Elvin-Lewis report that *antineoplastic* principles are suspected as being present in the plant's tannins and that it was scheduled to be looked at by scientists in this regard.*(Lewis & Elvin-Lewis 1977:135)* Whether this actually occurred, and with what results, I have been unable to discover.

Utilitarian Uses

Richardson says that the little bunchberries are widely rumored to be **EXCELLENT BAIT** for little fish*(Richardson 1981:53)*—useful to know in a survival situation, as catching little fish gives one bait for bigger fish!

BURDOCK
(Arctium minus; A. lappa)

Description

[See photos #24-25 in rear of book.]

Burdock is a biennial plant. During its first season, it appears only as a low-lying herb bearing long-stalked, heart-shaped leaves. These are quite large—at times growing to 18 inches! Other leaf features are as follows: They are dark green, wavy-edged, possessed of a peculiar net-like surface pattern, and woolly underneath. A characteristic rank smell at this stage of growth helps distinguish this plant from similar-looking species (i.e., rhubarb gone wild).

In its second season, a branching stem forms, eventually topping out at 3-5 feet. Stem leaves are alternate, ovate in shape, and smaller than the lower, first-season leaves. Like those basal leaves, however, the stem leaves are also veiny and slightly velvetlike to the touch (esp. the bottom side). Leaf stalks are hollow in *A. minus* and filled with a white pith in *A. lappa*.

Composite flowers—in the form of tubular florets—appear in the middle of the second season and are borne singly, or in clusters, at the upper part of the stem and branches. The flowerhead is almost an inch wide, very short-stalked, reddish violet to amethyst in color, and bristly. It emerges from a spiny, ball-like bract (reminiscent of thistles, to which the plant is related). During autumn, these hook-tipped bracts transform into the infamous, dry, brown burrs that so tenaciously stick to clothing and fur.

Range & Habitat

Found throughout Canada and in the upper 3/4 of the USA. In our area, *A. minus* is scattered throughout about 60% of the counties, while *A. lappa* is rare. The preferred habitat is waste places and other disturbed areas, old fields, pastures, vacant lots, roadsides, and neglected farmlands.

Seasonal Availability

Roots: Late spring/early summer of first year to spring of second year.

Chief or Important Constituents

Root: polysaccharides(inulin [to45%!]; mucilage), tannins, lappin, arctigenin, polyacetylenes(14 of them!), non-hydroxyl acids, volatile oil. Leaves/stem: arctiol, arctiin, taraxasterol, fukinone, a sesquiterpene lactone(arctiopicrin). Seeds: 15-30% fixed oils, a bitter glycoside(arctiin), chlorogenic acid. The plant's nutrients include vitamins (A, C, thiamine, and riboflavin in the seeds), phosphorous, potassium, sodium, & large amounts of chromium, magnesium, silicon, zinc, and iron.

Food

The cooked roots of first-year plants are edible and nutritious, being sweetish to the taste. (Roots of second-year plants are thin and woody, since they shrink in size as the plant begins to stalk.) They should be secured in late spring/early summer for best results. Unfortunately, they are often entrenched several feet deep, where they tenaciously cling to the soil, so that they must be dug out with a firm spade at the least (Euell Gibbons suggested a posthole digger as being ideal!). It requires some practice to get the hang of it. One needs to be careful when the root pops out—the sheer force of the affair can powerfully fling loose soil toward the eyes!

Once successfully unearthed, burdock roots are usually peeled, cut into chunks, and boiled in several changes of water (i.e., throwing off two waters, and doing the final boiling in a third). They should be cooked in a glass or enamel pot if one wants to keep the chunks from turning dark (caused by anthocyanin pigments in the root darkening upon reacting with ions from a metal pot). The addition of a little baking soda helps break down the fibers.

Very early spring leaves—especially of *A. lappa*—are enjoyed by some as a potherb (cooked 20 mins. in two waters, with a pinch of baking soda in the first).

Health/Medicine

Burdock was adopted by the American Indians for medicinal purposes after it arrived in America as a weedy import. It was enthusiastically used by the Eclectics and found official status in the *U.S. Pharmacopoeia* from 1831-42 and 1851-1916, as well as in the *National Formulary* from 1916-47. It is still official medicine in Germany, France, Belgium, and the U.K and remains a favorite among herbalists of many heritages. Moreover, it has attracted the attention of phytochemists (see below).

Both taproot and seeds (the latter especially) have long been celebrated as an excellent *diuretic*.*(Bever & Zahnd 1979:139-96)*, which was a favorite use of this plant made by the Eclectic physicians and by the Iroquois Indians.*(Herrick 1995:229)* And both—but especially the seeds—are somewhat *laxative* as well, making them useful for *CONSTI-PATION*. (Seeds—first available in late summer—can be decocted or tinctured for use.)

The Cowlitz Indians utilized a decoction of the root to treat *WHOOPING COUGH*.*(Gunther 1970:50)* Other herbal traditions in America have similarly employed it for *COUGH, ASTHMA*, and *PULMONARY COMPLAINTS*. Ojibwe Indians found it invaluable for a hard, dry cough and also drunk it after a coughing spell.*(Densmore 1974:340-41)*

Burdock is renowned as a potent *blood cleanser* (receiving official recognition as such in several European pharmacopoeias), especially for helping the kidneys to eliminate harmful uric and other acids from the blood, so that this plant is much appreciated by those afflicted with *RHEUMATISM* and *GOUT*.*(Moore1993:292)* (Interestingly, long ago, the Iroquois had used a burdock-leaf poultice to treat rheumatic parts of the body, while the Delaware had drunk a decoction of the root for same.-*Tantaquidgeon1942:31; Herrick 1995:229*) *Anti-inflammatory* principles in the plant (demonstrated via the rat-paw edema test-*Lin et al 1996:127-37*) are undoubtedly of value here as well and were widely made use of by various Amerindian tribes. For example, the Iroquois found that a mashed-leaf poultice worked wonders on a **BEE STING**, while a warmed leaf poultice eased pain from a **strained back**, and a root decoction eased the pain from **sore muscles.** *(Herrick 1995:229)* Moreover, herbalist Michael Moore finds the root decoction to be of great aid to those afflicted with **IgE-TYPE ALLERGIES**. *(Moore1993:292)*

The taproot is further known as a specific for **DRY ECZEMA** and **PSORIASIS**, being listed in the *British Herbal Compendium* for internal and external use in these conditions. Perhaps its most famous use, however, is for **BOILS** (gleaned from Amerindian use-*Tantaquidgeon 1942:56; Smith 1933:49; Shemluck 1982:312*) and the sometimes-resultant widespread condition therefrom, **FURUNCOLOSIS**. The noted herbalist Dr. Edward E. Shook described in graphic detail how he had witnessed the cure of a horrible case of furuncolosis by means of a decoction of burdock root taken internally along with a poultice of the leaves, sprayed with eucalyptus oil, as a topical treatment.*(Shook 1978:49)* Whether Shook knew it or not, the Delaware Indians had also treated boils with a burdock-leaf poultice.*(Tantaquidgeon1942:56)* Burdock is likewise employed for **DERMATOSIS** (note: not the same as dermatitis) and many other **CHRONIC SKIN PROBLEMS**.*(Bruneton1995:151; Moore1979:45)* The Iroquois found it invaluable for big **PIMPLES** on the face and neck.*(Herrick 1995:229)* The seeds (harvested in the fall after the burs have turned brown) are held by many herbalists to be the most efficacious part of the plant for this purpose, especially for **ADOLESCENT ACNE,** because it is thought that they react well with the male hormone. A decoction, or tincture, of the seeds is taken internally and sometimes also applied locally.

Accumulated clinical and experimental evidence suggests that burdock **enhances liver** and **gallbladder function.***(Chabrol &Charonnat1935:131-42)* What is more, it has exhibited **hepatoprotective** properties against CC14-induced hepatotoxicity.*(Lin et al 1996:127-37)* Bitter principles in burdock also **stimulate gastric secretions,** improving a **SLUGGISH APPETITE** and **POOR DIGESTION**.

Research on the part of German and other scientists has revealed that burdock's root—especially when fresh and not dried—has **antifungal** and **antimicrobial** properties, due at least in part to three of its fourteen polyacetylenes.*(Vincent 1948:669; Schulte 1967:829-30; Duke1985:54; Bruneton1995:151)* These, concentrated in the roots, were shown to possess activity against gram-negative bacteria.*(Moskalenko1986:231-59)* The sesquiterpene lactone present in the leaves, arctiopicrin, also has antibiotic activity, but against gram-positive bacteria.*(Bever & Zahnd 1979:139-96)* Together, the leaves and flowers showed activity against both gram-positive bacteria(*Bacillus subtilis, Mycobacterium smegmatis, Staphylococcus aureus*) and gram-negative bacteria(*Escherichia coli, Shigella flexneri, Shigella. sonnei*).*(Moskalenko1986:231-59)* (The leaves' effects against *Staphylococcus aureus* may partly elucidate their previously-mentioned healing of boils, which are usually caused by this gram-positive bacterium.)**Antiviral** properties are present as well; recently, an **anti-HIV effect** was demonstrated *in vitro*.*(Duke 1997:269)*

Japanese studies*(Morita et al 1985:925-32; Morita et al 1984:25-31; Ito et al 1986:55-60)* suggest that the root is **antimutagenic**. Additional research has isolated **antitumor** principles in it.*(Dombradi & Foldeak 1966:173-75; Foldeak & Dombradi 1964:91-93; Dombradi 1970:250)* The phytochemical arctiin, occurring in the seeds, has been shown to exert a protective effect against carcinogenesis of the breast, colon, and pancreas in rats.*(Hirose et al 2000:79-88)* Interestingly, the Potawotami Indians had used burdock to treat **CANCER** many decades before these functions were verified by scientists.*(EthnobotDB)* Likewise, scientist Jonathan Hartwell, in his mammoth study of plants used against cancer by indigenous peoples, devoted a sizable section to the traditional use of burdock.*(Hartwell 1968:71)*

Burdock is also one of four herbs in Canadian nurse Rene Caisse's Essiac® formula, purportedly based upon information given to her by Ojibwe Indians, and this formula has some documented evidence in its behalf for aiding cancer victims (*see* history and references in the entry for "Sheep Sorrel"). Burdock is likewise a component of the famed Hoxsey Formula, which, according to one limited study following the lives of cancer victims who had used it, appears to have some life-prolonging activity.*(Austin 1994:74-76)* A subsequent investigation by a scientist from the University of Texas who has examined cancer patients treated at the Hoxsey clinic in Tijuana, Mexico, has also reported preliminary results of an impressive nature.*(Richardson 1998)*

Blood-sugar-lowering (*hypoglycemic*) ability has been demonstrated for burdock root.*(Silver & Krantz 1931:274-84; Lapinina & Sisoeva 1964:52-58;Bever & Zahnd 1979:139-96; Leung 1980:81)* This is interesting in that French herbalists have long recommended it to diabetics to lower their blood-sugar levels, especially in that it contains inulin (not to be confused with insulin), a carbohydrate that does not rapidly convert to glucose as does starch.

Energetically, according to Traditional Chinese Medicine, burdock deals best with *hot (yang)* type of conditions,*(Pedersen 1998:58)* dispersing the pathogen of "wind heat."

Cautions

* The sesquiterpene lactones present in burdock can cause **CONTACT DERMATITIS** in people who have become sensitized to this chemical.*(Rodrgiguez 1995:134-35)*

* As burdock has been shown to be a uterine stimulant,*(Farnsworth1975:533-98)* **IT SHOULD NOT BE USED DURING PREGNANCY.**

* Burdock leaves should never be used if found *near a road or factory*, as they readily absorb nearby pollutants. And if one plans on sampling young burdock leaves as a salad ingredient or a cooked repast, one needs to be certain **NOT TO CONFUSE THE PLANT WITH RHUBARB** (*Rheum rhaponticum*), which has similar-looking leaves (although burdock's are not as smooth and shiny looking as the latter's and they possess a netlike pattern on their surface). Rhubarb, as a garden escape, sometimes encroaches on wasteland, vacant lots, and fields. Its leaves are highly toxic—possessing high concentrations of oxalic acid, oxalates, and anthraquinones—and can be fatal if consumed. However, the ripe stalks of rhubarb are bright red and they lack the characteristic rank smell that one learns to associate with burdock.

BUTTER-AND-EGGS
[TOADFLAX; FLAXWEED; YELLOW SNAPDRAGON]
(Linaria vulgaris)

Description

[See photo #26 in rear of book.]

A perennial, European import that reaches a height of 1-2 feet and tends grow in colonies.

The leaves of this odd-looking plant are numerous, entire, and narrow—even grasslike. Growing 1 - 2½ inches long, they tend to alternate on the upper part of the plant, but are often opposite or whorled on the lower portion.

The curious, yellowish flowers are irregular and two-lipped—very snapdragon-like, in fact. The upper lip is two-lobed and erect, while the lower one is three-lobed and more spreading. This lower lip also displays a prominent, orange-colored inflation (thought to be a nectar guide for insects, as well as insuring that these visitors pick up pollen along the way!). There is a long, curved, descending spur at the flower's base. The flowers are clustered in dense, terminal, clublike racemes.

Range & Habitat

A ubiquitous weed, preferring dry and sunny places such as fields, roadsides, railroad embankments, waste places, and disturbed soils.

Chief or Important Constituents

Quinazoline alkaloids(peganine), iridoid monoterpenes(antirrhinoside), aurones, phytosterol, flavonoids(linarin, pectolinarin, linariin), citric acid, tannic acid, choline, and goodly amounts of vitamin C.

Health/Medicine

This pretty plant is possessed of *cooling* and *drying* energies. Physiologically, this takes the form of **anti-inflammatory** and **diuretic** properties. As to the former, the noted European herbalist Nicholas Culpeper had pointed out:"The juice of the herb, or the distilled water dropped into the eyes, is a certain remedy for all heat, inflammation and redness in them." As to the latter, he commented upon its beneficial effects in **DROPSY**.(Potterton 1983:78) And well over a century earlier, horticulturist John Gerard observed that a decoction of toadflax "doth also provoke urine, in those that piss drop after drop, unstoppeth the kidneys and bladder."(Gerard 1975:556) Scientific experiments have since verified the plant's traditional uses for **URINARY-TRACT DISORDERS** and

INFLAMMATORY—or otherwise *"HOT"—CONDITIONS*, revealing that toadflax does indeed possesses *diuretic, anti-inflammatory, mild antimicrobial,* and *diaphoretic* properties.*(PDRHrbs; Benoit 1976:168; Fitzpatrick 1954:530)*

It is especially renowned, however, as a superb *hepatic,* with varied applications that seem to reveal its power only to those who truly suffer *in extremis.* Here toadflax's best-known use is for *JAUNDICE.* Culpeper had said here:"The decoction of the herb, both leaves and flowers, in wine, does somewhat move the belly downwards, open obstructions of the liver, and help the yellow jaundice."*(Potterton 1983:78)* Modern southwest-American herbalist Michael Moore adds that a rounded teaspoon made into a tea and drunk three times a day rapidly brings rebound bilirubin levels back within normal parameters.*(Moore 1979:153)*

Owing to its *discutient* and *deobstruent* properties, butter-and-eggs is one of the few herbs that some herbalists will use with jaundice or other hepatic problems when they are characterized by obstruction, that is, where gallstones are in the bile path. (Other, cholagogue- or choleretic-type, herbs may simply entrench the stones in the bile duct, worsening the situation.) Here, it is used by way of infusion, to the tune of two tablespoons a dose, up to three times a day.*(de Bairacli Levy 1974:145)*

Since, as we've earlier seen, butter-and-eggs possesses *anti-inflammatory* properties*(Benoit 1976:168;PDRHrbs)*, probably due to its flavonoids, its virtues are also seen in connection with *CHRONIC INFLAMMATION OF THE LIVER*, including those episodes which periodically characterize *HEPATITIS.* Moore goes on record as stating that it appears to be the superior remedy for such incidents, at least among those plants indigenous to his area.*(Moore 1979:153)* In addition, he finds the plant helpful for"liver stress" from a poor diet, overindulgence in alcohol, or malabsorption. He cautions, however, that it can be used for just a short while by itself because of its potential for irritation, but allows it for a longer period when combined with other herbs. His preferred form and dosage seems to be a standard infusion dosed at two to four ounces and allowed up to three times a day.*(Moore1993:301; Moore 1979:153)* As earlier stated, though, he prefers to use toadflax in combination with other hepatic herbs (he specifically naming burdock, dandelion, yellow dock, and red root) rather than by itself.*(Moore 1979:153)*

America's Eclectic physicians of the late-nineteenth- and early-twentieth-centuries likewise valued toadflax for hypertrophy of the liver, as well as for the spleen. They preferred either a decoction of one ounce of toadflax to one pint of water or a strong tincture dosed at 1-10 drops. They further emphasized that the plant should only be harvested while in flower, thereafter being rapidly and carefully dried (see below, under "Cautions"), and finally secured in airtight containers away from light.*(Felter & Lloyd)*

Because toadflax contains peganine, an alkaloid with peristaltic and spasmolytic effects, the herb has been found to be helpful for *INTESTINAL ATONY.*(Kresanek 1989:118)* The flowering stem's bile-stimulating properties also help to ensure a smooth bowel movement in those persons troubled by *CONSTIPATION* because of an insufficient flow of this fluid. On the other hand, the Iroquois Indians found that an infusion of toadflax leaves helped to firm up *DIARRHEA*.*(Herrick 1977:433)* Why the difference? One possible explanation is that the bile-moving properties may be largely resident in the flowers, whereas the Iroquois used the leaves, which are rich in astringent tannins that the flowers largely lack. Another possible reason is Moore's further explanation that the flowering stems only serve to stimulate bile flow *when it is insufficient*.*(Moore 1979:153)*.

Being thus *astringent*,*(Wren 1988:267)* toadflax also has a cherished reputation as a *HEMORRHOID* remedy,*(PDR Hrbs)* either applied by way of a fresh-plant poultice or as an ointment. The Eclectics often suggested the use of the ointment, made from one part of the fresh, bruised plant to ten parts of an emollient base such as hot lard or mutton tallow.*(Felter & Lloyd)* Another source suggests two parts fresh herb to five parts pork fat.*(Kresanek 1989:118)*

Butter-and-eggs is an herb that must be used judiciously and in the appropriate doses (see under "Cautions" below). It is best used as a tea of the fully dried herb. The strongest preparation I have seen, which seems to be at the outer limits of safety, is 2 tsp. of the dried plant infused in 3/4 cup of water and with the result *sipped very slowly throughout the entire day*.*(Lust 1974:417)* Higher doses of 2 to 3 ounces at a time would seem to be safely imbibed of a much weaker infusion (a scant tsp. infused in a cup of water), as long as the day's total intake of tea is kept to under 12 ounces.*(Moore 1994b:15)* Yet another source tells of using only a teaspoonful of the dried herb to two glasses of water, drinking only 1/3 to 1/2 glass of this.*(Kresanek 1989:118)* *The PDR for Herbal Medicines* says that the standard dose is 1-2 tsp in 2-4 cups of boiling water, "left to draw for 18 minutes," and drunk slowly throughout the day.*(PDRHrbs)* A 1:5 tincture of the flowering herb, with 60% alcohol, is also sometimes used, but should be kept to no more than 20 drops at a dose, well diluted, with no more than 3 doses per day.

Cautions

Toadflax should be *used with great caution by the layman*, if at all, due to an irritant—if not toxic—potential, which is probably due to the plant's bitter principles. The herb has also long been thought to be possessed of glycosides having the potential to cause cyanide poisoning, especially if the fresh or wilted plant—or an extract therefrom—is consumed internally and in too high an amount.*(Cooper & Johnson 1988:107)* (I have yet to see a lab analysis or study which confirms this, however.) Regardless, an irritant or toxic potential of some sort definitely exists (in fact, this plant used to be mixed with milk and used as a fly poison around barns!) and so **DO NOT EXCEED STANDARD DOSES.** Naturopathic writer John Lust cautions that as little as 20 drops of a strong tincture can give—and has given—rise to unpleasant effects.*(Lust 1974:417)*

BUTTERFLY-WEED
[PLEURISY ROOT]
(Asclepias tuberosa)

Description

[See photos #27 & 28 in rear of book.]

A native perennial growing 1-3 feet tall and certainly one of the loveliest plants in our area! Its single stem, rough and hairy, branches near the top. When damaged, it exudes a juice that is clear and not milky-white as in other *Asclepias* species.

The hairy leaves are ovate to oblong, stalkless, toothless, and mostly alternate. They grow 2-6 inches long.

The brilliant, orange flowers (sometimes more yellowish or reddish) occur clustered into flat, terminal heads that grow 2-3 inches long. Individual flowers have five, downward curved petals.

The ensuing seedpods are hairy, spindle-shaped structures, each growing to five inches. These, like the preceding flowers, are also grouped into clusters.

Range & Habitat

Throughout our area, in sandy soil such as occurs in fields, meadows, and prairies. Owing to its beauty, it is often a plant of choice in prairie restoration programs.

Chief or Important Constituents

A resin(inclusive of bitter principles such as asclepione), cardioactive glycosides, volatile oil, triterpenes(incl. alpha-amyrin and beta-amyrin), flavonoids(quercetin, rutin, kaempferol), beta-sitosterol, uzarin, caffeic acid, chlorogenic acid, gallic acids, tannic acid, amino acids, choline.

Health/Medicine

The dried rootstock of butterfly weed has been used and appreciated as a medicinal by both whites and American Indians for centuries. Most often referred to by healers under its alternate common name of pleurisy root, it was listed in both the *U.S. Pharmacopoeia* (1820-1900) and *National Formulary* (4th and 5th eds.) and was noted in the *U.S. Dispensatory* to be **emetic, cathartic, expectorant,** and **diaphoretic**.(Wood et al 1926:1293) The *British Herbal Pharmacopoeia* says that it is **antispasmodic** as well.(BHP1976:25) This makes sense when one considers that pleurisy root has historically been used for **DYSPEPSIA (INDIGESTION)**.(Millspaugh 1974:540) Many herbalists have also implemented it for **SPASMODIC ASTHMA.** (Shook 1978:301)

As to its ***expectorant*** properties*(cf. BHP 1976:25; Lust 1974:313),* it has been found help-ful for **BRONCHITIS***(Shook 1978:301),* especially when acute, hot, and dry.*(Moore 1994:4.3)* Here, though, the Natura physician R. Swinburne Clymer said that it should be given by tincture in hot water for best effects and also combined with skullcap (*Scutellaria* spp*) (see under that entry)*, with the dosage being 15 to 40 drops of the former and 2-20 drops of the latter (or of boneset [*Eupatorium perfoliatum*], if skullcap is unavailable), given every few hours until free perspiration is established, and then less frequently till the bronchials are clear; afterwards, followed up by cayenne, and finally by golden-seal.*(Clymer 1973:103)*

Butterly-weed is generally agreed to be most effective, however, for ***ACUTE, LOWER-RESPIRATORY-TRACT INFECTIONS****(Mills & Bone 2000:216),* where it serves to break up ***PULMONARY CATARRH.****(Shook 1978:301; Clymer 1973:103)* It is especially of benefit when combined with lobelia, cayenne, or gumweed (*Grindelia* spp.—*see under that en-try*).*(BHP 1976:25)* It is a specific when the catarrh is deep down and where the chest feels oppressed, but where the upper respiratory tract feels dry. The sufferer, says Clymer, also has *hot* and *dry skin*.*(Clymer 1973:104)*

Pleurisy root is superb for a non-productive ***COUGH*** that is acute and pulmon-ary or even when chronic, debilitating, and accompanied by indigestion.*(Moore 1994:4.5; Christopher 1976:222)* It has also been a favorite for ***PNEUMONIA.*** *(Shook 1978:301)* Georgian folk medicine has used it thusly, adding it first to whiskey.*(Bolyard1981:46)* It seems to be especially effective when the pneumonia is of the acute bronchial type, accompanied by vascular disturbance and fever, or when the person is recuperating but has difficult expectoration and the respiratory system is dry and tough.*(Moore 1994a:4. 13)*

This plant has also long revealed its benefits for ***TUBERCULOSIS.****(Shook 1978:301)* Interestingly, in the 1950s, this classic use was elucidated when it was discovered that butterfly-weed was powerfully active *in vitro* against a virulent human strain of TB.*(Fitzpatrick 1954:529)* As its name, pleurisy root, would imply, however, our herb is perhaps most specific for ***PLEURISY****(BHP1976:25;Clymer1973:102)* or for other types of ***INTERCOSTAL PAIN*** that occur upon inhaling*(Moore1994:4.5;Christopher1976:222)* and which are relieved by bending forward.*(Tilgner 1999:96)*

Unfortunately, pleurisy root is little used these days. This is in contrast to the widespread use and recognized benefits that it achieved in the nineteenth century. For example, medical doctor Jacob Bigelow, writing in the early 1800s, remarked that it had been utilized by practitioners in the "southern states in pulmonary complaints, par-ticularly in catarrh, pneumonia, and pleurisy; and has acquired much confidence of the relief of these maladies." Elucidating the nature of its healing properties, he continued: "It produces effects of this kind with great gentleness and without...heating... relieving the breathing of pleuritic patients in the most advanced stages of the disease."*(Bigelow1817-21:2:63)* Medical botanist R. E. Griffith, though critical of many botanicals popularly used in his time, yet noted that pleurisy root "has been employed with much benefit in [maladies] of the respiratory organs, and there is much ample testimony of its curative powers when judiciously administered for respiratory ailments." *(Griffith 1847:455)*

Pleurisy root is a very strong and reliable *diaphoretic,* and thus helpful for *FEVERS,**(Elliott1821-24:1:77;Shook 1978:301;Tilgner1999:96)* especially where the skin is moist. *(Moore 1994a:15.6)* Also, as nineteenth-century physician and botanist Laurence Johnson explained, the "diaphoretic effects have been found useful in acute pulmonary and bronchial affections and in rheumatism." *(Johnson 1888:231)* We have already delved into the traditional uses of this plant as an expectorant, above, but we must not fail to note here the thought that the diaphoretic effects may also be significant to the respiratory assist. As for Johnson's mention of the application to *RHEUMATISM,* such a use has also been extremely widespread and attested, both by whites and by the American Indians. *(Tantaquidgeon1972:37,116; Bolyard 1981:46; Shook 1978:301; Millspaugh 1974:540)*

As was true of so many plants, the Indians also found this one useful as a *vulnerary.* Our own area's Menominees, for instance, appreciated it for *WOUNDS CUTS,* and *BRUISES.* *(Smith 1923:25)* The Omaha Indians poulticed chewed roots of butterfly-weed on *SORES* and *WOUNDS.**(Gilmore 1991:57; Lewis & Elvin-Lewis 1977:341)* It has especially been revered among the Indians as a *SNAKEBITE* remedy. *(Speck 1942:13)*

As for yet other applications: Some herbalists have found butterfly-weed helpful for *OBSTINATE ECZEMA.* *(Millspaugh 1974:540)* Modern herbalist Michael Moore has found it useful for *HYPERTENSION* accompanied by a strong pulse. *(Moore 1994:5.9)* No doubt the herb's cardioactive glycosides and/or flavonoids are of assistance in this regard.

The usual dosage has been 1/2 to 1 teaspoon of the dried root, infused in one cup of water, or a tincture of 1-2 mls—either up to three times a day. *(Hoffmann 1986:218)*

Cautions

* Absolutely CONTRAINDICATED in *PREGNANCY* because it is highly estrogenic and oxytocic.

* *Do not use the fresh root*, which is toxic and will cause vomiting. Overdosing of the dried root will also cause unpleasant and even harmful side effects (Moore cautions that as little as one tablespoon per cup of water may cause nausea and vomiting-*Moore 1979:130*).

CATNIP
[CATMINT]
(Nepeta cataria)

[See photos #29 & 30 in rear of book.]

Description

A perennial herb reaching a height of 1½–3½ feet. Catnip possesses a distinctive musky-minty odor—especially when jostled or damaged. A member of the mint family, its stem is square and branching, as well as ridged and hairy.

Catnip's leaves are opposite, heart-shaped to arrowhead-shaped, coarsely toothed, and covered on the lower side with what is best described as a mealy sort of down, giving them a gray-green hue and a *very soft feel* to the touch—like velvet. Leaf length is 1–2½ inches.

In midsummer, a number of ½-inch-long flowers appear, crowded into 1½- to 2-inches-long, spike-like whorls at the end of the main stem and major branches. In color, the flowers are white to lavender to faint blue, but almost always bearing tiny purple spots. As is characteristic of mint flowers, they also possess two lips.

Range & Habitat

Catnip was brought over from Europe as a plant of cultivation, but escaped and is now widespread throughout North America as a wild plant. In our area, it is ubiquitous, except in a few counties in the upper central to eastern part of Minnesota and adjacent areas of northern Wisconsin. Its preferred habitat is disturbed ground—waste places, roadsides, trailsides, old homesteads.

Chief or Important Constituents

Volatile oil(nepetalactone [up to 90%, but some samples test around 40%], thymol, nepetalic acid, carvacrol, pulegone, camphor, humulene, carophyllene), a saponin, iridoid glycosides, tannin. Significant nutrients include vitamins A & C and the minerals manganese, chromium, iron, potassium, calcium, and selenium.

Food

I enjoy the velvety-soft leaves as a trail-nibble, putting one in each cheek and slowly chewing the juice out of them as I go my way (an idea I gleaned from one of Euell Gibbons' books). Catnip tea is appreciated by many, and can be made by pouring hot water over either fresh or dried leaves and steeping 10–20 minutes. (With aromatic mints such as this, however, the dried leaves are always preferable for tea.) This makes an excellent bedtime beverage—soothing, cooling, and relaxing (see more on this below).

Health/Medicine

At one time official medicine in America (the dried leaves were in the *U.S. Pharmacopoeia* 1842-82 and in the *National Formulary* 1916-50), catnip is one of the more interesting wild plants owing to the contrasting effects that it has on felines and humans. Its stimulating effect on cats is well known, but its *CALMATIVE EFFECT* on humans is not so well known in North America (it is in Europe, however). As such however, catnip is invaluable as a remedy for *TENSION HEADACHES* and for *INDIGESTION* induced by *NERVOUSNESS.* *(Felter 1983:281)* Then, too, a tea made from the leaves makes an excellent *SLEEP INDUCER* (or *SLEEP EXTENDER*) for those who find it difficult to catch their forty winks. This has been confirmed in non-feline animal studies. *(Harney et al 1978:369,373;Sherry&Koontz1979:68–71)* The soothing effect also combines with the herb's cooling energetics to provide a marvelous treatment for hot, inflamed conditions. Thus, herbalist and naturalist Tom Brown Jr. recommends a tea of half catnip and half boneset *(Eupatorium perfoliatum)* (*see under that entry*) for *SORE, STRAINED MUSCLES.* *(Brown 1985:66)*

The warm infusion is also a remarkable *diaphoretic/febrifuge* *(Felter1922:281;PDRHrbs)* and was utilized and appreciated as such by both our area's Ojibwe (Chippewa) and Menominee Indians. *(Densmore 1974:354-55; Densmore 1932:132)* The *antipyretic* action, it should be noted, has a special affinity towards fevers which are eruptive (measles, chicken pox, etc.) or intermittent (e.g., malarial). *(Holmes 1997:1:167)* As the Iroquois Indians experienced, too, a cup of catnip tea often proves helpful in yet other ways for those with colds and flu—by acting as a welcome *decongestant* in these situations. *(Herrick 1977:423),* The Menominees even applied a poultice of the leaves to the chest of a person afflicted with *PNEUMONIA.* *(Smith 1923:39)*

Catnip has classically been used by herbalists to *promote menstrual flow* and to ease *PAINFUL CRAMPS.* Of the former, the Eclectic physician Harvey Wickes Felter had noted: "When marked nervous agitation precedes menstruation in feeble and excitable women and the function is tardy or imperfect, this simple medicine gives great relief." *(Felter1922:281)* Regarding catnip's aid for cramps, the plant's nepetalactone is a good *antispasmodic* *(Leung 1980:137)* and this may be the responsible factor in providing relief.

Concordantly, this herb has long been used in formulas for *TREMORS/CONVULSIONS,* especially in infants to whom it has a special affinity. *(Holmes 1997:1:168)* It is thus most appropriately called a "spasmolytic" in the newest edition of the authoritative *Potter's Cyclopaedia* of herbs. *(Wren 1988:66)* Indeed, such an effect is undoubtedly largely responsible for its reputation as a *carminative*—especially valued with respect to *INFANT COLIC*, even as the Mohegan Indians had come to appreciate. *(Tantaquidgeon 1928:266)* Felter, too, had called catnip "a splendid quieting agent for fretful babies" and had listed its specific indications as follows: "Abdominal colic, with constant flexing of the thighs; writhing and persistent crying; nervous agitation." *(Felter 1983:281)* It was the use of catnip in such a situation in the late 1970s that first sparked my own interest in herbal medicine, when my infant son was miserably beset with painful colic and a lady we were staying with calmed him almost instantly and eased him into restful sleep with but a few tablespoons of catnip tea.

The tea is likewise the herb of choice for **INFANT DIARRHEA**, having been long ago used by the Iroquois Indians for this purpose.*(Herrick 1977:422)* Scientific evidence support this application, since an aqueous extract of catnip leaves has been shown to inhibit *E. coli*, and this has even led to the suggestion that the tea be used as oral replacement therapy for infant diarrhea.*(Crellin & Philpott 1989:140)*

The herb's curious affinity for youngsters further shows up in its splendid ability to relieve **BULL HIVES** on babies—a use long appreciated in Appalachian communities.*(Bolyard 1981:89;Crellin & Philpott 1989:140)* Catnip's efficacy as a nervine for **RESTLESS, AGITATED KIDS** is likewise most notable,*(Holmes 1997:1:168)* even as the Iroquois Indians had discovered.*(Herrick 1995:209)* In this latter regard, too, the beloved country herbalist Tommie Bass had cherished catnip for toddler frustrations during **TEETHING**.*(Crellin & Philpott 1989:140)* Herbalists have also long recommended a cup of catnip tea before bedtime for youngsters who are plagued by **NIGHTMARES**.*(Grieve 1971:1:74)* Finally, the Iroquois tapped into the herb's *antipyretic* effects, mentioned earlier, tailoring its application to **INFANTS** with **FEVER**.*(Herrick 1995:209)*

Utilitarian uses

Catnip is a first-class **INSECT REPELLENT**, owing mostly to its nepetalactone content. Dr. James Duke reports on a test of it involving 27 insects in which 20 were rebuffed.*(Duke 1985:325)* Another scientist, Thomas Eisner, detailed a fascinating array of experiments in which catnip oil repelled 13 out of 18 kinds of insects, acting like an impenetrable barrier around impregnated food, while controls were eagerly seized upon.*(Eisner 1965:1318)* Ants seemed especially dissuaded by the herb, a conclusion confirmed by another study performed two years later.*(Regnier et al 1967:1281)* Catnip has also traditionally been used to repel rats, although I am not aware of any studies that have confirmed this.

Cautions

* As is true of most mints, catnip should **NOT BE USED DURING PREGNANCY.** It is, as we've seen above, an emmenagogue and contains pulegone, which is the chemical in pennyroyal that contraindicates that relative of catnip for use during pregnancy.

* When making a tea from wildcrafted catnip for infant use, the tea must be carefully strained so as to remove any constituents; paper coffee filters are useful here. One needs to be sure that no solid particles remain.

CATTAIL
(Typha latifolia; T. angustifolia)

Description

[See photos #31 & 32 in rear of book.]

The familiar cattail is a tall (4-10ft.) perennial growing in thick stands or colonies in marshes, swamps, and on the perimeters of ponds and lakes. Its long stem is stout, tough, round, and smooth.

The plant's leaves are basal, enwrapping the stem on its lower portion. They are long, flat (straplike), and pointed. Because they also spiral several times, an onlooker gazing at cattail on a sunny day sees what appear to be alternating light-and-dark bands on the green leaves.

As the top segment of the stem grows, a pair of six-inch heads—sausage-like in appearance and wrapped in a papery husk—appear. The lower one is a flowerhead composed entirely of a cluster of female flowers and the upper one is a male appendage that eventually grows rich with yellow pollen in early summer and then releases its load and shrivels away shortly thereafter, leaving behind a bare stem above the female head (which, meanwhile, has become brown just like a sausage).

Buried in the muck below is the cattail's horizontal rhizome, whitish in color.

Two main species predominate in our area: *T. latifolia* and *T. angustifolia*. The latter is thinner than the former and has a noticeable space on its stalk between its male and female heads.

Range & Habitat

A nearly ubiquitous plant throughout our area (though surprisingly unrecorded in a few Minnesota counties). The habitat is, of course, swamps, marshes, roadside ditches, and the perimeters of ponds and lakes.

Chief or Important Constituents

Quercetin-neohesperidoside. Nutrients include high amounts of potassium and moderate amounts of phosphorous. The pollen contains sitosterol, iso-rhamnetin, and nutrients such as protein and vitamin A. The rhizome contains more protein than corn or rice, more starch than a potato, and a good supply of fatty acids. The stem and seeds contain leucanthocyanidin. There are more omega-3 fatty acids in the seeds than in nearly any other known source.

Food

One of the most utilizable of wild plants, the cattail offers a variety of culinary delights for the adventuresome forager....

Initially available are the plant's shoots, which can be collected by firmly grasping the stem inside the enwrapping leaves as far down as possible and then pulling the stalk (some recommend twisting it first) from the ground (the roots and rhizome will usually stay behind). Once the prize has been unearthed, one simply cuts free the bottom twelve inches or so, which portion can be peeled to reveal a white core that can be eaten raw or cooked (simmered not more than 10-15 minutes). This delightful repast has been called "cossack's asparagus," although the taste is not really like asparagus but more like raw cucumber, watercress, or parsnips. As the shoots mature, they become less tasteful. Even at this stage, however, they can still be cut into segments and added to stews.

The next cattail segment available for food is what has been called "cattail corn-on-the-cob." This is in reference to the sausage-like spike situated at the top of the plant (the male appendage). As it begins to appear, it is green and wrapped in a papery husk. *At this stage*, it can be cut free and prepared as a potherb—immersed for five to ten minutes in rapidly boiling water (salted or not). Or it can instead be cooked for five minutes in boiling water and then let sit for ten more while the burner is turned off. Both the form and method of eating the prepared spike—and even to some extent its taste—reminds one of eating corn on the cob. In a survival situation, the lower (female) spike can, at this (green) stage in its development, also be prepared in similar fashion. It is only deemed a survival food, however, because (1) it has very little substance to it, (2) it is not nearly as tasty as the male spike, and (3) it is a lot messier to consume. Either spike, too, can be boiled for just a few minutes and then packaged and frozen for later use. Bear in mind that both spikes become inedible (with the exception of some of their constituents, which see below) after they lose their green coloring.

Next available is the golden pollen which saturates the male appendage in June, visibly appearing on it as if it had been dusted with yellow chalk. It can be collected into a bucket, bowl, or paper grocery bag by bending the cattail stem so that the pollen head rests against the side of the collecting implement and then gently beating or rubbing the pollen into the container. Thereafter, the pollen can be separated from the large amount of chaff (also unavoidably collected) by means of a flour sifter. One can dry it and store it for future use or simply use it right away as flour in a variety of recipes for bread, corn bread, cookies, or pancakes (it is most often used for the latter, and I have fond memories of collecting cattail pollen with my kids and working with them to make the pancakes). As the pollen tends to resist wetting, however, it is best used half-and-half with a domestic flour like wheat or corn. Used alone and simmered in water for half an hour, however, it finally thickens into a pretty good oatmeal substitute.

Available in late summer are the tiny seeds enwrapped in the female head, which can be acquired by removing the fluff from the brown sausage head, laying this downy material onto a flat and inflammable surface such as a rock, and then igniting it, which burns away the fluff and leaves the parched seeds all ready to eat! (You will usually need to winnow the remaining chaff first, by blowing on the mass or letting the wind do it for you.) The seeds can also be added to breads or pancakes.

In autumn, the starch from the rhizomes can be acquired by first digging out the latter with a firm stick or long stone. Then, one can simply chew it out after roasting the rhizome in the slow embers of a campfire, which produces a very pleasant and sweetish taste. (Never consume raw rhizomes, however. Having been submerged in swampy water for some time, they may harbor parasites or other harmful microorganisms). The starch can also be separated from the fibrous part of the rhizome and then used as flour for muffins, bread, etc. Two main methods for doing this have evolved among wild-foods connoisseurs....

The first preparation method calls for one to dry the rhizomes in the sun for several days or in an oven set at about 200 degrees F. for a few hours. Then, using two large stones or a grinder, one simply pulverizes the dried rhizomes. Next, the fibers are picked out by hand or sifted and then the resultant flour is used immediately in a recipe such as listed above or stored for later use. The second method of preparation involves washing and peeling the rhizomes and then cutting them into small segments and putting them into a large bowl of cold water. Next, the material needs to be vigorously worked over with one's hands, in an endeavor to separate the starch from the fiber with the aid of (clean!) fingernails and a sloshing of the rhizome segments around in the water. When the starch seems to have largely separated from the core segments, the latter is discarded. Next, one strains the resultant fluid through a coarse sieve to remove any fiber segments and the remaining liquid is allowed to sit for half an hour, during which time the starch will settle to the bottom of the container. Then the water and any remaining chaff on the surface are carefully poured off so that only the starch is left at the bottom of the bowl. The container is then refilled with clean water and the process repeated to further purify the starch. Afterwards, it is repeated yet one more time. Then the cattail "dough" is wrapped around a stick and cooked over a campfire or fireplace for a tasty treat. Or the material can be used in a recipe. Or it can be dried in the sun (or an oven) and stored for later use.

Health/Medicine

While most people are acquainted with some of the above uses of cattail as an edible plant, very few are aware of its medicinal applications. However, such uses are legion, as witnessed below....

The Kentucky herbalist Tommie Bass once remarked that since grasses and aquatic plants are generally *diuretic*, cattail as a grass-like, aquatic plant should likewise have such a property, although he was not personally aware of such a use for the plant. Interestingly, however, history has recorded a variety of applications of cattail for the genito-urinary system. For example, **URINARY GRAVEL** was treated by Oklahoma Delaware Indians with the root, *(Tantaquidgeon 1942:80-81; 1972:122-23)* while the French valued this prolific marsh plant for **GONORRHEA**. *(Porcher 1863:544-45)*

The tireless, early-American writer on medicinal plants, Constantine Rafinesque, highlighted some of cattail's other physiological functions in noting that the "roots" were "subastringent" and "febrifuge."*(Rafinesque 1830:250-51)* America's Eclectic physicians agreed, stressing the **astringent** quality and adding as well that the rhizome was **emollient**. As to cattail's astringency, the French have traditionally used it for **CHRONIC DYSENTERY**.*(Porcher 1863:544-45)* The Cheyenne Indians found that a decoction of the pulverized, dried roots and white base of the leaves was useful for alleviating **STOMACH CRAMPS**.*(Hart 1981:13)*

The plant's astringent and other properties render it of value as a **vulnerary**. Here, Ojibwe (Chippewa) Indians from our area pounded the rhizome for use as a poultice on **SORES**,*(Hoffmann 1891:200)* while Malecite and Micmac Indians merely greased a leaf and laid it on the sore.*(Mechling1959:246)* Algonquin Indians of SW Quebec poulticed the crushed rhizome on **WOUNDS** and **INFECTIONS**.*(Black 1980;132)* Potawatomi Indians implemented cattail for **BURNS, INFLAMMATIONS,** and **WOUNDS**.*(Smith 1933:85)* As to burns, pharmacist Henry Youngken Jr., an aficionado of Amerindian medicine, wrote of cattail's use by Omaha Indians in such a regard: "The root was powdered, wetted and spread as a paste over the scald. The ripe blossoms were then applied as a covering and the injured part bound, so as to hold the dressing in place."*(Youngken1924:499;Gilmore1991:12)* Naturalist and herbalist Tom Brown Jr. points out further that the sticky juice found between the leaves yields a powerful **anaesthetic** which, in a wilderness situation, can even be used to dull the pain of a toothache or tooth extraction!*(Brown 1985:91)*

Cattail pollen is an important medicine in Traditional Chinese Medicine: In its raw state, it is used to relieve blood stagnation as manifested in **MENSTRUAL** and **STOMACH PAINS**. When cooked, its properties reverse and it becomes a **hemostat**.

Cautions

* Cattail is easily recognized once it has flowered, but should be used cautiously by the amateur before flowering due to a possibility of confusing it with the poisonous wild iris (see under *Blue Flag*), which also often grows in marshlike environments and bears similar looking leaves. Should the two be found growing together, one will note that wild iris' leaves are more blue-green compared to cattail's medium green; also, iris shoots are flattened where they meet their base, whereas cattail's remain rounded.

* Do not harvest any cattail growing near roads, as the plant collects pollutants from auto exhaust like a magnet.*(Erickson and Lindzey 1983:550-55)* Do not eat cattail rhizomes without cooking them first, due to possible parasitical contamination from having been submerged in swampy environs.

* Use of cattail pollen as a medicine, or heavy consumption of it as a food, should be **AVOIDED DURING PREGNANCY**, due to an emmenagogue effect. Also, those with pollen allergies should use caution in any use of it.

CHICKWEED

(Stellaria spp.)

Description

[See photos #33 & 34 in rear of book]

A small- to medium-sized annual or perennial (depending on the species), with white, starlike flowers. Chickweed often grows in clumps or mats. Several species occur in our area, two of which are described below....

Stellaria media (common chickweed) is a common, sprawling, puny-looking weed with weak, delicate stems. It reaches anywhere from 4-16 inches in length. Small pairs of toothless, pointed, egg-shaped leaves grow to about an inch in length along the threadlike stem. Lower leaves are often stalked, but upper leaves are frequently sessile. The stem of this species has *one* line of hairs running down its length, which curiously *switches sides at each node*.

Stellaria aquatica [alt: *Myosoton aquaticum*], "water chickweed" or "giant chickweed," is a perennial, and more upright (with a stronger, and entirely hairy, stem), growing to 24 inches high. Its likewise opposite leaves are also egg-shaped, but more accurately described as ovate since they are much larger—growing at times to four inches and are also often quite wide (to 1½ inches) besides. They are mostly or entirely sessile, although occasionally the bottom leaves are stalked. The stems are entirely hairy. The flowers appear on the forked, upright branches.

Stellaria blossoms open only on sunny days. All species have white flowers (*S. aquatica's* are biggest) with five sepals and five petals that are notched so deeply it looks as if chickweed really has ten petals! This is especially the case with *S. aquatica*, because while *S. media's* petals are *shorter* than its sepals, *S. aquatica's* petals are *longer* than its sepals.

Range & Habitat

S. media is found in most of Canada and throughout the U.S., including ubiquitously in our area. It springs up in lawns, gardens, and waste places. *S. aquatica* is also found in moist woods, though usually along streams or ditches and in partial shade.

Seasonal Availability

Amazingly, chickweed is available year-round, even during winter! This is because it is noted for its ability to survive under mild winter snows, which is facilitated by the mechanism of its leaves closing over its delicate flower buds and developing leaflets.

Chief or Important Constituents

Steroidal saponins, glycosides, esters(hentriacontanol and cerylcerotate), coumarins and hydroxycoumarins, octadecatetraenic acid, carboxylic acid, mucilage, rutin. Nutrients include protein, unsaturated fatty acids(linoleic, gamma-linolenic, and oleic), saturated fatty acids(palmitic and stearic), vitamins(A, several B vitamins, C [very high, 0.1 to 0.15%!], and K1[phylloquinone]), minerals(including calcium and potassium salts, magnesium, manganese, silicon, zinc, phosphorous, and an extraordinary supply of iron—one of the highest known in the herb kingdom!)

Food

Chickweed is delicious picked straight from the wilds and eaten raw, an activity that I engage in often and enjoy very much! It is extremely nutritious, bursting with many vitamins and minerals. Thus, it also makes a fine ingredient to include in a healthful "green drink" ala one's juicer or as an ingredient in a wild salad or simply enjoyed as a trail-nibble.

This scrumptious plant can also be prepared as a potherb. As such, it should be boiled in salted water for 2-8 minutes. However, one needs to pick double the amount of what one plans to eat, as it cooks down in size quite a bit. Many equate the taste of the cooked plant with that of spinach. It is often added as a vegetable to wild-game stews.

Chickweed's fascinating ability to survive under winter snows allows one to gather it in the winter if snowfall is not too deep and one knows where to find this herb. I enjoy winter foraging and always strive to make chickweed a menu item at this time. Those not so inclined, however, can try this: Cover known chickweed patches with oak leaves (about 6 inches deep) after the first frost, but before the ground freezes. Lay any flat covering (such as a board) over the leaves to compact them. When hungry, lift up the board and rake back the leaves to reveal fresh chickweed greens for the pickin'!

Health/Medicine

Chickweed provides a much-appreciated *anti-inflammatory* effect. Internally, *ULCERS, ARTHRITIS, RHEUMATISM*, and *GOUT* are benefitted thereby. There is a protracted history to its application in these regards, but with arthritis it has had only varying success. In this regard, herbalist Alexandra Donson has found it most effective for arthritis that is chronic, degenerative, and hemophilic.*(Donson 1982:57)*

URINARY-TRACT INFLAMMATION is also helped by this fragile herb, especially by its fresh juice. Externally, a tea for washing *INFLAMED, SORE EYES* has been frequently utilized. Also, a crushed-plant poultice (esp. when warmed) has long been used for exterior inflammation of almost any kind, including *ECZEMA, PRURITIS, HEMORRHOIDS, ARTHRITIC JOINTS, SKIN SORES*, etc. Such anti-inflammatory action may help explain chickweed's common use as a *vulnerary* for many kinds of *WOUNDS* (incl. *CUTS* and *SCRAPES*), pioneered in America by the Iroquois, and no doubt other, Indian tribes.*(Herrick 1977:317)* The anti-inflammatory effect here may be due to the saponins, which perhaps help solubize the toxins in the abscesses and rashes.*(Pedersen 1998:73)* A possible explanation for the increase in effectiveness when the plant is warmed is presented by Heinerman: "A volatile oil present in the herb permits a chemical change in the basic compound structure of the... glycosides under conditions of warmth and heat. It other words, chickweed's best virtues are brought out when subjected to an influence of heat, either internally or externally."*(Heinerman 1979:186)*

This plant is revered as a good *demulcent* and *expectorant*. Here it appears that the mucilage is primarily responsible for the former and the saponins for the latter. In addition, mild *antibiotic* activity against a human strain of *TUBERCULOSIS* has been found.*(Fitzpatrick 1954:531)* No wonder that the herb has been praised by many for its aid in dealing with *COUGHS, BRONCHITIS,* and *HOARSENESS*, even as numerous American Indian tribes had earlier discovered.*(Scully 1970:212-13)* Tom Brown Jr. adds that the congestion-relieving properties are especially appreciated because they perform this function *without drying up the passages* (as is often done by stronger herbs).*(Brown 1985:98-99)* Indeed, this plant is *preeminently* a *moistening* herb.*(Holmes 1997:1:458-58)*

Another of chickweed's assists lies in its *depurative* ability. This herb is thought to help rid the liver and kidney of wastes and is eagerly looked to by many each spring as a "tonic" along such lines. Thus, too, it is traditionally used for *ACNE*. Aside from cleaning blood, however, our herb is also a blood builder, being exceptionally high in iron and thus used by herbalists to help offset *ANEMIA.* Here, too the high vitamin-C content helps to assure absorption of the iron. In fact, the vitamin-C content is so pronounced that chickweed has also traditionally been used to offset *SCURVY.*.

A treasured cooling (*refrigerant*) herb, chickweed is known for its mild *antipyretic* action. Such a cooling action may also elucidate its value in treating the "heat" conditions highlighted above. In addition, herbalist Peter Holmes recommends it for *HEAT EXHAUSTION* and *DEHYDRATION*.*(Holmes 1997:1:458)* The Chinese themselves use the root of a related species, *S. dichotoma* (which they call yin chai hu), saying that it clears deficiency heat and cools the blood, so that they utilize it for night sweats, fever, malaria of a yin-deficient character, bleeding, and debility.*(Belanger 1997:504; Fan 1996:229; Yanchi 1995:60)*

This plant is further an important, though underrated, *laxative* (esp. the roots). Brown tells an interesting story of its anticonstipational power, relating how he accidentally overtreated a case of *CONSTIPATION* with it, bringing on three days of diarrhea!*(Brown 1985:97-99)*

The versatile weed has frequently been cited as an aid to help offset *OBESITY*, although with only anecdotal evidence so far. However, it may possess an actual ability to eliminate fat (as some advocates have alleged for it), owing to the form and content of its saponins.*(Pedersen 1998:72)* Whether owing to these chemicals or some other factors, clinical use suggests that it may do this partly by *stimulating the thyroid*.*(Holmes 1997:1:459)*

CHICORY
[SUCCORY, WILD ENDIVE, BLUE SAILORS, RAGGED SAILORS]
(Cichorium intybus)

Description

[See photos #35 & 36 in rear of book.]

This European perennial import, a relative of the cultivated endive (even being of the same genus), is erect, 1-3 ft tall, and possessed of a milky sap that exudes from the stem when damaged. The plant emerges from a very large and fleshy rootstock.

Chicory's initial appearance is as a basal rosette of spatula-shaped leaves that grow 3-6 inches long. These have edges that are cut, lobed, or toothed and they are often curled as well. Once the stem sprouts, smaller leaves—oblong to lanceolate—take form on it, clasping it tightly with their bases.

The flower heads, occurring along the upper stem from May to October, are sessile, generally two at a place, 1-1½ inches across, and usually of a beautiful blue color (and thus the plant's alternate common name of "blue sailors"), though sometimes they are white or pink. The petal tips are square with notches that make them appear ragged (and thus the other common name of "ragged sailors"). The flowers open only in the first half of the day and close around noon. The great botanist Carolus Linnaeus found that the process was so reliable in his area—opening at 5:00 AM and closing at 12:00PM—that he could even use the plant as a clock!

Range & Habitat

A ubiquitous, weedy plant that is most frequently found on roadsides, railroad rights-of-way, waste areas, and fields. Most common in the bottom half of our area.

Chief or Important Constituents

Root: sesquiterpene lactones(including lactucin, lactupicrine, and intybin), taraxasterol, coumarins(umbelliferone, cichoriin, esculetin), essential oil, fatty oil, tannin, inulin(15-58%), and pectin. Flowers: cichoriin, esculetin. Leaves: flavonoids (incl. hyperoside), unsaturated sterols or triterpenoids, catechol tannins, tartaric acid, glycosides. Nutrients incl. vitamins such as A(4000 IUs per 100g), C, K, thiamine, niacin, choline, and folic acid(50 mgs. per 100 grams); minerals such as potassium(a whopping 430 mgs. per 100 grams; for the root, an also impressive 290mg per 100 grams), iron, calcium, magnesium, and phosphorous.

Food

Chicory's early spring shoots have long been valued as a salad bitter, especially the whitened, developing leaves situated just below the visible leaves. After the plant flowers, however, its leaves become too bitter for the palate. However, at this time it can still be prepared as a potherb. Boiling it for ten minutes in two waters is usually preferred and this seems to effectively reduce the intense bitterness to an acceptable level. Blanching the plant before harvesting—such as by covering it for a week or more with a box or pail—can even further reduce the bitterness.

Blanched chicory leaves can be obtained throughout the winter, too—by "forcing" them. This can be accomplished as follows: Gather some plants in autumn, right before the ground freezes hard. Cut the leaves back to within an inch of the root. Then place the plants in a container filled with sand (or sandy soil) and put the whole thing away in a very dark place (a closet or basement is often ideal). Water your secreted chicory every few days. If you've performed your task with skill, and if you garnish a little luck besides, you'll discover very pale leaves emerging in just a few weeks!

Chicory crowns are also edible. They can be boiled for five minutes and then served with butter for a real tasty treat. These crowns, along with the lovely blue flowers, also make decent salad ingredients. The flowers can also be made into jelly.

The roots, dug early in spring, also have an edible portion: a small, white center. However, the most revered use of the roots is as the world's most famous coffee substitute: This is done by digging the roots anytime from midsummer until the following spring, then thoroughly scrubbing them and cutting them into long, thin strips and finally into quarters. Next, one allows these segments to dry in a warm place for a few days (or roasts them on a cookie sheet in a 250-300° oven for 2-4 hours) until they snap easily and are dark brown inside. Finally, the roots are ground and brewed the same as with regular coffee, using 1½ tsp. chicory root for each cup of water.

Health/Medicine

This plant has many of the same benefits as its close lookalike and tastealike, dandelion (*see under that entry*). First, chicory resembles dandelion in being a *diuretic*. In this regard, plant scientist Jean Bruneton remarks that chicory enhances the ability of the kidneys to eliminate water—so effectively that it has often been used in weight-loss regimens.*(Bruneton 1995:78)* Another similarity with dandelion is chicory's reputation of being, as second-century Roman physician Galen once put it, "a friend of the liver." And indeed, this herb has been demonstrated to be an effective *hepatoprotective(Sultana et al 1995:189-92)* and *cholagogue,(PDRHrbs 745; IRCS Medical Science 1986:ab 14, 22)* being most especially used for **LIVER ENLARGEMENT** and **JAUNDICE**. The root's bitter sesquiterpene lactones, shown to possess *antibacterial* properties,*(Chem Abstr 59, 5535c, 1963)* are undoutedly most responsible for its celebrated *hepatic* assists.*(Vuilleumier 1973:42-93)*

Owing to its bitter properties, too, chicory serves as a valuable *tonic,* even as the *U.S. Dispensatory* once noted,*(Wood et al 1926:1256)* and thus has long been valued as a treatment for **DYSPEPSIA**. Interestingly, it is also thought to rid the GI tract of phlegm in those persons in which this material has plugged the villi and thereby prevented full and proper nutrient absorption. **SCLEROSIS OF THE SPLEEN** has also been treated with chicory, especially by the Germans.*(EthnobDB)*

Traditionally used in European herbology as a *refrigerant*, the famed herbalist Nicholas Culpeper found that chicory "helps the heat of the reins and of the urine," that it is "good for hot stomachs....heat and headache in children," and that it "allays swellings, inflammation." *(Culpeper 1983:186)* The plant's reputation for cooling goes back at least as far as the ancient Greek and Roman herbalists, who found it to be 'binding, cooling, and good for the stomach.' The Portuguese and Iraqi have also esteemed it for **FEVER**. *(EthnobDB)* Then, too, an aqueous distillation of the flowers has traditionally been used in Europe as a wash for **INFLAMMATION OF THE EYES**. *(Grieve 1974:1:199)* For countless generations, too, herbalists have recommended that the boiled leaves and flowers, lightly wrapped in a cloth, be applied to painful inflammations on the body. *(Lust 1974:156)* In this regard, naturalist and herbalist Tom Brown Jr. vividly relates how his Apache herbal mentor, Stalking Wolf, had used just such a poultice to heal his family dog, Butch, from a painful abscess developed from a fight with another dog. *(Brown 1985:93)* It certainly comes as no surprise, then, that a 1976 lab study found chicory to be significantly *anti-inflammatory*. *(Benoit 1976:164)*

Chicory has historically been used to treat **TACHYCARDIA** and **ARRYTHMIA**, especially in Egypt. *(Fetrow & Avila 1999:163; Tyler 1993:91)* A study on toad hearts conducted by Egyptian scientists revealed the presence of a substance in the root that depresses heart rate in similar fashion to that of the alkaloid quinidine from cinchona. *(Balbaa et al 1973:133-44)* Cherokee Indians even used it directly as a *nervine*. *(Hamel & Chiltoskey 1975:29)* In fact, chicory has long been thought to be mildly *sedative* and as such useful in offsetting the stimulating properties of coffee (with which, as noted above, it is so often mixed). *(Fetrow & Avila 1999:162)* A European study conducted in 1940 revealed that this relaxant effect accrued from the bitter principles lactucin and lactucopicrin in the herb's white latex, *(Forst 1940:1-25)* the same principles found in the latex from the related wild lettuce (*Lactuca* spp.), a classic sedative plant in herbal medicine.

British herbalists say that a chicory syrup (made by simmering equal parts of root juice with honey until a gooey consistency is reached) serves as a useful *laxative* in children, especially valued because it does not cause irritation. *(Grieve 1974:1:199)* (*Never* give honey to a child under a year old, however.) This preparation is used in a dose of one teaspoon, administered three times a day. The laxative effects have also been appreciated by the French, German, and Iraqis. *(EthnobDB)*

Chicory has traditionally been used by way of decoction, to the tune of one ounce of the root to one pint of water, boiled five minutes and then infused for 10-15 minutes more.

Cautions

* For some people, touching chicory may cause contact dermatitis, owing to a sensitivity to its sesquiterpene lactones. *(Malten 1983:232)*

*Excessive and prolonged use has been said to impair the proper function of the eye's retina. *(Grieve 1974:1:199)*

CINQUEFOIL
[FIVE-FINGER GRASS]
(Potentilla spp.)

[See photos #37 & 38 in rear of book.]

Members of the *Potentilla* genus usually have five-petaled flowers (often yellowish in color) and leaves with five, fingerlike leaflets (hence the common name of cinquefoil = "five-leaved"). A number of species occur in our area, including:

P. anserina (silverweed) is a hairy, prostrate perennial, with long (4-8"), featherlike, basal leaves divided into many(7-31), inch-long, toothed leaflets that are green above and *silvery-white beneath*. Its flowers are solitary—not grouped, as are those of most other species—and they appear on separate stalks. The fruits resembles dry strawberries.

P. arguta (tall cinquefoil) can reach a height of 1-3 feet and has a stem covered with brown, clammy hairs. Its 7-11 toothed leaflets are hairy on their undersides. The flowers are not yellow like those of most of the other species, but creamy white with yellow centers. They are grouped into loose clusters at the tips of the branching stems.

P. argentea (silvery cinquefoil) is an erect perennial, growing 5-12 inches high. Its five leaflets are wedge shaped and have edges that are rolled back. The undersides are downy. The stem and its branches are also woolly.

P. norvegica (rough cinquefoil) reaches a height of 5-36 inches and has a stout and hairy stem. This species has leaves with only three leaflets and its flowers have five pointed sepals (calyx lobes) that are longer than its petals.

P. palustris (marsh cinquefoil; purple cinquefoil) grows 6-24 inches and has a stem with a sprawling tendency. Its star-shaped flowers are large and crimson (or purple), and grow on slender stems. The leaves of this species have 5-7 toothed leaflets.

P. pensylvanica (prairie cinquefoil) tops off at about 1-2 feet tall and has pinnate leaves that have lobes resembling those appearing on the leaves of white oaks. Like some of the other species, above, it is downy all over.

P. simplex (common cinquefoil; old field cinquefoil; five fingers), one of our area's more common species, reaches a height of only 6-20 inches and has alternate, toothed, palmate leaves, with the fingerlike leaflets growing to 2½ inches long. Its flowers—borne singly like those of *P. anserina*, above—have rounded petals.

Other species in our area include *P. paradoxa*, *P. tridentata*, and *P. fructiosa*—the latter known as the shrubby cinquefoil and possessed of toothless leaves rolled over at the edges and stems that are reddish-brown and woody with thin, loose bark peeling off in shreds. Being shrubby in nature, it will be discussed in the forthcoming volume, *Edible & Medicinal Vines, Shrubs, and Trees of Minnesota and Wisconsin.*

Range & Habitat

Most of the above species are found throughout our area and thrive on dry, open soil in fields and on trailsides. *P. anserina* can be spotted on lake shores and riverbanks. *P. palustris* makes its home in marshes.

Chief or Important Constituents

P. anserina (and probably most other species): Tannins, flavonoids, hydroxy-coumarins(umbelliferone, scopoletin). Nutrients include calcium.

Food

The very young leaves of most, if not all, species are edible. As cinquefoil plants mature, however, their accumulation of large amounts of tannin and other compounds renders their herbage unpalatable.

The thick rootstocks of *P. anserina* (silverweed) are edible and possessed of a parnip-like taste. These can be eaten raw, and as such, are quite pleasant. They can also be scraped, sliced, and cooked in water for about 45 minutes. (Steaming is the method that best preserves the flavor.) Quite sustaining, they once supported inhabitants of the Hebrides and Scotch islands for months during a food scarcity. Roots of some other species are reputed to be edible as well.

Health/Medicine

Astringency is the *Potentilla* genus' most famous and pronounced property and elucidates the implementation of this plant by the Navajo-Kayenta Indians for *BURNS.*(Wyman & Harris 1951:26) This property further sheds light on the the custom of our area's Ojibwe (Chippewa) Indians of taking a root decoction as a soothing drink for *SORE THROAT.*(Densmore1974:342) Such a tea, say modern herbalists, is also helpful for *CHRONIC RESPIRATORY CATARRH* and especially as a syringed injection up the nostrils for *SINUS PROBLEMS.*(de Bairacli Levy1974:135)

Astringency probably largely explains why the cinquefoils are *styptic*, so that our area's Ojibwe Indians could implement the dry or moistened root of *P. arguta* as a topical application for *BLEEDING WOUNDS.*(Densmore 1974:32) Much earlier, the first-century Greek herbalist Dioscorides had concordantly said of the European species, *P. reptans*, that it was applicable to "a fluxing belly, & ye dysentery."*(Herbal 4.42)* Writing in the early 1800s, the American botanist Constantine Rafinesque further explicated the *Potentilla* genus' astringent applicabilities in noting that species such as *P. reptans*, *P. canadensis*, and *P. fructiosa* were "mostly used in weak bowels, hemorrhage,...menorrhea, & c."*(Rafinesque 1830:253)*

As to "weak bowels," the Cherokee Indians employed a tea of the roots of *P. simplex* for *DYSENTERY,* while our Ojibwe did the same with *P. arguta.*(Densmore 1974:350) The Iroquois utilized cinquefoil for *DIARRHEA* in general,(Herrick 1995:164) as did the Okanagan Indians.*(Turner et al 1980:127)* Such an application is still popular in many lands and has even received the approval of the German Office of Health, which recommends cinquefoil for "acute, nonspecific diarrhea."*(Pahlow 1993:162)* Both the astringency and the antimicrobial properties of the tannins are no doubt of value here and may also explain why Dioscorides found that an infusion of *P. reptans* was useful "to assuage tootheache, and...rotten ulcers that are in ye mouth"*(Herbal 4.42)* and why the Cherokees had success in using it as a mouthwash for **THRUSH**.*(Hamel & Chiltoskey 1975:26)*

There is also an **antipyretic** effect that can be drawn from a tea made from the roots, which application has been widespread—from European use (as noted by Nicholas Culpeper-*Potterton 1983:46*) to our own continent's Cherokee Indians*(Hamel & Chiltoskey 1975:29; Hutchens1991:128)* and colonial-American healers who implemented cinquefoil to treat **AGUE** or **INTERMITTENT FEVER,***(Rafinesque 1830:253)* which was then so rampant in marshy areas. It is of interest here that scientists on the German Commission E panel confirmed that the tannin content in cinquefoil is sufficient enough to 'reduce fever.'*(GermCommE)* It is most noteworthy, too, that *antiviral* properties are present in at least *P. arguta*, for a recent study found a root extract of it to completely inhibit **RESPIRATORY SYNCYTIAL VIRUS**.*(McCutcheon et al 1995:101-10)*

European herbalist Juliette de Bairacli Levy points out that the genus name for cinquefoil is from an ancient word meaning "powerful," and such she finds it, deeming it a reliable **nerve sedative**, even with applicability to **EPILEPSY**.*(de Bairacli Levy1974:135)* Culpeper had also used *Potentilla* for epilepsy, advising use of the plant's juice and stressing that it needed to be taken to the tune of four ounces at a time, with such a regimen continuing for at least a month, before any results could be expected.*(Potterton 1983:46)* Certain species of *Potentilla*, and most especially *P. anserina* (silverweed), appear to be possess a potent **antispasmodic** property, and perhaps it is this quality that is primarily at play in the alleged aid for epileptics. Then, too, the antispasmodic property may also explain why many species of *Potentilla* have traditionally been used for **MENSTRUAL CRAMPS.** The revered European herbalist Parson Kneipp even felt that he had cured a case of tetanus with a decoction of silverweed in milk!*(Fischer-Rizzi 1996:249)*

The folk medicines of Russia and Sweden have used *Potentilla* plants for both internal and external **INFLAMMATION**.*(Hutchens 1991:128)* Culpeper had also affirmed that it was "used in all inflammations."*(Potterton 1983:46)* Such traditional uses are interesting in view of a 1976 study which found a species of cinquefoil tested, *P. argentes,* to produce an anti-inflammatory inhibition of 13% in the carageenan-produced, rat-paw edema test*(Benoit 1976:164)* and a more recent study which found *P. erecta* to strongly inhibit the cycloxygenase pathway to inflammation as well as to moderately inhibit platelet-activating-factor-induced exocytosis.*(Tunon et al 1995:61-76)*

Cautions

Due to cinquefoil's high amount of tannins, neither the plant nor its extracts should be used for an extended period of time.

CLEARWEED
[RICHWEED]
(Pilea pumila)

Description

[See photos #39 & 40 in rear of book.]

A colony-loving, annual herb growing 3-24 inches high. It is related to the well-known stinging nettle (*see under that entry*) and therefore looks a lot like that herb. (It lacks the stinging hairs, however.)

The stem is translucent and juicy, while the leaves are opposite, smooth, lustrous, egg-shaped, coarsely toothed, and grow 1-5 inches long. They each have three conspicuous veins.

As with its relative, stinging nettle, the minute, greenish flowers are clustered in the axils. Clearweed's flowers are darker green than nettles,' however. They are also short, curved, and drooping.

Range & Habitat

Clearweed thrives throughout much of our area, often growing in large colonies. For habitat, it prefers three things: rich soil, moisture, and shade. Find this combination and the chances are that you will find clearweed!

Food

Foraging author Steve Brill recounts meeting a family who had been digging and eating clearweed all summer long under the mistaken impression that it was stinging nettle!*(Brill 1994 :243)* Prior to this discovery, Brill had always assumed that clearweed was inedible, in harmony with the impression given by most foraging authors. However, writing way back in the 1940s, Fernald & Kinsey had opined that clearweed might prove to be a genuine wild edible since a related species had long been heartily consumed in Asia.*(Fernald & Kinsey 1958:166)* Euell Gibbons, in personal experiments with our American species in the 1960s and 1970s, found this to be the case, but said that palatability depended upon our native *Pilea* being cooked only briefly, even as is necessary with its close relative, stinging nettle. He discovered the taste to be rather bland, but suggested that such was valuable for a vegetable medley in order to offset the taste of stronger greens.*(Gibbons & Tucker 1979:73)*

Health/Medicine

Little is known about this succulent plant. However, several Amerindian tribes made use of it, to wit:

The Cherokee Indians found that an infusion of clearweed made a fine pacifier for children with **EXCESSIVE HUNGER**.*(Hamel & Chiltosky 1975:53)* (This may suggest that this plant possesses an unusually high protein content, like its relative stinging nettle.-*see under that entry*) They also rubbed the herb's stems between **ITCHING TOES**. If this was **ATHLETE'S FOOT**, one wonders whether clearweed might possess some sort of *antifungal* properties.

The Iroquois Indians inhaled juice squeezed from the stem to alleviate **SINUS PROBLEMS**.*(Herrick 1977:308; Herrick 1995:133)* The herb's **anti-inflammatory** properties, demonstrated in a 1976 lab study*(Benoit 1976:164)*, could be a major factor in any ability it possesses to soothe troubled sinuses. However, as recurrent sinus infections have recently been demonstrated by scientists to have a strong fungal component, such a finding may lend further weight to the possibility of antifungal properties in this plant. In this regard, it is interesting that clearweed's semi-translucent stem—housing a clear, runny fluid—is reminiscent of jewelweed's *(see under that entry)*, a plant demonstrated to have powerful antifungal capabilities.

CLEAVERS
[CLIVERS; GOOSEGRASS; CATCHWEED BEDSTRAW]
(Galium aparine)

Description

[See photos #41-43 in rear of book]

An annual weed of 1-6 feet, possessing weak, square stems that frequently latch on to other plants (esp. bushes) for support, so that this herb is often appropriately referred to as a "climber." Cleavers frequently grows in colonies, forming clumps.

Whorls of 6 to 8 leaves appear every so often along the stem. These leaves are narrow and lance-shaped, grow ½ - 3 inches long, and are nearly sessile. They are rough all over, being covered with many, tiny, backward-hooked bristles (ditto with the stem, which has four rows of these tiny hooks).

Clusters of two or three tiny, white flowers appear on stalks arising from the leaf axils. Starlike and four-petaled, they are about 1/8 inch wide and they lack sepals.

Blossoms are replaced by twinned fruits in summer, which are also covered by hooked bristles and which therefore cling easily to clothes and fur. These burs are globular, being about 1/8 inch in diameter.

Range and Habitat

Cosmopolitan weed, found throughout the USA and in Canada from Newfoundland to British Columbia and Alaska. Preferred habitat is partly shaded areas in rocky woods and rich thickets, trailsides, roadsides, meadows, fence rows, and waste places.

Seasonal Availability

Spring to autumn, but seeds available only from late summer onward.

Chief or Important Constituents

Coumarins, iridoid glycosides(asperuloside, monotropein), benzyl isoquinoline alkaloids(incl. protopine), quinazoline alkaloids, numerous organic and plant acids (citric, caffeic, salicylic, *p*-hydroxy-benzoic, *p*-coumaric), *n*-alkanes, flavonoids(incl. luteolin), tannins and pseudotannins(incl. gallic acid), anthraquinones in the roots. Nutrients include high amounts of vitamin C.

Food

Cleavers can be eaten as a potherb, but must be boiled or steamed longer than other plants (more than 15 minutes) so as to reduce the toughness of the barbed stems.

The bristly, autumn fruits make an excellent, dry-roasted coffee substitute. (This is not surprising, in that cleavers is distantly related to coffee!) One proceeds by first baking the fruits in a 275-to-375-degree oven till brown, then grinding them with a grinder, blender, or mortar & pestle. Next, about 3/4 cup of the ground fruits is simmered in one quart of water until the water turns dark brown. Then, the caffeine-free coffee substitute is strained and served. In a camping or survival situation, one can improvise by simply grinding the seeds between rocks, placing the meal in a drink container, pouring boiling water over it, and steeping the brew for about fifteen minutes.

Health/Medicine

Cleavers is a powerful *antiscorbutic,* due to its high vitamin-C content; also a renowned *diuretic blood purifier*, working in conjunction with the lymphatic system. This herb is considered to be one of the finest *tonics for the lymphatic system*, and one of the few that is safe for children. Cleavers stimulates this system, helping it to eliminate toxins and wastes by excreting them through the urine. It is one of the best herbs for so-called "swollen glands" (i.e., *SWOLLEN LYMPH NODES*) and even helpful for more heavy-weight afflictions such as *TONSILITIS, LYMPHADENITIS,* and *SCROFULA.* (Holmes 1997:1:701) (Interesting, in the latter regard, is that a 1954 study found cleavers to exhibit significant activity against the *TUBERCULOSIS* bacteria.-Fitzpatrick 1954:530) In China, it is even used to treat *LEUKEMIA*.(EthnobDB)

Strong *anti-inflammatory* properties for the urinary tract have been noted for this herb, and it has a long history of use by Amerindian tribes (including Micmac, Ojibwe, and Penobscot) in this regard.(Moerman 1986:1:192) This property, coupled with its diuretic effect, has allowed cleavers to be frequently employed for the alleviation of *CYSTITIS, SCALDING* or *BLOODY URINE* (in these cases, it has usually been recommended to be taken about an hour before meals, Moore 1979:57), and even *URINE STOPPAGE*(Ellingwood1983:432)—uses particularly appreciated by the Flambeau Ojibwe.(Smith 1932:386) A nineteenth-century Eclectic physician had elucidated here:"The first use of Galium is to relieve irritation of the urinary apparatus, and increase the amount of urine....One of our best remedies. In dysuria and painful micturition, it will frequently give prompt relief."(Scudder 1870:126) Not surprisingly, then, the plant is also famed as an aid to the reduction of *GRAVEL*. Teaspoonful doses of the juice have also been employed in the East Indies to help treat *GONORRHEA*,(Tierra 1988:221), as was done in bygone days by the Micmac, Penobscot, and other American Indian tribes.(Moerman 1986:1:192)

A tea made from cleavers is held to be one of the best skin washes for *PSORIASIS, DANDRUFF,* and other *DRY-SKIN CONDITIONS*. (The *British Herbal Compendium* lists it as official medicine in this regard.) The Iroquois Indians had even used it as part of a topically-applied, compound infusion to soothe *POISON IVY RASH*.(Herrick1977:439;Herrick1995:219) The Eclectics found it indicated for "nodulous growths or deposits in skin or mucous membranes."(Felter&Lloyd) Scudder, as one of such, even points out that "a case of hard nodulated tumor of the tongue, apparently cancerous, is reported in the British Medical Journal as having been cured with it."(Scudder 1870:126)

A renowned **refrigerant** herb, cleavers is appreciated in most conditions that require cooling. For example, it makes a soothing wash for **SUNBURN**. Also revered as a **febrifuge** ("impresses the temperature greatly, stimulates the excretion of all urinary constituents and the fever is shortened by its use,"*Ellingwood 1983:432*), a use frequently employed by Mexico's Mazatecs. Its refrigerant nature, plus its **astringent** factors (high tannin content, etc.), render this herb useful as a poultice for **BURNS** and **SCALDS**. Michael Moore suggests soaking slowly-healing burns in a tub of cleavers tea; he also adds that such a bath is helpful for **INFLAMED STRETCH MARKS**.*(Moore 1979:57)*

Astringent factors are likewise responsible for its traditional use in halting **DIARRHEA**. Seventeenth-century European herbalist Nicholas Culpeper noted that a cleavers poultice worked as a **styptic**,*(Potterton 1983:48),* undoubtedly due to the astringent tannins and pseudotannins, but also probably to the asperuloside content of the plant. At any rate, various Amerindian tribes (including the Micmac and Penobscot) implemented the herb to stop the "spitting up of blood," which sometimes occurred after a messenger's long run.*(Moerman 1986:1:192)*

Modern herbalists hold cleavers to be helpful with **LIVER PROBLEMS** and thus they sometimes include it in an herbal regimen for **HEPATITIS** (Moore notes that it is not irritating to the liver in such a case, *un*like most other hepatics.-*Moore1979:57*) The Mazatecs found it useful for **JAUNDICE**.

Herbalists of earlier times found cleavers to be helpful in the reduction of **OBESITY**; John Gerard, writing in the late 1500s, commented: "Women do usually make pottage of clevers with a little mutton and otemeale, to cause lankness and keepe them from fatnesse."*(Gerard1975:963-64)* Culpeper concurred.*(Potterton1983:48)* Some modern herbalists (e.g, the famed country herbalist Tommie Bass-*Crellin & Philpott 1990:155,235*) have revived this tradition, often combining (or interchanging) our herb with chickweed (*Stellaria* spp.), which was also implemented in bygone days for weight reduction.*(See under "Chickweed.")* The principle (or, principles) involved here has not yet been isolated, but possibly with *Galium* it is as Eleanor Viereck suggests—that the acids in cleavers may serve to speed up the metabolism of the body's stored fats.*(Viereck 1987:19)*

Truly, the herb's applications are legion: Culpeper maintained that cleavers juice droppered into ears relieved the pain of **EARACHE**.*(Potterton1983:48)* Numerous herbalists say that a cup of cleavers tea before retiring promotes sound, restful sleep; thus, it is recommended for **INSOMNIA**.*(Grieve1971:207)* A cleavers infusion applied to armpits has been held to be an effective **deodorant**. A study conducted in the 1940s found that our herb had a **hypotensive** effect when administered intravenously to dogs.*(Delas et al 1947:57)*

As is apparent from the above, cleavers is an interesting and widely applicable herb. However, its effects are definitely the most pronounced if the plant is used by way of infusion as opposed to tincture, etc.*(Scudder1870:126)* Here the fresh herb is preferable (if the dried herb is used, it must be infused for at least two hours.-*Gunn 1859-61:776*)

COLUMBINE
[WILD COLUMBINE; ROCK BELLS]
(Aquilegia canadensis)

Description

[See photos #44 & 45 in rear of book.]

A lovely native perennial, growing 1-3 feet tall.

Columbine's long-stalked leaves are compound, finding themselves divided into three leaflets and then subdivided into three more. They reach a length of 4 to 6 inches and they occur both basally and scattered along the stem.

The gorgeous blossoms, occurring singly on slender stalks arising from the leaf axils, are like drooping bells and grow 1-2 inches long. They are composed of five, yellow-and-red, funnel-shaped petals and five long, red, tubular sepals. The sepals are spurred and they end in a knob. Five yellow pistils, along with numerous stamens, extend beyond the length of the petals.

The erect, podlike fruits occur in clusters and contain tiny, black seeds.

Range & Habitat

Thrives in partial shade throughout our area in such environs as open woodlands, forest edges, rocky cliffs, and clearings.

Chief or Important Constituents

Cyanogenic glycosides. Roots contain alkaloids(aquileginine, berberine, magnoflorine, and others).

Health/Medicine

Pawnee Indians crushed and dried columbine seeds and steeped the resulting powder as a treatment for *FEVER*.*(Gilmore 1991:30)* Likewise, America's Eclectic physicians of the late 1800s and early 1900s found the roots to be *diaphoretic.*(Smith 1932:383)* Scientific research since these times has confirmed that at least the seeds possess this property and has also evinced that the leaves and stems are *antiscorbutic* and *diuretic.*(Mockle1955:41)* As to the latter, the Iroquois Indians steeped a handful of roots in one quart of water and imbibed one teacupful three times a day for *KIDNEY PROBLEMS.*(Herrick 1995:119)* Then, too, Mesquaki (Fox) Indians included columbine in a compound herbal formula for "when the contents of the bladder are thick." *(Smith 1928:238-39)*

The Cherokee Indians found columbine helpful for **HEART TROUBLE**. They also valued the herb for **FLUX**.*(Hamel & Chiltoskey 1975:30)*. The Mesquaki chewed the root for **STOMACH** and **BOWEL TROUBLES** and boiled the leaves and roots into a tea to cure **DIARRHEA**.*(Smith 1928:238)* Pillager Ojibwe Indians likewise considered columbine root to be invaluable for **STOMACH TROUBLES**.*(Smith1932:383)* Goisute Indians yearned for a decoction of columbine roots when "sick all over" or presenting **ABDOMINAL PAINS**.*(Chamberlin 1911)* The scientific panel known as German Commission E states that the European species (*A. vulgaris*) is a **cholagogue**.*(GermCommE)*

A few of the above-described ethnobotanical uses are perhaps significant in light of a 1954 study which found columbine to exhibit **antimicrobial** activity.*(Fitzpatrick 1954:530)* Concordantly, to control **HAIR-LICE INFESTATION**, some Amerindian tribes rubbed crushed seeds of this plant into the hair of one so afflicted.*(Foster & Duke 1990:136)*

Cautions

This plant is possessed of cyanogenic glycosides and is thus potentially poisonous. The recent *PDR for Herbal Medicines* notes that toxicity from ingestion of the leaf preparations has not been observed, presumably because the level of hydrocyanic acid that is releasable from the leaves is negligible.*(PDRHrbs)* However, it is important to bear in mind that there can be variation among species and that most authorities regard our American species, *A. canadensis*, as being more potent than the European species, *A. vulgaris*. Yet, the seeds of the latter are reported to have caused deaths in children.*(Lewis & Elvin-Lewis 1977:30)*

The above having been said, how do we explain the many and varied internal applications implemented by the American Indians, as detailed above? First of all, it should be noted that although we know that the Indians *did* in fact use columbine internally, we don't know the circumstances; their applications may have involved preparation techniques that somehow obviated any cyanogenic effects, such as using *dried* seeds or roots only (in that cyanogenic glycosides are usually inactivated upon drying).

All things considered, it seems prudent to urge that internal applications of *A. canadensis* not be implemented by the layman. If and when its utilization becomes necessary, it would be most wisely undertaken *only under the direction of a seasoned herbalist skilled in its use.*

COW-PARSNIP
[MASTERWORT; GIANT HOGWEED]
(Heracleum lanatum [H. maximum; H. sphondylium])

Description

[See photos #46 & 47 in rear of book.]

A huge plant, topping off at 5-10 feet, with a stem that is grooved, hairy, and hollow. This stalk can be up to two inches thick at the base and it emits a rank odor when bruised.

The alternate, palmate leaves are also huge (often reaching beyond a foot in length) and are possessed of three, oval leaflets. The base of the leafstalks has a large, inflated sheath, which is an important identifier distinguishing this plant from toxic plants which bear some similar features otherwise.

The tiny, white flowers are arranged in flat-topped clusters reaching a diameter of 6-8 inches. The notched petals are often tinged with purple.

Range & Habitat

This plant thrives throughout our region in wet or moist soil, including swamps, marshes, riverbanks, and lakeshores.

Chief or Important Constituents

The aerial portions contain pimpinellin, sphondin, imperatorin, psoralen, bergapten, angelicin, phellopterin, and xanthotoxin. 5-methoxypsoralen occurs in the fruits. The root contains vaginidiol, scopoletin, umbelliferone, apterin, ferulic acid, and para-coumaric acid.

Food

If you can get past this plant's rank odor and take a few precautions as to safety, you may find that you will be rewarded with some decent wild fixins'. The first-year root, as one instance, is quite edible. It can be boiled or roasted, peeled afterwards, and then consumed. The taste is somehwat reminscent of a rutabaga.

Blackfeet Indians peeled, roasted, and devoured the young stems.*(Hellson&Gadd 1974)* If instead boiled, as some prefer, the spring shoots need be cooked in two changes of water to lessen their otherwise rank taste and smell. Even thus properly prepared, a hint of rankness may remain. (But, this is the wilderness, bucko—not a candy shop!)

Woods Cree Indians peeled the leaf stalks of older plants and ate the naked petioles. They also peeled the main stem, split it, roasted it, and then scraped out the pith, which they proceeded to eat.*(Leighton 1985:40)* The older stems do indeed need to be peeled before being eaten, as the peel contains furanocoumarins which can cause blisters on the lips and skin.*(Szczawinski & Turner 1980:81; Leighton 1985:40)*

The maple-like leaves can be dried and burned, with the resultant ashes used as a salt subsitute.

Health/Medicine

This majestic and conspicuous plant has served a large array of Amerndian tribes as a valuable *vulnerary.* Eskimos, for instance, heated the large leaves and applied them to *SORE MUSCLES* and *MINOR CUTS*.*(Smith 1973:327)* So did Washington State's Quinault Indians.*(Gunther 1973:42)* Mesquaki Indians of our own area poulticed the (peeled?) stem onto *WOUNDS,*(Smith 1928:249)* while our area's Ojibwe Indians poulticed the pounded, fresh roots onto *SORES*.*(Smith 1932:390; Smith 1928:390)* The latter also used a decoction of the root as a gargle for *SORE THROAT* *(Densmore 1974:342-43)* and applied the boiled and pulped root to *BOILS* to force them to a head.*(Densmore 1974:350-51)* The latter application was appreciated by a wide range of Amerindian tribes: Cree and Gitksan Indians utilized the root for boils as well as for various *SWELLINGS*.*(Beardsley 1941:484; Smith 1929:61)* The Pawnee scraped the root, pounded it thoroughly, boiled it, and applied a compress of the decoction to these painful eruptions.*(Gilmore 1991:55)* Bella Coola Indians crushed and baked the roots and then applied them to the boils.*(Smith 1929:61)*

To relieve the *GUM DISCOMFORT* caused by *LOOSE TEETH*, the Coast Salish Indians of British Columbia sprinkled the powdered root onto the gums.*(Turner & Bell 1971:63-99)* However, Cree Indians appreciated cow-parsnip when they had a *TOOTH-ACHE per se*: they applied a one-inch-square portion of the root to the aching tooth, at the same time making sure not to swallow any of the juice (which they instead spat, since it was understood to be toxic).*(Beardsley 1941:491)* Sikani Indians applied the mashed roots to *NEURALGIC SWELLINGS*.*(Smith 1929:61)* Such was also widely used an an *anti-rheumatic* poultice. Here a mild *anti-inflammatory* property may derive from cow-parsnip root's scopoletin content.

It also appears that cow-parsnip (or at least the leaf) may possess *hypotensive* properties.*(Stuart 1982:73)* Here it is interesting that the Mesquaki Indians used the seeds to treat a *SEVERE HEADACHE*.*(Smith 1928:249)* Could this treatment have worked because it reduced the blood pressure in the head? Or was it rather because cow-parsnip is such a good *antispasmodic?*(Scudder 1870:138; Lust 1974:268)* As such, the plant tops were used by our area's Winnebago Indians as a smoke treatment for *SEIZURES*(Gilmore 1991:55)* as well as internally by the Eclectics (and other physicians before them) to treat *EPILEPSY per se*.*(Rafinesque 1830:227; Wood et al 1926:1330; Smith 1928:390; Lewis & Elvin-Lewis 1977:167)* A decoction of the root has also been used in various cultures for *ASTHMA*, as well as for other nervous disorders.*(Krochmal & Krochmal 1973:120; Smith 1928)*. The antispasmodic property would probably at least partly accrue from the root's scopoletin content.

Undoubtedly at least partly owing to this antispasmodic property, the root decoction was also used internally for **COLIC, FLATULENCE, INDIGESTION,** and **DIARRHEA,** both by various Amerindian tribes as well as by our nation's Eclectic physicians.*(Krochmal & Krochmal 1973:12; Rafinesque 1830:227; Smith 1928:390; Hellson & Gadd 1974:67,76)* It was even listed in the *U.S. Pharmacopoeia* from 1820 to 1863 in reflection of these uses. As late as 1926, in fact, the *U.S. Dispensatory* would make note of its *carminative* property.*(Wood et al 1926:1330)* The seeds are especially carminative, as well as being *stomachic* and *antinauseant.* Because cow-parsnip also has been experimentally shown to possess *antifungal* properties,*(McCutcheon et al 1994:157-69)* it is especially indicated when flatulence is due to an overgrowth of *Candida* organisms in the GI tract.

Our herb also has *stimulant,(Wood et al 1926:1330;Lust 1974:268)* *emmenagogue,(Stuart 1982:73; Lewis & Elvin-Lewis 1977:330)* and *purgative* properties, the latter of which the Thompson Indians especially appreciated, drinking a decoction of the root.*(Turner et al 1990:45)*

Finally, it should be noted that the seeds would appear to be *aphrodisiac.(Stuart 1982:73)* Although proof for such a property is always controversial and somewhat limiting because of the element of subjectivity, it is undoutedly significant that the roots of a related species growing in the far-off Himalayan kingdom of Nepal are similarly used.*(Lewis & Elvin-Lewis 1977:328)*

Cautions

* The plant (esp. in its blossom) looks somewhat like the deadly water hemlock (*Cicuta* spp.), which abounds in our area. However, cow-parsnip's leaves are much bigger and maple-like as compared to water hemlock's thinner, carrot-like leaves. (The poisonous water hemlock also has leaf veins that end *between* the teeth and not *in* the teeth as with not-so-deadly lookalikes.) *But be absolutely sure of identification before using what you think to be cow-parsnip!* Remember the forager's rule: "Better safe than sorry." (Or: "Risk takers belong in a casino, not on a foraging round.")

* Handling the fresh plant can produce severe, long-lasting dermatitis in susceptible individuals. Heavy exposure to sun afterwards may cause severe photosensitivity, including sunburn, blistering, and possible permanent purple pigmentation (good for sideshow acts perhaps, but not for a comfortable life in general).

* This plant is CONTRAINDICATED in *PREGNANCY* because of its *emmenagogue* potential. Its toxic potential would contraindicate its use during *LACTATION* as well.

CREEPING CHARLIE
[GROUND IVY; GILL-OVER-THE GROUND; ALE HOOF]
(Glechoma hederacea [Nepeta hereracea])

Description

[See photo #48 in rear of book.]

A colony-loving, ivy-like creeper that can grow to three feet long and may reach a height of eight inches. It bears the characteristic square stem of the mint family, to which it belongs, and it creeps owing to the fact that each of its stem nodes roots to the ground!

Like all mints, creeping Charlie's leaves are paired. In this particular case, they are also long-stalked, kidney-shaped, and possessed of scalloped margins. Growing ½ - 1½ inches long, they often display a purplish tinge.

Three to seven, blue-violet flowers bear the mint family's standard, two-lipped corollas, with this species having a three-lobed lower lip. There are four stamens.

Like many mints, creeping Charlie emits a powerful musky-mint odor when bruised—or sometimes even when just brushed!

Range & Habitat

Thrives in most and shady areas throughout our region, especially in lawns, waste places, fields, moist woods, and on roadsides.

Chief or Important Constituents

Glechomine, essential oil(incl. borneol, limonene, pulegone, linalool, myrcene, alpha-cadinol, p-cymene, alpha-terpineol), resin, bitter principles(sesquiterpenes, incl. glechomafuran), flavonoids(incl. apigenin, rutin, luteolin), diterpene lactone(marubiin), beta-sitosterol, caffeic acid, ursolic acid, rosmarinic acid, tannins. Stachyose is in the roots. Nutrients include choline, silicon, iodine, iron, sulfur, phosphorous, copper, zinc, potassium, molybdenum, and a very high amount of vitamin C.

Health/Medicine

Who would have suspected that this aggressive, much-cursed weed that can rapidly take over Minnesota and Wisconsin lawns is actually a valuable medicine? But such it is, and such are most common weeds. In fact, its uses are legion, and it has been vastly appreciated and utilized by European and American herbalists, including the American Indians.

First off, *The British Herbal Pharmacopoeia* notes that creeping Charlie is a specific for *tinnitus aurium*, that mysterious ringing in the ears that is increasingly affecting a wider and wider range of people to their extreme discomfort.*(BHP 1976:149)* Although we tend to think of *TINNITUS* as a modern condition, it really isn't: it's been with mankind for millennia! And, fortunately, the use of creeping Charlie to combat it has also been traditional for quite some time. Thus, way back in the late 1500s, John Gerard observed that "ground ivy is commended against the humming noyse and ringing sound of the ears, being put into them, and for them that are hard of hearing."*(Gerard 1975:201)* Herbalist David Hoffmann says that it can be used *internally* for tinnitus as well, being especially helpful when such is caused by catarrh.*(Hoffmann 1986:199)*

Aside from an **anticatarrhal** effect, the plant's moderate **anti-inflammatory** factors may be involved here,*(Mascolo et al 1987:28-31)* which accrue because of the high tannin content,*(Kresanek 1989:100)* the rosmarinic acid, the flavonoids, and the high vitamin-C content (and thus ground ivy has also been a revered **antiscorbutic**-*Wren 1972:141*). Understandably, then, the plant has long been used as a bath additive for *SCIATICA* and *GOUT*. Renowned European herbalist Nicholas Culpeper had advised taking it internally "in wine, drank for some time" for these same conditions.*(Potterton 1983:13)* Creeping Charlie's anti-inflammatory property has also been utilized by way of a poultice for *BRUISES* and *CONTUSIONS,(Stuart1982:70)* as well as internally for troubles in the respiratory system, especially *INFLAMMATION* of its *MUCOUS MEMBRANES*, including the *SINUSES*. *(Hoffmann 1986:199; Mills & Bone 2000:216)*

The plant has also proven its mettle with reference to *BRONCHITIS, COLDS,* and *COUGHS* (often being combined with coltsfoot, *Tussilago* spp.)*(Hoffmann 1986:199)* It has further been used against *TUBERCULOSIS,(de Bairacli Levy 1974:73)* and a 1954 lab study found it to be mildly active against a virulent human strain of such.*(Fitpatrick1954:530)* It decongests respiratory mucus with its essential oil and also dries it up owing to its astringent property.*(Ody 1993:138, 156-57)* Such a property, attributable not only to its tannins but also to its rosmarinic acid,*(Newall et al 1996:154)* lends its use as well to combatting *SORE THROAT,(Lust 1974:212)* *HEMORRHOIDS,(BHP1976:149)* and *DIARRHEA,(Kresanek 1989:100; BHP 1976:149;Lust 1974:212)* Even *FLATULENCE* is vanquished via ground ivy,*(Potterton 1983:13)* undoubtedly owing to the plant's volatile oil.

Glechoma is also a digestive *TONIC,(Lewis & Elvin-Lewis 1977:390)* owing mostly to its bitter principles. It is helpful for overall *DIGESTIVE TROUBLES*, in that it stimulates digestive juices.*(Kresanek 1989:100; Lust 1974:213)* It also serves to resolve *GASTRITIS(BHP 1976:149)* and *CATARRHAL ENTERITIS.(Lust 1973:213)*

Diuretic properties can be found here as well,*(Lewis&Elvin-Lewis1977:390; Lust 1974:212; Wren 1972:141; Stuart1982:70; Potterton1983:13)* making it helpful for *GOUT*, as we have noted above. Creeping Charlie has also proven its merit for *CYSTITIS(Hoffmann1986:199;BHP 1976:149)* and for certain kinds of *VENEREAL DISEASE.(Carse1989:104)* It is **antilithic** to *GALL-STONES,(Kresanek1989:100)* and has historically been used by country folk for *JAUNDICE*.

Our herb's uses do not stop here, however. Kentucky folk use has long mixed a small amount of ground ivy with catnip*(see under that entry)* for relieving"bull hives" on children (characterized by red spots on the face and a cross disposition).*(Bolyard 1981:82-83)* Appalachian herbalist Tommie Bass revered a tea of the fresh or dried leaves for **NERVOUS HEADACHES.***(Crellin & Phillpot 1989:1:238)* European herbalism has long appreciated that the plant helps to expel a**RETAINED PLACENTA** *(de Bairacli Levy 1974:73)* Herbalist Susun Weed writes of her own and others' experiences using the plant to expel the afterbirth in both humans and animals, at a dose of one cup a day or one-half to one teaspoon of the tincture.*(Weed 1986:70)* Scientific research has also confirmed *antitumor* effects, attributable to the plant's content of ursolic and oleanolic acids.*(Tokuda et al 1986:279-85; Ohigashi et al 1986:143-51)*

Probably ground ivy's most unusual application has been for offsetting **LEAD POISONING,** for which an infusion of the fresh plant has been used, drunken a wineglassful at a time and taken fairly frequently.*(Grieve 1971:2:443;Bolyard 1981:83; Carse 1989:104)* Amazingly, herbalists have used it for many years to great success with painters who have manifested this horrible illness. Many have assumed that the plant's high vitamin-C content explains its provings in this regard,*(e.g.,Dobelis et al 1986:204)* and a recent monograph in the *Journal of the American Medical Association* powerfully underscored the value of ascorbic acid in reducing levels of lead in the blood.*(Simon & Hudes 1999:2289-93)* Yet, other plants with an even higher content have failed to match its track record. I would guess, then, that at least one other chemical in the plant must enhance the vitamin-C effect, or otherwise act as an auxilliary agent.

Typical Usage: Tincture:1/4-1/2 fl dr; *Infusion*: Heaping teaspoon per cup, imbibing a total of 1-2 cups a day,*(Drescher 1940:115),* but dosed at 1/2 cup, 3-4 times a day.*(Carse 1989:104)*

Cautions

* Absolutely contraindicated in **PREGNANCY** and **LACTATION!** The plant has also traditionally been contraindicated in **EPILEPSY.**

* Stay within recommended doses or the plant can be poisonous. It contains an irritant oil that is definitely toxic to horses.*(Fuller & McClintock 1986:181-82)* Some authorities say that potentially harmful effects are increased after the first frost.

CULVER'S ROOT
[BLACK ROOT; CULVER'S PHYSIC; LEPTANDRA]
(Veronicastrum virginicum [formerly: *Leptandra virginicum*]*)*

Description

[See photos #49 & 50 in rear of book.]

A tall (2-7 ft.), striking, native perennial that often branches at the top.

The leaves are lanceolate, sharply toothed, and grow 2-6 inches. They are short-stalked and whorled at intervals along the stem in groups of 3-8. The very bottom leaves are often in pairs, however.

The tiny (1/4" diameter) flowers are creamy white, tubelike, and possessed of 4-5 fused petals and an equal number of pointed, green sepals. Two conspicuous stamens extend beyond the length of the petals. These have yellowish to reddish-brown tips.

Many of the flowers are arranged into 1-5, vivid, terminal, spiked clusters that *taper at the tip.* These slender racemes really hold one's attention, owing to their beauty and unusual design. Because of such visual splendor, this native plant is often an item of choice in prairie restoration programs

Range & Habitat

Thrives in colonies in moist meadows, mesic prairies, open savannas, rich woods, thickets, and alongside railroad tracks. Scattered throughout our area, but more heavily concentrated in the lower two-thirds.

Chief or Important Constituents

Bitter principle(leptandrin), resin, volatile oil, phytosterol(verosterol), saponin, citric acid, cinnamic acid, para-methoxy-cinnamic acid, mannite, d-mannitol, tannin.

Health/Medicine

Culver's root (leptandra, black root) is a classic and most effective *hepatic*. It is most specific for ***TORPIDITY OF THE LIVER,*** esp. when associated with ***HEAD-ACHE****(Clymer 1973:147; Ellingwood 1983:312)* The Eclectic physician Finley Ellingwood had noted here: "It certainly increases the discharge of bile and stimulates and gently improves the function of the liver." *(Ellingwood 1983:312)* For such purposes, it was at one time official in the *U.S. Pharmacopoeia* (from 1820-40, 1860-1916) and in the *National Formulary* (from 1916 to 1955). However, Culver's root can resolve a situation not only with too little bile but with too much as well, serving to rid the body of the excess, even as the Iroquois Indians had discovered.*(Herrick 1995:217)* This is because, when one is"all biled up," as the British say, stimulating bile flow will also stimulate its elimination. Interestingly, the cinnamic acid found in Culver's root has indeed been shown in lab experiments to act as a *choleretic*, thus shedding a spotlight on the herb's traditionally understood effects on the flow of bile.*(Galecka 1969:479-84; Das 1976:456-58; PDRHrbs)*

Culver's root has likewise proven its mettle for *CHOLECYSTITIS* and *NON-OBSTRUCTIVE JAUNDICE*(the latter especially when accompanied by pale and dry skin, a thick-coated tongue, and moderate liver pain-*Moore 1994:9.9*). It can also be of aid in *HEPATITIS*(Tilgner 1999:56) and it is specific for this condition when such is accompanied by hot skin, cold feet, abdominal congestion, a white-coated tongue, bad breath/bad taste, and pain in the right hypochondrium that is referred to either shoulder.*(Moore 1994:8.7)* There is also malaise, while the conjunctivae of the eyes are yellowish. Peter Holmes elucidates that, from the standpoint of Traditional Chinese Medicine (TCM), leptandra is indicated for syndromes reflecting "liver Qi stagnation with cold."*(Holmes 1997:1:397)*

Leptandra has quite famously been used as a *laxative.* *(PDRHrbs)* It is a specific for *CONSTIPATION* with symptoms of liver congestion.*(BHP 1979:225)* Although the fresh root is violently purgative, the dried root is a gentle laxative that removes excessive mucous from the GI tract and stimulates liver function to soften stool and increase peristalsis.*(Smith 1999:28)* Herbalist David Lytle says that, in his experience, the most effective dose of the tincture is 2 drams (1/4 ounce), mixed with an equal part of neutralizing cordial. He says that with this dosage, the stools are softened without watery discharge, cramps, or other discomforts.*(Lytle1992:103)* Even the dried root, however, is *cathartic* in high doses (one gram and over).*(Fetrow&Avila1999:80)* This latter effect was sometimes sought by the American Indians, including the Seneca *(Lewis&Elvin-Lewis1977:284)* and our own area's Menominee and Ojibwe (Chippewa) Indians.*(Smith1923:54-55;Densmore1974:346-47)* The latter accomplished such a task by steeping five roots in a quart of water to make a tea.

Lytle finds that Culver's root is also a valuable digestive *tonic.* The Eclectic physician John Scudder elaborated here: "The Leptandra exerts a gentle stimulant influence upon the entire intestinal tract, and its associate viscera, and in medicinal doses strengthens functional activity. Its action in this direction is so persistent that it might be called a gastro-intestinal tonic."*(Scudder 1870:151)* Herbalist Ed Smith notes that it is helpful with *ATONY OF THE GI TRACT*, even serving to restore function.*(Smith 1999:28)* Southwest-American herbalist Michael Moore finds it of value in *DYSPEPSIA,* primarily when prompted by indigestion of fats and proteins.*(Moore 1994:10.9)* Fetrow & Avila state that the herb has been shown to have an inhibitory effect on the gastric enzyme system, thus resulting in an *antisecretory* and consequent *antiulcer* effect.*(Fetrow & Avila 1999:80)*

The Mesquaki (Fox) Indians found Culver's root helpful for *URINARY GRAVEL* and even to stymie *FITS*.*(Smith 1928:247)* Our area's Dakota and Winnebago Indians discovered that it was valuable in treating *SNAKEBITE*.*(Andros 1883:118)*

Typical tincture dosage: 2-4 drops*(Clymer1973:147;Tierra 1992:208)* **Typical tea dosage**: 1/2 to 1 fluid dr,*(Christopher 1976:206)* 3-4 tbs every 3-4 hours, till effective.*(Christopher 1976:306)*

Cautions

　　* Contraindicated in ***PREGNANCY*** and ***BILE-DUCT OBSTRUCTION*** (incl. ***GALLSTONES)***

　　* *Fresh root is **toxic**.* Use only the *dried* root (aged one year is best and safest)

DANDELION
[LION'S TEETH; PISS-A-BED]
(Taraxacum officinale)

Description

[See photo #51 in rear of book.]

It is highly doubtful that a description of this plant is needed, but for the sake of consistency, here goes: A biennial or perennial herb growing 2-18 inches tall, with a thick taproot and an unbranched, hollow stem that exudes a milky, latex-like juice when broken or cut.

Leaves are entirely basal, forming a rosette. They are oblong, 3-16 inches long, ½-5½ inches wide, and pinnately lobed so that the teeth are coarse (although, when young, the leaves are almost entire). Like the stem, the leaves exude a milky juice when damaged.

The blossom is a composite flower, opening only on sunny days. It is solitary and sits at the tip of the stalk. Its strap-shaped ray flowers are a golden yellow and each possessed of a five-notched tip (actually, the flower petals!). There are 100 to 300 of these ray flowers per flower head.

Conspicuous bracts rest underneath the flower head. These are brown to green, thin, pointed, and reflexed downward.

The eventual fruits are dry, one-seeded structures, each attached to a feathery tuft. Together, they form a conspicous, globular head—like a delicate, silky ball. Eventually, they release themselves to the slightest wind and are blown hither and thither.

The dandelion can be differentiated from a number of lookalikes by four characteristics taken together: (1) a milky sap; (2) reflexed bracts, (3) an unbranched flower stem, and (4) leaves that are only slightly pubescent as opposed to being blatantly hairy.

Range & Habitat

Cosmopolitan weed, growing throughout the USA except in the extreme southeastern portions. The preferred habitat consists of fields, pastures, roadsides, wasteland, disturbed sites, and lawns (in which latter habitat they should be treasured, not cursed; cultivated, not exterminated; and utilized, not discarded).

Seasonal Availability

For medicinal uses, the root is collected in the fall when the inulin content is highest. For food use, the leaves, crown, and roots are collected in spring or fall, per personal preferences as to bitterness.

Chief or Important Constituents

The flower contains beta-amyrin, lecithin, carotenoids(chrysanthemumxanthin, violaxanthin, lutein, cryptoxanthin, flavoxanthin). The leaf contains sesquiterpene lactones(eudesmanolides), triterpenes(taraxol, taraxerol, beta-amyrin), phytosterols (beta-sitosterol), phenolic(hydroxycinnamic)acids(chicoric, chlorogeni, caffeic), organic acid(tartaric), coumarins(aesculin, cichoriin), and flavonoids(luteolin, apigenin). The root contains sesquiterpene lactones(both eudesmanolides and germacranolides), essential oil(largely triterpenes such as taraxol and taraxerol), taraxacerine, taraxicine, phenolic acids, organic acids(tartaric), fatty acids(oleic, linoleic, linolenic, stearic, palmitic, myristic, lauric), phytosterols(beta-sitosterol, stigmasterol), inulin(to 40% in autumn-harvested roots), flavonoids(luteolin, apigenin), coumarins(aesculin, cichoriin), mucilage, levulin, tannin, pectin, and the enzyme tyrosinase. Nutrients in dandelion include vitamins A(very high amounts—up to 7,100 IUs per cup of fresh greens), B1, B2, B3, choline(only in the root), C, D, and minerals such as calcium, magnesium, zinc, selenium, boron(only in the leaf), and silicon in moderate amounts; there is an especially high amount of iron, manganese, phosphorous, and the electrolytes potassium & sodium.

Food

People are often amazed when I tell them that the commonest of weeds, as opposed to exotic wildflowers, provide the very best in food and medicine. And there could be no greater example than the accursed dandelion! Indeed, all parts of the plant except for the flower stalk are edible, nutritious, and very tasty. In fact, the Apache Indians valued dandelions so highly that they would scour the surrounding countryside for days in order to satisfy their prodigious appetite for this herb!

Dandelion leaves are wonderfully tasty as a salad herb, especially if they are taken from well-shaded plants from rich soil. The leaves of sunned plants, however, are usually overly bitter. So here are some foraging tips: It searching for dandelions in an area new to you (in other words, not from your own lawn), look for the leaves that are not deeply toothed, since this is a visual clue that these are the ones that consistently get the most shade during the day. Also, give prime attention to those plants lacking flower heads (i.e., existing merely as rosettes), which also indicates that they get a lot of shade, since sun-drenched plants will almost variably sprout a stalk

Rather than merely harvesting leaves, it is often preferrable to take the whole plant, not only because you may want to utilize other parts of it as well (*see below*), but because dandelions are often gritty and the whole plant is easier to clean than simply the leaves. Such a cleaning can be accomplished by rinsing the plants under the faucet or often even more effectively by dunking them up and down in a container of water.

Dandelion greens can be enjoyed as a cooked vegetable, too. Here they can be steamed in a non-aluminum pot in just enough water to cover them, for about 3-5 minutes. Some foraging manuals urge that the usual cooking procedure for greens—immersing plants in already boiling water—should be reversed for dandelions, and that they should be started in cold water which is then heated to a boil. It is felt that this method most effectively removes any excess bitterness, especially if a water-change—again starting with cold water—is implemented half way through the boiling period. Bear in mind, however, that dandelion's bitterness has health benefits (see below), so that removing all aspects of this should not usually be one's goal.

The greens can also be frozen for winter use, but should first be given a two-minute blanching. And yet, dandelions—like nettles (*see under that entry*), chicory*(see under that entry)*, and some other plants—can also be "forced" through the winter in one's house, so that one has access to a fresh crop throughout the cold season. How so? Here's what to do: Gather some whole plants in autumn, right before the ground freezes hard. Cut the leaves back to within an inch of the root. Then place the plants in a container filled with sand (or sandy soil) and put the whole thing away in a very dark place (a closet or basement is often ideal). Water your secreted dandelion remnants every few days. If you've performed your task with skill, and if you garnish a little luck besides, you'll discover very pale leaves emerging in just a few weeks! "Blanched" like this, they are less bitter than outdoor dandelions.

But, you're not through yet! Cut them back as initially, keep watering them every few days, and you may find yourself harvesting another 2-3 crops before winter's end! It's a lot of fun, even though it doesn't work with every plant. If you like the taste and texture of these blanched dandelions, you can also blanch the ones growing outside during the warmer seasons, too: Simply cover them with a plastic pail (or some other structure) that will block the sunshine, yet allow them space to grow and thrive. In about a week, they will be paled to readiness.

The roots of dandelions are likewise edible, nutritious, and delicious. Moreover, they are a renowned survival food, having at one time kept alive the sizable population of a Meditteranean island after locusts had destroyed the cultivated food crops. Those from older plants should be peeled before being consumed, however, and such are especially tasty after being cooked. Here they can be baked (at 375 degrees F. for about 30 minutes) or boiled (in two waters, for about 20 minutes, with a pinch of baking soda added to the first water). After being cooked, they usually taste even better if they are chilled, in which condition one is tempted to employ them as a salad ingredient.

The roots can also be oven-roasted until dry (this takes about four hours) and then ground and brewed into a caffeine-free coffee substitute. Many, though, simply prefer them raw, chopping them up and mixing them into salads in that state. (Some persons react with flatulence to the raw roots, however, as they contain inulin, a carbohydrate that does not digest very easily. Still, for diabetics, this is ideal, since inulin does not as readily convert to glucose as does starch, and so consequently blood-sugar levels are not so rapidly raised. *See more on this below*, in the medicinal section.)

The flower heads are also edible and very sweetish to the taste. To remove them from their stalks, simply twist them off (as opposed to pulling them off or cutting them off). They are most often dropped into pancake batter to add a hint of sweetness to the resultant flapjacks. They may also be eaten raw, as a trail nibble or in salads, though some find them far too sweet when consumed in this fashion. (Those with known reactions to pollen should exercise caution when contemplating any consumption of flowers.) Or they can be steamed for five minutes, stirred, and then let stand for a few minutes prior to serving.

In the opinion of many, the prize dandelion part is the crown, located just under the surface of the ground where the stem and leaves meet the taproot. Cut this structure from the plant and cook it till tender (about 3-5 minutes), then top it with your favorite dressing. (You can make your own "wild" dressing from vinegar and the seed-pods of peppergrass—*see under that entry*.) Unlike the leaves, the crown doesn't become excessively bitter with age. Developing buds, hidden inside the crowns of new dandelions, are also cherished. These need be boiled for just a few minutes and then can be buttered and served.

If your taste buds wind up being delighted and your interest piqued to the point that you desire to become a dandelion connoisseur, simply contact Dr. Peter Gail of Goosefoot Acres Center for Resourceful Living of Cleveland, Ohio, who has an abundance of literature on dandelions, including even a newsletter devoted to the plant's appreciation. As of 2001, you can write to him at: Dr. Peter Gail, P O Box 18016, Cleveland OH 44118. Check out his website, as well, at www.goosefootacres.com (also at www.edibleweeds.com).

Health/Medicine

Possessed of *cooling* and *drying* energies, this "weed" so castigated by North American civilization has so many healing benefits for mankind that the dogged attempts to eradicate it from suburban lawns prove to be nothing short of criminal!

First, in view of its bitter sesquiterpene lactones, dandelion is a revered *tonic* for **POOR DIGESTION,** especially when attributable to **HYPOCHLORRHYDRIA.** Potawatomi Indians, among other native American tribes, greatly treasured its benefits in this regard.*(Smith 1933:54)* Pillager Ojibwe Indians found a root tea helpful for **HEARTBURN.***(Smith 1932:366)* Modern British herbalists feel that dandelion is especially indicated for **DYSPEPSIA** when it exists in combination with constipation.*(BHP 1976:1:197)*

A classic herb used by various cultures to enhance liver function, dandelion's high choline content, in conjunction with its sesquiterpene lactones, appears to be highly contributory to its revered role as a **hepatic**. Indeed, countless thousands throughout history have benefitted from dandelion's gentle but effective properties in this regard. In the mid-nineteenth century, for instance, the American physician Asahel Clapp reported that dandelion was widely used at that time for **CHRONIC LIVER DISEASE,***(Clapp 1852:806)* and it has a long pedigree of use in this regard in Europe.

Various clinical trials during the last fifty years have confirmed that dandelion can be most helpful for liver-related conditions such as **HEPATITIS, CIRRHOSIS, CONGESTIVE JAUNDICE, CHRONIC LIVER CONGESTION**(esp. related to **PRE-MENSTRUAL SYNDROME,** where estrogens are not clearing properly) and **CHOLE-CYSTITIS.***(Kroeber 1950:122-27; Faber 1958:423-36; Susnik 1982:323-28; Sankaran 1977:621-26; BHP 1976:1:197)*

Dandelion's assistance with these problems has been shown, through a variety of animal studies, to be attributable to *cholagogue* and *choleretic* abilities possessed by the plant, even as herbalists had long surmised.*(Bohm 1959:376-78; Bolyard 1981:168; Benigni et al 1964:593; Chabrole et al 1931:1100; Popowska et al 1975:491; Buesemaker 1936:512; Chabrol & Charonnat 1935:131-42)*

In view of the above information, too, dandelion has understandably proven to be invaluable for **CHRONIC CONSTIPATION** associated with inadequate liver function and poor overall digestion.*(Winston 1999:39)* Such a *laxative* effect, appreciated as well by the Delaware Indians,*(Tantaquidgeon 1972:39)* seems at least partly attributable to the plant's essential oil.*(Pedersen 1998:78)*

The greatly despised lawn "weed" is also a treasured *depurative* for **ECZEMA** and **ACNE,** including the juvenile forms.*(Scott 1990:56)* Eclectic physicians esteemed it as "the herb" for **AUTO-INTOXICATION***(Ellingwood 1983:326),* of which acne, eczema, and **CANKER SORES (APHTHOUS ULCERS)** were often seen as signs*(cf. Moore 1994a:10.6)* Dandelion's effect for auto-intoxication seems largely due to its bitter flavonoids, but detoxification appears also to take place owing to the content of mucilage and inulin, which respectively absorb toxins and regulate the beneficial intestinal microflora which generate their own toxins to kill harmful bacteria.*(Pedersen 1998:79; Winston 1999:39)* Such a regulatory function with respect to the microflora may perhaps at least partly explain the herb's help in vanquishing canker sores, since research has connected these aphthous ulcers with lowered quantities of the specific B-vitamins manufactured by the colon's bifidobacteria.*(Porter et al 1988:41-44; Palopoli & Waxman 1990:475-77; Wray et al 1978:418-28; Mills & Bone 2000:173)*

Dandelion's regulatory effect upon the intestinal microflora may also explain why a clinical study revealed that the herb (in combination with lemon balm, calendula, fennel, and St. John's wort) helped sufferers of chronic, nonspecific **COLITIS,***(Chakurski et al 1981:51-54)* in that reduced quantities of beneficial intestinal microflora have been found in colitis sufferers.*(Hentges 1983; Murray & Pizzorno 1998:593)* Dandelion's aid toward improving human microflora activity, coupled with a study which showed direct activity against *Candida albicans*(*see Duke 1985:476*) suggests that it can be an important weapon in any ongoing battle against **CHRONIC YEAST INFECTIONS**.

Additionally, dandelion may directly fight colitis' inflammation, since *anti-inflammatory* effects have been verified for the plant,*(Mascolo et al 1987:28-29)* due at least partly to its flavonoids.*(Pedersen 1998:78)* This makes it valuable as well for other irksome inflammatory conditions, such as **MASTITIS**, where the Chinese often successfully use a compress of a dandelion-root decoction.*(Yanchi 1988:64)* Still others*(Hoffmann 1986:190)* utilize dandelion to treat **MUSCULAR RHEUMATISM**, which application has had a long tradition in Europe and has even been recommended in the *British Herbal Pharmacopoeia.(BHP 1976:1:197)*

Both dandelion's leaves and its root are *diuretic* (not for nothing have the French long called dandelion *piss-a-bed!*), but studies show that the leaves are much stronger in this regard.*(Racz-Kotilla et al 1974:212-17)* The reason(s) for dandelion's diuretic effect is/are not known for a certainty. Good arguments can be presented for such an action owing to: (1) its bitter flavonoids,*(Pedersen 1998:78)* (2) an ability to inhibit sodium reabsorption by the kidneys,*(Tilgner 1999:57)* and (3) its high potassium content and/or (4) the action of its sesquiterpene lactones.*(Duke 1997:83; Blumenthal et al 2000:80)* Unlike pharmaceutical diuretics, however, it literally replaces potassium lost through diuresis by means of its own rich contribution of that mineral to the system.*(Racz-Kotilla et al 1974:212-17)* The diuretic effect, and perhaps other actions of the plant, helps to eliminate metabolic toxins such as uric acid (responsible for *RHEUMATISM* and *GOUT*).*(Ellingwood 1983:326; McQuade-Crawford 1996:142)* Dandelion is also commonly used by herbalists to alleviate *EDEMA* relative to cardiac disorders, as well as the accompanying *HYPERTENSION*.*(Winston 1999:38)*

The diuretic effect, combined with the hepatic properties outlined above, makes dandelion the herb of choice for *PMS-H*, the subtype of premenstrual syndrome where a woman gains five or more pounds of water-weight just prior to her menstrual period. Here the plant's diuretic effect leaches out the excess fluid, while the hepatic effect helps the liver to process the estrogens most efficiently. This use of dandelion has probably been my most frequent application for clients, and it usually works in admirable fashion. Further with rerference to its estrogen-processing boost, herbalist Amanda McQuade-Crawford finds dandelion helpful for *FIBROIDS*,*(McQuade-Crawford 1996:142)* which seem to owe their existence to a hyperestrogen condition in the body.

Because dandelion contains generous amounts of the coumarin aesculin, the active agent in horse chestnut *(Aesculus hippocastanum)*, it can improve the quality of the veins, tonifying their structure,*(Williams et al 1996:121-27)* so that it proves to be of assistance for *VARICOSE VEINS/HEMORRHOIDS*.*(Brooke1992:103; McQuade-Crawford 1996:142)* It has also shown *anti-platelet-aggregating* action.*(Neef et al 1996:S138-40)* Moreover, dandelion has traditionally been used to offset *DIZZINESS*(Brooke1992:103;McQuade-Crawford1996:142)* and *SLEEPINESS*,*(de Bairacli Levy 1974:57)* but whether this has to do with any stimulation of circulatory function remains unclear.

Dandelion has been used by various cultures for *PULMONARY COMPLAINTS*. Mesquaki Indians, for example, used the herb for "pain in the chest" after the usual remedies had failed.*(Smith 1928:218)* Interestingly, too, a test-tube study conducted in the 1950s even demonstrated some effects agains a virulent human strain of the tuberculosis organism.*(Fitzpatrick 1954:531)*

Extracts of dandelion have inhibited growth of *CANCER* cells and have produced antibodies to tumor polypeptides.*(Kotobuki Seiyaku1979:14530m;idem,1981:10117;Salvucci 1987:930-36)* Even a simple hot-water extract has shown marked *antitumor* activity.*(Baba et al 1981:538-43)* The oriental species, *T. japonicum*, has likewise shown anticarcinogenic activity.*(Takasaki et al 1999:606-10)* Probably dandelion's sesquiterpene lactones are a prime factor in such activity, even as they have been shown to be crucial in the tumor-fighting capabilities of another plant previously monographed in the present work, boneset *(Eupatorium perfoliatum) (see under that entry)*. Dandelion has further been experimentally shown to restore damaged nitric-acid production by the body—an important factor in immune function.*(Kim et al 1998:283-97)*

Our herb has demonstrated *hypoglycemic* effects on lab animals.*(Farnsworth 1971:52-55; Yamashita et al 1984:491-96).* This suggests that it could have potential for treating diabetes, although another study revealed that it only lowered blood-sugar levels in normal animals and *not* in diabetic ones.*(Akhtar et al 1985:207-10)* To what degree the plant's own inulin content (discussed above, in the culinary section) contributes to this effect remains unclear.

The latex in dandelion's stem and leaves has been a revered topical folk remedy for *WARTS*, applied three times a day for ten days, after which the wart supposedly blackens and falls away. My experimentation with this treatment has revealed that it is not as effective as the latex from milkweed *(Asclepias syriaca)* (see under that entry), another folk remedy for warts, but it is still often worth a try.

Dandelion root was official in the *U.S. Pharmacopoeia* from 1831-1926 and in the *National Formulary* from the first edition until 1965. It was listed as *tonic, diuretic,* and *aperient*. The root is still official in the British, German, Austrian, and Czech pharmacopoeias.

Cautions

* The plant's sesquiterpene lactones have occasionally produced *CONTACT DERMATITIS* in persons handling this plant who are sensitive to these substances.

* Never collect dandelions from lawns or other cultivated areas where herbicides or chemical fertilizers are suspected as being used.

* Persons with *ACTIVE ULCERS* should be aware that bitter foods like dandelion leaves and roots will stimulate production of hydrochloric acid, which excess may further erode the ulcer.

DOCK
[WILD BUCKWHEAT; CURLY DOCK; YELLOW DOCK]
(Rumex crispus)

Description

[See photos #52-54 in rear of book.]

A familiar perennial that begins life in spring as a rosette of long (4-12"), wavy, curly-edged leaves. The summer plant shoots up to 1 to 5 feet tall, developing an upright, smooth, ribbed stem that becomes rigid and hollow as the plant fully matures. Branch stems may or may not appear. If they become present, it is only toward the top of the plant.

The stem leaves are dark green, oblong to lanceolate, pointed at the tip, and grow alternately along the stalk. They are about half as long as the basal leaves and *have strongly curled and wavy margins.* They show a strong vein down the central axis, with a number of veins branching off from it that angle toward the edge and then curve back to join other branch veins. Finally, a papery, straw-colored sheathing membrane can be found around the stem where the leaves attach. It is sometimes slimy to the touch.

In early summer, an abundance of tiny green flowers (with a slightly reddish or purplish tinge) are densely clustered along the upper part of the stem in wandlike fashion. In late summer, these flowers transform into brown, winged, three-angled achenes, reminiscent of tobacco, and present a more striking appearance than the previous green flowers.

Dock's taproot is long and carrot-like. It is noticeably yellowish-red when cut open (and thus the alternate common name, "yellow dock").

Range and Habitat

This plant, castigated as a "weed," is almost as ubiquitous as the dandelion, and is common to wasteland, vacant lots, ditches, and fields throughout North America.

Chief or Important Constituents

Roots contain resins, a volatile oil, anthraquinone glycoside derivatives (including nepodin, chrysophanic acid, emodin, physcion), rumicin, fatty acids(stearic, palmic, and erucic), sugars(fructose, dextrose), tannin, pigments. Above-ground portions contain flavonoids(avicularin, quercetin, quercitrin, hyperoside), calcium oxalate, oxalic acid, chrysophanic acid, tannin, and nutrients consisting of vitamin A(very high—a 100-gram portion contains 12,900 IU's), vitamin C(a 100-gram portion contains 119 milligrams), B vitamins (niacin, riboflavin, and esp. thiamine), phosphorus(41 milligrams per 100-gram portion), potassium(338 milligrams per 100-gram portion), and iron(1.6 milligrams per 100-gram portion); also some manganese, selenium, sodium, and zinc.

Food

This abundant weed is widespread and easily recognized, which is fortunate considering that it allows for a variety of repasts.....

The curled leaves can be consumed raw *sparingly* (see important cautions below), but some are too bitter (especially during certain times of the year), and any plants growing in soil known to be high in nitrates should not be eaten. Well-chosen plants, however, yield delightfully sour leaves, similar in taste to the closely related sheep sorrel *(R. acetosella) (see under that entry)*.

Dock also makes a delicious cooked vegetable. To prepare it, boil in two changes of water—about five minutes in the first and about ten in the second. (The extra pot of water is necessary only if the leaves are bitter.) Recipes appear in several weed cook-books. One recipe I especially enjoy is "Clam Soup with Dock," provided in Karl Knutsen's informative foraging manual.*(Knutsen 1975:37)*

Unfortunately, dock is one of those very tender wild veggies that is difficult to keep by freezing. Successful efforts relate to careful preparation. One recent foraging guide*(Young 1993:26)* suggests cutting the dock leaves up into smaller pieces, putting them into a pot with one cup of water for each quart of dock, bringing the water to a boil, and then steaming the contents for three minutes, stirring occasionally. Thereafter, it is suggested that the segments should be ladled into 8-ounce plastic (or glass) containers with 1-inch headspace and then cooled in a pan of ice water (insuring that the ice water does not get into the containers). Finally, it is urged that the containers be capped and frozen.

The fully dehusked and winnowed seeds can be ground into flour for bread, but the vision of work involved is too prohibitive for most persons.

Health/Medicine

Possessed of *cooling* and *drying* energies, yellow dock has long been regarded as a superb *alterative* and *hepatic.* Not surprisingly, then, southwest-American herbalist Michael Moore notes that a tea made from dock root can often be of service for***BILIOUS PROBLEMS*** (it was used as such by the Delaware-Oklahoma Indians-*Tantaquidgeon 1942:28, 78*) and particularly for the ***POOR DIGESTION OF FATS***. His recommended dose is a teaspoon of the chopped root, boiled (not steeped) in water, taken twice a day (see cautions below*).(Moore 1977:166)* Delaware and Chickasaw Indians likewise used yellow dock for another hepatic problem: *JAUNDICE.(Tantaquidgeon 1972:33; Swanton 1928:267)*

The Cherokee Indians found that dock, rather curiously, possesses both *astringent (antidiarrheal)* and *laxative (anticonstipational)* properties.*(Hamel & Chiltoskey 1975:32)* The tannin-rich leaves, stems, and fruits are astringent and thus antidiarrheal (Western Eskimos cured persons afflicted with severe cases of diarrhea by feeding them dock leaves and stems as a potherb in the morning before breakfast and also before bedtime-*Oswalt 1957:24*), while the root is usually thought of as laxative owing to its content of anthraquinone glycosides and because of the plant's favorable effect upon the liver's activation of bile. Yet, because it, too, contains tannins, there is a balancing effect therefrom, which results in yellow dock being much less powerful and irritating a laxative than other anthraquinone-containing plants like rhubarb (even though dock has a greater anthraquinone content than that plant does), cascara sagrada, or senna. Some cultures, such as the Tarahumara Indians,*(Kay 1996:240)* have even coaxed an astringent effect from the root so as to allow it to serve as an antidiarrheal as well. James Duke, a former economic botanist with the USDA, asks whether the body, in its wisdom, might be able to choose which effects it desires, depending upon the need at the time.*(Duke 1985:415)*

Yellow dock also serves as a wonderful *depurative*, and it was made use of in this fashion by numerous tribes of American Indians, including the Cherokee, Delaware (Oklahoma), Mohegan, Paiute, Rappahannock, and Shoshone.*(Moerman 1986:1:421–22)* The Eclectic physician Finley Ellingwood held it to be "a renal depurative and general alterative of much value when ulceration of mucus surface or disease of the skin result from impure blood."*(Ellingwood1983:378)* Even the nineteenth-century physician and botanist Laurence Johnson, so generally critical of plant remedies then in vogue, acknowledged that yellow dock's "properties render it useful in a variety of chronic affections, such as scrofula, obstinate cutaneous diseases... syphilis, etc., in which an alterative and depurative effect may be desired for a long time."*(Johnson 1884:238)* Charles Millspaugh highlighted the popular use of a dock-root ointment as a *discutient* for "indolent glandular tumors."*(Millspaugh 1974:576)*

One of dock's most renowned uses, however, is as a salve for *SKIN RASHES*, for which it was employed by the Cherokee Indians*(Hamel & Chiltoskey 1975:32)* and by European and American herbalists. The specific indication for its use is skin rash accompanied by signs of liver stagnation and constipation.*(BHP1976:1:171)* The chrysophanic acid is thought to be the most important constituent toward the dermatological assists, and Duke notes that as late as 1977 chrysarobin was used in orthodox medicine as a topical agent for skin disorders, proving particularly effective against *PSORIASIS*.*(Duke 1985:415)* (Chrysarobin is also widely regarded as a fungicide.) Undoubtedly, the tannins and perhaps other constituents are important in this regard.

In fact, yellow dock is a classic herb for *ITCHING* of almost any kind.*(McQuade-Crawford 1996:182; Harrar & O'Donnell 1999:81)* Our own area's Ojibwe (Chippewa) Indians applied the powdered root via a wetted cloth to itchy areas on the skin.*(Densmore 1974:350-51)* Concordantly, a mid-1980s lab study found that leaves of the related species, *R. nepalensis*, significantly reduced itching, which was attributed to *anti-cholinergic, antibradykinin*, and *antihistamine* properties.*(Aggarwal et al 1986:177-82)*

Topical application of yellow dock has also been heralded for parasitic skin infections such as *SCABIES, RINGWORM*, and *URTICARIA*, and here it is thought that the consistuent rumicin is largely responsible, having an *antiparasitic* effect.*(Stuart 1982:128;Duke1985:415;Krochmal&Krochmal 1973:193)* An *antibacterial* effect from the anthraquinones may further explain its aid in these or other skin conditions.*(Anton & Haag-Berrurier 1980:104-12)* One study has shown effects against *Escherichia coli, Mycobacterium smegmatis, Shigella sonnei* and *S. flexneri,* and *Staphylococcus aureus*, although this was from an extract of the aerial portions, not the root, and attributed to the plant's essential oil.*(Miyazawa & Kameoka 1983:45-47)* Another study on the leaves has shown a moderate effect against a virulent human strain of the tuberculosis organism.*(Fitzpatrick 1954:530)*

It is not clear whether it has to do with the mechanisms for dermatological aid outlined above, but crushed leaves of this weed—also those of jewelweed, mullein, and plantain (*see under those entries*)—have long been held to be efficacious for relieving the **STING** produced from **NETTLES** and even, to some extent, the **RASH** accrued from **POISON IVY**. I cannot vouch for poison ivy, but my own experience has convinced me that rubbing with the leaves of this plant does relieve the sting from nettles.

A large variety of Amerindian tribes employed dried, powdered dock root or leaves as a sprinkle for **CUTS** and **WOUNDS**, including the Flambeau Ojibwe*(Smith 1932:381)* and Indians of the Nevada region.*(Train et al 1957:87)* Teton Dakota Indians crushed the plant's leaves and poulticed them onto **BOILS** so as to suppurate them.*(Gilmore 1991:25)* Undoubtedly the tannins largely account for dock's long-standing reputation as a **BURN** remedy, a function gratefully seized upon by various Amerindian tribes.

Astringent properties undoubtedly account as well for dock's aid in helping to heal **SPONGY GUMS.** *(Millspaugh 1974:576)* Thus, too, a cold infusion of the leaves swished in the mouth sometimes helps to offset the discomfort caused by **CANKER SORES**, which use has been appreciated by Navajo Indians.*(Mayes & Lacy 1989:33)* Mexican-Americans sometimes gargle a root infusion to soothe a **SORE THROAT**.*(Kay 1996:240)*

Dock has long been a treasured remedy for **ANEMIA.** It indeed contains some iron, but not nearly as much as one would expect for an herb that is supposedly the best thing around for this condition. And although it may take months to get the job done, yellow dock does indeed seem do so quite efficiently, as I can testify from much experience in using it. How, then, does it work? Some have suggested that it serves to liberate stored iron in the liver. This seems a reasonable explanation, especially in view of dock's demonstrated hepatic properties (as detailed above). I am not aware of any confirmatory evidence at this point in time, however.

Dock has potential against some heavyweight afflictions, too: An Egyptian study found that it may be useful in the war against **AIDS**, since an extract of its fruits inhibited HIV reverse transcriptase.*(el-Mekkaway et al 1995:641-48)* Emodin, a quinone found in the roots of our own *R. crispus*, has also shown pronounced **antitumor** activity.*(Harborne & Baxter 1993:499)*

In view of the material we have thus far considered, it should come as no surprise that yellow dock was listed in the *U.S. Pharmacopoeia* from 1863-1905 and in the *National Formulary* from 1916-36, initially as an **alterative** and later as a **laxative** and **tonic**. It is, of course, also official in some European nations, including the U.K.

Yellow dock is available commercially in the form of capsules, tincture, and tea. Capsules most effectively concentrate the minerals, and so to best utilize dock's iron-building properties, this would seem to be the form of choice, at least as far as actual intake of iron (although, as we've seen, dock's content of that mineral is not all that phenomenal and any other iron-building mechanisms it may possess have not yet been fully explicated).

Cautions

* *As a wild food*: Because dock contains a significant amount of oxalic acid and its salts, which can hinder calcium absorption in the body by chelating with this mineral, it should **NOT BE CONSUMED AS A REGULAR STAPLE**, but only every once in awhile at most. Further underscoring this caution is that the calcium oxalate crystals that can be formed from the combination as mentioned above are excreted through the kidneys and, if bunched up in quantity, can cause mechanical damage there. The potential problem is usually not a concern for occasional consumption, as long as it is in moderation. Then, too, cooking tends to break down the oxalate. But that heavier consumption of dock presents a genuine danger is shown by the fact that animals have died from eating it, probably because they ate so much of it that their systems could not process the oxalates, and perhaps as well because amounts of nitrates toxic to these creatures were present.*(Panicera et al 1990:1981-84; Duke 1985:415)* In the late 1980s, too, a man with insulin-dependent diabetes died from consuming a large amount of dock soup (equating approximately 500 to 1000 grams of the plant).*(Farr et al 1990:1981-90)* It was estimated that his total consumption of oxalic acid was somewhere between 6 and 8 grams, which is in accord with the mean lethal dose (set at 5 to 30 grams). Any human consumption should therefore be with care, as outlined above, and in moderation.

* *As a medicinal*: Not to be taken internally during *PREGNANCY*, owing to laxative effects which could precipitate a miscarriage. As the purgative anthraquinones can also make their way into the breastmilk, dock is also contraindicated during lactation. Owing to its content of oxalic acid and oxalates, it should also not be used by those with *KIDNEY FAILURE, DIABETES*, or *ELECTROLYTE ABNORMALITIES*, nor be used concurrently with drugs known to lead to *HYPOCALCEMIA.*

EVENING PRIMROSE
[SUNDROPS; GERMAN RAMPION]
(Oenothera biennis)

Description

[See photos #55-57 in rear of book.]

A hairy, biennial plant growing to six feet.

It starts off as a basal rosette of thick, wavy-edged leaves, sometimes with a dab of red here and there. Growing to eight inches long, these leaves each display a prominent midrib that is white or red.

In the summer of the second year, the flowering stem appears, crowded with many, alternate leaves of lanceolate shape.

The showy, yellow flowers are four-petaled and possessed of a characteristic, cross-shaped stigma (*see close-up illustration*). They occur clustered at the end of the stem. The flowers open only in the afternoon and close before dawn of the next day, and thus the plant's common name.

The fruit consists of thickish, horn-like capsules jutting upward, the skeletons of which remain throughout the winter months, making the plant easily recognizable even during this season.

Range & Habitat

Waste places, fields, meadows, railroad embankments, riversides. Throughout our area, except in the very northern strip of Minnesota.

Chief or Important Constituents

Oenotherin, mucilage, resin, bitter principle, phytosterols, tannins and pseudo-tannins(ellagic acid and gallic acid), flavonoids(quercetin, kaempferol, delphinidin), caffeic acid, digallic acid, neochlorogenic acid, o-coumaric and p-coumaric acids. Nutrients include calcium, magnesium, and potassium in significant amounts. Seed contains tryptophan, phenylalanine, beta-sistosterol, and a fixed oil of which about 70% is cis-linolenic acid, 9% gamma-linolenic acid(GLA), and the rest a mixture of oleic, palmitic, and stearic acids. Nutrients in the seed include boron, copper, iron, sodium, zinc, and a variety of free amino acids such as arginine. The leaves are high in vitamin C.

Food

The plant's taproots, known as German rampions, are edible and can be collected from first-year plants in the fall or spring. Eaten raw, they have a powerful, turniplike afterbite. Fortunately, proper cooking renders them more palatable. This should be done in three different waters, for a total cooking time of about half an hour. (Europeans often boil them for two hours, however!) The roots should be peeled, either before or after being cooked.

Some say the taste is like parsnips, while to others it is more like turnips or salsify. Be forewarned that the roots will have an unpleasant, peppery, biting quality if undercooked or if harvested too early in the fall or too late in the spring. However, the peppery quality can still be useful as a spicy ingredient in soups or stews, and so the cut-up roots are sometimes added to such. The pungent leaves can also be added—in moderation—to these dishes, if desired.

The rosette's central crown also makes a decent cooked vegetable, as master forager Euell Gibbons was fond of explaining. *(Gibbons 1973:97)*

Health/Medicine

In 1901, Professor of medical botany B. B. Smythe listed evening primrose's root as possessing the physiological properties of an *alterative,* an *astringent*, and a *demulcent*. *(Smythe 1901:202)* The demulcent property at least partly explains why the plant has long been used as a soothing *COUGH* remedy. *(Lust 1974:187)* Because our herb is also *antispasmodic*, it has traditionally been employed for *WHOOPING COUGH, HIC-COUGH* and *SPASMODIC ASTHMA*. *(Mockle 1955:65; Smith 1932:376; Wood & Ruddock 1925)* To make a cough medicine, southwest-American herbalist Michael Moore says to chop up the fresh or dried root and then boil it slowly in twice the amount of honey, thereafter giving a tablespoon every 3-4 hours, as needed. *(Moore 1979:74)* (Never give honey to a child under a year old, however.) Naturopathic doctor John Lust, in his interesting and detailed book on herbal remedies, wrote instead of making a syrup from the flowers, which he suggested for whooping cough and asthma. *(Lust 1974:187)*

Herbal authorities Wood and Ruddock spoke of evening primrose as being "an efficient remedy as a nervine and sedative to quiet nervous sensibility, well adapted to neuralgia." *(Wood & Ruddock 1925)* As to the sedative effect mentioned, which Michael Moore notes as being stronger in some people than in others, *(Moore 1979:74)* perhaps this is at least partly due to the high tryptophan content.

Evening primrose is also a much-appreciated *vulnerary*. *(Millspaugh 1974:60)* The Flambeau Ojibwe Indians employed the "whole plant soaked in warm water to make a poultice to heal bruises." *(Smith 1932:376)* Constantine Rafinesque, an early American authority on medicinal plants, detailed an application popular in his day wherein the "leaves" were "bruised and applied to wounds." *(Rafinesque 1830:247)*

This plant has a long history of use as an external application in "infantile eruptions."*(Millspaugh 1974:60)* Writing in the mid-1800s, Dr. R. E. Griffith wrote: "Some years since, hearing of the efficacy of a decoction of the plant in infantile eruptions, I made a trial of it in several cases of an obstinate character, which had resisted other modes of treatment and became satisfied that it was highly beneficial; and this opinion has been confirmed by subsequent experience."*(Griffith 1847:304)*

The Cherokee Indians used a tea of the plant for **OBESITY**.*(Hamel & Chiltoskey 1975:33)* They also utilized a hot-root poultice as an application for **PILES,** while the Iroquois employed an infusion for the same affliction—boiling one root of it with one root each of wild mint *(Mentha arvensis)* and self-heal *(Prunella vulgaris)* (*see under those headings*) in one quart of water, then drinking two cups of the brew and using the rest for a wash.*(Herrick 1995:175)*

Evening primrose has also had a place in treating gastrointestinal problems. Charles Millspaugh, in his classic work on American medicinal plants, wrote of a certain Dr. Winterburn as one who found the plant helpful for "gastric irritation and chronic exhaustive diarrheas."*(Millspaugh 1974:60)* A Canadian researcher likewise detailed the plant's use in his country's folk medicine to "soothe gastrointestinal disturbances."*(Mockle 1955:65)* Dr. Griffith wrote of it as having been "a favorite emollient in ulcers."*(Griffith 1847:304)* The Montagnais Indians implemented evening primrose for **PAIN IN THE BOW-ELS.**.*(Tantaquidgeon 1932:267)* Another authoritative source lists it as being helpful for "pains of the lungs, stomach, heart, liver, bowels and womb."*(Wood & Ruddock 1925)* British herbalists have traditionally used it to offset **DYSPEPSIA**, theorizing that it finds success in this regard by improving liver function.

Traditional forms and dosage for most of the above conditions have been to infuse one teaspoon of evening primrose in one cup of water and then imbibe one cup a day, a mouthful at a time. A tincture of the plant has been used to the tune of 5-40 drops a day, as needed.*(Lust 1974:187)*

The seed oil, available commercially and rich in the essential fatty acid known as gamma-linolenic acid(GLA), has been the subject of over 300 studies. GLA supports the production of a series of prostaglandins(PGE-1's) in the body which modify inflammatory processes, lower blood pressure, and inhibit platelet aggregation (thereby reducing blood clotting, which in turn can help prevent **HEART ATTACKS** and **STROKE**). This being so, the seed oil has been put to good use in the treatment of a large variety of inflammatory conditions, including the following: **ATOPIC ECZEMA** (numerous clinical trials in a variety of research centers have shown effective results, and some more so than with the use of steroids-*see the summary of studies in Mills & Bone 2000:367*), **MIGRAINES** (Dr. James Duke, a former economic botanist with the USDA, suggests that the phenylalanine content may be significant here-*Duke 1997:235*), **RHEUMATOID ARTHRITIS** (a Scottish study showed improvement in 60% of patients, most of whom were even able to quit their anti-arthritic drugs-*Mabey 1989:89*), **PMS PAINS** (numerous studies have shown effectiveness here, including three double-blind ones-*seeHorrobin1983:465-68;Brush1982*),

MULTIPLE SCLEROSIS (because EFA's are crucial to the health of the myelin which protects the nerves and which degenerates during MS), ***SJOGREN'S SYNDROME*** (two good studies—one at Copenhagen and another at Glasgow—have shown effectiveness with this condition-*Campbell&MacEwen1982,Horrobin1990:1-45*), ***SCLERODERMA,*** *(Horrobin 1984:13-17)* and ***ASTHMA.***

Other conditions shown to be helped by evening primrose oil include: ***ADHD, PARKINSON'S DISEASE*** (one study showed improvement with this disease's characteristic tremors in 55% of people taking the equivalent of two teaspoons a day for several months-*Duke 1997:353*), ***SCHIZOPHRENIA,*** *(see the studies cited in Mills & Bone 2000:369)* the pain and nerve damage of ***DIABETIC NEUROPATHY*** (although Duke recommends that the crushed seeds be used instead of the extracted seed oil, since little tryptophan is available in the latter.-*Duke 1997:348;Mills & Bone 2000:367*), and ***BLOOD-SUGAR-HANDLING PROBLEMS/ALCOHOLISM*** (which two conditions, since the pioneering research of endrocrinologist John Tintera, have been known to be directly related-*Tintera 1966:126-50*)*(see the research summary in Mills & Bone 2000:364-65, 368-69, 370-71*). A Scottish study showed that evening primrose oil even aided in the regeneration of livers damaged by alcohol.*(Mabey 1988:89)* Looked at from a TCM perspective, the oil would be a cooling "yin tonic," favorably affecting the liver and kidney meridians.*(Tierra 1998:137)*

The oil is widely used to bring back a healthful sheen to dry hair and skin and it is also sometimes rubbed into the vaginal area by menopausal women experiencing dryness in this region. (However, its effect, while often quite helpful, doesn't hold a candle to regular, pulsepounding sex, which can dramatically improve blood flow to this region and thereby enhance its tone and lubrication!)

I personally use the commerical softgels of EPO quite a bit with clients with reference to the majority of the abovementioned conditions. Here my experience has been that this is a very effective measure in most of these complaints—especially PMS cramps/breast tenderness and infantile eczema. (It is usually the treatment of choice in the latter situation, especially if the baby has not been breastfed or where the complaint develops after weaning from breastfeeding; this occurs because the GLA in breastmilk is not then available and because the baby's body does not possess, or cannot efficiently use, the enzyme responsible for converting *cis*-linolenic acid from foods into GLA). I have also witnessed the oil to enhance the quality of life with clients afflicted with MS. The effective dose ranges from a total of 2 to 6 grams a day and these softgels should be taken at the beginning—or in the middle—of meals, to help minimize any burpback.

Cautions

* Evening primrose oil should **not** be utilized by persons who have ***EPILEPSY,*** as it may worsen this condition.

* As EPO is somewhat of a uterine stimulant, it should be ***used cautiously by pregnant women***, if at all.

FALSE SOLOMON'S SEAL
[SOLOMON'S PLUME; WILD SPIKENARD]
(Smilacina racemosa; S. stellata)

Description

[See photos #58-61 in rear of book]

The number of Solomon's seal (*see under that heading*) look-alikes can be quite bewildering to the neophyte, though one eventually learns to distinguish between the more common varieties. The two species listed in the heading above, both perennials, are each commonly referred to as "false Solomon's seal" by various sources, so that one should learn their respective scientific names and distinguishing features.

The first plant, *Smilacina racemosa*—variously called "false spikenard," "wild spikenard," or "Solomon's plume"—is the larger of the two species (1 to 3 feet tall), and the one that most resembles true Solomon's seal, being generally of the same height and bearing similar-sized, oval, alternating leaves that are 3 to 6 inches long and have prominent veins. The leaves, however, are not as waxy-looking as with true Solomon's Seal. Most noticeably, the berries do not dangle under the plant's stem from the leaf axils as they do in the true Solomon's seal, but are borne in a branched, pyramidal cluster at the tip of the stalk, replacing tiny, creamy-white flowers. (True Solomon's-seal flowers are greenish-yellow bells and are much larger than *S. racemosa's* flowers.) At first, the berries are white and gold, later a blotchy pinkish-red, and finally pure red (thus, never blue like those of true Solomon's seal). Finally, the arching stem of *S. racemosa* is constructed of a curious zigzag design (giving rise to yet another common name for this plant: "Solomon's zigzag"), in contrast to the non-zigzag stem of true Solomon's seal.

The other species, *S. stellata*, is smaller (1 to 2 feet high) and more graceful in appearance. The leaves—more spikelike and upswept looking—usually have a slight bluish hue to them, and their bases appear almost to wrap themselves around the plant's stem, in that they are sessile (that is, stalkless). The six-petaled flowers—giving rise to the alternate common name, "star-flowered false Solomon's seal"—are, unlike those of *S. racemosa*, in unbranched clusters. They are thus fewer (3 to 15), but individually larger in size, than those of their cousin. Otherwise, this species shares the feature of the placement of the berries (and earlier, the flowers) with the larger *Smilacina* species. The berries change color as they grow: first, a blotchy sort of red; second, greenish with reddish brown stripes; and finally—in August and September—an attractive, dark-ruby-red. Eight or more seeds can be discovered in each berry.

Range and Habitat

Rich deciduous woods, coniferous woods (especially *S. stellata*), thickets, in clearings at edges of woods, sandy banks and along streams and rivers. Range is southern two-thirds of Canada and the top half of the eastern US, dipping more deeply south in the western states. Throughout our area's woodlands.

126

Chief or Important Constituents

Very little research has been done on either species, although *S. racemosa* is known to contain asparagin, saponins, mucilage, sitosterol, and the cell-proliferant allantoin. The cardioactive glycoside convallarin has been reported as occurring in the rhizomes. Berries of both *Smilacina* species contain large amounts of vitamin C. Much chemical work remains to be done on these plants.

Food

The ripe berries of both *S. racemosa* and *S. stellata* are edible. In that they each sometimes persist into late autumn, they should be committed to memory as being possible rations for survival predicaments. The berries of *S. stellata* are the best tasting—the flavor being strongly like molasses. Those of *S. racemosa* are slightly bitter, posessing a only a mild molasses taste. (As a point aside, in early New England, the berries of the *Smilacina* genus were called "treacle berries" because their taste was adjudged to resemble an English molasses by that name.) In small amounts, the berries are also gently laxative, so obviously they should not be consumed in large amounts in the raw state. Many say that they taste best simmered, however, which reduces the laxative principle.

The early spring shoots of *S. racemosa* may also be eaten raw or as a potherb (boiled in water—preferrably salted—for 10 minutes). The rootstocks are toxic as is, but can be prepared in such a way as to make them edible. The process is involved and depends upon the availability of lye, so it will not be detailed here.

Health/Medicine

Although neither species is used much by modern herbalists, available records reveal that the American Indians made use of both species in numerous ways. This has led some herbalists to begin looking at this genus more closely—as indeed they should!

Available records inform us that the Paiute Indians of the west used to dry, slice, and powder the roots of *S. stellata* for use as an emergency **styptic**. They found from experience that such a powder, thrown onto a **WOUND**, would clot it instantly.*(Kindscher 1992:282; Train et al 1957:93)* The Washoe Indians did the same, believing the root to be also **antiseptic** for **BLOOD POISONING**.*(Train et al 1957:93)* On the other hand, the S. Ojibwa Indians preferred *S. racemosa* as a **stypic**—employing, however, a poultice composed of the fresh, smashed leaves.*(Hoffman 1891:199)*

Because this genus possesses anti-inflammatory properties, an important use of *Smilacina* rootstock—fresh or dried/powdered—has been as a poultice, not only for wounds and sores, but also for **BURNS, SPRAINS, BOILS, SWELLINGS, INSECT BITES/STINGS, POISON IVY RASHES**, and other **INFLAMMATIONS**.*(Zigmond 1981:64; Train et al 1957:92)* The latter has even included **INFLAMED EYES** and **EARS**. As to the application for the eyes, the Shoshone Indians would soak the rootstocks in water, mash them, and apply the resultant liquid (presumably strained) to the inflamed peepers.*(Train et al 1957:93)* As to Amerindian applications for the ears, the Paiute Indians would sometimes handle this by forcing pulped material of *S. stellata* through a cloth (thereby straining it) into the afflicted ear.*(Train et al 1957:93)* **RHEUMATISM** was another sort of inflammation for which the plant was found helpful; here the Thompson Indians drank a decoction of the leaves two to three times a day.*(Turner et al 1990:129)* The Gitksan Indians preferred a decoction of the "roots" (presumably dried),*(Smith 1929:53)* while the Gosiute Indians pounded such for topical application.*(Chamberlin 1911:382)* Then, too, the plant's leaves and twigs were also used by Maritime Indians to heal ***ITCHY RASHES****.(Chandler et al 1979:62)*

Michael Moore, an American herbalist who has shown interest in the *Smilacina* genus for healing, has found the dried root of *S. racemosa*, prepared as a decoction (a tsp. of the dried rhizome, finely chopped, in a cup of water and boiled 15+ minutes), useful for ***FRONTAL HEADACHES*** caused or accompanied by ***INDIGESTION***; also to soften and expectorate ***RESPIRATORY MUCUS***. (Our area's Menominee Indians had also used the plant for catarrh, but chose instead to grind up the rootstock, mix it with water, and heat it to release steam that was inhaled by the sick person.*-Smith 1923:41*) Moore also uses the rootstock decoction for the ***INFLAMMATORY STAGES*** of ***LUNG*** and ***THROAT INFECTIONS****.(Moore 1979:77)* Interesting here is that the Delaware-Okl. Indians employed *S. stellata* for "scrofula"*(Tantaquidgeon 1942:80)* and further that a 1954 study examining hundreds of American wild plants for antibacterial effects found this same species to inhibit a human tuberculosis pathogen in a 1:20 dilution.*(Fitzpatrick 1954:531)*

Delaware-Oklahoma and Shoshone Indians decocted the root of *S. stellata* for treating ***VENEREAL DISEASES****.(Tantaquidgeon 1942:80; Moerman 1986:1:458)* A quite different application affecting the reproductive system was employed by the Nevada Indians, who had their women drink a half a cup of the leaf tea daily for one week as a most effective ***contraceptive****.(Trease et al 1957:92;de Laszlo & Henshaw 1954:626-31)* The Costanoan Indians employed the root of *S. racemosa* instead.*(Bocek 1982:28)* The Thompson Indians used a decoction of the latter for women during menstruation or after childbirth.*(Turner et al 1990:129)*

Because the rhizome contains convallarin, it was used in early America as a substitute for digitalis when the latter could not be secured. Gentler in its effects because of having much less an amount of the glycoside than foxglove, *Smilacina* was nevertheless felt to be more effective as a continuous treatment for ***DROPSY****.(Smith 1928:231)* The Thompson Indians had also employed it for ***HEART*** problems.*(Turner et al 1990:50, 129)*

Interestingly, the Potawatomi Indians implemented a smudge of the rhizome of *S. racemosa* on hot coals to revive a person in a ***COMA****.(Smith 1933:63)* One cannot help but wonder if such a procedure might find some use today with so-called "hopeless" cases.

Cautions

    * The ***FRESH, BELOW-GROUND PORTIONS*** of *Smilacina* are **TOXIC** if ingested, unless treated per the procedure alluded to in the text above.

    * ***Overconsumption*** of the berries **WILL RESULT IN DIARRHEA.**

FIREWEED
[GREAT WILLOW HERB]
(Epilobium angustifolium)

Description

[See photos #62-64 in rear of book]

A lovely native perennial, possessed of a smooth stem, and growing 2-6 ft. high.

The leaves are narrow and lanceolate—even willowy (giving rise to the alternate name of "great willow herb"). They grow alternately along the stem, varying from 2 to 8 inches long and from 1 to 1½ inches wide. They are shiny green on their topsides only, toothless or minutely toothed, and nearly stalkless.

Beautiful, four-petaled, pinkish-purple flowers are stalked on a showy, spikelike raceme. *(Note for the novice*: the number of petals is a quick means of distinguishing this plant from the similar looking purple loosestrife. *See entry under that name.*) Like mullein and blue vervain (*see earlier entries under those names*), fireweed winds up never having its flower spike entirely in bloom. This is because the blossoms begin opening at the bottom of the stem, with the process moving upward—so that one is likely to encounter fruits at the bottom, blossoms in the middle, and buds at the top!

This being as it is, it is not difficult to conceive why many—myself included—find this plant to be one of the most attractive of our native flora. Here, too, fireweed seems to induce a sense of deep serenity, thus being a wonderful plant to simply behold (and thus insure one's dose of sublimeness for the day).

By late summer, the red, slender, bean-shaped seedpods—ever angling upward—split into four segments, folding back in spirals to reveal white, fluffy seeds with long, silky hairs (reminiscent of down) which, becoming wind-borne, whiten the surrounding landscape. Prior to being released, they give the plant a very shaggy look, making fireweed easily recognizable even during this stage of its existence.

Range & Habitat/Seasonal Availability

An inhabitant of roadsides, open woods, clearings, dry soils and fields; also of rich, wetter soils along streams. It invariably seems to spring up in areas that have been burned over, since the seeds need a high temperature in order to germinate. In fact, fireweed grows so quickly after a fire that some persons claim to have witnessed the plantlings emerge before the smoke has fully cleared! Grows in the southern four-fifths of Canada and throughout much of the northern third of the USA—dipping deep into the southern states in the west and into mountainous areas of Georgia in the east. Flourishes in the upper three-fourths of our area, except for the extreme western part of Minnesota.

Chief or Important Constituents

Tannins, egallitanin(oenothein B), beta-sitosterol, flavonoids(incl. kaempferol, quercetrin, quercetin, and myricetin-3-O-beta-d-glucuronide), chlorogenic acid, ursolic acid, palmitic acid, and nutrients(including sizable amounts of vitamins A and C).

Food

Very young leaves and flower buds can be eaten raw in small amounts. Buds and leaves of any age can also be boiled and eaten, although one or more waters may need to be thrown off. Western Eskimo tribes would even add the leaves to stews.*(Oswalt 1957:22)* Otherwise, the leaves—either fresh or fully dried—can be made into a tea that is very distinctive in taste and so delectable that I often curse myself because I never seem to gather enough to carry me through the winter! Most agree that the tea is best if the leaves are gathered before the plant flowers.

The pith inside the stalks can be eaten raw. In fact, peeling the stalks (a knife— or even fingernails—will do the job) and partaking of the surprisingly sweet pith can be quite a pleasant and relaxing activity. In the opinion of many, however, boiling the young stems provides the tastiest treat. Older stems can be boiled and eaten, too, but they should first be stripped and cut into chunks. As with the leaves, above, however, one may wind up having to throw off one or two waters.

Health/Medicine

A tea made from the leaves is reported to be *tonic* to the human system. Fireweed is also a powerful and revered *antispasmodic* (long used in folk medicine for ***WHOOPING COUGH, HICCOUGH, and ASTHMA***-Grieve 1971:2:848), ***astringent*** (due to its tannin content), and ***anti-inflammatory*** (due to several factors, including the plant acids and the flavonoids, especially 3-O-beta-d-glucuronide, which is well evidenced as to its antiphlogistic ability-*Hiermann et al 1991:357-60*). The latter two physiological functions make it immensely practical as a wash for ***MOUTH SORES*** and as a gargle for ***SORE THROAT***—uses heartily implemented by the Snohomish and other Amerindian tribes.*(Gunther 1973:41; Moerman 1986:1:163)*

The abovementioned physiological functions also enable fireweed to serve ably as an aid in ***GASTRITIS*** (an application appreciated by the Kayenta Navajo Indians-*Wyman&Harris1951:32*), certain ***STOMACHACHES, PILES*** (used internally and especially topically after steeping the flowers and leaves in olive oil), ***BOWEL HEMOR-RHAGE*** (a use implemented by the Cheyenne Indians-*Grinnell 1972:181*), ***ULCERATIVE COLITIS***(*Moore1993:138*),***GASTROENTERITIS***(esp. in children-*Fearn1923:583*) and ***CHRON-IC DIARRHEA*** (especially when pasty and green or yellow in color-*Moore 1993:138*).(The antidiarrheal use was a favorite of the Eclectic physicians, who recommended it with "a dry, red tongue, with...the abdomen...contracted and the evacuations...very painful.... There is enfeeblement and many disturbing colicky pains; discharges [that] are frequent and feculent, not watery."*(Bloyer 1899:276; Felter & Lloyd)* In addition, herbalist Michael Moore recommends the tea as a wash for infants who have ***INFLAMMATION*** in the region of the *ANAL OPENING.(Moore 1993:138)* A pediatric utilization was also implemented by the Blackfoot Indians, who administered an enema of an infusion of the inner cortex and root of the plant to babies who had *DIFFICULTY ELIMINATING.(Hellson & Gadd 1974:66)*

Widely appreciated as *vulnerary*, a poultice made from the grated root has been found to be useful in treating *ULCERATED SORES* (a use implemented by the Cree and Thompson Indians-*Leighton1985:38;Turner et al 1990:235*) and *BURNS* (owing at least partly to the tannin/egallitannin content). (The latter use is especially of interest because, as Canadian herbalist Terry Willard notes, this plant also heals the burns *of the earth*, i.e., it grows in burned-over ground-*Willard 1992a:146*) Our own area's Ojibwe Indians poulticed the leaves on *BRUISES,* as well as on *SLIVERS* to draw them out,*(Densmore1974:352-53)*, while their relatives of the Flambeau band poulticed the pounded root to draw out inflammation from a boil or carbuncle.*(Smith 1932:376)* The Bella Coola Indians also used the roasted/smashed root as a poultice for *BOILS.(Smith1929:60)* The Iroquois Indians used fireweed as a remedy for *INTERNAL INJURIES* from lifting and as a poultice for *SWOLLEN KNEES.(Herrick 1977:389-90)*

The plant is thought to have an affinity to *the genito-urinary system*, having long demonstrated its value for *UTERINE HEMORRHAGE, MENORRHAGIA, LEUKOR-RHEA*, and *INFLAMED URINARY TRACT* (as manifested by scalding urination) to various cultures. (The latter was a much-used application on the part of the Iroquois Indians.-*Herrick 1977:389-90; Herrick 1995:174*) *Epilobium* especially has a cherished reputation in folk medicine for *PROSTATIC SWELLING*. Recent scientific research*(Ducrey et al 1997:111-14)* has confirmed that oenothein B, a chemical occurring in significant amounts in this species,*(Ducrey et al 1997:111-14; Levisse et al 1996:490-92)* inhibits the enzymes 5-alpha-reductase and aromatase, thought by many to be responsible for prostatic swelling (the former enzyme by metabolizing testosterone into dihydrotestosterone and the latter by converting testosterone into 17-beta-estradiol).

It has also been evinced that oenothein B is a powerful *antiviral*, significantly reducing replication of *HIV(Okuda et al 1989:117-22)* and strongly inhibiting *HERPES SIMPLEX (HSV-1).(Fukuchi et al 1989:285-98)* The Skokomish Indians of Washington state even used fireweed to combat *TUBERCULOSIS,(Gunther 1973:41)* and a recent Finnish study found it to indeed possess powerful *antibacterial* properties, attributed to its polyphenols.*(Rauha 2000:3-12)* Earlier research found that both a tincture and an aqueous extract of the whole plant showed activity against several microorganisms, including *Staphylococcus aureus, S. albus*, and *Candida albicans.(PDRHrbs)* The activity against *S. aureus* is of interest because the Bella Coola Indians used to roast fireweed root in ashes, mash it, and then apply it to *BOILS,(Smith 1929:60)* which are often caused by this very bacterium. And regarding the reported activity against *Candida albicans*, it is of interest here that several, discerning American herbalists that I am aware of have long used fireweed for *CHRONIC YEAST PROBLEMS*.

This botanical is usually used as a tea because of its *very* pleasant taste. The Eclectic physicians recommended that one ounce of leaves be infused in one pint of water, with a dosage set at 2 to 4 fluid ounces, 5 or 6 times a day.*(Felter & Lloyd)*

FLEABANE

(Erigeron spp.)

[See photos #65-66 in rear of book.]

A weedy member of the *Compositae (Asteraceae)* family of plants that is often confused with aster (*Aster* spp.) *(see under that entry)*, but blooms in early- to mid-summer as opposed to late summer when aster spreads its petals. The *Erigeron* genus is represented in Minnesota by about a half-dozen species, the chief of which are as follows:

E. annuus (annual fleabane, daisy fleabane) is a native annual, growing to 4 ft tall, and possessed of a stem with spreading hairs. Its alternate leaves are wide, ovate to lanceolate, coarsely toothed, and grow to about five inches long. Their bases clasp the stalk. The composite flowers have a yellow center with many (70-120), thin, tightly clustered ray flowers that are white or lavender in color.

E. canadensis [*Conyza canadensis*] (horseweed) is another native annual, but can reach a height of five feet (2-3 ft. being average). Its stem is bristly and unbranched in the lower half, but branches above. The alternate leaves are dark green, hairy, and minutely toothed. They grow 1-4 inches long and are short-stalked or stalkless. Horseweed's flowers—composed of yellow disks surrounded by inconspicuous, greenish-white ray flowers—are situated on narrow, pointed bracts and occur in panicled clusters on stalks arising from the leaf axils.

E. philadelphicus (common fleabane, Philadelphia fleabane, daisy fleabane) is a hairy perennial or biennial, growing ½ - 2½ feet tall. Its alternate, spatula-shaped leaves are toothed and grow to four inches long. They are stalkless and they conspicously clasp the stem. The flower heads have a yellow center disk surrounded by up to 150 thread-like rays that may be colored lavender, pink, or white.

E. strigosus (rough fleabane, daisy fleabane) is an annual or biennial plant that reaches a height of three feet and is possessed of a stem covered with hairs that lie flat. Its leaves are alternate, narrow, and toothless (or barely toothed). Its flower heads are the smallest of the *Erigeron* species listed here (only 1/2 inch across), with 40 or so white ray flowers surrounding a yellow center.

Range & Habitat

The fleabanes are widespread weeds of fields, meadows, roadsides, waste places, and disturbed ground. Found throughout our area in such locales.

Chief or Important Constituents

Essential oil(incl. sesquiterpenes and terpenes such as limonene, myrcene, beta-pinene, linalool, dipentene, terpineol), flavonoids(incl. apigenin), plant acids(vanillic, caffeic, succinic), gallic acid, tannic acid

Food

Miwok Indians ate the raw, pulverized leaves and tops for food. The flavor is onion-like.

Health/Medicine

The various species of fleabane are all *astringent* and *stypic* to a large extent.*(Millspaugh 1974:319)* In fact, they are among our area's very best plants in these regards! Both the American Indians and white herbalists have used one or another of these species accordingly for health problems where bleeding or other tissue laxness has become a concern. For example, the Thompson Indians toasted the leaves, mixed them with grease, and used them as a salve for *SORES, SWELLINGS,* and *WOUNDS*; they also used the fresh plant for same.*(Turner et al 1990:46)* Kayenta Navajo Indians applied a lotion containing horseweed (or simply the crushed leaves of such) to *PIMPLES*.*(Wyman & Harris 1951:47, 50)* Miwok Indians placed the chewed roots in tooth cavities to alleviate *TOOTHACHE*.*(Strike & Roeder 1994:58)* A decoction of the upper part of *E. canadensis* has traditionally been used for *SORE THROAT,(Krochmal & Krochmal 1973:93)* *INTERNAL HEMORRHAGE* (esp. *UTERINE HEMORRHAGE*),*(de Bairacli Levy1974:68-69)* *HEM-ORRHOIDS* (especially those which are congested, as opposed to being in-flamed),*(Moore1989:23)* *CHILDBIRTH HEMORRHAGE,(Johnston1987:56)* *BLEEDING UL-CERS,(Moore 1994a:10.15)* *BLOODY DYSENTERY,(Moore 1994a:10.9)* and *ULCERATIVE COLITIS*.*(Moore 1989:23)* Cree Indians valued a decoction for plain 'ol *DIARRHEA*.*(Holmes 1884:303; Youngken 1925:172)* The Houma Indians employed a species of fleabane for *LEUK-ORRHEA*, drinking a decoction of the root as hot as possible.*(Speck 1941:64)* Modern southwest-American herbalist Michael Moore finds the herb useful for *IRRITABLE BOWEL SYNDROME* when it has a well-established cholinergic phase and is charac-terized by episodes of diarrhea that are painful and aching.*(Moore 1989:23)*

The extracted oil from *E. canadensis* was official in the *U.S. Pharmacopoeia* throughout the latter half of the nineteenth century for hemorrhages. Its terpenes would seem to be the chief hemostatic agents.*(Stuart 1982:58)* The *U.S. Dispensatory* even noted the oil's ability to stymie *HEMOPTYSIS*.*(Wood et al 1926:1293).* The Eclectic physician Finley Ellingwood recommended erigeron oil for "menorrhagia with profuse flow of bright-red blood" and for "dysmenorrhea with blood clots, bloody lochia increased by movement, epistaxis, haemoptysis, hematuria, haematemesis...in all passive hemorrhages where there is no fever or constitutional irritation."*(Ellingwood1983:352)* Herbalist Ed Smith markets a compound of the oils of erigeron and cinnamon in tincture form, based upon an original formula developed by Ellingwood. It not only works well as a hemostat, but it also smells very nice! (However, see "Cautions," below, re: sensitization to the oil on the part of some persons.)

The various fleabanes are *diuretic*, too, even having been recognized in the USP as such (from 1820-1882 for *E. canadensis*, from 1831-1882 for *E. philadelphicus*, and from 1831-82 for *E. heterophyllum*). *E. philadelphicus* would appear to have the most pronounced ability in this regard. Early-American botanist Constantine Rafinesque informs us that it was once demonsrated to "have increased the daily evacuation of urine from 34 to 67 ounces."*(Rafinesque 1828:1:166)* Thus, Lakota Indians utilized the plant's ability in this regard to help adults who had **DIFFICULTY URINATING**.*(Munson1981:234)* Philadelphia fleabane's diuretic capabilities have also made the plant useful for **DROPSY, DYSURIA, STRANGURY,** and **URETHRITIS**.*(Millspaugh 1974:319; Jackson 1876:235)* "It has been used beneficially in diseases of the urinary organs and in dropsies," acknowledged nineteenth-century physician and botanist Laurence Johnson, an authority generally critical of botanicals then in use.*(Johnson 1888:175)* The herb's combination of astringent and diuretic actions has proven useful as well for **URINARY STONES***(Jackson 1876:235; Barton 1900)* and for certain kinds of **KIDNEY DISEASE**. *(Potterton 1983:78)* European herbalist Juliette de Bairacli Levy says she has found it helpful for **SCALDING URINE** and for **ENURESIS**.*(de Bairacli Levy 1974:69)*

The plant is also "a powerful emmenagogue,"*(Millspaugh 1974:308)* and at least *E. canadensis* was used by our own Ojibwe Indians for "female weakness."*(Densmore 1974:356-57)* Maidu and Houma Indians also utilized fleabane for **MENSTRUAL PROBLEMS**.*(Strike & Roeder 1994:58; Speck 1941:62)* Here the whole plant, rather than merely finding itself applicable for menorrhagia as does the oil (per above), achieves abalance between astringency and other properties and hence is capable of a more comprehensive effect on the female system. Thus, Thompson Indians discovered that the decoction quickly eliminated **CRAMPS** and **BACKACHE** associated with **MENSTRUATION**.*(Turner et al 1990:180)* The Cherokees found the plant helpful when there was **SUPPRESSED MENSTRUATION**.*(Hamel & Chiltoskey 1975:35)* Modern herbalists maintain that fleabane seems especially indicated when **AMMENORRHEA** occurs in conjunction with a late menstrual period.*(Heatherle 1998:88)* The *Erigeron* genus has historically been used to treat **VENEREAL DISEASE**, too.*(Lewis & Elvin-Lewis 1977:333)*

Early-American physician Benjamin Barton said that the combination of the herb's diuretic effects with a seeming *sudorific* effect explained the reports made to him by sufferers that fleabane was useful not only for gravel, but also for **GOUT**.*(Barton 1900)* The Cherokees also used it for this exceedingly painful condition.*(Hamel & Chiltoskey 1975:35)* Both horseweed *(Conyza)* and *E. strigosus* have demonstrated *anti-inflammatory* activity,*(Benoit et al 1976:165; Lenfield et al 1986:268-69)* shedding further light on the plant's reported effects with inflammatory conditions like gout. This also explains how the Kayenta Navajo Indians could use a hot poultice of *E. canadensis* on an **EARACHE***(Wyman & Harris 1951:47)* and perhaps why Kawaiisu Indians could implement a decoction of the root as a wash for **HEADACHES,***(Strike & Roeder 1994:58)* while the Cherokees could utilize a poultice of the fresh plant for same.*(Hamel & Chiltoskey 1975:35)* The Okanagan-Colville Indians found that they could get headache relief, however, simply by drinking an infusion of the leaves and blossoms.*(Turner et al 1980:83)* As for the sudorific property mentioned by Barton above, this has yielded use of the plant to treat **FEVERS**, even as the Pillager Ojibwe Indians discovered.*(Smith1932)* The Miwok Indians utilized both Philadelphia fleabane and another species, *E. foliosus*, to treat **FEVERS** with **CHILLS**.*(Strike & Roeder 1994:58)*

Writing about *E. canadensis* in the *Canadian Pharmaceutical Journal*, John R. Jackson observed: "An infusion of the powdered flowers is considered antispasmodic, and is used in cases of hysteria and affections of the nerves."*(Jackson 1876:235)* The Cherokee Indians even used *E. philadelphicus* for **EPILEPSY**.*(Hamel & Chiltoskey 1975:35)*

E. canadensis has shown experimental **hypoglycemic** activity.*(Lewis & Elvin-Lewis 1977:218)* This is interesting in that, way back in the 1870s, Eclectic physicians had found this species useful for **DIABETES**.*(Scudder 1870:119)*

A related European species, *E. linifolius,* has demonstrated **antifungal** activity.*(Nene & Kumar 1966:363)* It is undoubtedly not without significance that *E. canadensis* has long been used for **RINGWORM** in Africa.*(EthnobDB)*

Tommie Bass, an Appalachian herbalist of note, used to recommend—before his passing from the scene—a washing of **LICE**-infested hair with fleabane tea.*(Crellin & Philpott 1990:153)* It has classically been used for fleas as well, even as its common name suggests, and this by a burning of the plant to drive them away!*(Vogel 1970:305)*

Cautions

 * Certain sensitive persons may develop **DERMATITIS** from handling this plant—especially the leaves, which are saturated with an essential oil inclusive of sesquiterpenes.

 * Fleabane is CONTRAINDICATED during **PREGNANCY**, due to *emmenagogue* and other properties possessed by the plant.

FRAGRANT GIANT HYSSOP
[ANISE HYSSOP]
(Agastache foeniculum; A.scrophulariaefolia)

Description

[See photos #67& 68 in rear of book.]

A pretty native perennial, growing 2-5 feet tall, and possessed of stems that are squarish, smooth, and branching above. The entire plant is faintly anise scented (strongly so, when its leaves are crushed).

The leaves are opposite, shiny, egg-shaped, coarsely toothed, and tapered at their apex. They are downy on their undersides, with the top sides being hairless but rough. The leaves of *A. foeniculum* (blue giant hyssop, lavender hyssop) are small (1-3 inches long and 2 inches wide) and short-stalked, whereas those of *A. scrophulariaefolia* (purple giant hyssop) are larger (2-6 inches long) and have longer petioles. The latter species also has purple stems with whitish hairs.

The small, purplish-blue flowers appear in whorls on dense, terminal, spiked clusters that are 1-4 inches long. Each flower bears four stamens occurring in two protruding pairs, with the pairs crossing and one of them curving upward. Many, tiny, black seeds evolve from the flowers.

Range & Habitat

Thrives in dry, open areas: prairies, open upland woods, sandy and well-drained soil. Prefers sun, but will tolerate some shade. *A. foeniculum* occurs throughout our area in these environs, but *A. scrophulariaefolia* thrives only in our southern half.

Chief or Important Constituents

Essential oil consisting of methylchavicol, alpha-limonene, beta-carophyllene, germacrene B & D, eugenol, camphene, myrcene, geraniol, beta-pinene, linalool, pulegone, and other terpenes.

Food

The leaves of this anise-scented plant make both a nice trail nibble (the Indians called it *wahpe yatapi*, meaning "leaf that is chewed") and a pleasant tasting tea. The whole plant can be dried, powdered, and stored for use as a licorice-like flavorant and sweetener in baking, in the tradition of certain Amerindian tribes.*(Gilmore 1919:113)* The seeds of some species (e.g., *A. urticifolia*) have also been reported as being edible.*(Kirk 1975:80)*

Health/Medicine

To treat **BURNS**, our area's Ojibwe (Chippewa) Indians utilized a poultice of the leaves or stalk of this plant. It was felt that it would "prevent blister and take out the fire." *(Densmore 1974:352-53)* They also employed fragrant giant hyssop for chest pain and for the cough of an "internal cold with tendency to pneumonia." *(Densmore 1974:340-41)* The Cheyenne Indians likewise appreciated that a cooled infusion of this plant was useful for **CHEST PAINS**, including those from **EXCESSIVE COUGHING**. *(Grinnell 1905:42)* Likewise, the Woods Cree Indians found this herb to be invaluable in treating the **COUGHING UP OF BLOOD**. *(Leighton 1985:26)*

Interestingly, Traditional Chinese Medicine (TCM) uses the related species, *A. rugosa* (referred to as "huo xiang"), for "oppression in the chest" as well. *(Reid 1992:106)* They understand *Agastache* to be one of the best remedies to fight off what they call "damp summer heat." They therefore use it to offset **HEAT STROKE** and **SUMMER COLDS**. *(Reid 1992:106)* Likewise, halfway across the world, the Cheyenne had used the American species similarly, rubbing the powdered leaves on the body of a person engripped by a **HIGH FEVER**. *(Hart 1981:27)* When I myself get a summer cold, which is rarely, I always rely on this herb, chewing the leaves straight from the fields where the plants grow. (Their delicious, licorice-flavored taste makes this a most pleasant experience, although they are so dry that I always bring a bottle of mineral water with me to keep them from getting caught in my throat on the way down!) Amazingly, this treatment has never failed to vanquish the cold for me most speedily, sometimes even on the very day in which I have consumed the leaves!

The Cheyenne, it is of interest, also appreciated fragrant giant hyssop to offset a **DISPIRITED HEART**. *(Hart 1981:27)* I find this fascinating in that the related mint called lemon balm *(Melissa officinalis)*, having a chemical profile to its essential oil similar to our own herb (they both contain linalool, limonene, eugenol, etc.), is used by contemporary herbalists to offset melancholy. (I use the latter extensively and have found it to be wonderfully effective in most cases, especially for teenage girls.)

Other uses for *A. rugosa*, especially by Chinese practitioners, include topical application of the tea to **VAGINAL YEAST INFECTIONS** and internally for **NAUSEA** and **VOMITING**. *(Yanchi 1995:78; Belanger 1997:582-83)* Recent published studies have also confirmed an **ANTI-HIV ACTIVITY** in the roots of this species. One of these studies found the active agent to be rosmarinic acid, *(Kim et al 1999:520-23)* while another isolated two diterpenoid compounds as the crucial chemicals. *(Min et al 1999:75-77)*

Caution

As a member of the mint family and thus possessed of an essential oil inclusive of pulegone and other powerful chemicals, this plant should **NOT BE USED DURING PREGNANCY.**

GOLDENROD
(Solidago spp.)

[See photos #69-71 in rear of book.]

A native perennial, represented in our area by almost 20 different species. The *Solidago* genus is marked by alternate leaves and golden flowers (even as the common name implies—they often being situated in a "rod," too). Our area's more important species are as follows:

Solidago canadensis (Canada goldenrod) grows 2-4 ft tall and has a stem that is hairy on the upper, leafy part. Like some other goldenrods, it commonly has insect galls on its stem as well. Its sharply toothed leaves grow to six inches long and display three main veins. The small, yellow flowers are clustered terminally in long plumes, with the tip of the chief cluster nodding to one side.

S. flexicaulis (zigzag goldenrod) tops off at 1-3 ft and is characteriozed by a zigzag stem. Its alternate, ovate leaves are pointed and coarsely toothed. They grow 1-3 inches long and are attached to short leafstalks. This species is unusual in that its yellow flowers are arranged in small (1-2"), rounded, clusters in the axils of the upper leaves.

S. gigantea (late goldenrod) is 5-6 ft tall, with a smooth, shiny stem that is green or purplish-red. Like *S. canadensis*, this species also has an arching flower stalk and commonly insect galls on its stem as well.

S. hispida (hairy goldenrod) has, as its common name implies, hairy stems. Its elliptical leaves are blunt and taper toward their base. Here the flower heads are narrow, ascending, and colored orange-yellow.

S. missouriensis (Missouri goldenrod) grows 1-2 ft tall and is possessed of a smooth stem. Its stem leaves are narrow, toothed, and somewhat erect. The leaves on this species are markedly three-veined, as with *S. canadensis*, above.

S. nemoralis (gray goldenrod) reaches a height of only 1-2 ft. Its slender stem is densely hairy, as is the rest of the plant. This species has large basal leaves and small, narrow stem leaves. There is also a pair of tiny leaflets in the axils of the upper leaves. The golden flowers are arranged in an arching plume that nods to one side.

S. rigida (stiff goldenrod) grows 1½- 4 ft tall. It has short hairs that cover the entire herbage. This species has rough basal leaves that reach a length of 10 inches and round, clasping, stem leaves. There are numerous flower heads appearing in wide clusters (several inches across).

S. speciosa (showy goldenrod) reaches a height of 1-3 ft. Its stout stem is smooth on the lower end, but rough above. The leaves are elliptical and barely toothed. Both the basal leaves (growing up to 10 inches long) and the lower stem leaves are stalked, while the upper stem leaves lack stalks. The terminal flower cluster looks like a fat plume, although sometimes it is more pyramidical in shape.

Other species common to our area include *S. ptarmicoides* and *S. uliginosa*

Range & Habitat

Most of the species in this genus prefer dry, sunny ground such as can be found in fields, prairies, and meadows. Very common throughout our area in such locales.

Chief or Important Constituents

Phenolic glycosides(incl. leiocarposide), flavonoids(chiefly rutin, but also quercetin, quercetrin, astragalin, campherol), carotenoids, hydroxybenzoates, essential oil(incl. borneol), clerodane diterpenes(incl. solidagolactones and elongatolides), poly-acetylenes, pseudo-tannins(polyphenols), polysaccharides, saponins, organic acids, tannins.

Health/Medicine

Goldenrod is a medicinal plant with a long tradition of use in numerous healing communities. It has, as one example, been treasured by native Americans as a **vulnerary** Thus, the Constanoan Indians used a decoction of the leaves of *S. californica* as a wash for **SORES** and **BURNS** and they also sprinkled toasted and crumpled leaves on **WOUNDS**.*(Bocek 1984:255)* Our own region's Ojibwe (Chippewa) Indians implemented the pulverized and moistened root as a poultice for a **BOIL**, the flowers as a poultice for a **BURN**, and the stalk or root as a poultice for a **SPRAIN** plagued by severe swel-ling.*(Densmore 1974:348-49, 362-63)* Russian healers have traditionally applied powdered flowers to slow-healing wounds.*(Hutchens 1991:143)* Herbalist and naturalist Tom Brown Jr., who was mentored by an Apache-Indian medicine man, relates of how, when he was a youngster, his mentor had once eased a bevy of yellowjacket stings that he had acquired with a salve made from goldenrod tea and tallow, on top of which the medicine man had poulticed some chewed leaves. Brown says that this treatment took away his pain almost instantly.*(Brown 1985:123))*

The soothing effects achieved with these conditions have no doubt been due at least in part to the plant's tannins and pseudotannins, which explain as well why Cree Indians could effectively utilize a decoction of the root for **DIARRHEA**.*(Turner et al 1990:50)* Such **astringent** properties, acknowledged even in the pharmaceutical literature,*(Wood et al 1937:1588)* would also explain goldenrod's many applications to the oral cavity. Thus, a decoction of *S. virgaurea* has been used in England to help fasten **LOOSE TEETH**, while the flowers have been chewed to combat **PYORRHEA**. *(Lewis & Elvin-Lewis 1977:261)* In the Ozarks, various species of *Solidago* have been placed on **TEETH** with **CARIES**.*(Lewis & Elvin-Lewis 1977:258)* Miwok Indians found that a decoction of the plant held in the mouth relieved **TOOTHACHES**. *(Strike & Roeder 1994:148)* The Cherokees swished goldenrod tea in a

SORE MOUTH.*(Hamel & Chiltoskey 1975:36)* Zuni Indians chewed the crushed blossoms and swallowed the juice to soothe a **SORE THROAT**.*(Stevenson 1993:26)* The Blackfeet relished goldenrod tea for **SORE THROAT, THROAT CONSTRICTIONS,** and **NASAL CON-GESTION**.*(Hellson & Gadd 1974:74)*

Astringency would also explain goldenrod's demonstrated assistance for **MINOR INTERNAL HEMORRHAGES,** but only partly, since the plant has also been shown to decrease the fragility of capillaries, no doubt owing to its flavonoids.*(Kresanek 1989:180)* At any rate, the Cherokee Indians were grateful for its aid when afflicted with **BOWEL HEMORRHAGE**.*(Hamel & Chiltoskey 1975:36)* Ojibwe Indians also employed a cooled decoction made from one root of *S. rigiduscula* and one quart of water to **CHECK HEMORRHAGE COMING FROM THE MOUTH** when a tribe member had been wounded.*(Densmore 1974:352-53; cf. 340-41)* In fact, one or another species of *Solidago* has been famed as a wound herb in numerous cultures throughout history, and a prominent European herbalist informs us that the ancient Saracens would even refuse to go into battle without this herb at their side!*(de Bairacli Levy 1974:72)* The famous horticulturist John Gerard perhaps summed it up best: "[Golden rod] is extolled above all other herbes for the stopping of bloud in wounds."*(Gerard 1994:109)*

Our herb's **vulnerary** uses might be partly elucidated by at least three studies which have demonstrated powerful **anti-inflammatory** properties for a variety of goldenrod species (e.g., *S. canadensis, S. rugosa, S. flexicaulis, S. virgaurea, S. gigantea*).*(Benoit et al 1976:165; el-Ghazaly et al 1992:333-36; Leuschner 1995:165-68; Okpanyi et al 1989:698-703)* Such a property has since been acknowledged by the scientific panel known as German Commission E.*(GermCommE)* This effect, which seems largely due to the plant's content of leiocarposide*(Metzner 1982:869)* and perhaps as well to its flavonoids, unboubtedly sheds light on why the Zuni Indians could imbibe an infusion for practically any sort of **BODY PAIN***(Stevenson 1993:26)* and why many herbalists have valued goldenrod for **CHRONIC ECZEMA,** as well as used it as an **anti-allergenic**.*(Stuart 1982:138)* While some may balk at this last use and claim that goldenrod, by way of its pollen, *produces*—rather than stymies—allergic reactions, many researchers challenge this, noting that goldenrod's pollen is not wind-borne so that it is most unlikely to be a nasal allergen as commonly believed. These researchers point out that ragweed blooms at the same time as does goldenrod and that the former is most likely to be the true culprit when goldenrod is implicated.*(Wunderlin & Lockey 1988:3064-65)*

A rat study has also found this herb to be **antispasmodic**.*(Leuschner 1995:165-68)* Perhaps this property explains why our Ojibwe Indians could soothe **STOMACH CRAMPS** by poulticing the heated root on the stomach*(Densmore 1974:344-45)* or why the Cherokee Indians could find relief from **FACIAL NEURALGIA** by holding the root tea in their mouths.*(Hamel & Chiltoskey 1975:36)*

A tincture of goldenrod is also **diuretic,***(Chodera et al 1985:199-204 Leuschner 1995:165-68; Kresanek 1989:180)* which effect has been confirmed by several lab studies*(Leuschner 1995:165-68; Chodera et al 1991:35-37)* and attributed again to the herb's leiocarposide content*(Duke 1997:208)* as well as to its array of flavonoids and saponins.*(Fetrow & Avila 1999:299)* The plant's historic uses for **RHEUMATISM** and **GOUT** are thus explained through a combination of these anti-inflammatory and diuretic properties.*(Pahlow 1993:171; Kresanek 1989:180)*

Goldenrod has also been used for **UTI INFLAMMATION** and **STONES**.*(Pahlow 1993:171; Kresanek 1989:180; GermCommE)* Maud Grieve's herbal even asserts that in 1788, a ten-year-old boy who had been drinking the infusion for months wound up passing large amounts of gravel, including fifteen large stones and fifty others a bit larger than a pea in size!*(Grieve 1971:1:361)* The Eclectic physician Finley Ellingwood listed the herb's specific indications for the urinary system as follows: "Difficult and scanty urination, where the urine is of dark color, and contains a heavy sediment. Where there is nephritis, either acute or chronic. It is useful where there is suppression of urine in infants, or retained urine, which causes general depression, with headache. Urinary obstructions, from any character."*(Ellingwood 1983:452)* Modern herbalists, like Oregon's Ed Smith, often find it helpful for **KIDNEY PAIN, ASSOCIATED BACKACHE,** and even **ULCERATION OF THE BLADDER.***(Smith 1999:35)* Herbalist Peter Holmes regards it as as genuine trophorestorative to the kidneys.*(Holmes 1997:1:190)* However, goldenrod should probably not be used for edema associated with kidney or heart failure since its unique diuretic effect causes mainly water, and not salts, to be excreted.*(Tilgner 1999:70)*

The genito-urinary system as a whole would seem to benefit from goldenrod's power. A decoction of *S. spathulata* was drunk by Thompson Indians for **SYPHILIS**.*(Turner et al 1990:48)* Goldenrod's saponins are active against *Candida* organisms, those tiny but common bodily inhabitants that can aggressively get out of control and produce vaginal yeast infections.*(Bader et al 1987:140; GermCommE)* It is also a classic **emmenagogue,** used in Bello-Russia for **AMENNORHEA**.*(Hutchens 1991:143)* The Cherokees likewise treasured it for "female obstructions."*(Hamel & Chiltoskey 1975:36)* Our own area's Ojibwe Indians implemented it to alleviate a **DIFFICULT LABOR**.*(Densmore 1974:358)*

Goldenrod is also a cherished digestive **tonic,***(de Bairacli Levy 1974:72)* having often demonstrated its benefits for conditions like **GASTRIC CATARRH** and **EXCESSIVE FERMENTATION**.*(Kresanek 1989:100)* Then, too, the Thompson Indians used a decoction of *S. spathulata* to stimulate the appetite.*(Turner et al 1990:44)* Physician and botanist Laurence Johnson noted that "golden-rod is gently stimulant, diaphoretic, and carminative. It has been used in domestic practice to produce diaphoresis, to relieve colic, and to promote menstruation."*(Johnson 1888:176)* The Houma Indians drank a decoction of *S. nemoralis* for **YELLOW JAUNDICE,***(Speck 1941:66)* while the Iroquois treated this same condition with an infusion of *S. juncea*.*(Herrick 1995:237)*

The **diaphoretic** property that Johnson mentioned, attested as well in the pharmaceutical literature for our herb,*(Wood et al 1926:1484),* is probably connected with the plant's essential oil. This property, in conjunction with the plant's saponins—which are possessed of an expectorant effect—probably explains why goldenrod is widely regarded as one of the first herbs to consider for **UPPER RESPIRATORY CATARRH,** whether of an *acute* or a *chronic* nature.*(Hoffmann 1986:197)* Diaphoresis no doubt also largely explains the plant's traditional use for *FEVER*, such as implemented by the Cherokee,*(Hamel&Chiltoskey1975:36)* Forest Potawatomi,*(Smith1933:53)* Delaware,*(Tantaquidgeon1972: 122-23)* Iroquois,*(Herrick1995:237)* and Ojibwe Indians.*(Densmore 1974:354-55)* From the standpoint of energetics, goldenrod's *cooling* and *drying* nature would squarely impact on these conditions.*(Holmes 1997:1:188-90)*

All in all, this common weed dotting Minnesota and Wisconsin fields is a true treasurehouse for suffering humanity!

Cautions

* Contraindicated in *PREGNANCY* because of the herb's *emmenagogue* property.

* As this plant can absorb high levels of nitrates from the soil, harvest only from soils known to be low in nitrates (and thus never from areas where fertilizers have been used.)

GUMWEED
[GUMPLANT; GRINDELIA; ROSINWEED]
(Grindelia squarrosa)

Description

[See photo #72 in rear of book]

The *Grindelia* genus is composed of some thirty species, but only one fluorishes in Minnesota, *G. squarrosa*. It is a low, branching perennial or biennial with many flowerheads. This species grows 1-2 feet tall.

Gumweed has stiff, alternate, toothed, oblong leaves that are faintly aromatic and grow 1-2 inches long. They are each covered with translucent glands that look like tiny dots. They upper leaves clasp the stem, while the lower ones disappear as the plant begins to flower.

Yellow, dandelionlike flowerheads (each 1-1½ inches wide) are situated on top of rough burs. And gumweed is appropriately named, since these burs are composed of five to six rows of *extremely gummy*, cuplike bracts, having pointed tips that curl downward and then recurve outward. Even the flowerheads are sticky to the touch—and before them, the buds!

Range and Habitat

Gumweed grows from British Columbia to our area and south to California and Texas. Originally native to the U.S. west of the Mississippi, it is now rapidly spreading eastward. In Minnesota, it has been recorded from all quarters but can currently be confirmed in only about half of the counties. Occurs less frequently in Wisconsin. Found in open, dry areas such as waste places, roadsides, fields, and prairies.

Chief or Important Constituents

Resin(20% of plant!) consisting of grindelic acid and other diterpene acids, saponin(incl. grindelin), a bitter alkaloid(grindeline), polyynes(incl. matricarianol and matricarianol acetate), essential oil(chiefly borneol), flavonoids, and tannin

Health/Medicine

Gumweed yields *antispasmodic* and *expectorant* properties and has thus been employed as a respiratory aid, mainly for spasmodic conditions like *ASTHMA*(Hoffmann 1986:199), *WHOOPING COUGH,*(Hoffmann 1986:199) *SLEEP APNEA,*(Moore 1994:4.1) and *BRONCHITIS*.(Smith 1999:37) The herb relaxes and opens bronchial tubes and helps expel phlegm (at least partly owing to its saponin content). In that its resinous nature also helps to moisten the respiratory tract, it is held by herbalists to be especially useful for *DRY, HACKING, UNPRODUCTIVE,* and *SPASMODIC COUGHS.* *(Moore1994:4.5; Smith 1999:37)*

Gumweed's uses for the respiratory tract were pioneered by the Shoshone, Paiute, Crow, Flathead, and other Indian tribes. *(Vogel 1970:313;Kindscher1992:120)* Indians of the Nevada region found the young and particularly gummy portions of the plant invaluable for treating *PNEUMONIA*. They would drink one-half cup of the decoction at a dose, and that while still quite hot. *(Train et al 1957:54)* The Ponca Indians even implemented it to treat *TUBERCULOSIS.* *(Gilmore 1991:81)* Interestingly, herbalists have usually insisted that only the top one-third of the plant be used when treating respiratory problems. Noted herbalist Michael Moore stesses that a tea is preferable to a tincture. *(Moore 1979:80)*

The antispasmodic properties have also lent the plant for use in *POSTPARTUM PAIN* (The Crow Indians used an infusion of the flowers here-*Shemluck 1982:333*), *TOD-DLER COLIC* (utilized by the Teton Indians-*Gilmore1991:81*), *INDIGESTION,* and *HEAD-ACHE*. The latter was an application especially valued by America's Eclectic physicians of the late nineteenth and early twentieth centuries. Finley Ellingwood, as a representative of such, noted that gumweed was the herb of choice when there was "headache accompanied with dizziness, and some nausea" and also when the headache was "persistent, day after day, and there is dullness, drowsiness,...dizziness,...lassitude, and the patient tires easily. A dull headache is present when he awakes in the morning, and with some exacerbations continues all day." *(Ellingwood 1983:320)*

Verified *anti-inflammatory* properties (via the standard, rat-paw-edema laboratory test-*Benoit 1976:164)* may help explain the Indians' application of this plant as a *vulnerary* and *dermatological aid*, for which they poulticed it on *WOUNDS, CUTS, SORES, BROKEN BONES*, and *BURNS*. *(Shemluck1982:333;Strike&Roeder1994:67)* Other dermatological uses ala the anti-inflammatory effects can be coaxed out of this plant as well, so that it has been found useful as a topical application for itchy skin conditions like *IMPETIGO, VAGINITIS,* and *ECZEMA* (especially the chronic, vesicular kind). *(Wood et al 1926:537; Hutchens 1991:147; Pedersen 1998:136)* Its most famous implementation, however, has been as a remedy for *POISON IVY RASH,* which use was established by various Indian tribes. *(Vogel 1970:313;Shemluck 1982:334;Strike&Roeder1994:67)* This application was brought to the attention of the medical profession in 1863 by Dr. C. A. Canfield of California, who was greatly enamored of the Indian treatment. *(Vogel 1970:313)* Later, it was incorporated into the official American pharmacological textbooks; thus, the *U.S. Dispensatory* of 1926 (21st ed.) noted: "It has considerable use...as a local application in the treatment of *rhus poisoning*." *(Wood et al 1926:537-38)* Though this application was later abandoned by the medical profession (which, however, has yet to come up with anything comparable!), it is still utilized by some of today's herbalists. Various forms have been used here: (1) the flowers have been crushed or pounded and then poulticed onto the rash; (2) a wash has been made from a tincture mixed with a little water; and (3) a wash has been made from a 15-minute decoction (which is said to be as effective as pharmacological preparations, i.e., fluid extracts, of the plant). *(Moore 1979:80;Weiner 1980:109)*

The anti-inflammatory properties may also partially elucidate gumweed's traditional aid for **GENITO-URINARY AFFLICTIONS**, widely appreciated by the American Indians.*(Train et al 1957:55; Moerman 1986:1:208-09)* Herbalist Peter Holmes finds it called for in the TCM syndrome known as "kidney Qi stagnation," manifested by cystitis, malaise, fatigue, and skin rashes.*(Holmes 1997:1:511-12)* Moore notes that the tincture is preferable to the tea for the urinary afflictions, suggesting one-quarter teaspoon in water every four hours for an adult.*(Moore 1979:80)* Various Indian tribes—including the Navajo, Shoshone, and Montana—even implemented gumweed to treat **VENEREAL DISEASES.***(Shemluck 1982:333; Moerman 1986:1:208-09; Mayes & Lacy 1989:48)* Here a decoction of the plant was imbibed, varying in dose from small amounts to one-half cup, or even two cups, a day by some tribes; it was usually felt that a length of time was required for this regimen to prove successful.*(Train et al 1957:54)* (See *Cautions*, below, regarding extended use of the tea.)

The Karok Indians used a decoction of *Grindelia robusta*, the western species, to kill **HAIR LICE**.*(Strike&Roeder1994:67)* Recent investigations have verified that highly viscous substances (such as mayonnaise, an old folk treatment) can smother hair lice, so it is conceivable why treatment with the highly resinous gumweed might have worked in this regard. And there may still be other factors involved, e.g., the plant's essential oil.

The plant has long been a cherished folk remedy for two types of *CANCER*—of the **STOMACH** and of the **SPLEEN**. In fact, the plant is held to be a spleen tonic *par excellence*.*(Ellingwood 1983:320)* and is specific for a congested, swollen spleen.*(Smith 1999:37)*

Finally, the herb is understood to be a ***heart relaxant***—capable of slowing a rapid heartbeat and moderating blood pressure. Holmes points out that it is most effective for what Traditional Chinese Medicine (TCM) calls "heart Qi constraint," manifested by a nervous, rapid heartbeat.*(Holmes 1989:1:202)* Moore notes that it is not always dependable in this regard, however.*(Moore 1979:80)* Holmes and other herbalists point out that gumweed can also sometimes act as a ***circulatory stimulant*** to help relieve a situation such as acute pulmonary **EDEMA**.*(Holmes 1997:1:511-12)*

The dried leaf and flowering tops of this herb were official medicine in the *U.S. Pharmacopoeia* from 1882-1926 and appeared in the *National Formulary* from 1926 all the way up to 1960. Gumweed is still official in some European pharmacopoeias. Because of its tough, resinous nature, a decoction has been preferred by most natural healers over an infusion. Tinctures and fluid extracts have also been employed.

Cautions

Martindale's Extra Pharmacopoeia, 28th ed. of 1982, warns that *large internal doses* may **irritate the kidneys**. This is presumably because of the large amount of resin in the plant. Older literature also linked diarrhea and gastric irritation with large doses.*(PDRHerbs)* Then, too, in western soils, where gumweed prefers to grow, the plant can take up too much selenium from the soil, making extended consumption (at least of the tea) potentially toxic.

HAWK'S BEARD
(Crepis tectorum; C. runcinata)

[See photo #73 in rear of book.]

Description

This is a weedy annual or perennial, growing 1-3 ft. tall, and with a slender stem filled with a milky juice. It possesses both basal leaves and *very* narrow stem leaves.

The dandelionlike flowerheads are yellow and about 1/2 inch in diameter. They are arranged in spreading clusters. These eventually give way to tufted achenes—similar to those produced by dandelions when they go to seed. Two species occur in our area:

C. runcinata is a perennial growing 2-3 feet high and having 1-6 flowerheads per plant. Its basal leaves are spatulate to oblong, 2-6 inches long, and have smooth midribs. Its sporadic stem leaves are oblanceolate to elliptic.

C. tectorum, an annual, is possessed of 3-50(!) flowerheads. Its stalkless stem leaves are thinner than in the above species—oblanceolate to linear. The midribs on its lanceolate, 4-to-6-inch-long basal leaves are rough to the touch.

Range & Habitat

Throughout our region in open, disturbed soils.

Food

I could find no record of either of our two native species ever having been used as food. However, the peeled stems of *C. acuminata* were eaten as greens by the Karok Indians*(Schenck & Gifford 1952:389)* and the leaves of *C. glauca* were consumed by Goisute Indians.*(Chamberlin 1911:367)*

Health/Medicine

Meskwaki (Fox) Indians viewed *C. runcinata* as a powerful topical medicine, strong enough to open not only a **CARBUNCLE** but even a **CANCEROUS GROWTH** so that it could be easily excised.*(Smith 1928:213)* The application here was simply as a poultice of the entire plant.

Homesickness and lonesomeness were viewed as amenable to an infusion of the young plants of this species by the Keres Indians of the western U.S.*(Swank 1932:40)* Western tribes also used an indigenous species, *C. scopulorum [C. modocensis]*, as an eyewash, employing a decoction of the root.*(Train et al 1957:41)* **CAKED BREASTS** were treated by a poultice of the mashed plant of this species and of *C. acuminata*.*(Train et al 1957:41)* Even the latex of *C. scopulorum* was utilized by these Indians, who applied it to **BEE STINGS** and **INSECT BITES**, for which it seems to have greatly reduced the discomfort.*(Train et al 1957:41)*

HAWKWEED
(Hieracium spp.)

[See photos #74-75 in rear of book.]

A weedy, perennial member of the *Compositae (Asteraceae)* family, possessing ray flowers with squared-off tips and hairy herbage. The several important species in our area include the following:

H. aurantiacum (orange hawkweed, Devil's paintbrush) grows ½ -1½ feet tall and possesses a slender, very hairy stem that is conspicuous in that it is entirely leafless. There is, however, a cluster of long basal leaves which, like the stem, are also very hairy, aside from being entire and spatula-shaped. Inflorescence is in the form of a terminal, spreading cluster of 2-10 reddish-orange flowerheads, each about 3/4 inches across. There are 10-20 ray flowers, but no ostensible center disk flower. The ends of the rays are abruptly squared. The green bracts cupping the flowerheads are covered with black, gland-tipped hairs.

H. kalmii [*H. canadense*, misapplied] (common Canada hawkweed) reaches a height of 1-3 ft. and possesses a stem sprouting many stalkless, clasping leaves that can range from being elongate/elliptical to oval. They are usually coarsely toothed and grow to about an inch long. The basal leaves, 2-5 inches long, disappear by flowering time, which sees the opening of 4-10 yellow, dandelionlike flowerheads having 20-30 ray flowers.

Other species in our area include *H. pilosella* (mouse-eared hawkweed), with white bloom on the underside of its leaves, as well as *H. longipilum, H. scabrum*, and *H. scabriusculum*.

Range & Habitat

Thrives in dry, sunny places: Open woods and fields, roadsides, clearings, pastures, disturbed soils. Its range is the mid-to-northeastern part of Minnesota and into adjacent areas of Wisconsin.

Chief or Important Constituents

Hydroxycoumarin(umbelliferone), essential oil, tannins.

Health/Medicine

This plant has **diuretic** and **astringent** properties, even as noted in *the U.S. Dispensatory* for 1926.*(Wood et al 1926:1332).* As such, a western species, *H. fendleri*, was used by the Navajo Indians for **VENEREAL DISEASE, BLADDER STONES, PELVIC PAIN, ANURIA,** and **HEMATURIA**.*(Wyman&Harris1951)* The Rappahonnock Indians chewed our *H. scabrum* or prepared a tea from it to treat **DIARRHEA**.*(Speck 1942:27)* The Cherokees likewise valued a species of hawkweed in their area, *H. venosum* (veiny hawkweed), for **BOWEL TROUBLE**.*(Hamel & Chiltoskey 1975:37)*

These Amerindian uses paralleled the usage of the European species of this plant, *H. murorum*, by famed 17th-century herbalist Nicholas Culpeper. He had written of that species: "The juice in wine helps digestion, dispels wind, hinders crudities abiding in the stomach, and helps the difficulty in making water....The decoction of the herb and Wild Succory with wine cools heat, purges the stomach, increases blood, and helps diseases of the reins and bladder."*(Culpeper 1983:91)* Culpeper had also found hawkweed invaluable for **RESPIRATORY COMPLAINTS**: "The decoction of the herb in honey digests phlegm, and with Hyssop helps the cough." *(Culpeper 1983:91)* Culpeper's predecessor in British herbalism, John Gerard, claimed that the hawkweeds were "good for the eye-sight, if the juice of them be dropped into the eyes."*(Gerard 1975:300)* This idea has been perpetuated in herbal practice ever since.

H. venosum (veiny hawkweed), referred to above, has been renowned as a remedy for the **BITES** of **RATTLESNAKES**. (In fact, one common name for this species is "rattlessnake weed.") So convincing was the evidence accumulated by the end of the first quarter of the nineteenth century that a Pennsylvanian physician, Richard Harlan, felt compelled to weigh that evidence in experimental fashion: He did this by subjecting animals to serpentine attacks and then watching to see if his form of administration of the remedy would take effect! Such a crude (and cruel!) methodology did not, however, prove out the treatment—which is understandable given the plentitude of variables involved!*(Vogel 1970:223)*

Hawkweed was also implemented by the Cherokee Indians as one of several plants in a formula to treat what they called "deer disease."*(Mooney1890:44-50)* This is left uninterpreted in ethnobotanical texts, but one wonders whether this might have been what we now know as the deer-tick-transmitted lyme disease? It is of interest here that hawkweed has pronounced **antimicrobial** activity (although it has not been tested specifically for spirochetes): A mid-1950s screening of plants for antibacterial activity found the fresh flowers of *H. praeaetum* to strongly inhibit *Mycobacterium tuberculosis*, while its leaves and those of *H. venosum* mildly inhibited it.*(Fitzpatrick 1954:530)* (Of interest here is that the Iroquois had earlier used hawkweed against TB!-*Herrick1977:480*) Then, too, of possible significance for the "floating arthritis" that lyme disease can produce, a mid-1970s study found *H. pratense and H. florentinum* to produce an **anti-inflammatory** inhibition of 24% and 19%, respectively, in the carageenan-produced, rat-paw edema test.*(Benoit 1976:164)*

148

HORSEMINT
[DOTTED MINT]
(Monarda punctata)

Description

[See photo #76 in rear of book.]

A gorgeous native perennial, growing to two feet tall, and possessing a square stem. The latter features suggests a membership in the mint family, which it indeed has.

The paired leaves are lanceolate, 2-4 inches long, and with dotted glands on their undersides. The upper leaves are whitish-green, often approaching lavender.

Flowers emerge in the upper leaf axils, in dense whorls. They are yellow with purple fleckings and are surrounded by whitish to pale-purple, leaf-like bracts.

Range & Habitat

Flourishes in the eastern two-thirds of the USA—in sunny, alkaline, sandy soil of dry fields. In Minnesota, it is found in Anoka county down through the southeastern counties and over into most of Wisconsin.

Chief or Important Constituents

Essential oil known as monarda oil, composed of thymol(60 percent!), cymene, carvacrol, thymohydroquinone, d-limonene, camphene, heptanal, pulegone, and a trace of formaldehyde. This plant is the *highest known source* of both thymol and carvacrol! *(Duke 1994:12)*

Food

Dried horsemint leaves may be made into a tea, but the herb is so pungent and odoriferous that a tea made solely from it proves too overpowering for most palates! Then, too, due to the irritating nature of the monarda oil, my thought is that horsemint is best served in a blend with a demulcent herb. I also like mixing in some other mints.

Health/Medicine

Like most other mints, horsemint proves useful for expelling gas from the colon (i.e., it is a *carminative*) and it was recognized in the *U.S. Dispensatory* (e.g., 1926 and 1937 eds.) for such activity. These editions also highlighted its*stimulant* quality, which may explain why Carolina moms have tended to use it as an aid to*POSTPARTUM RECOVERY.(Morton 1974:97)* Indeed, too, the Mesquaki (Fox) Indians even used horsemint as a *nasal stimulant* to rally a fellow Indian who was slipping into death! *(Smith 1928:225, 226)* Some modern herbalists also use a stimulating cup of horsemint tea to dispel*NASAL CONGESTION,* even as have the Mesquaki and Nanticoke Indians.*(Moerman 1986:1:298)*

Horsemint, like some other mints, has a valuable *anti-emetic* effect, which was appreciated by the Eclectics.*(Felter & Lloyd;Ellingwood 1983:281)* Even earlier, Dr. Jacob Bigelow, a prominent physician practicing in the very early 1800s, was most enamored of this property, among several of the plant's other effects.*(Vogel 1970:338)*

Horsemint is the best-known source of thymol, a very powerful *fungistatic* and *bacteriostatic* chemical that is the major component of its essential oil historically known as "monarda oil." Toxicologist David Spoerke notes that because 50% of this oil is excreted in urine, it might serve as a urinary *antiseptic*, although he feels that the proteins would limit the bacteriostatic potential.*(Spoerke 1990:91)* Clinical experience of herbalists suggests otherwise; likewise, the Eclectics had found it to be highly useful in "urinary disorders."*(Felter & Lloyd)*

Owing to the thymol content, too, the essential oil was formerly used as an *anthelmintic*, especially for *HOOKWORM;* but the dosage required was so high as to be dangerous.*(Weiner 1980:23)* (However, such use of the oil may have been unnecessary, as simple infusions of the plant were popularly used as anthelmintics by physicians in colonial America-*Christianson1987:73*) Monarda oil was also used, in times past, as a *rubefacient* for *RHEUMATIC* joints*(Vogel 1970:338; Bolyard 1981:88), NEURALGIA,(Felter & Lloyd)* and—what is especially of interest—*DEAFNESS (Bolyard 1981:88; Vogel 1970:338)* But as it tended to be quite severe in its effects if left on too long—often producing blistering!*(Felter & Lloyd)*—it has been largely abandoned.

Grieve observes that horsemint has been widely utilized as both an *emmenagogue* and a *diaphoretic.(Grieve 1971:546)* As for the effect upon menstrual flow, the Eclectic writers Felter and Lloyd had noted: "The warm infusion...has acquired some reputation as an emmenagogue."*(Felter & Lloyd)* The recent *PDR for Herbal Medicines* likewise lists "dysmennorrhea" as an indication for the use of this odoriferous plant.*(PDRHerbs)* Horsemint's sweat-inducing ability has been noted by numerous sources and strongly contributes to its renowned *antipyretic* property, which was gratefully implemented by Indian tribes as diverse as the Delaware,*(Tantaquidgeon 1972:73, 130)* Mohegan,*(Tantaquidgeon 1942:54)* Creek,*(Taylor 1940:54)* and Navaho-Ramah*(Vestal 1952:42)* whenever fever broke out. The antipyretic properties are what originally impressed a Dr. E. A. Atlee, who brought this herb to the attention of the medical profession in 1819, demonstrating it to be useful for various low forms of *FEVER* as well as topically in a liniment for *NEURALGIA* and *RHEUMATISM,(Crellin & Philpott 1989:252)*

In a fascinating article published in 1994, Dr. Jim Duke pointed out that horsemint contains four ingredients (i.e., thymol, carvacrol, pulegone, and limonene) demonstrated by Austrian scientists to prevent the breakdown of acetylcholine in the brain—usually thought to be a factor in Alzheimer's Disease. Since at least some of these compounds reportedly pass through the blood-brain barrier, Duke asks whether a shampoo made with horsebalm might prove useful for *ALZHEIMER'S* patients.*(Duke 1994:12-13)*

Cautions Because of the high *thymol* and *carvacrol* content, and the classic use as an *emmenagogue*, this herb should *be EXCLUDED FROM USE DURING PREGNANCY*

JERUSALEM ARTICHOKE
[SUNCHOKE]
(Helianthus tuberosus)

Description

[See photo #77 in rear of book.]

A lovely perennial plant growing to 12 feet (7 to 9 being average). The stem is hairy and branches considerably toward the top.

The long-stalked leaves are oblong to oval in shape, sharply pointed, and coarsely toothed. They grow 6-10 inches long and 2-4 inches wide. They are usually trident-veined (i.e., with two branching veins from the midrib forming what looks like a trident), starting at the base of the leaf. As with the stem, stiff hairs envelop the leaves, so that a rough, bristly feeling is experienced when running one's hand along them. Interestingly, too, while the upper leaves are alternate, the lower leaves are often opposite.

The plant's gorgeous, yellow-orange blossoms, appearing at the tip of the main stem and its branches (up to 18 blossoms per plant!), open in September and are 2-3 inches across. They are composite—each consisting of a flat, yellowish, central disk and numerous (10-20) ray flowers.

The rhizome has numerous, large tubers, each possessing opposite buds.

This plant is a sunflower (*Helianthus* genus) and thus resembles other sunflowers to a large degree. This being as it is, it is often frustrating for the amateur to try to distinguish it from other sunflowers. Here, however, the following pointers can be borne in mind: Jerusalem artichoke has (1) blossoms that open later in the year than do those of most other sunflowers; (2) more branches—and consequent flower heads—than most other species; (3) longer leaves than most other species; and (4) winged leafstalks.

Range & Habitat

A lover of wet soil and disturbed ground, Jerusalem artichoke grows pretty much throughout our area.

Seasonal Availability

Tubers are available in October (after the blossoms have fallen) as well as in the winter (in still unfrozen ground) and spring (before the shoots start popping up again).

Chief or Important Constituents

Carbohydrate(inulin). Nutrients include high amounts of iron.

Food

Of course, the famous tubers are edible. Locate them by scouting out patches in September when the plants are flowering and then returning to harvest them in late fall (after the first frost, when the flowers have disappeared), winter, or early spring before the shoots start sprouting again. (Those dug in the spring are sweetest because the sugar content is highest at this time.)

Peeled of their unpleasant tasting rind, many feel that the tubers taste like Chinese chestnuts. There's only one problem: Eating too much of them induces flatulence! (For persons with sensitive systems, even eating just a little gets them going like motorboats!) Cooking, however, greatly reduces the "gassy" nature of the tubers and renders the taste a lot like potatoes.

Sunchokes can be stored in the frig for up to two weeks, but then spoil rapidly, becoming mushy.

Health/Medicine

Jerusalem artichoke has **tonic, diuretic,** and **hypoglycemic** properties.*(Porcher 1849:798-99;Lewis & Elvin-Lewis 1977:218)* Not surprisingly, then, its fleshy tubers have proven useful in **DIABETES.** This is because, as we have seen above, they contain inulin, a carbohydrate that is not a starch and so does not raise blood sugar levels. (Being largely indigestible, however, inulin often—as noted above in the culinary section—produces flatulence, especially if the tubers are eaten uncooked.)

A tincture of Jerusalem artichoke's blossoms has been used for **BRONCHITIS,** and especially when there is chronic bronchial dilation.*(Angier 1978:150)*

Cautions

* Be absolutely sure of identification, since the tubers resemble those of several highly poisonous plants!

* Avoid eating raw Jerusalem artichokes if sensitive to inulin (e.g., if eating raw onions gives you gas).

JEWELWEED
[TOUCH-ME-NOT; SILVERWEED]
(Impatiens capensis [biflora]; I. pallida)

Description

[See photos #78-82 in rear of book.]

A succulent, annual herb that reaches a height of 2-5 feet. The branched stem is smooth, characteristically *translucent*, and *very* juicy.

The ovate leaves have rounded teeth and grow 1½ - 3½inches long. They are densely covered with small hairs. (In fact, if submerged in water, they glisten, from which fact the plant's alternate name of "silverweed" derives.) The uppermost leaves are alternate, while the lower leaves are sometimes opposite.

Snapdragonlike flowers appear on slender stalks arising from the leaf axils. These flowers are bilaterally symmetrical, with three petals emanating from a bell-shaped corolla ending in a spur. The flowers of *I. capensis* are orange with reddish-brown spots, while those of *I. pallida* are pale yellow. As a further point of distinction, the spur of *I. capensis* extends backwards while that of *I. pallida* extends downward.

Range & Habitat

Flourishes in moist ground (streamsides, perimeters of marshes and swamps, wet woods) throughout our area.

Chief or Important Constituents

Lawsone, tannin, minerals(selenium, calcium oxalate), and probably 2-methoxy-14-napthoquinone

Food

This is an edible and tasty plant if used when under six inches tall and properly prepared: To do this, separate the stems from the rest of the plant and cut them into string-bean-size pieces. Then boil these for 10-15 minutes in two changes of water. When done, shake all of the water off before consuming. (However, the water, while having extracted mildly toxic principles from the plant, can still be saved and frozen for skin applications—see below.) Many wild-foods authors stress that jewelweed is too powerful for the system if consumed alone and should therefore have its potency moderated by being mixed with other vegetables before being eaten. One can do this after cooking as outlined above or alternately, after only five minutes of simmering, the stem segments can be mixed into oriental stir-fry dishes.

When jewelweed seeds are ripe, they are edible and can be eaten as a trail nibble, tasting a bit like walnuts. However, they are not just placidly gathered. Rather, because they are rapidly discharged from the fruits when the plant is brushed, their efficient collection requires some manual dexterity and a large collection device (a shopping bag is ideal)! Yet, before the seeds can be eaten, their jettison coils must be removed.

Health/Medicine

The crushed, juicy plant has been an age-old topical treatment for a wide variety of skin ailments. The Delaware Indians applied it to *BURNS*.*(Speck 1915:320)* But, *RING-WORM, IMPETIGO, ATHLETE'S FOOT*, and *CONTACT DERMATITIS* have also been successfully combatted with topical use of this plant. A powerful antifungal chemical known as 2-methoxy-1, 4-napthoquinone has been confirmed in a related species, *I. balsamina*, and may in fact be present here, shedding light on the above applications. The Mesquaki (Fox) and Potawatomi Indians also rubbed the juice on *NETTLE STINGS* to bring relief.*(Smith 1928:205; Smith 1933:42)* Dr. James Duke reports that a chemical extracted from the plant called lawsone is the active agent in this nullification process, which he has confirmed via personal experimentation with the extracted chemical on himself and volunteers who subjected themselves to nettle stings.*(Duke 1997:262)* *RASH* and *ECZEMA* were also treated by the Omaha Indians with jewelweed.*(Gilmore 1991:49)*

The most celebrated topical use of this plant, however, has been for *POISON IVY RASH*. The Potawatomi Indians reverenced the use of this plant for this miserable affliction.*(Smith 1933:42)* So did the Cherokee Indians, who preferred rubbing the juice from seven flowers on such a rash.*(Hamel & Chiltoskey 1975:41)* Although two published studies*(Guin & Reynolds 1980:287-88; Long et al 1997:150-53)* have declared such treatment ineffective, others have shown efficacy: One 1957 study, following a physician's use for 115 patients, noted its effectiveness for an astounding 108 of these.*(Foster & Duke 1990:136)* But while there is conflicting evidence for jewelweed's effect on a full-blown poison-ivy dermatitis, evidence for positive *prophylactic* efficacy is not generally contested. According to Rutgers University chemist Robert Rosen, Ph.D., the active ingredient in jewelweed's prevention of poison ivy-induced rash is lawsone, which binds to the same receptor sites on the skin as poison ivy's active agent, urushiol, and beats the latter to those sites if applied quickly after initial contact with the resin. The net effect is that the urushiol is "locked out." Duke adds that the greatest concentration of lawsone is in the reddish protuberances at the base of jewelweed's stem.*(Duke 1997:359)*

As for other topical uses, the ethnobotanist Huron Smith tells us that he watched Potawatomi Indians boil down an infusion of the plant into a thick mass to apply it to *SPRAINS, BRUISES*, and *SORES*.*(Smith 1933:42,116)* Naturalist and herbalist Tom Brown Jr. finds jewelweed to be soothing to *INSECT STINGS, BLISTERS*, and *SUN-BURN*.*(Brown 1985:132)* Mixed with lard or some other base oil or paste, jewelweed juice has been used in a variety of cultures to soothe *HEMORRHOIDS.*(Grieve 1971:2:450)*

> *Caution:* Jewelweed's high content of minerals such as selenium and calcium oxalate—each of which can be toxic in excess—makes frequent consumption unwise.

JOE-PYE WEED
[GRAVEL ROOT; QUEEN OF THE MEADOW; PURPLE BONESET]
(Eupatorium purpureum; E. maculatum)

Description

[See photos #83-85 in rear of book.]

Joe-Pye weed is a tall (2-7 feet) plant with an erect, rigid, unbranched stem.

It is possessed of lanceolate leaves that are rough above and downy below. These are positioned in whorls of three to five at intervals along the stem.

The flowerhead is in the form of clustered blossoms resting at the tip of the stem. The blossoms are solely disk flowers and are pinkish-purple in color (thus the alternate common name of "purple boneset," in distinction from its white-flowered relative, boneset—*see under that entry*).

A fluffy seed-head replaces the flowers in autumn.

Several species exist in North America, but those in our area are *E. maculatum* (spotted Joe-Pye weed) and *E. purpureum* (sweet Joe-Pye weed). The former has a flat-topped inflorescence and a purple—or purple-spotted—stem. The latter has a dome-shaped inflorescence, purple spots on the stem at the leaf nodes only, and leaves that exude a strong, vanilla-like odor when crushed.

Range & Habitat

Joe-Pye weed thrives in marshes, ditches, shores, damp meadows, and open and wet woods. *E. maculatum* flourishes throughout our area, while *E. purpureum* has been mainly recorded east of the Mississippi and thus is sparse in western Minnesota.

Chief or Important Constituents

A yellow flavonoid(euparin), cistifolin, an oleoresin(eupurpurin), volatile oil, sesquiterpene lactones, and an unsaturated pyrrolizidine alkaloid(echinatine).

Health/Medicine

Joe-Pye weed has served most prominently as a reliable *diuretic*(Mockle 1955:90) and was listed as such in the 1820 to 1830 editions of the *U.S. Pharmacopoeia*. Dr. Gunn, in his famous household medical guide of the latter half of the nineteenth century, said here: "The root is the part used, and is regarded by botanic physicians as a valuable diuretic....highly esteemed in dropsical affections...and affections of the kidneys and urinary organs....[Implemented are the] bruised roots in water...given in doses of from half to a teacupful three or four times a day."*(Gunn 1859-61:846)*

Joe-Pye weed is more than just a diuretic, however—it is more broadly a **urinary** and hence it is renowned as being a specific for **hematuria***(Tierra 1982:220; Willard 1993:171)* as well as being a very powerful **lithotriptic** (hence its alternate common name of"gravel root"). As to the latter use, however, Canadian herbalist Terry Willard notes that, in his experience, it sometimes has eliminated kidney stones too quickly, causing a lot of pain. He therefore recommends its use for this purpose only when accompanied by a demulcent herb such as marshmallow root (*Althea* spp.) and further stresses that Joe-Pye weed should not be taken in large quantities.*(Willard 1994:6:9)*

Explicating further on its diuretic/urinary applications, turn-of-the-century botanical writer Charles Millspaugh wrote—in reflecting upon comments in the materia medica of Eclectic physician John King—that the root of Joe-Pye weed seems"to exert a special influence upon chronic renal and cystic trouble, especially when there is an excess of uric acid present."*(Millspaugh 1974:309)* Therefore, a decoction has classically been used by herbalists for **JOINT STIFFNESS** and **LUMBAGO** when known or suspected to be caused by uric-acid deposits.*(Hutchens 1991:255)* Such a tea was in like accord used by the Cherokee Indians for **GOUT** and **RHEUMATISM**,*(Hamel & Chiltoskey 1975:42)* while our own area's Ojibwe (Chippewa) Indians preferred a lukewarm decoction as a wash for those painful joints.*(Densmore 1974:348-49)* A recent study*(Habtemarian 1998:683-85)* identified the chemical cistifolin as being one active agent in preventing rheumatic inflammation. It was found that this chemical inhibited integrin-mediated cell adhesion.

Joe-Pye weed is also a **nervine**,*(Wren1972:140;Tierra1982:220)* and as such it is especially thought to be helpful for genitourinary conditions of an"irritable" nature. Said the Natura physician R. Swinburne Clymer: "It is of much value in irritable conditions of the bladder and in kidney conditions accompanied by aching in the small of the back. In irritated conditions of the female organs, it is most helpful."*(Clymer 1973:158)* Western-American herbalist Ed Smith elucidates that for irritability of the female organs, one should especially think of the **UTERUS,** including **ATONY** or **DISPLACEMENT** of such as well as **CHRONIC ENDOMETRIOSIS**.*(Smith 1999:37)* Not surprising in view of the latter is that Joe-Pye weed is a fair **emmenagogue,** especially applicable for **PAINFUL AMENNORHEA***(Ody 1993:57)* It is also often helpful for **PELVIC INFLAMMATORY DISEASE**.*(Mabey 1988:48)*

But it is hardly specific to women! British herbalist Mary Carse also finds the decoction valuable for **SPERMATORRHEA** in men, dosed at one teaspoon every three waking hours.*(Carse 1989:119)* Others would list it for **PROSTATE PROBLEMS** as well, especially for a boggy prostate.*(Smith 1999:37; Willard 1993:171; Christopher 1976:254)* One champion of the latter application was the Alabama herbalist Tommie Bass, who had cited several examples of his successful use in this regard, including one case of a client who had prostate trouble so bad that his doctor had urged an operation. This man proceeded to drink two or three cups of the tea a day for awhile and then submitted to his physician's exam, upon which he was thrilled to get a clean bill of health as well as his doctor's strong urging: "I don't know what you have been doing, but keep doing it." *(Crellin & Philpott 1989:274-75)*

Joe-Pye weed has also historically been used to help deal with *DIABETES*. Lab experiments have verified *hypoglycemic* activity for *E. purpureum*. *(Lewis & Elvin-Lewis 1977:217-18)* Bass, having used both this species and his indigenous *E. serotinum* for this condition, found that both species helped diabetes considerably, although he felt that the latter was the more effective of the two. *(Crellin & Philpott 1989:274)*

The leaves were utilized by various Amerindian tribes as a *vulnerary*. Thus, Potawatomi Indians implemented the fresh leaves as a poultice for *BURNS*. *(Smith 1933:52)*

A decoction is made with 20g of the root to 600ml water. *(Ody 1993:138)* Bass, in his appealing, "down-home" style, declared: "One quart of roots makes a gallon of medicine. The roots can be dried in the sun and stored in paper sacks. Boil one hour and strain it and keep it in the refrigerator or it will sour.... If it don't turn your stomach, it is simple and safe." *(Crellin & Philpott 1989:274)* In view of the fact that this plant contains toxic pyrrolizidine alkaloids, however, it should not be used long-term. (See further under "Cautions," below.)

Cautions

* Because this plant contains unsaturated pyrrolizidine alkaloids, which can be toxic to the liver in excess, it is *contraindicated for long-term use*. Even with short-term use, one needs to be careful not to exceed standard doses.

* Absolutely contraindicated in *PREGNANCY* and *LACTATION*. It should also not be given to young children.

KNOTWEED
[PROSTRATE KNOTWEED; KNOTGRASS; DOORWEED]
(Polygonum aviculare; P. arenastrum)

Description

[See photos #86-87 in rear of book.]

A sprawling annual, reaching anywhere from 4-24 inches and possessed of jointed stems marked by silvery or brownish sheathes at the juncture points.

Knotweed's leaves are alternate, hairless, oblong, and pointed (both at the tip and at the base). They grow ½ - 1½ inches long. Each leaf has a prominent center vein.

The flowers, opening only on sunny days, are tiny and yellow, pink, or green. These give way to small, three-angled, reddish-brown seeds.

Range & Habitat

A cosmopolitan urban weed, found in hard trampled areas such as at the edges of parking lots, lawns, sidewalks (and in between the slabs), etc. Also thrives in dooryards and waste places.

Chief or Important Constituents

Polygonic acid, oxymethylanthraquinone, flavonoids(rutin, avicularin[quercetin-3-arabinoside], quercetin, quercitrin, kaempferol, avicuroside, delphinidin, isovitexin, vitexin), hydoxycoumarins(scopoletin, umbelliferone), salicylic acid, plant acids(caffeic and chlorogenic), polyphenols (catechin), tannin, mucilage. Nutrients include vitamin C and silica.

Food

Eating raw knotweed can cause intestinal disturbances,*(Dobelis et al 1986:228)* but it is edible once it is cooked. It is popular as such in Asia and was also at one time in Europe.*(Couplan 1998:128)*

The seeds can be ground into flour.

Health/Medicine

Knotweed was an esteemed herb of the ancients: Back in the first century of our Common Era, the scholarly and skilled herbalist Dioscorides described this plant and its powers in detail, noting first that it"hath a binding, refrigerating faculty." He next went on to spotlight its usefulness for *EPISTAXIS (NOSEBLEED)* and as a pessary for"ye womanish flux," and finally commented that a poultice of the leaves was used when there was "casting up of blood."(*Herbal*, 4.4)   The learned naturalist Pliny the Elder, a contemporary of Dioscorides, similarly related that the juice when taken with wine was capable of allaying *EPISTAXIS* as well as *HEMORRHAGE* in any part of the body.(*HN*, 27.114-17)   Celsus, a Roman physician writing just a few years later, alluded as well to knotweed's powerful *styptic* properties(*De Medicina*, 5.1; 6.7.4) Then, too, the ancient physician Soranus treated *UTERINE HEMORRHAGE* with a knotgrass plaster or sometimes even with an actual injection of the juice.*(Jashemski 1999:80)*

The intervening centuries have verified that knotweed is indeed both a powerful ***hemostat*** and ***astringent.*** Thus, owing to its demonstrable anti-hemorrhagic effects, Choctaw Indians could imbibe a strong knotweed tea with the firm conviction that it would prevent miscarriage.*(Campbell1951:286)* Russian scientists have scientifically verified the herb's ability to stop uterine bleeding and it is part of the official medicine in their country, with a commercially prepared liquid extract available (but only by prescription!).*(Zevin 1997:93)*

The herb has proven helpful for both *DIARRHEA* and its most severe form, *DYSENTERY.**(Willard 1992:148; Lust 1974:245)* Understandable, too, in view of its astringency, is that the tea has also been recommended by the scientists composing German Commission E for *SORE THROAT* and for *LARYNGITIS.**(GermCommE)* The*PDR for Herbal Medicines* notes its traditional use as a hemostat, acknowledges its astringency, and further points out that it is an acetylcholinesterase inhibitor (this is the enzyme that breaks down acetylcholine in the brain and which seems to be responsible for Alzheimer's Disease.)*(PDRHrbs)*

INFLAMMATION of the *MOUTH* and *THROAT* is an indication for knotweed's use according to the*PDR for Herbal Medicines.* A 1995 study revealed that the plant does indeed have anti-inflammatory properties, inhibiting both the cyclooxygenase path to inflammation as well as platelet-activating-factor-induced exocytosis.*(Tunon et al 1995:61-76)* Earlier research had likewise shown *antiphlogistic* properties.*(Kresanek 1989:54; Duke 1995:389)* Interestingly, Dioscorides had long ago advocated it for *STOMACH INFLAMMATION(Herbal,4.4)*, while, centuries later, Ramah-Navaho Indians halfway across the globe in western America would use it similarly.*(Vestal 1952:23)*

Knotweed would seem, indeed, to be a good herb for the GI tract in general. Pliny had drawn attention to its common implementation for *DIGESTIVE AILMENTS.*(HN 27.114-17) Knotweed has also been used in folk medicine for *STOMACH CANCER,* as well as for malignancy of the breast and kidneys.*(Duke 1995:389)*

Pliny had held it to be effective for *COLDS,*(HN 27.114-17) while the present-day German Commission E outlines its application for *RESPIRATORY COMPLAINTS*, inclusive of *BRONCHITIS*.*(GermCommE)* The *PDR for Herbal Medicines* lists its indications as "cough" and "bronchitis" and notes its traditional usage as an **anticatarrhal**.*(PDRHrbs; cf. Kresanek 1989:54)* Knotweed's effectiveness with pulmonary complaints may at least partly be because of its content of silicic acid, which strengthens connective tissue.*(Chevallier 1996:251)*

Knotweed was used in ancient Pompei for **HEART TROUBLES.** In a fascinating continuity of tradition, it continues to be used in this fashion in the same area today.*(Jashemski 1999:80)* Other cultures in Europe have employed it to moderate **HYPERTENSION.** *(Duke 1995:389)*

A treasured **diuretic,***(Kresanek1989:54;Willard1992b:148)* knotweed is best known to herbal students as a **urinary** aid, especially as one capable of carrying off **GRAVEL.***(Kresanek 1989:148)* The Natura physician R. Swinburne Clymer had noted, indeed, that its"chief influence is on the bladder" and that it is regarded as a **lithotriptic** of last resort, often being combined with horsetail (*Equisetum* spp.) for best effect.*(Clymer 1973:91)* Dioscorides, noting that it "moves ye urine also manifestly," had also highlighted its benefits for **STRANGURY**(*Herbal,* 4.4)

Traditional Chinese Medicine(TCM) uses knotweed (under the name of"bian xu") as a medicinal plant as well. It is viewed as an herb that helps clear *dampness* and *heat.* The Chinese, like Western herbalists, thus employ knotweed as a urinary aid (esp. for **PAINFUL** or **DRIBBLING URINATION**), as a hemostat (for a **BLOODY** or **MUCOID VAGINAL DISCHARGE** as well as for **POSTPARTUM BLEEDING**), and as an astringent for **DIARRHEA** and **DYSENTERY**.*(Belanger 1997:588;Yanchi 1995:106)* As to the latter affliction, Chinese lab research has established that knotweed is powerfully effective against **BACILLARY DYSENTARY.** In one study, involving 108 people who were afflicted with the condition and concurrently treated with a knotweed paste that was ingested daily, 104 recovered within 5 days!*(Chevallier 1996:251)* The Chinese also herald knotweed for conditions such as intestinal worms, weeping eczema, fungal skin conditions, and certain kinds of jaundice.*(Belanger 1997:588; Yanchi 1995:106)*

Caution:

CONTACT DERMATITIS can develop in some persons who touch this plant.

LADY'S THUMB
[SPOTTED KNOTWEED; HEARTWEED; HEART'S-EASE; MILD ARSSMART; DEAD ARSSMART; PINKWEED]
(Polygonum persicaria)

Description

[See photos #88-89 in rear of book.]

This perennial relative of both knotweed (*see under that entry*) and smartweed (*see under that entry*) grows 6-30 inches tall and has a jointed stem that is reddish or pinkish hued.

The leaves of lady's thumb are lanceolate, 2-6 inches long, and marked by a *dark, three-angled blotch* in the middle of their blades. In that this is a *Polygonum* species, papery sheaths occur at the leaf nodes which in this case are fringed with short bristles.

The blossoms are pink to purple and occur in terminal, elongate clusters that measure one or two inches long.

Range & Habitat

Thrives throughout our area in gardens, waste places, marshes edges, damp clearings, cultivated ground, and roadsides.

Food

Boiled for five or ten minutes, the young leaves are quite tasty, being somewhat reminiscent of spinach. Many wild-foods connoisseurs also add them to soups, stews, casseroles, etc.

Health/Medicine

Lady's thumb is an esteemed *diuretic* and has a history of use by various native American and other cultures for *GRAVEL*(Foster & Duke 1990:160) and *SCANTY URINE*(Angier 1978:161) For these conditions, a teaspoonful is steeped in a cup of bubbling water and sipped cold during the day, to the tune of one or two cupfuls every 24 hours (Angier 1978:161)

The tea is also highly *astringent* and has thus been implemented to treat *DIAR-RHEA, ULCERS,* and *APHTHOUS SORE MOUTH.*(Rafinesque1830:66;Mockle1955:38) Its astringency probably largely explains as well its folk use for *ECZEMA*(Potterton 1983:18) and *POISON-IVY RASH*(Foster & Duke 1990:160) Such a property may further account for the long-time European and Ojibwe-Indian application for **STOMACH PAIN**.*(Mockle 1955:38; Densmore 1974:344; Foster & Duke 1990:160)* Lady's thumb has also traditionally been used for **JAUNDICE** and for various other **LIVER DISEASES.** *(Silverman 1990:75; Potterton 1983:18)*

For ages, the plant has been appreciated as a *vulnerary.*(Mockle 1955:38)* Here it has been poulticed on **PAINFUL AREAS,** providing pain relief as an *anodyne.*(Foster & Duke 1990:160).* A leaf-tea has a long history of use as a foot soak for**RHEUMATISM,***(Foster & Duke 1990:160; Potterton 1983:18)* including among the Iroquois Indians, who boiled one plant in a gallon of water and soaked the inflamed feet in the cooled bath, thereafter bandaging them, and then removing the bandage on the next morning. Such a process was repeated on for one more day.*(Herrick 1995:144)*

Lady's thumb has had a history of use in Canada as an *anthelmintic.*(Mockle 1955:38)* It has also been used in folk medicine by a variety of cultures for **CANCER**.*(Hartwell 1970:373)* Certain Indian tribes also found the plant helpful for**HEART TROUBLES,***(Herrick 1995:145; Foster & Duke 1990:160)* and hence its alternate common name of "heart's-ease." A mid-1950s study found it to be active against the tuberculosis organism*in vitro.*(Fitpatrick 1954:530)*

As for preparation and dosage, herbal authority John Lust says to steep one teaspoon of the herb in one cup of water and then drink one to two cups a day.*(Lust 1974:247)*

Cautions

* Raw lady's thumb possesses an acrid juice—though not as acrid as that of some of its smartweed cousins—which can irritate mucous membranes (eyes, lips, mouth, throat, etc.). A number of foragers have reported that the very young leaves are free from this caustic nature, but eating them raw is still not advised.

* One should definitely avoid rubbing one's eyes when harvesting—or otherwise handling—this plant!

LAMB'S QUARTERS
[FAT HEN; GOOSEFOOT; PIGWEED]
(Chenopodium album)

Description

[See photos #90-91 in rear of book.]

A weedy, annual or perennial plant, growing 1-4 feet tall, and producing ascending branches as it reaches maturity. The stems are often red-streaked.

Lambs' quarters possesses alternate leaves, 1-4 inches long, having prominent, greenish-white, center veins. These sprout from long leafstalks. The young leaves are broad and triangular to diamond-shaped—much like a goose's foot (hence the alternate common name of "goosefoot"). They are blunt-toothed to wavy-edged and possessed of a mealy-white look to their undersides. This mealy-whiteness also appears prominently on top of the new, uppermost leaves, so that they look like they have recently been dusted with a fine, white powder (the species name, *album*—meaning "white"—alludes to this). As the plant ages, the leaves—especially the upper ones—become more linear and their edges smooth out. The uppermost leaves often become sessile at this time as well.

Many minute, sessile flowers appear in small, dense clusters arranged as spiked panicles at the ends of the stem and branches and in the leaf axils. These are replaced in late summer by papery fruits that encompass many small, black seeds.

Range & Habitat

Cosmopolitan weed, found throughout most of North America. Lamb's quarters prefers disturbed, well-worked soils, e.g., illkept gardens and lawns, waste places, roadsides, grain fields, etc.

Chief or Important Constituents

Oxalic acid, oxalate, hydroxycoumarin(scopoletin), an alkaloid(chenopodine), ferulic acid, imperatorin, oleanolic acid, campesterol, stigmasterol, n-triacontanol, traces of ascaridole. The roots contain polypodine B, ecdysteroids, and beta-ecdysone. Nutrients include very high amounts of vitamins A(11,600 IU's per 100 grams), C(80 mgs per 100 grams), thiamine, riboflavin, and niacin. A large number of free-form amino acids are present as well.

Food

Lamb's quarters was a staple food of early man and was used quite extensively as a cultivated crop in Europe until just recently. This abundant weed is edible raw or cooked.

Many prefer lamb's quarters raw—especially when the leaves are young—as a trail nibble. The plant is quite a forager's dream: one can keep clipping off the upper few leaves and return mere days later for a new growth! (See "Cautions" below, however, before endeavoring to "pig out" on this pigweed.)

To prepare the plant as a potherb, it needs only to be boiled or steamed in salted water for 7-12 minutes (North Carolinans always throw off one or more water in doing so, but most feel that this is not necessary.) The leaves lose bulk during cooking and so must be used in a large quantity in order to get any substantial results, but the results do reward one's efforts: This is certainly one of the tastiest of the weedy potherbs described in the present work! (I rank it only behind the leaves of marsh marigold as a wild-potherb delight.) In addition, lamb's quarters has a good spectrum of vitamins and minerals. As if all of this is not enough, Eskimos have found that cooking the leaves with beans helps to offset the intestinal gas that can accrue when eating the latter! (Burritos, anyone?)

The plant's leaves and stems even keep well in the freezer. Simply blanch them for two minutes, then cool. Next, drain or dab dry. Finally, put them in freezer bags and store them in the freezer.

The seeds also provide food. They can be obtained by collecting the seed heads and then beating them against a rock to release their precious cargo. An incredible amount can be collected from each plant—at least 50,000 (one authority counted 70,000 on a single plant!). The seeds can then be boiled and made into mush, or simply roasted. They can also be ground (best done with a blender, as they slip between grinder blades; or boil till soft, then crush, dry, and grind in a grinder) and made into a buckwheat-like meal for pancakes, biscuits, or bread. The great Napoleon once even lived on bread made from these seeds when his supplies were low!

Health/Medicine

This plant evidently possesses powerful *anti-inflammatory* (probably partly due to the scopoletin content) and/or *astringent* properties, in that a large variety of American Indian tribes utilized the bruised leaves as a poultice—or a tea therefrom as a compress—for *BURNS* (incl. *SUNBURN*), *HEADACHE, WOUNDS, INFLAMED LIMBS, PILES, SORE THROAT*, and *EYE INFLAMMATION*. *(Wyman & Harris 1951:20; Herrick 1977:316; Leighton 1985:351; EthnobotDB; Asalkar 1990:196)* The Mesquaki Indians even called upon a root infusion of lamb's quarters to deal with *URETHRAL ITCHING*. *(Smith 1928:209)* This herb was also used by white settlers of California as a poultice for *ARTHRITIS* and *RHEUMATISM,* *(Westrich 1989:99)* while the juice obtained from the boiled greens was applied to soothe a *TOOTHACHE*. *(Westrich 1989:99)*. The Mendocino Indians of the same region reverenced the larger and older leaves as being very helpful for a **STOMACH-ACHE**,*(Chestnut 1901:346)* while Iroquois Indians out east implemented a cold infusion of this herb to deal with **DIARRHEA**. Their manner of implementation was to steep a small bundle of whole plant in a quart of water for five minutes, then allow the infusion to cool, and finally to drink freely therefrom.*(Herrick 1977:315)* •

On the other hand, other cultures (e.g., the Chinese, Sudanese, and Spanish-*EthnobotDB*) have coaxed *laxative* properties out of this plant (probably more so from the seeds). Lamb's quarters has also traditionally—and most successfully—been used for **PINWORMS**.*(EthnobotDB)* Like its famed cousin, wormseed (*C. ambroisides*), it contains ascaridole, a chemical that is deadly to parasites. Even though our species has this chemical in only trace amounts—unlike wormseed's huge quantity of it—its presence is nevertheless undoubtedly the chief factor in its time-honored **anthelmintic** reputation.

Ethnobotanist Huron Smith informs us that Potawatomi Indians noticed that lamb's quarters possessed **antiscorbutic** properties and therefore made a special effort to include it in their diet.*(Smith 1933:47)* In view of the incredibly high vitamin-C content since uncovered in this plant, their wisdom has been confirmed. Lamb's quarters also has a history of use in treating the unsightly and embarrassing skin condition known as **VITILIGO,** which is currently thought to be due to dysfunction of the adrenal glands. Here again, the plant's high content of vitamin C may be significant, in that the adrenal glands store and utilize more vitamin C than does any other part of the body!

Zulu tribes in Africa have put our herb's seeds to use in dealing with an **ENLARGED SPLEEN, BILIOUSNESS,** and **LIVER PROBLEMS**.*(Asalkar 1990:196)* The Chinese have likewise used it for troubles with the liver.*(EthnobotDB)*

Arabian peoples have a long tradition of using lamb's quarters for **TUBERCULOSIS**.*(EthnobotDB)* Interestingly, a 1954 study found it to definitely inhibit a virulent human strain of TB.*(Fitzpatrick 1954:530)*

Cautions

* **PREGNANT WOMEN** should not consume the seeds of lamb's quarters, which can be abortive.*(Asalkar 1990:196)*

* Do not collect in known, high-nitrate soils or where cultivated crops grow.

* Do not consume in quantity or regularly over a period of time, as this herb contains oxalic acid in sufficient quantity as to be harmful with extended use. Some serious poisoning once occured in Europe when lamb's quarters was consumed in high amounts owing to a food shortage following a war.*(Cooper & Johnson 1984:35)* It is safe as an *occasional* trail nibble (in moderation) and the safety margin increases when it is cooked (which tends to break down the oxalic acid), allowing for periodic use as a meal in this form. But, even cooked plants rife with oxalates have caused deaths (see under "Dock"), so care should always be exercised with this plant.

MARSH MARIGOLD
[AMERICAN COWSLIP]
(Caltha palustris)

Description

[See photo #92 in rear of book]

This interesting perennial plant is not a true marigold (contrary to what its name would suggest), but a buttercup instead. Growing to a height of 8–24 inches, it possesses a stem that is thick, stout, succulent, branching, and hollow—the latter feature enabling the leaves to ably float on the water (the favored habitat of this species).

The leaves themselves are glossy (almost fluorescent), heart-shaped to kidney-shaped, crispy green, wavy-edged, and with small, rounded teeth. Altogether, as they are situated on the plant, they present a saucer-like appearance. They occur clustered at the base and scattered alternately along the stems. Stem leaves are smaller (3 inches across) than basal leaves (5-6 inches across) and lack the long stalks characterizing the latter. A helpful identifier of this plant is the occurrence of a papery sheath where the leaf stalks join the stem.

The gorgeous golden blossoms—composed not of true petals, but of five to seven sepals—emerge in May, often opening and closing with the appearance and disappearance of the sun. They resemble those of their relatives, the buttercups, but are larger—being approximately 1–2 inches across.

Range and Habitat

Occurs throughout the northern two-thirds of the US (including at least 85% of our area) and parts of Canada (Labrador, Newfoundland, and adjacent areas—up to the Arctic Circle). Always grows in—or very near—water (swamps, marshes, ditches, streams, and ponds).

Seasonal Availability

The blossoming plants are limited to May in our area, but leaves can be harvested for a month or so afterwards (if having been positively identified earlier, in their flowering stage).

Chief or Important Constituents

Triterpenoid saponins(oleanolic acid, hederagenin, hederagenic acid), triterpene lactones(caltholide, palustrolide, epicaltholide), coumarins, ranunculin, sitosterol, very small quanitites of isoquinoline alkaloids, calthaxanthin. Nutrients include iron.

Food

Although marsh marigold contains a glycoside that makes it impossible to eat raw (as such, it would blister the mouth badly), this principle is destroyed by drying and/or boiling (it dissipates at 180°F), so that this plant can be used as a potherb. Ideally, one should gather the stem and leaves before the flowers actually open (although very shortly after flowering is allowable). These plant parts *must* be boiled in *several* (three or more) changes of water (entire boiling time should be 20 to 30 minutes or more—the lower limit for younger plant parts and the higher limit for segments from an already flowered plant). Unfortunately, the whole process is very difficult to time out just right. As overcooked plants are tasteless and undercooked ones leave a sting in the throat, first-time efforts are often frustrating. This, then, is my method: I get two pots and fill them with water. I then turn on the burner under one to start it boiling. Two to three minutes later, I turn on the burner under the other one. When the first pot is boiling, I immerse the plant parts therein. About two to three minutes after the water in the first pot has returned to a boil, I transfer the plant contents to my second—now boiling—pot. Immediately, I pour out the liquid of the first pot into the sink under running water, then quickly rinse out this first pot, re-fill with clean water, and put back on the burner to get it boiling. As soon as it is boiling, I transfer the plant contents from the second pot back to this one and finish cooking in this pot for the remainder of the boiling period (based on age of plant as given above).

Marsh marigold lends itself to a variety of recipes, some of which can be gleaned from the better foraging manuals. However, I prefer the plant simply boiled as above, topped with a dab of butter and some sea salt.

Not only is this herb quite tasty when properly cooked (many—including myself—swear that it makes the best of all potherbs!), but it is rich in iron as well.

Health/Medicine

Marsh marigold has been extensively used in folk medicine for a long time. However, its use has not been entirely without risk (see below under "Cautions"), and other herbal aids can often substitute at less risk. Below is presented a summary of uses.

First of all, *Caltha palustris* has been used in folk medicine for *GENITO-URINARY PROBLEMS*, primarily *DYSMENORRHEA* (Bhandari et al 1987:98) and *STOPPAGE OF URINE* (the latter by our area's Ojibwe [Chippewa] Indians, who used the leaves and stem--Densmore 1971:348)

Marsh marigold has also been employed as medicine for respiratory afflictions. Herbalists of both yesterday and today have found that this plant, properly prepared, can serve as a useful aid for *BRONCHIAL* or *SINUS INFECTION*. (Bhandari et al 1987:98; Moore 1979:106) Thus, our area's Ojibwe (Chippewa) Indians boiled two roots in a scant teaspoon of water, removed it from the fire when it boiled, strained and cooled it, and then drank it all at once as a powerful *expectorant.* (Densmore 1974:340-41; Foster & Duke 1990:88) They also combined a *Caltha* tea with maple sugar to form a *COUGH* syrup that was also popular among colonists.*(Foster & Duke 1990:88)* Finally, they employed a poultice of the root—dried, powdered, and moistened—for application to *SCROFULA* (tuberculosis) sores.*(Densmore 1974:354)*

Then, too, marsh marigold contains useful *antispasmodic/anticonvulsive* proper-ties. The famed herbal writer Maud Grieve relates how this came to the fore by quoting the accidental discovery of Dr. William Withering, the famed 18th-century physician who had also (re)discovered digitalis, as follows: "On a large quantity of the flowers of [marsh marigold] being put into the bedroom of a girl who had been subject to fits, the fits ceased," so that "it would appear that medicinal properties may be evolved in the gaseous exhalations of plants and flowers."*(Grieve 1971:519)* The anticonvulsive properties have been sought through an infusion of the dried flowers, though nothing as dramatic as Withering's aromatherapeutic experience has been recorded.*(North 1967:117; Jordan 1976:147)*. Still, some beneficial effects have been achieved, and, interestingly, Turkish folk medi-cine has also long employed this plant for *SPASMS*.*(EthnobDB)*

As to other historical uses: A warm-leaf poultice has been carefully applied and watched, by skilled herbalists, as a counterirritant for *RHEUMATISM* and *FACIAL PARALYSIS*.*(Westrich 1989:81; Moore 1979:106)* A drop of the caustic juice on *WARTS*, applied once daily, is an old folk remedy to rid them from the body.*(Gibbons 1970:141-42; Dobelis et al 1986:245)*

Finally, saponins isolated from marsh marigold have been found to be potent *molluscicides**(Bhandari et al 1987:99)* suggesting that the plant may prove useful for the *SNAIL INFESTATIONS* responsible for serious and widespread disease in third-world nations

---

Cautions

*This herb is **highly toxic if used raw**, producing **severe blistering***! It used to be felt that helleborin, a glycoside found in the related genus *Helleborus* (same family as *Caltha*), was the principle responsible for marsh marigold's blistering effects (and indeed, R. N. Chopra had reported it as found in the roots of the present plant, along with veratrin-*Chopra et al 1956:47)*. But research since that time*(Bruni et al 1986:1172-73)* has revealed that the heavy-handed culprit is actually protoanemonin, an irritant oil formed by enzymatic action from the glycoside ranunculin present in marsh marigold and in more toxic members of the buttercup family. The transformation to protoanemonin is initiated when the plant is damaged in any fashion (such as by foraging or through in-gestion). Since protoanemonin poisoning is not a pleasant thing to experience, cooking plants carefully per instructions provided above is strongly advised. Both cooking and drying destroy the toxic irritant principle contained in this plant (by converting the protoanemonin to innocuous anemonin). In view of this, some have employed a tea made from the fully-dried plant in order to benefit from one or more of the therapeutic benefits listed above. But even this is not without its hazards. Michael Moore cautions that any such tea should comprise no more than a scant teaspoon of the dried herb per cup of water, and that significant quantities should not be drunk, nor should smaller quantities be imbibed for more than a few days, since **KIDNEY** or **LIVER INFLAM-MATION** could accrue to some individuals due to the residual acridity.*(Moore 1979:106)*

MILKWEED
[SILKWEED; MILK PLANT]
(Asclepias syriaca)

[See photos #93-94 in rear of book.]

Description

A native perennial that can reach a height of five feet, but which usually tops off at 2-3 feet. The stem is stout, erect, hairy, and unbranched.

Milkweed's fleshy, dark-green leaves are opposite, oblong, smooth-edged, and nearly stalkless. They each possess a conspicuous midrib and are a bit hairy. In length, they grow anywhere from 4 to 10 inches. Width is 2-4 inches.

The gorgeous, pleasantly-scented, lilac-colored flowers are arranged in dense umbels on the upper third of the plant—at the tip of the stem, as well as on stalks arising from the leaf axils. There are 25-140 flowers in a cluster. Individual blossoms have five, reflexed petals.

The eventual fruits are large, warty pods that grow to five inches. They finally split along one side, revealing the plant's silky seeds, which are then scattered hither and thither by the wind.

Milkweed is, of course, the host plant to the lovely monarch butterfly and its beautiful striped larva, which metamorphoses into the former via a striking green chrysalis dotted with gold that can sometimes be found hanging from this plant.

Range & Habitat

Dry, sunny, and open places—prairies, railroad embankments, meadows, roadsides, old fields, pastures, vacant lots, woodland margins.

Chief or Important Constituents

Proteolytic enzyme(asclepain), cardioactive glycosides?, resins, flavonoids, asclepion, caoutchouc, gum, salts(including those of acetic acid). Nutrients include high amounts of vitamins A and C.

Food

Milkweed yields some mighty good repasts, with one or another of its succession of structures available as food until the end of summer. The initial culinary possibilities arrive with the young, still-unfurled shoots (i.e., with the leaves still largely clasping the stem), which can be hand rubbed to remove their natal wool and then cut and cooked like asparagus, in two waters. (Even when the leaves are fully opened, many wild-food enthusiasts urge that the topmost ones can still be gathered and cooked.)

Once the grayish-green flower-bud clusters appear and are about one inch across, they can be harvested and cooked like broccoli.(See specs. below) They can also be blanched and frozen for use during the winter months. As for the eventual blossoms, when the Swedish botanist Peter Kalm was traveling in North America and recording observations on its plants, he observed that "the French in Canada make a sugar of the flowers, which for that purpose are gathered in the morning when they are covered with dew. This dew is pressed out, and by boiling yields a very good brown, palatable sugar." *(Kalm 1966:387)*

In late summer, the flowers yield to the warty seed-pods. These, when still young (up to 1½ inches long and firm to the touch—not elastic, as they will eventually become when they develop their silk-ridden seeds) can be boiled in salted water to which a pinch of baking soda has been added. (See specs. below.) The texture and taste of the finished product is reminiscent of okra. If cooked instead with meat, these pods will tenderize it due to their enzyme content. (Here the Sioux Indians capitalized on this phenomenon by cooking the pods with buffalo meat!)

Numerous other tribes of American Indians relished one, or all three, of the above plant parts for food.*(Kindscher 1987:56-57)* A bitterness which may hint at a subtoxic element in the sap occurs in northeastern plants but does not seem to be much of a problem here in the midwest, probably because, as Wisconsin forager Sam Thayer opines, our milkweed does not interbreed with other closely related species, since these are not common to our area.*(Thayer 2001:2-4)* Yet, traditional wisdom on the part of wild-foods connoisseurs has urged that the leaves and pods should always be cooked in at least three waters and that the first boiling—never be from a cold-water start but always from already-boiling water—should be for at least four minutes before the water is changed. The second and third boilings are allowed to go only one minute each (or until the water has returned to a boil after placing the pods therein). Then the final boiling is done for about 10-15 minutes. The buds, however, are allowed to be cooked in three waters in quicker succession, that is, for only about seven minutes total (the first bath for two minutes, the second for one, and the last in salted water for about four minutes).

Health/Medicine

This native plant has a long tradition of use in our nation as a medicinal and it was even in the *U.S. Pharmacopoeia* from 1820-1863. It is probably best known to the casual herbal student of today as a treatment for **WARTS,** on which its milky juice has been rubbed by many, including the Catawba, Rappahonnock, Iroquois, and Cherokee Indians.*(Speck 1944:44; Speck 1942:30, 32; Herrick 1995:199; Hamel & Chiltoskey 1975:44)* Some of these Indian tribes also implemented it for **RINGWORM**.*(Speck 1942:30, 32)* Others (including residents of the Appalachian region-*Bolyard 1981:45; Crellin & Philpott 1997:251*) have used it for **POISON IVY RASH**, although here the application, while often proving helpful, can sting quite a bit!

In one of his popular books on wild plants, naturalist and herbalist Tom Brown Jr. relates his interesting experiences with all three of the aforementioned applications, except that his fungal malady tested was *ATHLETE'S FOOT* and not ringworm. (Appalachian folk medicine, however, has traditionally used milkweed for sores and cracks between the toes.*-Bolyard 1981:45*) For athlete's foot and poison-ivy rash, Brown informs us that he has broken the stems into cold water, strained them out, and then used the white liquid as a topical splash, leaving it on for a few moments and then washing it off.*(Brown 1985:137-38)* For warts, Brown relates that as a lad he had applied the juice to a wart twice a day for seven days, with successful results, although it took an extra week for the wart to dry up and fall off. My own use has had generally good results, although I have stressed application of the fresh juice *three* times a day for at least *ten* days. This age-old treatment for warts may be effective because the juice contains the proteolytic enzyme asclepain, which, as Dr. James Duke points out, might not only serve to soften the warts but perhaps to inhibit the virus as well.*(Duke 1997:453-54; cf. Moore 1979:106)*

Milkweed has a good reputation as an *expectorant(Grieve 1971:1:64-65)*, being *anti-catarrhal.(Mockle 1955:74)* Thus, a modern southwest-American herbalist, Michael Moore, finds that it softens bronchial mucus and even dilates the bronchioles a bit.*(Moore 1979:106)* Early American physicians even used powdered milkweed roots an an*ASTHMA* remedy. For example, one physician of the early 1800s deemed it an expectorant and said that it was quite useful in asthma and in *CATARRHAL INFECTIONS OF THE LUNGS.(Clapp 1852:847)* The early American botanical scholar Constantine Rafinesque wrote that it also inhibited the *pain* of asthma, thus serving effectively as an *ano-dyne.(Rafinesque1828:76)* As to preparation and dosage, Moore writes of using a teaspoon of the chopped root boiled in a cup of water and drunk hot, but warns that anything above this amount could cause nausea.*(Moore 1979:106)* Modern British herbalist Mary Carse finds a hot infusion of the flowers to be most effective for the respiratory problems described above. She gathers them when dewy in the early morning and boils them down into a syrup, which she then strains and uses for a *COUGH.(Carse 1989:131)*

Our herb is also a celebrated *diuretic(Gunn 1859-61:829; Rafinesque 1828:76)*, increasing both the volume and the solids of urine.*(Millspaugh1974:134)* It was thus used by various Amerindian tribes for kidney troubles, including *BRIGHT'S DISEASE.* Here the treatment was for the root to be cut into small pieces and made into a tea, of which three swallows were to be taken three times a day.*(Taylor 1940:52)* Interestingly, at least one tribe required that the patient not ingest any salt while under treatment.*(Swanton1928:667)* Then, too, the Cherokee Indians valued milkweed for urinary *GRAVEL.(Hamel & Chiltoskey 1975:44)* Herbalist Mary Carse finds it to be soothing in general to the genito-urinary tract, as well as analgesic to a backache caused by kidney problems; she even lauds it for *GALL-STONES.(Carse 1989:131)* Michael Moore says that it will especially aid chronic kidney weakness typified by a slight, nonspecific ache in the middle back that is most notice-able in the morning.*(Moore 1979:106)*, Both Carse and Moore agree that it is best used as an infusion for kidney problems.*(Carse 1989:131; Moore 1979:106)* Moore suggests a tablespoon of the chopped root boiled in a pint of water, with one-half cup drunk four times a day.*(Moore 1979:106)* Carse says to use one teaspoon of the powdered root in one cup of boiling water, then drink one-half cup, cold, three times a day, for *EDEMA* and for gallstones. (She cautions, however, that, owing to the plant's toxic potential, the tea is too risky to use with children.-*Carse 1989:131*) Early-American physician John Gunn said to take one-half pound of the dried root, bruise it, decoct it in six quarts of water, and then boil the mess down to two quarts, thereafter taking one-half teacupful three or four times a day.*(Gunn 1859-61:829)*

Milkweed's diuretic effect, in combination with a *cardiotonic* effect, has yielded use of the root tea for *DROPSY,* both by various Amerindian tribes*(Hamel & Chiltoskey 1975:44; Lewis & Elvin-Lewis 1977:193)* and by early American physicians.*(Gunn 1859-61:829;Rafinesque1828:76)* The diuretic action, combined with a *diaphoretic* effect,*(Millspaugh1974:134;Hutchens1991:196)* has likewise proven useful for *RHEUMATISM.* *(Johnson1884:230-31;Hutchens1991:196)*

This meadow-loving plant has had a variety of other applications as well: It was used by our area's Ojibwe(Chippewa) Indians as a *galactagogue.*(Densmore1974:360)* Early-American healers employed it as an *emmenagogue.* *(Gunn 1859-61:829)* Various Indian tribes (including the Ojibwe and the Mohawk) even used it as a female *contraceptive.* *(Rousseau 1945:59; Densmore 1974:320)*

And there's more! Milkweed leaves were used as a *vermifuge* by the Mohegan Indians.*(Tantaquidgeon1972:128-29)* The Natchez Indians implemented them for *SYPHI-LIS,*(Swanton 1928:667)* while the Cherokee Indians drank a tea made from the roots for various *VENEREAL DISEASES.*(Hamel & Chiltoskey 1975:44)*

All in all, this common plant of fields and "wasteland" seems capable of yielding a plethora of benefits to man. My personal feeling is that it is much too neglected by herbal professionals nowadays.

Cautions

* Do not confuse milkweed with dogbane (*Apocynum* spp.), a similar-looking plant that is toxic. When both plants are mature, they are easily distinguished: At this stage, dogbane, unlike milkweed, has a *branching* stem and it also contains *tiny, white, bell-shaped flowers* which appear in *cymes* rather than in *umbels* as with milkweed's. When young, however, the two plants closely resemble each other. The distinguishing features at this stage are as follows: Milkweed has hairy stems (which, however, can only be seen under magnification) and a very prominent marginal vein on the leaf. Dogbane's stems are smooth, tough, thinner than milkweed's, and often red-tinged.

* The *Asclepias* genus has toxic potential owing to the cardioactive glycosides typically found in its species, although it is not known to what extent—if any—such occur in *A. syriaca.* I wish to stress, then, that internal use of milkweed in therapeutic does and applications (i.e., not just prepared as food) would most wisely be done under the supervision of an herbal professional.

MONKEY-FLOWER

(Mimulus ringens; M. glabratus)

Description

[See photos #95 & 96 in rear of book.]

This herb grows 1-4 feet high. Its opposite leaves are stalkless and in fact clasp the stem. Coarsely toothed, they are oblong to lance-shaped and reach a length of 2-4 inches.

Several blossoms can be found on slender stalks sprouting from the leaf axils. These flowers, having four stamens, are irregular and possessed of two lips. The upper lip is erect and has two lobes while the lower lip is puffy and three-lobed. Fascinatingly, the corolla structure, viewed in its entirety, gives the impression of a monkey's face and ergo the plant's common name.

Two species occur in our area, to wit:

M. ringens (purple monkey-flower; square-stemmed monkey-flower; Allegheny monkey-flower) has a square stem and several purplish blossoms each growing to about an inch long. Two yellow spots can be found on the inside portion of the corolla of this species' flowers, the throat of which is nearly closed.

M. glabratus var. fremontii (yellow monkey-flower) has a smooth stem and pale yellow flowers that grow to about half an inch long. In this species, the corolla's throat is wide open.

Range & Habitat

Monkey-flower can be found growing in open, wet places such as streambanks, lakeshores, ditches, marshes, and wet meadows. It appears scattered throughout most of our area, but is less common in the far western portion.

Food

There are over 150 species of *Mimulus* worldwide, and a number of these have been utilized for food. The leaves have most often been eaten—either as a trail nibble or as an ingredient in salads. It is usually agreed that the greens taste best if gathered before flowering, but the plant is of course difficult for the novice to identify at this time.

The leaves of our own area's species are edible. My opinion is that they have a strong, but not unpleasant, taste. Indeed, I look forward to finding this plant each summer so as to add some taste variety to my wild-foods outings. But a species of the western states, *M. guttatus* (another species known as "yellow monkey-flower"), is by far the most popular food species of *Mimulus* in America, having been especially appreciated as a salad ingredient by the Indian and white inhabitants of the Rocky Mountain region.*(Coffey1993:217)* Its leaves have a salty taste and several Indian tribes have therefore used them in recipes to enhance the overall flavor. On certain occasions, too, the leaves were even torched and the ashes used directly as salt.*(Strike&Roeder1994:92)* Yellow monkey-flower's stems and leaves are also enjoyed as a steamed potherb, often being added to soups or casseroles.*(Schofield 1989:201)*

Health/Medicine

Appreciated as a ***vulnerary***, the bruised or chewed leaves and stems of various *Mimulus* spp. can be applied as a poultice to minor wounds such as ***BITES, SCRATCHES***, and ***CUTS***, *(Schofield 1989:202)* as did the Maidu Indians.*(Strike&Roeder 1994:92-93)* The effects here may be partly due to ***astringent*** properties evidently possessed by the plant, in that other Indian tribes found *Mimulus* useful for ***DIARRHEA, HEMORRHAGE***, and as an wash for ***SORE EYES***. *(Strike&Roeder 1994:92-93)*

URINARY PROBLEMS were treated by the Costanoan Indians with an infusion of the related species, *M. aurantiacus*.*(Bocek 1984:253; Strike&Roeder1994:92-93)* A decoction of the leaves and flowers was valued by the Tubatulabal Indians for dealing with ***STOMACH DISCOMFORTS***, while the Maidu used such to treat ***NERVOUS DISORDERS***.*(Strike&Roeder 1994:92-93)* The latter also found monkey-flower helpful to reduce ***FEVER***.*(Strike&Roeder 1994:92-93)*

The roots of *M. ringens* were highly esteemed by the Iroquois, who used them as part of a compound infusion for ***EPILEPSY*** and as part of a compound wash to counteract the miserable rash produced by ***POISON SUMAC (Rhus vernix)***.*(Herrick 1977:435; Herrick 1995:213)*

Mimulus flower essence is used by flower-essence practitioners to heal emotional states such as "the fear of known things."

MOTHERWORT
(Leonurus cardiaca)

[See photo #97 in rear of book.]

A sparsely branched, perennial herb, growing 2-4 feet tall. As a member of the mint family, it possesses that family's characteristic square stem.

As also do all mints, it bears opposite leaves, which in this case are noticeably held out quite stiffly. These leaves are long-stalked and toothed. Unlike those of other mints, however, they are deeply cut (lobed). While the stem leaves have three pointed lobes, the lower leaves are maple-like—that is, with three to five lobes, each growing up to four inches long.

Lilac to pale-lavender flowers occur in spiny clusters situated at leaf axils along the stem. (The spininess occurs because the sepals are sharp-tipped.) Each flower is two-lipped, with a fuzzy upper lip.

Range & Habitat

Motherwort occurs throughout most of North America and entirely throughout our area. It can be found growing in fields, waste places, fencerows, trailsides, and old homesteads.

Chief or Important Constituents

Alkaloids(leonurinine, leonurine, leonuridine, turicine, betonicine, stachydrine), iridoid glycosides(incl. leonurid[ajugoside]), cardiac glycosides(bufanolides), essential oil(incl. alpha-humulene, alpha-pinene, beta-pinene, limonene, linalool, marrubiin), resin, flavonoids(apigenin, genkwanin, hyperoside, kaempferol, quercetin, quercitrin, iso-quercitrin, rutin), organic acids(malic acid, citric acid), plant acids(oleanolic acid, caffeic acid, ursolic acid), saponin, pseudotannins(catechin, pyrogallol), tannin. Nutrients include vitamins A, C, & E, and choline.

Health/Medicine

Motherwort is perhaps best known as an *emmenagogue,* (Willard 1993:287; Ody 1993:74) helping to alleviate **MENSTRUAL PAINS,** (Ody 1993:64) and being indicated primarily when the flow is scanty due to cold and when there is irritability and unrest, sometimes accompanied as well by pelvic/lumbar pain. (Ellingwood 1983:483; Moore 1994:13.2) Traditional Chinese Medicine also finds it valuable for **DYSMENORRHEA** and **AMENORRHEA** caused by impeded blood circulation. (Yanchi 1988:119) It can also often relieve a menstrual headache, with one authority advising 8-15 drops of the tincture every three hours during an attack and otherwise three times a day as a preventative. This is advised in conjunction with a tincture of partridgeberry*(Mitchella repens),* dosed identically.*(Clymer 1973:188)* Motherwort has also been used by some as a **partus preparator,** during the last couple weeks of pregnancy (used 2-3x a day), where, owing to its alkaloidal content, it tones the uterus so that it can achieve coordinated contractions during labor.*(Kong et al 1976:373-82; Newall et al 1996:197; Mabey 1988:68; McIntyre 1994:97; Mowrey 1986:138)*

This striking plant is also one of the outstanding**nervine** herbs available to herbalists, and one which I personally use quite a bit for family, friends, and clients. First, it is very helpful for **NERVE PAIN,** as in **SCIATICA***(Clymer 1973:97)* and **SHINGLES** (or other **HERPES-FAMILY ERUPTIONS***-McQuade-Crawford1996:161)* Secondly, like skullcap (*Scutellaria* spp.) (*see under that entry*), it is also effective in obviating**NERVOUS-NESS** and **ANXIETY**, especially when associated with disturbed sleep, feeble digestion, debility, heart problems (see below), or any of the conditions described above.*(Ellingwood 1983:483)* Both the Cherokees and the Iroquois used it for nervous complaints,*(Hamel & Chiltoskey 1975:45; Rousseau 1945b:98)* and east-european research has confirmed that motherwort is not only **relaxant**, but also **hypotensive** and **antispasmodic.***(Arustamova 1963:47-52; Isaev & Bojadzieva 1960:145-52)*

Then, too, motherwort can also—sometimes remarkably—resolve menopausal **HOT FLASHES,** especially when these are associated with nervous tension and palpitations.*(McQuade-Crawford 1996:161)* In fact, as for **PALPITATIONS,** our herb is probably the single best one on the planet for these, not only when due to nervousness but also when following from organic disease.*(Willard 1993:287)* In fact, motherwort is a classic herb for **HYPERTHYROIDISM**, helping to stabilize the **TACHYCARDIA** occurring with this condition, and often being combined with lemon balm (*Melissa officinalis*) and bugleweed (*Lycopus* spp) (*see under that entry*) to treat this functional disorder from all angles (the latter two herbs serving to lower production of thyroid hormones). Our herb is even approved for this use in the *German Commission E Monographs* prepared by German scientists.*(GermCommE)*

Motherwort's effects on palpitations are accomplished via two pathways: the nervous system and the cardiovascular system. In that first regard, the German phytotherapist-physician R. F. Weiss commented: "My own investigations have shown that there is indeed a medicinal action, mainly for functional heart complaints due to autonomic imbalance. It appears to be predominantly sedative, similar to valerian. It is necessary, however, to take the drug for a long period, over months."*(Weiss 1988:186)* Secondly, it is a proven **cardiotonic,** at least partly due to its glycoside content,*(Mowrey 1986:138)* so that it is indicated for **CARDIAC DEBILITY.***(BHP 1976:1:133)* Chinese studies with rats show that it relaxes pulsating myocardial cells, improves circulation (both coronary and microcirculation), and inhibits platelet aggregation, thus performing several vital functions that could thwart situations that can lead to **HEART ATTACKS.***(Yanxing 1983:185-88; Zhang1982:267)* Additional work by the Chinese found that motherwort prevented **STROKES** in animals.*(Zhang et al 1988:254-56; Kuang et al 1988:37-40)* Yet further Chinese research with animals found that the plant decreased platelet aggregation, blood viscosity, and the volume of fibrinogen in the blood.*(Zou 1989:65-70; Peng 1983:41)* Our herb also has proven **antioxidant** effects.*(Bol'shakova 1997:480-83)*

Motherwort has been found to alleviate **ALBUMIN** in the **URINE.** *(Willard 1993:288)* It would seem to eliminate congestive material from the system in general, thus even proving of aid for **RHEUMATISM**.*(Clymer1973:97)* Its demonstrated *anti-inflammatory* ability is no doubt helpful as well with this latter condition.*(Fetrow & Avila 1990:440)*

A recent study found virucidal activity with this herb: An aqueous extract was shown to inactivate tick-born encephalitis (TBE) almost completely *in vitro*—as well as *in vivo* with mice. Such a tea increased survival rate and prolonged average longevity.*(Fokina et al 1991:18-21)*

Traditional Chinese Medicine (TCM) uses not only the plant (in *Qi contraint* patterns—esp. of the heart and uterus),*(Holmes 1998:2:572-73)* but also the seed, which is said to influence the liver meridian and thereby brighten the vision (in that the eyes are understood to be related to this meridian). The seeds are even used to help heal **CONJUNCTIVITIS.** *(Ziyin & Zelin 1994:135; Holmes 1998:2:572-73*)

Cautions

* Some persons can develop **CONTACT DERMATITIS** from touching this plant. This is due to the plant's volatile oil. The effect will flare with sunlight.

* *Not* to be used during **PREGNANCY,** due to its emmenagogue and parturitive effects. Also, avoid during menstruation if the flow is heavy *(MENORRHAGIA).*

* Should be used with caution and under supervision if simultaneously taking pharmaceutical anticoagulants and/or digitaloids; otherwise, it is contraindicated with these drugs.

MOUNTAIN MINT
(Pycnanthemum virginianum)

Description

[See photo #98 in rear of book.]

Possessed of a powerful 'minty' fragrance, this plant grows 1-3 feet high. Like all mints, it has a square stem, which in this case is blanketed with many tiny hairs.

Growing to 3 inches long, the opposite leaves are lanceolate to ovate, rounded at their base, and whitish on their undersides. The uppermost leaves often look as though they have been dusted with talc on their surface. The leaves are toothless and nearly stalkless on our *P. virginianum*, while they are noticeably toothed and stalked on species outside of our area.

The many, small flowers, white to lavender, are two-lipped (the upper lip being single and the lower lip being three-lobed and spotted with purple) and appear in dense, flat-topped clusters. These cymes appear in the leaf axils or terminally on the stem and branches. They are cupped underneath with hoary bracts.

Range & Habitat

Mountain mint can be found on streambanks and in open woods, thickets, wet meadows, and moist prairies. It thrives in the lower three-quarters of our area.

Chief or Important Constituents

Essential oil(consisting of pulegone, carvacrol[approx. 25% or more], geraniol, dipentene, and other terpenes).

Food

The dried leaves make a pleasant, minty tea as well as one darn good mint julep! There are a variety of opinions on how to best make the latter: Stewart and Kronoff suggest gathering about a third of a cup of fresh leaves and then mixing them in a blender at low speed along with six tablespoons of sweetener (they say sugar, but health nuts like me prefer rice syrup, fructose, or honey). Next, they advise adding two cups of shaved ice and eight ounces of bourbon and then blending again, but this time at high speed, until the mess has a soft-frozen consistency.*(Stewart & Kronoff 1975:131)* Others would crush the leaves and drench them with the bourbon, then steep the mess for several hours, and finally strain and serve over shaved ice.*(Phillips 1979:142)* Either way, it's hard to go wrong—this drink is as good as its reputation!

Health/Medicine

Mountain mint has been valued as a treatment for **COLDS** by many peoples, including the Cherokee Indians.*(Hamel & Chiltoskey 1975:45)* The Lakota Indians used it to deal with a bad **COUGH**.*(Rogers 1980:50)* The Cherokees also poulticed this plant's leaves on the head for **HEADACHE**,*(Hamel & Chiltoskey 1975:45)* while the Choctaw Indians drank an infusion of the plant to achieve the same goal, as well as poured it on the head.*(Taylor 1940:55)* A useful *antipyretic,* mountain mint was much appreciated by both the Cherokees and our area's Ojibwe (Chippewa) for this purpose.*(Hamel & Chiltoskey 1975:45; Densmore 1974:354)* Such an effect is successfully achieved because, as the *U.S. Dispensatory* noted, "its [*Pycnanthemum*'s] hot infusion is diaphoretic."*(Wood et al 1926:1441)*

Like most mints, *Pycnanthemum* is a valuable *carminative,* and was appreciated as such by the Cherokees.*(Hamel & Chiltoskey 1975:45)* This, in fact, remains its most well-known use today and is probably my most oft-used application for clients.

It was felt by the Cherokees that a tea of mountain mint's leaves was useful for **HEART TROUBLE**.*(Hamel & Chiltoskey 1975:45)* The plant was even used as a *stimulant* by early American physicians and folk healers.*(Smythe 1901:204; Burlage 1968:94)* Concordantly, the Mesquaki (Fox) Indians appreciated a tea of this plant for those who were "all run down."*(Smith 1928:226)*

Like other mints (e.g., catnip, horsemint, motherwort—*see under those entries*), mountain mint has a history of use as an **EMMENAGOGUE**. *(Smythe 1901:204; Burlage 1968:94; Foster & Duke 1990:68)* But it is not exclusively a female herb, even as the Indians astutely recognized. Thus, when a Cherokee brave had an **INFLAMED PENIS,** he would soak his organ in a receptacle filled with a tea of a *Pycnanthemum* species.*(Hamel & Chiltoskey 1975:45)*

A final application is worth mentioning: James Duke, a skilled ethnobotanist and prolific author on medicinal plants, finds mountain mint to be an excellent topical *mosquito repellent! (Duke 1997 :291)*

Caution

Contraindicated in **PREGNANCY** due to the pulegone and carvacrol content.

MULLEIN
[AARON'S ROD; FLANNEL LEAF]
(Verbascum thapus)

Description

[See photos #99-101 in rear of book.]

A striking biennial, sometimes reaching a height of ten feet!

At the beginning of its existence, this plant's large, hairy (even *velvet-like*) leaves form a spring rosette that is easy to identify even at this stage. Come winter, the rosette survives or dies back under the snow. But next spring, the plant begins to shoot up a tall (4 to 10 feet!) center stem topped with a club-like spike bearing many flowers. Full height is reached by late summer, and the stalks persist throughout the winter.

The stem is round, fibrous, sometimes winged, and tends to branch toward the top. It contains a white pith. The clubhead is clustered with small(½–1 inch), yellow to yellow-orange, five-petaled flowers, which open only randomly, so that one inevitably finds the spike to possess a patchy appearance (part green with unopened flowers and part yellow with opened blossoms).

Leaves of the second-year plants occur clustered at the base and along the stem, often tapering into the stem toward the top. They are large, ovate, and densely covered with the same fine, bristly hairs characterizing the basal leaves. The hairs often give the leaves a slight grayish tint.

The plant has been ingeniously designed to retain moisture in the dry, sunny environment in which it thrives, trapping rainwater with the hairs on its upper leaves and then dripping it down to the lower leaves.

Range & Habitat

Mullein thrives in open, dry places—esp. fields, wasteland, and roadsides. It can be found throughout our area (although less so in the northern one-third of our region).

Chief or Important Constituents

Iridoid glycosides(aucubin, catalpol), coumarins, flavonoids(verbascoside, rutin, apigenin, hesperidin, luteolin), sterones, triterpenoid saponins(verbascosaponin), resin, traces of essential(volatile) oil, rotenone, caffeic-acid derivatives, bitter amaroid, malic acid, phosphoric acid, mucilage, calcium phosphate. The root contains sugars(raffinose, verbascose, stachyose, penteose, and sucrose). The flower contains thapsic acid and polysaccharides(arabinose, galactose, uronic acids). The seed contains fatty acids(oleic, palmitic, linoleic) and beta-sitosterol. Nutrients include vitamins(A, D, and B-complex), iron, magnesium, potassium, manganese, selenium, sodium, silica, zinc, and sulfur.

Food

Mullein's velvety leaves, when properly dried, make a marvelous tasting tea. (The leaves themselves should not be consumed—either raw or cooked—since they contain irritating stinging hairs and are also quite high in aluminum. As to that latter factor, steeping the leaves is regarded as safe, since minerals are not easily transported into an infusion unless it is left to steep for many hours.) However, the infusion needs to be strained through one or more coffee filters in order to trap the irritating hairs.

Some people enjoy nibbling on the little yellow flowers—sucking the juice out of them. But, one should always discard the mildly toxic seeds.

Health/Medicine

There are few herbs in my healing arsenal that I employ more frequently than mullein! This versatile herb, listed in the *National Formulary* from 1916-36, is, first of all, a wonderful *astringent* (owing to its tannin content) and *demulcent* (owing to its mucilage content and to some other factors). In accord with these physiological proper-ties, the late-first-century naturalist Pliny the Elder had noted mullein's popular utili-zation for *DIARRHEA*,(HN 26.44) while his herbalist contemporary Dioscorides de-scribed its use for "lask and fluxes of the belly."(*Herbal, 4:104*) The plant has also been recognized by twentieth-century herbalists as being one of the herbs of choice to treat adult *DIARRHEA.* (Shook 1978:195) (Herbalist Mary Carse recommends it even over the most renowned botanical for this affliction, blackberry root/leaves, urging one to switch to the latter only if mullein doesn't work.-Carse 1989:135) The herb is often helpful as well for another intestinal condition with which some persons are afflicted: *COLITIS.* (Keville 1991:202) I have seen it work wonders for this condition—although it is not a cure.

Mullein's astringency further explains its traditional usefulness for soothing *LARYNGITIS* and *SORE THROAT.* Its most famous application for the respiratory tract, however, has been as an *expectorant*,(Germ Comm E) proving successful here owing to a reflex action arising from an irritation of the gastrointestinal mucosa by its saponins. This is accompanied by a tonifying of the mucuous membranes so that production of fluids is increased and hence a cough is made more productive.(Hoffmann 1986:210; Fetrow & Avila 1999:444) Thus, Catawba Indians could prepare a syrup from the boiled root for children's *CROUP*,(Speck 1937:190) while the German physician-phytotherapist R. F. Weiss could come to regard mullein as one of the best botanicals to treat *CHRONIC BRON-CHITIS*.(Weiss 1960, 1988). It is most widely regarded as being specific for *a tight, hacking, dry, and non-productive COUGH*—not only when found in viral afflictions, but also when existing in the form of a *smoker's cough,* a *nervous cough,*(Moore 1994a:15.2) or as Dioscorides would have it, simply an 'old cough,'(*Herbal,* 4:104) The cough is also often worse when the sufferer is lying down.(Moore 1994a:4.5) I have used mullein for these indications hundreds of times and it has seldom failed me. In fact, it is one of the most reliable herbs around!

The hot tea has a special application for certain, stubborn *SINUS PROBLEMS*: Steeping it 5-10 minutes in a teacup with a plate on top and then moving one's nose close to the tea while removing the plate and breathing the vapors—followed by a straining and imbibing of the tea—is a favorite remedy of mine to alleviate impacted *SINUS CONGESTION* (or even ordinary *NASAL CONGESTION,* for that matter). In fact, I have made many converts to herbal medicine with this simple procedure.

Asthmatics often find some relief from such a treatment as well. In fact, a mullein steam treatment for *ASTHMA* was even practiced by the Micmac Indians.*(Lacey 1993:35)* On the other hand, several Amerindian tribes preferred to *smoke* the leaves for the same condition.*(Tantaquidgeon 1928:265;Speck 1917:310;Smith 1933:83;Smith 1923:53;Lewis & Elvin-Lewis1977:299-300)* A large variety of Amerindian tribes, in fact, used mullein for cold, chest, and lung troubles of many sorts.*(Moerman 1986:1:505-06)*

In India, too, mullein has a long history of use as a medicinal plant for pulmonary and other afflictions and it is very highly esteemed. Here fresh mullein leaves have been boiled in milk and the resultant drink imbibed for tuberculosis, often with excellent results. This success was shared to some extent on the American continent by the early settlers,*(Lewis & Elvin-Lewis 1977:299-308)* as well as by the Iroquois*(Herrick 1977:432)* and Salishan*(Teit 1928:293)* Indians. Early-American physicians found a hot mullein decoction helpful in maintaining body weight in the disease's early stages and were also enamored of its help in checking the constant cough and dyspnea.*(Millspaugh 1974:433)* In fact, the scientific name of the family to which mullein belongs, *Scrophulariaceae,* is related to the word scrofula, an old term for chronic swollen lymph nodes eventually identified as a form of TB. In harmony with this, in the early 1950s, a virologist who tested about 300 different plants as to their effects on a virulent human strain of tubercle bacillus found that mullein was one of 72 plant extracts that yielded complete inhibition of growth at a dilution of 1:40—a significant percentage.*(Fitzpatrick 1954:531)*

There are several possible reasons for its efficacy in this regard, but nobody knows yet for sure just why it helps. However, the plant has for some time been known to possess antimicrobial properties. Its chemical aucubin is definitely antiseptic. In the opinion of some, another of its chemicals, verbascose, may also at least partly account for its action against TB, while the plant's saponins may help disinfect the lungs.*(Heinerman 1979:158; Lepore 1988:181)* The flavonoids and tannins may also play a hand. The flower—though not yet the leaf—has been shown to possess *antiviral* action *in vitro* against *Herpes*-family viruses and influenza A & B.*(Slagowska et al 1987:55-61; Zgorniak-Nowosielska et al 1991:103-08; McCutcheon et al 1995:101-10)* Then, too, it has long been used for *SWOLLEN LYMPH NODES* in general.*(Carse 1989:135; Willard 1993:188)* Cherokee Indians have a history of poulticing scalded leaves onto swollen nodes and also placing them around the neck for *MUMPS,(Hamel & Chiltoskey 1975:45)* while the Iroquois preferred a smashed bunch of leaves as a poultice for this latter affliction.*(Herrick 1995:216)*

Mullein is a true miracle plant for the ears, too! The Iroquois Indians used to heat mullein leaves in sweet milk and poultice them onto *ACHING EARS.(Herrick 1995:215; Herrick 1977:432)* Likewise, a renowned use of the flowers is as the main ingredient in an old earache remedy: They are combined with olive oil and then sun-treated for a few days to produce a concentrated bottle of oil which, after being carefully strained through cheese-cloth, is stored in a cupboard, to be brought out as needed and used with an ear-dropper. (Herbalist Michael Moore points out that setting the mixture on one's basement water heater for several weeks will accomplish the same task-*Moore 1979:113*) This oil appears to stimulate secretion of cerumen in the ear*(Fetrow & Avila 1999:444)* and it certainly takes the edge off of the pain. Charles Millspaugh observed that this same oil can often be applied to **HEMORRHOIDS** with good results.*(Millspaugh 1974:432)*

Indeed, mullein flower definitely possesses mild *anodyne/analgesic* properties, making it somewhat useful for acute or chronic **PAIN** of almost any kind. The late, great herbalist Edward Shook once called mullein "the only herb known to man that has remarkable narcotic properties without being poisonous. It is the great pain killer and nervous soporific, calming and quieting all inflamed and irritated nerves."*(Shook 1978:199)* Here the first-century herbalist Dioscorides found mullein helpful for **EYE INFLAM-MATION, TOOTHACHE,** and **SCORPION STINGS**.(*Herbal*, 4:104) Centuries later, the Iroquois Indians would moisten the warmed leaf with vinegar and poultice it on the face of a person with a **TOOTHACHE.** Some modern herbalists find mullein to be par-ticularly valuable for the pain of **SCIATICA**.*(Donson1982:85)* America's Eclectic phys-icians appreciated that mullein was "mildly nervine, controlling irritation, and favoring sleep."*(Felter & Lloyd)* The flower tea is especially useful as a *sedative* for **INSOM-NIACS**.*(de Bairacli Levy 1974:101)*

Mullein has especially been applied locally to **SPRAINS, BRUISES**, and various **INJURIES,** including by Amerindian tribes as diverse as the Catawba,*(Speck1937:190)* Iroquois,*(Herrick1995:215-6)* Malecite,*(Mechling1959:246)* and Delaware-Oklahoma.*(Tantaquidgeon 1942:66,82)* Canadian Delaware Indians used a crushed, mullein-leaf poultice for **SWEL-LINGS**.*(Tantaquidgeon 1972:108)* The Rappahannocks did, too, but chose a different appli-cation for a swelling as opposed to a sprain: For the former, they boiled leaves in a gallon of water until three pints remained, then rubbed it on the swelling. But for sprains, they simply applied boiled leaves as hot as the person could stand!*(Speck 1942:28)* For swellings and **ABSCESSES**, Iroquois Indians sectioned a piece of leaf the size of the problem area and applied it just when the afflicted area was about to open.*(Herrick 1995:215)* When the swelling was related to an injury, however, they instead boiled a mullein plant in water and then washed the area with the decoction. Next, they mashed the warm leaves and tied them onto the swelling as a poultice. Finally, they imbibed what was left of the decoction.*(Herrick 1995:216)*

Richard Mabey suggests that mullein-flower oil be applied to **RHEUMATIC JOINTS**.*(Mabey et al 1988:113)* Likewise, Alexandra Donson gives mullein reverential in-clusion in her informative tome on therapeutic herbs for **ARTHRITIS** and **RHEUMA-TISM**.*(Donson 1982:85)* Appalachian folk medicine has long appreciated mullein tea and leaf poultices for both conditions.*(Bolyard 1981:133)* Then, too, many have found that a warmed-leaf poultice is a Godsend for the excruciating pain of **GOUT.** *(Bolyard1981:134;Morton 1974:156)* Indeed, lab tests have shown this common weed to possess *anti-inflammatory* proper-ties: Its glycoside verbascoside even inhibits 5-lipooxygenase, the enzyme responsible for the formation of inflammatory leukotrienes.-*Fetrow & Avila 1999:444; Dobelis et al 1986:259*)

Because the herb is revered as having an affinity to *hollow, boggy, body struc-tures lined with mucous membranes,*(Landys 1997:189) its main effect is seen with the lungs (as above) and with *the bladder*, where it acts to support the structure. Thus, a traditional remedy for **URINARY OBSTRUCTION** has been an infusion of mullein, cleavers (*see under that entry*), and wild strawberry (*see under that entry*).(Bolyard 1981:134) The Eclectic physicians treasured the herb for "urinary irritation with painful micturition," finding it most "useful in allaying the acridity of urine which is present in many diseases."(Felter & Lloyd) The herb's mucilage, tannin, and other anti-inflammatory principles render it invaluable for **CYSTITIS/UTI'S**.(Mabey 1988:113) Southwest-American herbalist Michael Moore explains that a quarter-teaspoon of the crushed root in a quarter-cup of water, taken before retiring, will strengthen the bladder, helping to alleviate **BED-WETTING**.(Moore 1979:113) Furthermore, New Jersey herbalist David Winston has found a tincture of mullein seeds helpful as an amphoteric for weak kidneys.(Winston 1998) As for the root, he finds that useful for facial nerve pain, such as occurs with**BELL'S PALSY** and **TRIGEMINAL NEURALGIA**.(Winston 1999:46)

When historian-archaeologist Wilhelmina Jashemski was excavating sections of the devastated city of Pompei in 1966, some of her workmen spotted mullein plants and excitedly began digging them from the soil. When she inquired as to the purpose for their elation, she was informed that they were gathering the plant's roots to help offset **LIVER PROBLEMS**, including **JAUNDICE**. Her workmen proceeded to boil the roots and drink the resulting decoction after meals, to the tune of a liter a day for six days.(Jashemski 1999:90)

To dispel a **FEVER**, the Nanticoke Indians took fresh mullein leaves, dipped them in vinegar, and bound them to the forehead, wrists, back of the neck, and soles of the feet. (Tantaquidgeon 1972:98) The Iroquois implemented a decoction for same.(Herrick 1995:216)

Utilitarian Uses

An application for mullein that is often welcome to hikers with sores on their feet is to use the large, velvety leaves as insulating pads. Although the leaves eventually break, and the effect therefore diminishes, I have found that it is still worth the effort. Another utilitarian use of mullein is to employ the dead or dying clubheads as torches—a use often put to them by ancient native Americans.

Cautions

* Because hairs on the leaves collect not only moisture (for the plant's survival in its arid habitat), but inadvertently also dust, pollutants, etc., the leaves should be rinsed and shaken before being dried for tea. (Of course, too, one should never collect mullein from roadsides.)

* As mentioned above, the tea must be carefully strained, so as to remove the irritating hairs.

*Use cautiously in **PREGNANCY**.

NUT GRASS
[CHUFA; NUTSEDGE; UMBRELLA SEDGE]
(Cyperus esculentus; C. odoratus)

Description

[See photo #102 in rear of book.]

A perennial herb with a triangular, smooth, unbranched stem. When flowering, it can reach a height of three feet, but usually tops off at 1-2 feet.

Leaves are grasslike, light green, rough edged, and with a prominent midrib. Basal leaves rise about six inches from the ground, but can grow as high as the flower cluster. Two to six smaller leaves form an involucre just below the flower cluster. Very occasionally, there occur stem leaves at the nodes between these two groups.

Flowers are contained in an umbel, looking like an inverted umbrella (and thus the alternate common name of "umbrella sedge"). They are composed of five or six rays that in turn are subdivided again. (The appearance is that of straw-colored spikelets, flattened on top.) The eventual fruits are small, yellowish, triangular nutlets.

This plant is known and named for its dark, round, half-inch tubers, which are borne from scaly, underground stems. They are possessed of a milky juice.

Several species occur in our area, including:

Cyperus esculentus (chufa; yellow nut-grass) grows 6-30 inches tall. It has a yellowish-green stem and bears yellowish to yellowish-brown flowers. The subtending bracts are longer than the inflorescence.

C. odoratus (fragrant nut-grass) is similar, but, as its name implies, *fragrant*.

C. rotundus (purple nutsedge; coco grass), a species native to the Eurasian continent as well as thriving in the southwestern USA and Mexico, has popped up in a lone Minnesota county. It has purplish flowers shorter than its subtending bracts and its tubers possess scales that are shiny and reddish.

Range & Habitat

Nutgrass thrives in moist and sandy fields, roadsides, lakesides, streambanks. It can be found in the southern three-fourths of our area.

Chief or Important Constituents

The tuber contains a potent volatile oil. It also contains a fixed oil, protein, and large amounts of carbohydrates.

Nut-grass tubers, which taste like sweet nuts as the name implies, have been used by many cultures throughout history and have even been found in Egyptian tombs dating as far back as 2400 B.C.!

The tubers of our yellow nut-grass, or chufa (note the species name, *esculentus*, referring to the esculent potential of the species) must usually be dug from the ground and not merely pulled, as otherwise they tend to break off and remain in the soil. Once secured, they should be peeled of their dry, tough rind. They can then be eaten raw or tossed into salads.

A delicious, milky drink can also be made from them by soaking them in water for two days and then mashing them in fresh water, adding sweetener, and straining out the solid matter. The measurements used in this recipe are one-half pound of tubers to four cups of water, but that is for the cultivated variety and it is doubtful that today's forager would round up anywhere near enough tubers as half a pound! Therefore, one could convert the one-half pound to eight ounces and the four cups to thirty-two fluid ounces and then divide according to what has been actually uprooted. (For example, two ounces of mashed tubers could be mixed with eight fluid ounces of water.)

The tubers can also be steamed or roasted till dark brown and then peeled and consumed. Alternately, they can be roasted, ground, and brewed as a coffee substitute (about one tablespoon to one cup of water) or used as flour. (The roasting must be done in a slow oven and with the door left slightly ajar to let moisture escape; otherwise the tubers will not become brittle enough to grind.)

The base of the stem can also be cut free and eaten raw, being quite tasty. These "hearts" can alternately be tossed into stews. Nutgrass' seeds can also be collected and then dried, ground, and used as flour.

Health/Medicine

Paiute Indians poulticed the smashed tuber of *C. esculentus*, mixed with tobacco leaves, onto **ATHLETE'S FOOT**.*(Clarke 1977:162)* (Success here was probably due to nut-grass' volatile oil, in that such oils are often antifungal to a marked degree.) Wild-foods pioneer Euell Gibbons considered this species' tuber to be a superb nutritive vegetable and quite helpful to digestive function as well.*(Gibbons 1970:264)* The tuber of *C. odoratus* also has **stomachic** properties,*(Porcher 1849:850-51;Potterton 1983:82)* while *C. rotundus* is usually reverenced as an **hepatic** herb, rebuilding liver function when it is deficient.*(Landis 1997:335)*

The famed European herbalist Nicholas Culpeper extolled the **carminative** virtues of the native European perennial, *C. longus* (sweet cyperus; umbrella plant) and also remarked that it was **diuretic**.*(Potterton 1983:82)* This species, however, does not occur in our area. Nor does *C. articulatus,* which has **vermifuge** properties, according to the early American physician Francis Porcher.*(Porcher 1849:850-51)* However, it would not be surprising if the various species of *Cyperus* interchangeably shared all of the physiological functions described above, owing to their common possession of volatile oil.

PEARLY EVERLASTING
[COTTONWEED; INDIAN POSY; LIFE EVERLASTING]
(Anaphalis margaritacea)

Description

[See photos #103 in rear of book.]

A native perennial, growing 1-3 feet tall. The stem is tall and erect and branches at the top. The entire plant is covered with a loose, cottony substance and bears an unusual, musky sort of scent.

The many alternate, stalkless, toothless leaves are linear to lanceolate, 3-5 inches long, and often curled. Their tops are greenish-white, while their bottoms are more whitish.

The blossoms are golden, globe-shaped disk flowers, about 1/4 inch wide, and surrounded by stiff, shiny, pearly-white bracts. Arranged in flat-topped clusters (compound corymbs) at the end of the stem, the flowers are unisexual—with male and female blossoms growing on separate plants.

Range & Habitat

This plant thrives in sunny and dry areas such as pastures, roadsides, waste places, and old fields. It occurs in northeastern Minnesota and in adjacent parts of Wisconsin.

Seasonal Availability

Flowerheads are available in late summer (they should ideally be gathered while still compacted).

Chief or Important Constituents

Essential oil(incl. the monoterpene anaphalin), resin, flavonoids, tridecapentainen, trans-dehydromatricariaester, 5-chlor-2-5,6-dihydro-2h-pyran, phytosterol, tannin,

Health/Medicine

This herb's alternate common name, life everlasting, derives from its ability to survive for up to a whole year after being picked (and thus this plant is quite popular in floral arrangements). This unusual 'survival' aspect of the plant has been thought to carry over to humans (and animals) who use the plant, in the sense of imparting *vitality* and *longevity*. Thus, the Northern Cheyenne of Montana rubbed the dried and powdered flower bundles all over their arms and legs before going into battle so as to ensure strength and vitality in warfare. They also applied these topically to their horses for the same reasons.*(Grinnell 1905:42; Grinnell 1972:187; Hart 1981:18)*

However, the plant's most celebrated and oft-utilized property is its *astringency*. Here, then, it has sometimes been employed topically for **BURNS**, especially in Quebec.*(Coffey 1993:240)* The Tete de Boule Indians also greatly appreciated this application.*(Raymond 1945:119)* As to manner of preparation and application, modern southwest-American herbalist Michael Moore recommends briefly simmering the dried flower-heads in a small portion of water, then cooling them to lukewarm and applying them to the burn, thereafter holding the herb in place with a moist cloth.*(Moore 1993:19)*

DIARRHEA and **BOWEL HEMORRHAGE** have likewise been treated with this powerfully scented herb.*(Millspaugh 1892:89; Rafinesque 1828-30:224; Herrick 1995:228; Grieve 1971:2:477)* Then, too, a decoction of pearly everlasting applied to **HEMORRHOIDS**, popular among certain Indian tribes,*(Lewis & Elvin-Lewis 1977:293)* has brought soothing relief to many. An infusion has also been traditional as a swish for **MOUTH SORES** and as a gargle for **SORE THROAT**.*(Callegari&Durand 1977:14)* The Iroquois appreciated an infusion of the plant as a wash for **SORE EYES**.*(Herrick 1995:228)*

BRUISES have been treated by topical application, even as older editions of the *U.S. Dispensatory* have noted.*(e.g.,Wood et al 1926:1319 ;cf. Grieve 1971:2:47)* The preferred manner of implementation here is as a compress made from an infusion of the plant. The early nineteenth-century botanical scholar Constantine Rafinesque wrote that pearly everlasting was also useful by way of a wash for "tumors, contusions, sprains."*(Rafinesque1828-30:224;cf.Millspaugh 1892:89)* Canada's Kwakiutl Indians, too, have long valued it for **SWELLINGS.** *(PhytochemDB)* It appears, then, that the plant has some *anti-inflammatory* properties*(Moore 1993:19)* aside from a simple astringency. Its utilization by modern herbalists for **ALLERGIC ASTHMA**, as well as for other **SWOLLEN MEMBRANES**—where it effectively decreases **EDEMA**—reinforces this conclusion.*(Moore 1993:19)*

Because this plant has *laryngeal* and *expectorant* properties,*(Chai 1978:70)* it has traditionally been implemented for **PULMONARY CATARRH**, even as the *U.S. Dispensatory* for 1942 noted.*(cf. Grieve 1971:2:477)* The Cherokee Indians appreciated this application, drinking a tea of the plant—or even smoking it—to achieve the desired effects. They did the same to alleviate **BRONCHIAL COUGHS**.*(Hamel &Chiltoskey1975:48)* The Montaignais Indians used it by way of decoction for **COUGHS** and **TUBERCULOSIS**.*(Speck 1915:314)* The Mohegan Indians utilized an infusion of pearly everlasting for **COLDS**.*(Tantaquidgeon 1915:319)* The Baja California Sur still use it for coughs, colds, and bronchitis, especially as part of a compound infusion taken before retiring for sleep.*(Kay1996:153)* For **ASTHMA**, the Mohawk Indians combined an infusion of pearly everlasting flowers with a decoction of the root of mullein *(Verbascum thapsus)* (*see under that entry*).*(Rousseu 1945:63)*

This plant was used by the Baja California Sur for ***FEVER***,*(Kay 1996:153)*while the Thompson Indians implemented a decoction of pearly everlasting's root particularly for ***RHEUMATIC FEVER***.*(Turner et al 1990:50)* Energetically, it is thought to be a *cooling* and *drying* herb, indicated when the tongue is red-tipped and moist.*(Moore 1993:19)*

Steam inhaled from *Anaphalis* tea splashed onto hot rocks was appreciated by the Cherokee as a cure for certain forms of ***HEADACHE*** and also for ***BLINDNESS*** caused by intense sunlight.*(Hamel&Chiltoskey 1975:48)* Washington's Quileute Indians likewise steamed tribesman afflicted with ***RHEUMATISM***.*(Gunther1973:48)* Our own area's Ojibwe (Chippewa) Indians used a steam of this plant—after powdering it and sprinkling it on live coals—to ease ***PARALYSIS***, often in combination with a steam of wild mint (*Mentha arvensis*) (see under that entry).*(Densmore 1974:362-63; Smith 1932:362)* As a point aside, I can personally appreciate why the wild mint might have been added, since pearly everlasting smells (and *tastes,* for that matter) terrible! (I therefore personally use it only when absolutely necessary, and then in combination with other herbs so as to dull the taste as much as possible!)

In view of the horrific smell and taste, I don't find it surprising that this herb also has some ***vermifuge*** properties, even as Rafinesque had pointed out.*(Rafinesque 1828-30:224)*

PENNYCRESS
[FIELD PENNYCRESS; STINKWEED; FANWEED]
(Thlaspi arvense)

Description

[See photos #104 & 105 in rear of book.]

An annual herb with a smooth stem that grows 6-24 inches tall.

This plant has both basal leaves and stem leaves. As to the former, they begin as a rosette of lanceolate, coarsely-toothed blades that reach a length of four inches. When the stem shoots up, it begins to show alternate, wavy-edged leaves that grow 1½ - 2½ inches long. These are sessile and in fact clasp the stem.

Flowers are four-petaled, as is typical of the mustard family to which this weed belongs. They are also white, 1/4-inch wide, and appear in short, spiked clusters that reach a length of 1-2 inches.

The striking seed pods (fruits) are round (looking like a penny, and thus the plant's common name), papery, deeply notched at the tip, and filled with tiny, black seeds. They are at first green, but eventually become tan.

Range & Habitat

A common weed throughout our area. Flourishes in dry, sunny places like old fields, waste places, gardens, and roadsides.

Chief or Important Contituents

An irritant oil (allyl isothiocyanate), a glycoside (sinigroside). Nutrients include substantial amounts of vitamin C.

Food

According to Myra Jean Perry, who did her Master-of-Science thesis at the University of Tennessee on wild-food use by the Cherokee Indians, this tribe much appreciated pennycress leaves as a food. Having munched on them several times myself, I can testify that the taste—while strong—is not all that bad, especially when mixed in with other greens to make a salad. The smell, however, is somewhat unpleasant and bears witness to its alternate common name of stinkweed. (That "stink" becomes increasingly strong after the plant has been picked.)

Pennycress can also be prepared as a potherb. The shoots and leaves are usually boiled in two or three waters, with the total cooking time being 15-25 minutes if one wants to reduce their zing*(Elias & Dykeman 1982:82)* or only five minutes if one wants to retain the strong, onion-like flavor. *(MacLeod & McDonald 1977:56)*

The seedpods can be dried, ground, and used as a pepper substitute. Such can also be mixed with olive oil and vinegar to make a tasty salad dressing. *(Medve 1990:61)*

Health/Medicine

For a **SORE THROAT,** the Iroquois Indians valued an infusion of this plant as a gargle and also drunk as a tea. *(Herrick 1977:341; Herrick 1995:155)* The Ramah Navajo used the related *T. fendleri* as a cold infusion for **ITCH**—both as a tea and as a lotion. *(Vestal 1952:29)*

Pennycress was once an important ingredient in "theriac," the famed poison antidote attributed to Mithridates Eupator, king of Pontus in the first century B.C., who is said to have made himself invulnerable to poison by taking increasingly large doses of this alleged antidote.

In our present day and age, the Chinese utilize this herb, which they call *bai jiang cao,* for clearing heat, revitalizing blood congealed by that heat, and dissipating pus from the body. Applicable conditions include*: APPENDICITIS, TUMORS AND SWELLINGS, ACUTE CONJUNCTIVITIS, EDEMA, HEPATITIS, LUNG ABCESS, SWOLLEN BREASTS, ENDOMETRIAL PAIN, POST-PARTUM PAIN, BLOOD STASIS* due to heat, *DYSMENORRHEA, BOILS, TONSILITIS, ERYSIPELAS*, and *INFLAMED BITES* and *STINGS*. It is also used in *INFECTIOUS DISEASES,* where it excels with *MUMPS. (Belanger 1997)*

Caution

It is thought that this plant has toxic potential, since cattle have been poisoned from eating it in abundance. Dr. James Duke, a former economic botanist with the USDA, warns that internal bleeding could possibly result from overconsumption on the part of humans. *(Duke 1992:194)* Experts in Traditional Chinese Medicine (TCM) —who probably have the most experience with this plant among herbalists of the present day— warn that it may cause nausea and dizziness. *(Belanger 1997)*

PENSTEMON
[LARGE-FLOWERED BEARD-TONGUE; WILD FOXGLOVE]
(Penstemon grandiflorus; P. gracilis)

Description

[See photo #106 in rear of book.]

A lovely native perennial of field and prairie. The major species, *P. grandiflorus* (large-flowered penstemon, large-flowered beard-tongue), reaches a height of anywhere from 2 to 4 feet, while *P. gracilis* (slender penstemon, lilac-flowered beard-tongue) reaches only 6 to 18 inches.

The leaves of both species are opposite, 1-3 inches long, and short-stalked (often even clasping the stem). Those of *P. grandiflorus* are egg-shaped, smooth-edged, waxy, thick, leathery, and bluish-green, whereas those of *P. gracilis* are narrow and toothed.

Two to six flowers appear just above the upper leaf nodes on the stem. The large-flowered pentstemon is so named because its gorgeous, lavender, bell-shaped blossoms reach a length of about two inches, whereas the smaller species bears flowers growing only to about three-quarters of an inch. The blossoms flare at their ends into two upper lobes and three lower lobes. They possess five stamens, one of which is sterile and somewhat hairy at the tip, giving rise to the alternate common name of "beard tongue."

So striking is the large-flowered pentstemon, with its lilac colored flowers and its bluish sheen, that it never fails to catch the eye, even when nestled in among dozens of other flowering plants in a field or thicket! It is truly *beauty personified*.

Range & Habitat

Thrives in sandy soil in fields, prairies, woods, and thickets. Both species predominate in the middle section of Minnesota. Less common in Wisconsin.

Chief or Important Constituents

Cardioactive glycosides(acetyldigitoxin, lanatoside B, gitoxin)

Health/Medicine

"Penstemon in, pain out!" might likely have been a slogan of the Indians: The Navajo, for instance, discovered that the plant was invaluable for *TOOTHACHE.*(*Wyman &Harris1951:42-43; Hocking 1956:162)* To relieve a *STOMACHACHE*, the Kiowa Indians found a decoction of penstemon roots to do the trick.*(Vestal & Schultes 1939:51)* Our region's Dakotas Indians likewise utilized an analgesic application of the boiled roots: the decoction was relied upon as a remedy for ***PAINS IN THE CHEST***.*(Kindscher 1992:267)* (If these were angina pains, as seems likely, then perhaps the plant's cardioactive glycosides were a factor in offsetting them.)

Then, too, various Indian tribes of the southwestern USA have used penstemon as a ***vulnerary***, especially for ***BURNS, WOUNDS***, and ***BITES***.*(Wyman&Harris1951:42-43;Hocking 1956:162)* The Navajo combined an infusion of the blossoms of a *Castilleja* species with one of penstemon to create a tea for topical use on ***CENTIPEDE BITES***.*(Weiner1980:33)* California's Costanoan Indians implemented the western species, *P. centranthifolius*, as a poultice for ***DEEP, INFECTED SORES***.*(Bocek 1984:254)* Modern California Indians mash, or blend, a *Penstemon* species and then mix it with olive oil as an external application for ***CHAPPED HANDS, RASHES***, and ***COLD SORES***.*(Westrich 1989:97)* In this regard, modern southwest-American herbalist Michael Moore provides an excellent recipe for a penstemon salve for ***SKIN*** and ***EXTERNAL MUCOUS-MEMBRANE (lips, anus, etc.) IRRITATION*** or ***INFLAMMATION***: He says to pulp fresh penstemon (via a hand mill, blender, food processor, or juicer), combine it with equal amounts of a base oil (almond, apricot kernel, olive oil, etc.), and then place the concoction in a warm place for a week. When that time has elapsed, Moore says to strain or express it through a cloth, add some beeswax, and then warm the combination till the wax melts. Finally, he urges one to transfer the mess into a widemouthed pot, thereafter stirring it and allowing it to set.*(Moore 1979:123-24)*

The astute reader may recall that this sort of salve preparation is reminiscent of a mullein-oil preparation mentioned earlier under that plant's entry (i.e., combining the plant with olive oil, heating it for a week, etc.) This is not surprising in that penstemon is so closely related to mullein that the two plants would only naturally have similar uses via similar preparations! In fact, a careful look at penstemon's applications as outlined above will remind one closely of the spectrum of mullein's uses: both plants are held to possess ***anti-inflammatory, analgesic,*** and ***antimicrobial*** properties.

Caution

Penstemon is a member of the same family as foxglove, and thus caution should be exercised in any internal use owing to a potential for possible poisoning from the cardioactive glycosides (which, however, are most concentrated in the flowers and even there occur in much less a concentration than in foxglove).

PEPPERGRASS
[POOR MAN'S PEPPER; PEPPERCRESS;VIRGINIA PEPPERWEED]
(Lepidium densiflorum; L. virginicum)

[See photos #107-09 in rear of book.]

Description

Peppergrass—represented in our area by the two species, *Lepidium densiflorum* and *L. virginicum*—is one of several "cresses" found here, to be distinguished from the cow (or, field) cress *(L. campestre)* and two introduced species, *L. perfoliatum* and *L. sativum* (the latter being the popular and cultivated gourmet salad ingredient called "garden cress"). It should also be distinguished from pennycress, *Thlaspi arvense (see under that entry)*, which is similar in appearance, but is bigger and has seed pods that are four times larger than those of peppergrass.

This herb can be either an annual or a biennial. It grows 1½ feet high, but is usually recognized in the spring when it is one of the few weeds "up and running" and about 3-6 inches high.

Peppergrass starts as a basal rosette of deeply cut (lobed) leaves, which are about two inches long and possessed of double-toothed margins. The terminal lobe is noticeably larger than the side lobes. The eventual stem leaves are smaller, narrower (even willow-like), pointed at their tips, and very slightly toothed. They taper to a very short leaf-stalk or simply clasp the stem, appearing one at a node and alternating up the stem. One feature distinguishing *L. densiflorum* (greenflower pepperweed) from *L. virginicum* (Virginia pepperweed) is that the leaves of the former are slightly pubescent whereas the leaves of the latter appear to be without hairs (they are there, but are so small as to not usually be visible to the human eye without a magnifying glass).

Inflorescence is in the form of narrow, elongate clusters of small, white or green flowers appearing at the tip of the stem and its branches. These blossoms have only four petals and four sepals—characteristic of the mustard family to which this plant belongs. In *L. densiflorum*, the flowers are more greenish than whitish and the petals are shorter than the sepals (often the petals are even absent), while in *L. virginicum* the flowers are more whitish than greenish and the petals are as long as—or longer than—the sepals.

Like some other plants bearing terminal racemes, this herb frequently has both flowers and fruit growing at the same time, with the fruits appearing on the lowest part of the flowering stems while the flowers are at the top. The fruits are flattened, winged, orbicular seed pods, growing 2½ - 4 mms. wide, and having a shallow notch at their apex. Each fruit contains only two seeds.

Range and Habitat

A cosmopolitan weed in the USA and southern Canada, inhabiting roadsides, vacant lots, fields, and trail edges.

Chief or Important Constituents

Glucosinolate compounds. Nutrients include extremely high amounts of vitamins A and C, as well as goodly amounts of vitamins E and some B vitamins; also iron and protein.

Food

In their famed crosscountry expedition of the early 1800s, the American explorers Lewis and Clark came across a "sand-bar extending several miles, which renders navigation difficult, and a small creek called Sand Creek on the south, where we stopped for dinner, and gathered wild cresses or tongue-grass"(from their official journal). How might they have prepared and eaten this cress? In several interesting ways: The leaves add a tangy zest to an otherwise bland salad. The pungent seed pods are renowned as providing the "poor man's pepper" (ala the plant's alternate common name). As such, they are excellent in camp stews, added shortly before the stew is finished. The pods can also be chopped and mixed with vinegar and salt to use as a dressing for wild game. Native California Indian tribes even parched and ground the seeds and combined them with water to make a mush.*(Strike&Roeder 1994:81)*

Health/Medicine

Early American physicians and healers found peppergrass to have ***antiscorbutic, deobstruent***, and ***stimulant*** properties. In the 1800s, the medical practice also discovered that the plant has ***diuretic*** properties and so it was utilized at this time to help deal with ***DROPSY***.*(Clapp1852:739ff)* Likewise, the Lakota Indians employed a peppergrass infusion for ***KIDNEY PROBLEMS***.*(Rogers1980:41)* Then, too, traditional Chinese medicine uses seeds of the related species, *L. apetalum*, for ***WATER-RETENTION*** in the ***CHEST*** and ***ABDOMEN***.*(Reid 1992:140)* Interestingly, *Lepidium*'s diuretic properties were recently confirmed, via a testing of *L. latifolium* on rats.*(Navarro et al 1994:65-89)*

An earlier study with this same species revealed significant ***anti-inflammatory*** powers in the standard rat-paw-edema test.*(Benoit et al 1976:166.)* The significance of this may be gauged by the fact that the Keres and other Indian tribes of the west used to rub the crushed plant on ***SUNBURN*** to heal it rapidly*(Swank 1932:51)* and that America's Eclectic physicians of the late 1800s and early 1900s found it prudent to utilize this herb to treat the pains of ***RHEUMATISM***.*(Smith 1933:550)* In fact, application of the plant or its extracts to a variety of ***SKIN PROBLEMS*** proved to be of considerable popularity in early America.*(Clapp 1852:739ff)* Here the plant was most highly regarded as a treatment for ***POISON IVY RASH***—a use pioneered by the Menominee, Maidu, and other tribes. These Indians utilized a wash made from an infusion of the plant (or sometimes simply the bruised plant itself, as a poultice).*(Smith 1923:33; Strike &Roeder 1994:81)* The Cherokee Indians, on the other hand, implemented a bruised peppergrass poultice to quickly draw out a ***BLISTER***.*(Hamel &Chiltoskey 1975:48))*

Peppergrass has also been put to use in treating ***INHIBITED SEXUAL DESIRE*** (thus adding a little "pepper" to one's love life!). The Lumbee Indians used it most effectively in this regard*(Croom 1982:83-84)* and a few herbalists have since taken the clue. The effect may be due to the herb's stimulant property.*(Crellin&Philpott 1989:333)* Such a property no doubt contributed to its reputation as a reducing aid for ***OBESITY***, which application was popular with the Mahuna Indians.*(Romero 1954:66)*

The plant has been used in white folk medicine as a shampoo ingredient to offset ***HAIR LOSS***.*(Crellin &Philpott1989:333)* This was undoubtedly gleaned from the Cahuilla and Maidu Indians of California, who did likewise.*(Strike&Roeder1994:81)* There may be some sort of actual regenerative effect here, since the related species, *L. sativum*, has been shown to rapidly heal ***BROKEN BONES,*** in line with traditional Arabic use.*(Ahsan et al 1989:235-39)*

The herb evidently has a healing effect on the respiratory system, no doubt due to its pungency and to its vitamin C content, if not also to other factors. Thus, the Cherokee Indians used a poultice of the plant to help heal ***CROUP.*** *(Hamel&Chiltoskey 1975:48)* An early-nineteenth-century physician, Dr. Asahel Clapp, noted that it was commonly used in his time for ***CHRONIC COUGHS*** and that it had ***expectorant*** properties.*(Clapp 1852:739)* (Traditional Chinese Medicine likewise uses the related species, *L. apetalum,* as an expectorant.-*Reid 1992:140*) A compound infusion including *L. virginicum* and wild celery (*Apium graveolens*) was even implemented by the Houma Indians to help manage ***TUBERCULOSIS***, although it was felt that this application needed to be continued for some time.*(Speck 1941:64)* Then, too, the Isleta Indians found that chewing several peppergrass leaves helped to dissipate certain kinds of ***HEADACHE***.*(Jones 1931:34)*

Cautions:

    * The infusion can be ***emetic***.

    * One should exercise ***caution in applying any member of the mustard family to the skin***, as *irritation, redness, or blistering* may occur—especially if left on too long.

PIGWEED

[GREEN AMARANTH; RED AMARANTH; ROUGH PIGWEED; REDROOT PIGWEED; WILD BEET]
(Amaranthus retroflexus)

Description

[See photo #110 in rear of book.]

A coarse, stout-stemmed annual growing to six feet (usually 2-4 ft.). Older plants have a red root and often stems that are red (or red *striped*) as well.

This herb is possessed of long-stalked, dull-green leaves that are egg-shaped to lanceolate, long-pointed, prominently veined, and possessed of wavy margins. They commonly grow 2-4 inches long (though they can reach 10 inches!) and alternate along the stalk. Quite often, at least some of them are reddish-tinged.

The greenish flowers are densely crowded into spikes that appear at the tip of the stem and on short stalks from the leaf axils. This inflorescence is usually dioecious, that is, possessed of male and female flowers on separate plants. (Occasionally, both genders can be found on the same plant.)

The eventual fruits are flattened and possessed of small, shiny, black seeds—to the tune of over 100,000 per plant!

Range & Habitat

Sparsely scattered throughout our area, in wasteland and especially cultivated ground or areas with livestock (as it prefers manured soils).

Chief or Important Constituents

Saponins; mucilage. Nutrients include high amounts of vitamins A (6100 IU per 100g) and C (80mgs per 100g) as well as minerals such as iron(3.9 mgs per 100g) and calcium (215mg per 100g).

Food

The young, tender leaves and stalks (from plants less than six inches tall) can be sparingly added to salads. Or they can be prepared as a potherb—boiled like spinach for 10-20 minutes.

The plentiful seeds can be boiled into mush (two parts water to one part seed, cooked until the water is absorbed) or ground into meal and thereafter used half-and-half with wheat flour or cornmeal in pancakes and other recipes. Before being ground, however, they are best roasted (for 1-1½ hours in a 350-degree oven, stirred every so often).

Health/Medicine

Although pigweed is revered as a wild food, its medicinal applications have been little known. These, however, are many and mighty. First, it is *astringent*,*(Chai 1978:69; Lust 1974:95)* making it useful for **SORE THROAT, DIARRHEA,** and as a douche for **LEUKORRHEA***(Krochmal & Krochmal 1973:35; Chai 1978:69; Lust 1974:95),* It is also **hemostatic** (primarily due to its astringency),*(Chai 1978:69; Lust 1974:95)* and thus has a cherished reputation for curtailing external bleeding,*(Chai 1978:69; Lust 1974:95)* as well as **BOWEL HEMORRHAGE** and **MENORRHAGIA.** *(Krochmal & Krochmal 1973:35; Lust 1974:95; Angier 1978:34-35)*

The plant is also a valuable *alterative.* *(Tierra 1992:339)* Externally, it has been used as a wash for **ULCERS**, as well as for **HIVES, ECZEMA,** and **PSORIASIS.** Effects with these conditions are undoubtedly at least partly because of its high concentration of saponins, allowing the wash to serve as a purifying **disinfectant**.*(Stuart 1982:23; Krochmal & Krochmal 1973:35; Angier 1978:34-35)*

As for dosage, naturopathic writer John Lust recommends using 1/2 to 1 teaspoon of the tincture, while for an infusion he suggests adding one teaspoon of the crushed leaves to one cup of water, then drinking the tea when cold.*(Lust 1974:95)*

Caution: Plants growing in cultivated areas or western soils can accumulate dangerous amounts of nitrates

PINEAPPLE-WEED

(Matricaria matricarioides [alternately:Chamomilla suaveolens])

Description

[See photo #111 in rear of book.]

A shrimpy, annual weed that is closely related to chamomile and even often called "wild chamomile" (though that is a misnomer). This is a sprawling and branching herb, but reaches only 2-6 inches. It owes its common name to the delectable pineapple scent that exudes from it when it is crushed, trampled, or otherwise damaged.

This herb is possessed of alternate, finely-dissected leaves, growing ½ -2 inches long. The leaves look a lot like those of parsley, but are even thinner.

Small (1/4-inch wide), spherical, yellow-green flowerheads appear singly at the tip of the stem and its branches. Each flowerhead is composed of many, tiny, tubular, disk flowers, but totally lacks any sort of ray flower (in contrast to the true chamomile, which does possess ray flowers). These heads rests on a cup of overlapping bracts.

Dry fruits (achenes) start to develop from the flowerheads in late summer, each containing only one seed. These fruits are tan to gray in color.

Range and Habitat

This is a cosmopolitan weed in North America, except for the upper 1/3 of Canada. It has been naturalized either from the western USA or, as some allege, from the wasteland of northeast Asia (in either case, probably due largely to the tires of man-made vehicles, in that the seeds readily adhere to their corrugated surfaces!). Now a common "urban weed," growing between sidewalk slabs, at the edges of driveways, and on roadsides. Also likes to grow at the edges of fields and paths, in barnyards, on railroad embankments, and in waste places.

Chief or Important Constituents

Not much work has been done on this species, but it is known to contain helaniol (a triterpene alcohol).

Food

The flowerheads can be infused to make a pleasant, pineapple-scented tea.

Health/Medicine

This herb has unfortunately been far overshadowed by its cousin of the same genus, chamomile, and is thus vastly underutilized by today's herbalists. However, it has most of chamomile's valuable properties—plus a few more!—and it is much more readily available in that it is a common weed. The significance of such ready availability may be seen in light of the fact that this genus suffers in potency when dried and stored for any length of time.

Interestingly, while little used today by white healers, pineapple-weed has a protracted history of use by Indian tribes, especially by those of Alaska and Canada's northern territories, who still treasure it as an important ingredient in their materia medica. Thus, the Dena'ina Athabascans boil the herb and give a cup of the tea to mothers after childbirth, as well as a few drops to the newborns, feeling that this herb"cleans them both out and helps the mother's milk to start."*(Kari 1977)* This use as a *galactagogue* is also still appreciated by Indians living in the area of Port Graham, Alaska.*(Schofield 1989:304)* Additionally, Aleut elders drink pineapple-weed tea to offset ***STOMACH UPSET*** and ***COLIC***.*(Smith1973)* The West Eskimos imbibe it for ***INDIGESTION***.*(Moerman1986:1:285)*

These *stomachic* and *carminative* functions are the herb's most celebrated uses, and they were widely implemented by Amerindian tribes of times past as well. Thus, California's Cahuilla and Diegueno Indians found the plant to be of great aid for ***STOMACH UPSET, COLIC***, and ***DIARRHEA***.*(Westrich1989:101;Strike&Roeder1994:91)* The Salish Indians utilized pineapple-weed for the same discomforts,*(Shemluck1982:341)* while the Costanoans and Western Eskimos used it primarily for ***STOMACH PAIN*** and ***INDIGESTION***.*(Bocek1982:27; Oswalt 1957:22-23)* The Maidu appreciated an infusion of the flowers for ***GASTROINTESTINAL CRAMPS*** and a decoction of the leaves and flowers for ***DIARRHEA***.*(Strike&Roeder1994:91)* Pineapple-weed is the first herb that I myself consider for most of the above-listed G.I. problems, in that I have found it to be quite dependable.

This plant is also a pretty good *emmenagogue,* especially helpful for ***MENSTRUAL CRAMPS*** (perhaps owing to its chemical helaniol-*Akihisa et al 1996:1255-60*)**,** and it can sometimes deliver a retained placenta—both of which functions the Flathead and Diegueno Indians had come to appreciate.*(Shemluck 1982:341;Strike&Roeder1994:91)* Pineapple-weed is a fair *febrifuge,* too, for which the Maidu, Costanoan, and Coast Miwok Indians were grateful.*(Strike&Roeder1994:91)* The Diegueno likewise tapped into its *antipyretic* properties, reserving a simple flower tea for youngsters and a whole-plant infusion for adults.*(Strike&Roeder1994:91)* (This may have been simply owing to patient compliance, since the flowers are sweet and pleasant but the leaves are somewhat bitter.)

Finally, the European phytotherapist R.F.Weiss noted, in his wonderful opus on botanical medicine,*(Weiss1988:122,24)* that pineapple-weed flowers have been clinically verified as being a powerful *vermifuge.* (Weiss was especially impressed with the work of the French phytotherapist LeClerc, who had experimented heavily with this herb on soldiers.) He pointed out that it is most effective with ***THREADWORMS***, but that ***ROUNDWORMS*** and ***WHIPWORMS*** are also susceptible to its properties as long as the remedy is continued for a period of time. It perhaps comes as no surprise, then, that Flathead Indians found this herb to be a good ***INSECT REPELLANT***. *(Shemluck 1982:342)*

Caution: Use with caution during ***PREGNANCY*** (low doses only, if at all)

PLANTAIN
[SOLDIER'S WOUNDWORT; INDIAN WHEAT; WAYBREAD]
(Plantago major)

Description

[See photos #112 in rear of book.]

A perennial plant growing 4-20 inches high, with a short rootstock.

The herb starts as a basal cluster of bright green, spadelike leaves that are smooth-edged, finely haired, and blatantly veined in parallel fashion—a total of seven veins running down the length of the leaf. Each leaf bears a thick petiole that is also very long (in fact, about as long as the blade).

Eventually, a long and leafless center-stalk forms, bearing a flower spire. This whole structure can reach eight inches. The greenish-white flowers, small and inconspicuous, are arranged on the top portion of the stalk in tightly clustered fashion, so that the entire structure gives the impression of a long pipe cleaner.

In autumn, the flowers transform into fruits (seed pods), each bearing 12-18 seeds that are very tiny and angular.

Range & Habitat

A cosmopolitan weed, occurring throughout the USA and Canada. The preferred habitat is illkept lawns, streetsides, between sidewalk slabs, pastures, meadows, cultivated fields, waste places, and along roadsides. Brought by European settlers to America, plantain was named "white man's foot" by the native Americans because it seemed to follow the white man wherever he traversed the new land.

Chief or Important Constituents

Iridoid glycosides(aucubin, catalpol), flavonoids(apigenin, luteolin, baicalein, plantagoside, scutellarin, hispidulin, nepitrin, homoplantaginin, aucuboside), phenolic [hydroxycinnamic] acids(salicylic, rubichoric[asperuloside], vanillic, syringic, caffeic, ursolic, p-coumaric, chlorogenic and neo-chlorogenic), oleanolic acid, planteolic acid, organic acids(citric, fumaric), essential fatty acids(linoleic, linolenic), amino acids, enzymes(invertrin, emulsin), allantoin, mucilage, tannin. Nutrients include goodly amounts of vitamin A, as well as other vitamins (C & K), zinc, potassium, and silica.

Food

I enjoy young plantain leaves raw in salads. (They can also be munched "as is," but have a strong taste that is best offset by mixing them with other greens.) But this common broadleaf also makes a good cooked vegetable. Here the leaves are usually soaked in salted water for a few minutes and then boiled or steamed for anywhere from 3-15 minutes (depending upon whether you like them crispy or tender) in very little water (just enough to cover them). They can then be eaten plain or combined with cheese or lemon juice in a variety of dishes. Alternately, one can simply chop and add the leaves to omelots, casseroles, and stews. Euell Gibbons once related of a couple he knew who even blended, cooked, and strained plantain to feed to their infant child, owing to the fact that the plant is so rich in nourishing factors. But, not to leave adults feeling neglected, he then proceeded to give big folk a mouth-watering recipe for plantain soup!*(Gibbons 1966:163)*

The seeds can be dried and stone-ground into flour for use in breads and pancakes. This was evidently a common practice in ancient times, as the seeds have been found in the stomachs of mummified bog people from fourth-century Europe.

Health/Medicine

While nothing short of torture could force me to name a favorite herb, I could perhaps be coaxed into providing a list of most-commonly-used herbs, and that list would definitely include plantain. In fact, plantain's *cooling* and *drying* energies allow it to be used in such multifarious ways that a monograph such as the present one could only begin to scratch the surface. But, let's try...

First, we should note that plantain's medicinal wonders have been venerated from antiquity: Dioscorides, a first-century-A.D. herbalist who may also have been a medical officer under Nero or Claudius, noted (in his *Herbal,* 2.153) that plantain had a "drying, binding facultie" and he went on to attest to its even-then-time-honored uses for *SORES/ULCERS* (including *CANKER SORES*) and *BLEEDING* (esp. of the *GUMS* or of the bowel [*FLUX*]). Celsus, a medical practitioner or scholar who lived in the Roman empire when Jesus trod the earth, said that if "bleeding comes from the throat, or from more internal parts,... the patient should sip... plantain...juice."(*De Medicina* 4.2:4-7) Galen, the famed Greek physician practicing in Rome in the second century, said that "among all remedies that curtail bleeding, remove heat, and astringe ulcers, there is none greater than plantain, or at least none that can surpass it."

The plant is a fine *astringent*, which would largely explain the uses explicated by these ancient healers. Plantain's astringency would also at least partly explain its time-honored application for *SORE NIPPLES(Bolyard 1981:113;Krochmal & Krochmal 1973:172)* and *PILES/HEMORRHOIDS*, the latter use being recommended in the *British Herbal Pharmacopoeia.(BHC 1976:1:163)* For these troublesome, embarassing afflictions—which are actually varicose veins of the anus—American herbalist Susun Weed prefers an ointment consisting of extracts of plantain and yarrow*(Achillea millefolium)(see under that entry),* which she has found efficacious both for offsetting the pain and for shrinking the hemorrhoids—sometimes in mere days!.*(Weed 1986:32)*

Astringency also explains the tradition of using plantain leaves and roots for **DIARRHEA.***(de Bairacli Levy 1974:112)* (Celsus had even found it helpful for **DYSENTERY.**-*De Medicina* 4.23) The Russian Ministry of Health has further recommended plantain for chronic **COLITIS.***(Hutchens 1991:221)* In modern Pompei, where this latter application is also appreciated, five leaves are heated in 1/4 - 1/2 liter of water and this tisane is drunk throughout the day.*(Jashemski 1999:78)* Extensive clinical experience in Russia has revealed that the plant is helpful as well for **GASTRIC INFLAMMATION,***(Hutchens 1991:221)* probably at least partly because the herb's proteolytic enzymes act as a mild vasoconstrictor.*(Ody 1993:86; Moore 1979:129)* Most significantly, however, a Romanian study published in 1990 found that a polyholozidic fraction extracted from the leaves and seeds exhibited a "statistically significant gastroprotective action...in two experimental models."*(Hriscu et al 1990:165-70)* Not surprisingly, then, plantain has also long been a revered aid for both stomach and surface **ULCERS,**(esp. the juice-*Moore 1979:129*) which application has been used by some of the native peoples of North America.*(Lacey 1993:90)*

Another traditional application of the juice, tea, or tincture has been as a topical treatment for **ACHING TEETH**.*(de Bairacli Levy 1974:112; Brown 1985:174)* British herbalist Mary Carse even recommends drops of the strong decoction in **TOOTH CAVITIES** until these can be professionally repaired.*(Carse 1989:143)* Further on this, an early American physician once wrote to another physician about the amazing benefits of the fibrous strings in the herb's leafstalk and how they can actually vanquish the pain of a carious tooth if placed in the ear on the same side as the aching molar!*(Millspaugh 1974:420)*

A swish of the tea or diluted tincture is also most welcome for **SORE** or **INFLAMED GUMS** or **MOUTH**, and this treatment is even endorsed by the scientists who authored *The Complete German Commission E Monographs.(GermCommE)* I have found it to be especially appreciated by those suffering from oral inflammation due to **CARRAGEENAN SENSITIVITY,** a largely unknown but important problem in modern society, caused by food industrialists tampering with nature in order to make the cheapest possible food stabilizer. What is involved here is that the food industry takes the seaweed known as irish moss or carrageenan and reduces its molecular weight to make a stabilizer that is added to creamy foods, sauces, and toothpastes in order to maintain a consistent, lump-free smoothness. What these commercialists don't tell us is that molecular-weight-reduced carrageenan is also used in scientific laboratories to *inflame rat paws so that anti-inflammatories can be tested against the swelled appendages!* People with disturbed intestinal microflora (beneficial bacteria like *Lactobacillus acidophilus* and *Bifidobacteria* spp.) seem to be especially sensitive to the commercial carrageenan's inflammatory potential.*(Murray&Pizzorno1998:593)* Some researchers even suspect carrageenan as being a causative—or contributing—factor to Crohn's disease and ulcerative colitis—two, devastating, inflammatory-bowel diseases on the increase in today's advanced nations.*(Murray & Pizzorno 1998:593)* Interestingly, though, lab research has found plantain to inhibit carageenan-induced inflammation in rat paws!*(Shipochliev et al 1981:87-94; Mascolo et al 1987:28-31)*

Not surprisingly, such an ***anti-inflammatory*** effect for plantain has also been otherwise well demonstrated,*(Murai et al 1995:479-80; Lambev et al 1981:162-69; Guillen et al 1997:99-104)* which action appears to be due to a combination of the plant's iridoids, flavonoids, and hydroxycinnamic acids.*(Bruneton 1995:100; Ringbom et al 1998:1212-15; Maksyutina 1971:824)* Its soothing aid as a wash to ***SORE EYES*** may also stem from this property (as well as its astringency), and a plantain infusion has even been used topically to good effect in ***BLEPHARITIS*** and ***CONJUNCTIVITIS****.(Stuart1982:114;Levy1974:112)* Numerous peoples have long praised the relief accomplished by binding the leaves to the soles of ***ACHING FEET*** after a long walk,*(Kay 1996:216)* (a use that I have employed on several occasions, so that I can add my own affirmative testimony).

Amerindian tribes such as the Algonquians, Flambeau Ojibwe, and Canada's Delaware found that crushed plantain leaves rapidly healed bruises when applied topic-ally.*(Speck1917:329;Tantaquidgeon1972:108;Smith1932:381)* Various Amerindian tribes also employed the crushed leaves and juice for ***CUTS, SORES,*** and ***SWELLINGS****.(Tantaquidgeon 1972:108;Herrick 1995:211; Train 1957:80)* Indeed, plantain's most famous use has certainly been as a ***WOUND*** herb, not only because of possessing the capability to curtail the flow of blood (as delineated above), but also because of its having the power to disinfect the wound. Here the fresh plant has been shown to be effective against *Staphylococcus* infections,*(LRNP 1998; Caceres et al 1991:193-2080)* while the juice has proven itself against *Bacillus subtilus.(Lin et al 1972:76-78)* The ***bacteriostatic/bactericidal*** properties are largely attributed to its aucubin content, which, when modified by an enzyme in the plant that is released when it is damaged (e.g., bruised or crushed to make a poultice), becomes the active bactericide, aucubigenin.*(Tyler 1985:101;Holmes 1998:620)* However, because this enzymatic ac-tion is nullified by heat, use of the raw herb or a cold infusion is necessary to achieve the antimicrobial effect.*(Blumenthal et al 2000:309)* Thus, in a test-tube study conducted in the 1950s, a cold infusion of plantain made from a 48-hour sample yielded moderate effects against a virulent human tuberculosis organism, although the roots were more active than the leaves.*(Fitzpatrick 1954:530)* Finally, Russian scientists who examined plantain and a variety of other plants known to be antimicrobial discovered that ***antioxidant*** properties may be strongly at play in the microorganism-inhibiting powers of these herbs.*(Bolshakova et al 1998:186-88)*

Through mechanisms not yet entirely understood (although the proteolytic en-zymes and plant acids would seem to be at least partly responsible), plantain serves as an ***antitoxin par excellence***, proving capable of drawing out—or neutralizing—the poison or infection involved in ***WOUNDS, BOILS, POISON IVY RASH, BITES, STINGS, ABSCESSES, BLOOD POISONING,*** and the like.*(de Bairacli Levy 1974:112)* The mysterious "drawing" power possessed by this lowly weed is so powerful that Ponca Indians even found that applying heated leaves to ***THORNS*** or ***SPLINTERS IN THE FOOT*** could easily remove them!*(Gilmore 1991:63)* Pliny the Elder, that inveterate purveyor of plants and all things natural, further underscored this drawing power by relating that for ***SHINGLES*** "plantain is thought to be the sovereign remedy, if it is incorporated with fuller's earth [itself possessed of great "drawing" power]."(HN 26) (There may be ***nervine*** effects involved with the application for shingles as well, since Alma Hutchens relates that she has had good success using 2-5 drops of plantain tincture every ten minutes to dull the pain of *NEURALGIA*.-*Hutchens 1973 :219*)

There is so much information available relative to plantain's use for these toxic conditions that we would do well to consider the applications for each of them separately. First, then, as to *BOILS*, Appalachian folk use has long implemented wilted or bruised plantain leaves to bring these nasty, painful afflictions to a head.*(Bolyard 1981:112)* It has often been observed by persons so afflicted that such a poultice will cause the core to be expelled in a mere 12-36 hours!

Second, as to *INFECTED CUTS*, *BLOOD POISONING* and *ABSCESSES*, a number of Amerindian tribes found plantain to be a Godsend for these conditions*(Herrick 1995:211; Train et al 1957:79)* Ethnobotanist Huron S. Smith, in a 1930s study of the Ojibwe Indians, related that he himself had "cured a badly swollen and lacerated hand, which swelled to three times its normal size, probably because dirt from a sewer was ground into it, with the simple leaf bound upon the hand."*(Smith 1932:381)* And herbalist Rosemary Gladstar has testified that plantain poultices and tea taken internally saved a friend badly wounded and blood-poisoned owing to a wilderness emergency.*(Schofield 1989:308)* Such instances are not atypical, as two clients of mine could well testify: One was a man who had a tennis-ball-sized abscess in his mouth, but who was having no luck with penicillin, and so was in great desperation owing to his pain and discomfort. Another was a young woman with a swelling just about the same size. Plantain tincture, applied repeatedly as a compress, fixed them both up just fine, to their utter amazement. And although I have witnessed plantain's successes now with a good number of people over the years, I am *still* amazed at the astonishing results it can obtain; this is one *incredible* herb!

Fourth, as to bites and stings, numerous Amerindian tribes (eg., the Algonquian, Cherokee, Flambeau Ojibwe, Mohegan) used the rough, lower sides of fresh plantain leaves to draw out the poison from the *BITES* of *SNAKES* and *INSECTS*.*(Tantaquidgeon 1972:74; Smith 1932:380; Hamel & Chiltoskey 1975:50; Speck 1917:319)* (This application was widespread among both North American Indians and the white settlers.) Cherokee Indians treated *YELLOWJACKET STINGS* with a poultice of wilted plantain leaves.*(Hamel & Chiltoskey 1975:50)* Application to snakebite was especially renowned and utilized by many tribes as well as by white settlers. Ethnobotanist Frances Densmore even described, in some detail, an incident wherein an Ojibwe-Indian man saved his snakebitten wife from certain death with a moistened plantain-root poultice.*(Densmore 1974:353)* And although it sounds like something right out of the pages of Ripley's "Believe It or Not!," in the year 1750 the South Carolina legislature even purchased a snakebite recipe consisting chiefly of plantain roots from a black slave named Caesar!*(Crellin & Philpott 1989:345; Millspaugh 1974:420)* (Other sources say that the individual concerned was an American Indian.-*Grieve 1971:2:641*)

The Iroquois Indians ground up plantain leaves and applied them to *SPIDER BITES*.*(Herrick 1995:210)* Even the bites of poisonous spiders have been brought to nothing with this amazing plant, such as, for example, the bites of violin (brown recluse) spiders: In a recent issue of *Medical Herbalism*,*(MH 9:4:20)* herbalist Sasha Daucus related her experiences in treating bites from these spiders with fresh plantain-leaf poultices. In one instance, her client, with an initially small and itchy bite that in ten days had developed into a nasty open gash, received an overnight plantain poultice and on the very next day went on to expel a tiny black core from the wound, proceeding to heal rapidly from that point onward. This is not surprising in view of plantain's "drawing power" and what we saw above relative to a plantain poultice often causing the core of a boil to pop out. Daucus explains that her favorite method of application is to advise her arachnid-attacked clients to pick a fresh leaf, rinse it, chew it, and then apply it. She also allows a plantain oil (made from the dried leaf soaked in olive oil—one part leaf to two parts oil) to be used when fresh plantain leaf is unavailable, such as during the winter season.

My own experience in using plantain on **MOSQUITO BITES** is similar: I have found that chewing a leaf and then rubbing it thoroughly into the bites works rapidly to negate the toxin and the itch, although applying the tincture works just fine, too. The quick results (1-15 minutes) that one experiences can be truly awe-inspiring: what has kept one awake all night long now quickly quiets down and permits a restful sleep for the remainder of the night! And although I don't like to make guarantees, I can pretty much assure you that every outdoor enthusiast that you share this remedy with will never stop thanking you!

Third, as to **POISON IVY RASH**, Appalachian traditional healing has had good success with a poultice of plantain leaves*(Bolyard 1981:112)* even as the late, great folk herbalist from that region, Tommie Bass, found to be the case.*(Crellin & Philpott 1997:152)* My own experience with clients has likewise been most satisfying and successful. I have had about a dozen cases that I can remember and in most of them I advised wrapping the rashy areas in a gauze compress of the cooled infusion, which quickly thwarted the itching and dried up the rash. (I myself am among the 15-20% of the populace that is immune—quite fortunate, given the fact that I am often forced to tramp through so much of this stuff in my wilderness wanderings!) A physician's report published in a widely-known and respected medical journal also described the successful use of plantain leaf to permanently relieve the terrible itching associated with this condition in ten people among his family and friends who had contracted the rash.*(Duckett 1980:583)*

Not only poison-ivy rash and mosquito bites, though, but almost every kind of **ITCHING** yields to the power of plantain—something that I have witnessed on numerous occasions. Tommie Bass found the same to be true.*(Crellin & Philpott 1997:152)* Susan Weed even recommends a plantain salve for **DIAPER RASH**,*(Weed 1986:112)* something that I have used to good effect on numerous occasions. And her remedy for baby's **THRUSH** is another winner, namely, to soak plantain seeds overnight, skim off the slime appearing on top of the water in the AM, and apply this goo to the affected area.*(Weed 1986:110)*

A proven *diuretic* as well,*(Doan et al 1992:225-31)* used as such by both Europeans*(BHP 1976:1:163)* and American Indians,*(Mayes & Lacey 1989:82)* plantain has been greatly treasured for **URINARY GRAVEL, UTI'S, CYSTITIS**(esp. being indicated when accompanied by hematuria),**URETHRITIS, PROSTATITIS, INCONTINENCE**(esp. when painful or when producing a large quantity of urine from a relaxed bladder), and various other **URINARY COMPLAINTS.**(*McIntyre 1994:217; Moore 1979:129; BHP 1976:1:163; Hamel & Chiltoskey 1975:50; Green 1991:235, 244)* The diuretic effects no doubt contribute to its legendary aid for **GOUT** and **RHEUMATISM,** especially since the plant's aucubin has been experimentally shown to be contributory toward uric-acid excretion.*(Wren 1988:311)* Possibly the diuretic potential contributes somewhat to plantain's time-honored efficacy for **OBESITY** as well, although here the polyphenols in the leaves have been found by Russian research to be the biggest contributing factor.*(Maksiutina et al 1978:56-61)*

Many have appreciated plantain's aid for **COUGHS.***(Millspaugh 1974:420)* It is indeed a wonderful **expectorant,** owing chiefly to its saponins.*(McIntyre1994:217)* This wonderful herb is also strongly **anticatarrhal,***(GermCommE)* and thus useful for **CATARRHAL CONGESTION IN THE MIDDLE EAR, EAR INFECTIONS,** and **SINUS PROBLEMS** involving static catarrh.*(McIntyre 1994:217)* Clinical trials conducted by east-European scientists have even found that a plantain preparation significantly reduced the coughing, wheezing, irritation, and pain of **CHRONIC BRONCHITIS**. In one of the trials, "a rapid effect on subjective complaints and objective findings was obtained in 80 per cent" of the 25 patients studied over a period of one month.*(Matev et al 1982:133-37; cf. Koichev 1983:61-69)*

Plantain has long been used in Bulgaria to treat **CANCER OF THE SPLEEN**.*(Hutchens 1973:346)* Interestingly, scientific research conducted in the late 1980s revealed that when the herb's polyphenol complex (designated as plantastine) was introduced into the diet of rats, it proved capable of *cutting in half* the tumor yield chemically induced in the rats by the authors of the study, and it also reduced the toxic damage to their livers as a consequence of the carcinogenic drugs.*(Karpilovskaia et al 1989:64-67)* A study conducted a few years later by the Federal University of Sao Paulo in Brazil found that intracellular fluid of plantain injected subcutaneously into female, breeding mice largely prevented the formation of breast cancer in the rodents: While the malignancy occurred in 93.3% of the controls, it appeared in only 18.2% in the plantain-treated mice!*(Lithander 1992:138-41)*

Finally, a few miscellaneous findings about plantain that are of interest: First, a **cholesterol-lowering** effect has been demonstrated for this common weed.*(Maksyutina et al 1978:56-61; Ikram 1980:278-79)* Second, it has been shown to be **hepatoprotective**.*(Chang & Yun 1985:269)* Last but not least, our herb would seem to be a mild **antipyretic.***(Smith1932:381)* (Rappahannock Indians appreciated this property, binding bruised plantain leaves to the head to reduce **FEVERS**.-*Speck 1942:25-26*)

All in all, I'm sure you would agree that this is *one marvelous weed*!

Cautions

* When harvesting plantain leaves, be wary of contaminants, for not only does this herb prefer civilized habitats, but its shortness and broadness make it an easy target for animal urination, air pollutants, etc. It may thus be advisable to wash the leaves prior to using them.

* Plantain is not usually contraindicated during pregnancy. However, its seeds are a mild laxative and it has shown uterine activity *in vitro* (not *in vivo*, however), which factors at least suggests *caution in any internal use during gestation*. The seeds—and probably the plant as well—should also definitely not be used by those with *allergies to psyllium*, to which plantain is closely related.

PRAIRIE SMOKE
[TORCH FLOWER; LONG-PLUMED AVENS; OLD MAN'S WHISKERS; THREE-FLOWERED AVENS]
(Geum triflorum)

Description

[See photos #113-15 in rear of book.]

A perennial herb growing 6-16 inches high and thriving in large colonies owing to proliferation from creeping, underground stems. The plant's aerial stems are upright, soft, and hairy.

Prairie smoke possesses both basal leaves and stem leaves. The former are fernlike, being pinnately compound and having 7-19 leaflets that each grow to nine inches. The largest leaflets are toward the tip of the leaf. The stem leaves are few—often consisting of only one pair near the middle of the stem—and much smaller (3/4 to 1 inch long).

Urn-shaped flowers, colored russet-pink to ruddy-purple, appear in the spring, having five petals and five calyx lobes. These blossoms are situated in groups of three at the ends of the leaning stems, where they nod quite blatantly.

The fruits replace the flowers as early as the beginning of the summer. These are seed-like and tipped with prominent, plumelike hairs—the whole structure being some-what reminiscent of a feather duster. The plant's common name of "prairie smoke" derives from this eye-catching appearance.

Range & Habitat

Prairie smoke thrives in prairies, rocky hillsides, and open woodlands. It can be found in the eastern two-thirds of the USA (mainly in the northern half) and into most of Canada. In our area, it thrives in the southern, central, and western portions of Minnesota.

Chief or Important Constituents

Phenolic glycoside(gein), tannin

Food

Amerindian tribes anxiously awaited the sight of this herb's fernlike leaves in the spring, since it was one of the first edible plants to appear in the new year. Eagerly, they would then harvest, cook, and relish the bright pink rhizomes.*(Scully 1970:11)* The

Thompson and other Indians also made a culinary tea from this same part of the plant.*(Ward-Harris1983)* Virginia Scully remarks that the taste of such is pleasant and even compares it to that of sassafras tea.*(Scully1970:11)*

Likewise, an acidic, astringent, and quasi-chocolate-tasting beverage can be made from the fresh or dried roots of the related species, *G. rivale*, the water avens (or, purple avens), also appearing in our area. Here the root stalk is cut into pieces and boiled.*(Harrington 1967:360)* It is felt by many that the taste of this tea is most agreeable when prepared from the spring or autumn-gathered roots.

Health/Medicine

Medicinal uses of this plant relate chiefly to its **astringent** properties (the related *G. rivale* was listed with such a property in the *U.S. Pharmacopoeia* from 1820-82). The Blackfeet utilized these astringent properties by preparing a decoction of the plant's roots to: (1) bathe **SORE/INFLAMED EYES,***(McClintock1923:321;Goodchild 1984:143)* (2) swish as a mouthwash for **CANKER SORES,***(Hellson & Gadd 1974:66)* and (3) use as a gargle for **SORE THROAT.** *(Hellson & Gadd 1974:66; Goodchild 1984:136)*

The Blackfeet also mixed the root with grease and applied the concoction to **SORES, RASHES, BLISTERS**, and **FLESH WOUNDS.** *(McClintock 1923:321;Hellson 1974:275)* Similarly, the Thompson Indians bathed **BODY ACHES** and **PAINS** with a wash made from the rhizomes, as well as used it for steam treatment.*(Steedman 1928:466; Turner et al 1990:48)*

A tea of the plant's aerial portions was used by the Blackfeet as a **tonic.** *(EthnobDB;McClintock1923:321;Hellson&Gadd 1974:72)* The Thompson Indians also valued *G. triflorum* as a tonic, but preferred a decoction of the roots to the aerial portions.*(Turner et al 1990:44)* Our area's own Ojibwe (Chippewa) Indians used a compound decoction of the roots for **INDIGESTION** and even chewed them before undergoing feats of endurance.*(Densmore 1974:342, 364, 366)*

The Flathead Indians appreciated prairie smoke for **CHILLS.***(EthnobDB)*, while the Blackfeet esteemed it for **SEVERE COUGH.** *(EthnobDB;McClintock1923:321;Hellson1974:72)* The Ojibwe preferred a weak decoction of the boiled roots of the related *G. strictum* for **COUGH** and also implemented it to relieve **SORENESS** in the **CHEST**.*(Hoffmann 1885-85:200)*

The related yellow avens (*G. aleppicum*) and large-leaved yellow *avens (G. macrophyllum)*, both also found in our area, closely resemble *G. triflorum* in properties and applications, although they are quite different in appearance (having, as their common names suggest, yellowish blossoms). As for *G. aleppicum,* its roots were boiled by the Ojibwe Indians and a weak decoction derived therefrom was drunk for **COUGH** and **CHEST SORENESS.** *(Hoffman 1885:200)* On the other hand, *G. macrophyllum* was utilized by Snohomish and Quileute Indians, who poulticed the leaves on **BOILS**.*(Gunther 1973:37)* The Bella Coola Indians likewise applied the leaves to boils, after chewing or bruising them.*(Smith 1929:59)* The Quileute Indians also poulticed the leaves on **OPEN CUTS** and chewed them to facilitate labor (as a ***parturient).*** *(Gunther 1973:37)* As to another gyneco-logical use, Chehalis Indian women steeped yellow avens leaves to make a tea which they drunk as a ***contraceptive.*** *(Gunther 1973:37)* The Flambeau Ojibwe also used this plant as a female remedy. *(Smith 1932:384)*

The Thompson Indians discovered that a decoction of this plant's root was a godsend for serious, rash-causing diseases such as smallpox, measles, chicken pox, and the like. *(Steedman 1930:507)* Smallpox took a heavy toll among many Indian communities, but according to anthropologists James Teit and Elsie V. Steedman who studied the Thompson Indians in depth, none among this tribe who faithfully imbibed this beverage died during the final smallpox epidemic that swept through North America, which otherwise took a massive toll. *(Steedman 1930:476)*

PUCCOON
[GROMWELL; INDIAN PAINT; STONE SEED]
(Lithospermum spp.)

[See photos #116 in rear of book.]

Description

A native perennial with gorgeous, yellow blossoms and thin leaves. Several species are found in our area, including:

Lithospermum cansecens (hoary puccoon) has several, erect, green stems covered with soft, white hairs. It has alternating, lanceolate, mostly-stalkless leaves, about an inch long, growing only on the upper part of the stem. The corolla is orange-yellow and tubular, with five flaring lobes. This species is possessed of a thick, red root.

L. caroliniense (hairy puccoon; Carolina puccoon) has an orange-yellow corolla. Its stems and leaves are rough to the touch, being covered with stiff, spreading hairs—longer than in *L. canescens*, above. Its roots are black—not red, as in the above species.

L. incisum (narrow-leaved puccoon) has a lemon-yellow corolla, with the ends of its petals being fringed or toothed. The surfaces of its leaves are covered with stiff hairs that lay flat. As the common name implies, the leaves of this species are narrower than those of others.

L. latifolium (American gromwell) has ovate to ovate-lanceolate leaves that are 2-5 inches long and 1-2 inches wide.

Range & Habitat

Dry and open woods; fields; prairies. The various species occur throughout most of Minnesota, except rarely in the northern counties. Sparser in Wisconsin.

Seasonal Availability

Gathered in late spring and early summer when flowering

Chief or Important Constituents

Phytosterol, trans-dehydromatricariaester, tridecapaentain-en, resin, tannin(at least in some spp., incl. *L. cansescens*),

Food

L incisum was utilized by several Amerindian tribes as food: The Okanagan Indians of British Columbia relished the herbage as a potherb, while the Blackfoot and Thompson Indians boiled or roasted the roots and feasted on them.

Health/Medicine

In 1612, Captain John Smith wrote of *L. canescens* that it had a small root "which being dried and beat in powder turneth red; and this they [the Indian tribes of the region] use for swelling, aches, anointing their joints, painting their heads and garments."(cited in *Coffey 1993:191*) Our area's Menominee Indians also used this species, including it in a compound infusion to offset **CONVULSIONS**.(*Densmore 1932:128*) Other tribes used a tea of the plant as a wash for **FEVER** with **SPASMS.** (*Foster&Duke1990:136*)

The black-skinned roots of the hairy puccoon, *L. carolinense*, were reduced to a powder and used by the Lakota Indians on **CHEST WOUNDS**, including those occurring from bullets.(*Munson 1981:236*) The Cheyenne Indians found that rubbing the powdered plant of *L. incisum* onto **PARALYZED BODY PARTS** seemed to help renew them, causing a prickly sensation.(*Grinnell 1905:37-45*) Alternately, they employed the chewed, green plant. They also chewed this and spit it onto the faces of drowsy persons to awaken them!(*Grinnell 1972:2:185;Hart 1981:15*)

L. incisum was further implemented by the Hopi Indians for **HEMOR-RHAGES**(*Colton1974:331*) and by the Sioux and Teton Indians for **LUNG HEMOR-RHAGES** in particular.(*Densmore1918:269; Munson 1981:236*) The Ramah Navajo found it helpful for **COUGHS** and **COLDS,**(*Elmore 1944:71, 96*) and they also pulverized the roots and seeds to use as an eyewash.(*Vestal 1952:41*) They further employed a decoction of the roots as a compress for a **SORE UMBILICUS** on mother or child occurring after the umbilical cord was cut.(*Elmore1944:71*) The Zuni Indians appreciated puccoon for **SKIN INFEC-TIONS, SWELLINGS,** and **KIDNEY PROBLEMS.** (*Stevenson 1915:56;Camazine & Bye 1980:374*)

L. incisum was one of two species of puccoon used by Indian tribes as an ***oral contraceptive***: Here it was the herb of choice of the Navaho,(*Hocking 1956:161*) whereas *L. ruderale* (stoneseed)—fluorishing in states west of Minnesota—was the species employed by Indian tribes of the Nevada region.(*Vogel 1970:242-43*) The latter has been tested and found to be contraceptive in animals; it was suspected by the authors of the study that it inhibited gonadotrophins in the ovaries.(*Vogel 1970:242-43; Stone 1954:31-33*) Likewise, the European *puccoon (L. arvense)* is used as an oral contraceptive; it suppresses the menstrual cycle.(*Trease & Evans 1973:364*) Unfortunately, however, the plant seems to inhibit or shrink *most*—or *all*—of the endocrine glands, something that would definitely be detrimental to homeostasis!

Caution: At lease one species of puccoon, *L. officinale*, contains toxic pyrrolizidine alkaloids, which can be destructive to the liver.

PURPLE LOOSESTRIFE
(Lythrum salicaria)

Description

[See photo #117 in rear of book]

A common, tall-spiked, perennial import from Europe, growing 2-5 feet tall, and possessing a stout, squarish stem.

Leaves are in opposite pairs or whorled in threes. They are narrow and lance-shaped and grow 1½ - 4 inches long.

Flowers appear clustered on long wands at the top half of the stem. They are ½ - ¾ inches wide and rose to purple in color. Six (occasionally five) crinkled petals are seen upon investigation.

Range & Habitat

Swamps, marshes, borders of ponds, ditches, wet meadows—often found in large colonies whatever its habitat. Range is southern Canada to the northern USA Common in wetlands in our area, despite concerted efforts to control it. This plant can become a menace to other plants, which it tends to crowd out owing to its habit of growing in very dense colonies. This ricochets further into the animal kingdom, adversely affecting the diverse vegetable food-chain needed by many animals to survive.

Chief or Important Constituents

Tannin, gallotanic acid, ellagic acid, salicarin, flavonoids(orientin, vitexin), pectin, beta-sitosterol, chlorogenic acid, p-coumaric acid, resin, phthalides(phthalates), Nutrients include substantial amounts of vitamin A. Calcium occurs in the oxalate form.

Health/Medicine

Purple loosestrife's traditional use as a *BURN* remedy has been confirmed by chemical analysis, which reveals not only a high amount of tannins, but also the presence of the related ellagic acid. The tannin content also at least partly explains the plant's time-honored use as an *astringent* for *MOUTH SORES, SORE THROATS*, and—in combination with the pectin present—*DIARRHEA*, its most famous application.*(Bruneton 1995:329)* (The folk medicine of Spain and Turkey has especially long relied upon this plant as a major antidiarrheal-*EthnobDB*) Undoubtedly, too, the tannins are largely behind the plant's traditional use as a wash or douche for *LEUKORRHEA,(Tierra 1980:339)* its topical application to *BODY SORES/ULCERS,* and its implementation as an eyewash for *SORE EYES/OPHTHALMIA*.*(Tierra 1992:339)*

Further with reference to the eyes, herbalist Peter Holmes finds *Lythrum* useful for **VISION IMPAIRMENT** and **DISTURBANCES**, incl. **BLURRED VISION**— especially as it relates to the TCM syndrome known as *kidney-Qi stagnation*, characterized also by skin rashes, malaise, and painful/dark urine. *(Holmes 1997:2 :649)*

Because of strong **hemostatic/stypic** properties (the genus name, *Lythrum*, even derives from an ancient word for blood, in recognition of this physiological function), the plant will rapidly stop **BLEEDING**, *(Stuart1982:89)* being especially revered for halting **UTERINE BLEEDING**. *(Holmes1997:2:648)*

There is an **antibacterial** element present as well, due to the glycoside salicarin, *(PDRHrbs)* which proves especially useful in **FOOD POISONING**. *(Stuart 1982:89)* (Memorization of such a fact could prove valuable should food poisoning occur during camping/hiking, since purple loosestrife is usually easy to find in a wilderness setting owing to its magenta-colored tops and its habit of growing in large colonies on the perimeters of bodies of water.) The combined antibacterial and stypic properties make topical application of this plant an exceptional one for **WOUNDS** and **BITES**, *(EthnobDB)*, especially with inflammation and bruising—the chlorogenic acid in the plant making short work of the inflammation and the flavonoids eventually resolving any bruising. *(Stuart1979:89;Holmes1997:2:648)*

The anthocyanin-containing, flowering tops are official medicine in the tenth edition of France's *Pharmacopoeia*. They are used by the French to treat **PILES** and **VENOUS INSUFFICIENCY**. *(Bruneton 1995:329)* Such applications are also popular in Turkey and Iraq. *(EthnobDB)*

A series of animal studies (using rabbits, mice, and rats) conducted in the 1980s found an extract from purple loosestrife to be powerfully **hypoglycemic**, reducing blood-sugar levels in the creatures tested. *(Torres et al 1980:559-63; Lamela et al 1986:153-60; Lamela et al 1985:83-91)* This, of course, suggests a possible application for **DIABETES**.

Caution

Because this plant contains calcium oxalate and is also very high in tannins, it should *not be used internally on a regular basis*.

PURSLANE
[PUSLEY]
(Portulaca oleracea)

Description

[See photo #118 in rear of book.]

An annual weed with a sprawling or trailing inclination. The stems are smooth and red- or flesh-colored. The branches forming from the main stem allow this weed to reach a length of twelve inches, whereby it carpets the ground like a dense mat.

The inch-long leaves are teardrop-shaped to paddle-shaped, being noticeably widest at their tips. They are thick and succulent, almost rubbery. They occur most commonly in groups of five and they sometimes display a purplish tint.

The small, yellow flowers are borne singly in leaf axils or at the stem tips. They each possess five petals and open only during sunny mornings. Tiny, black seeds are scattered throughout the plant, contained in capsules resembling flower buds.

Range & Habitat

A ubiquitous weed that is common to gardens, waste places, edges of driveways, and eroded areas in general.

Chief or Important Constituents

DOPA?, dopamine?, norepinephrine?, sterols, resins, a glycoside, organic acids (malic, citric, oxalic), amino acids(glutamic acid, alanine, glutathione), omega-3 fatty acids in goodly amounts, sugars. Nutrients include high amounts of vitamins A(one sample has tested at a whopping 8,300 IUs per 100 grams!), folic acid, C(26 mgs. per 100 grams), E, and minerals such as iron(a whopping 3.5 mgs. per 100 grams!), potassium (494 mgs. per 100 grams!), sodium(45 mgs. per 100 grams), calcium(65 mgs. per 100 grams), magnesium(at nearly 2% of the dry weight), phosphorous, and lithium.

Food

Because of its low-lying form, this plant is often gritty and therefore should be washed before being prepared for consumption. Thereafter, it can be boiled in salted water for 3-10 minutes. In view of its salty and oily taste, it is delicious eaten just as it comes out of the pot, with no additions being necessary. However, it is also great served with crumpled bacon or mixed into casseroles and stews. What is more, it can be blanched for two minutes and then frozen for future use. Who could ask for more?

The seeds can be ground into meal and mixed with wheat flour to make bread-stuffs. Or they can be boiled and made into a porridge.

215

Health/Medicine

Purslane has proven to be a valuable vulnerary for *BEE STINGS, SNAKEBITE, SWELLINGS,* and *SORES*.*(Tierra 1992:199)* The Iroquois Indians were fond of implementing this plant as a poultice for *BURNS* and *BRUISES*.*(Herrick 1995:143)* It has also been used both topically and internally for *BOILS, ERYSIPELAS, DYSENTERY, FEVERS, HEADACHES*, and *"OVERHEATED BLOOD."*(Levy 1974:115-17; Reid 1992:102)* One can understand and appreciate this range of applications when one realizes that this plant is energetically considered to be a *refrigerant*.*(Mockle 1955:40)* Unfortunately, however, purslane is little used by Western herbalists these days; the Chinese, however, still appreciate and utilize its refrigerant properties. As with Western herbalists of yore, they feel that it is best used as a fresh herb or by way of decoction.*(Reid 1992:102)*

This much-despised garden weed is a useful astringent, whereby it has historically proven helpful for *HEMORRHOIDS, LEUKORRHEA,* and *DIARRHEA.* Herbalists have also found the fresh juice to provide marked relief from *PAINFUL URINATION (STRANGURY)* and *STOMACHACHE*.*(Grieve 1971:2:660)*

Former USDA scientist James Duke notes that purslane's nutritional profile might lend it for use in helping to prevent—or offset—*ASTHMA, DEPRESSION* (because of the lithium content), and *HEART DISEASE.* As to the latter, purslane has also been shown to be *HYPOTENSIVE* and *DIURETIC.*(Mockle 1955:40)* In the 1890s, Parke-Davis even marketed a fluid extract of the herb, stating that it was indicated for "chronic catarrhal affections of the genitourinary tract." *(Crellin&Philpott 2:1989:363)*

Since Roman times, the seeds have been esteemed as a *VERMIFUGE*. Recent clinical studies in China have verified that the juice is active against *HOOKWORM INFESTATION*; other studies there have demonstrated mild antibiotic effects.*(Chevallier 1996:253)* The latter results might at least partly explain purslane's effectiveness against dysentery, boils, and erysipelas as mentioned above.

If, as certain Chinese scientists claim to have discovered, purslane actually contains the neurotransmitters norepinephrine and dopamine (as well as the latter's precursor, DOPA),*(Chevallier1996:253;Tierra1992:199)* this suggests possible application in those conditions wherein these neurotransmitters are deficient in humans (e.g., Parkinson's Disease). However, any therapeutic benefits here remain yet to be thoroughly investigated and proven. It is perhaps of significance, however, that a study published in 1987 found purslane to be a powerful *muscle relaxant,*(Okwusaba et al 1987:85-106)* demonstrative of a palpable effect on the neuromuscular system.

Cautions

* Extended use of purslane in medicinal strength is contraindicated during *PREGNANCY*, as it affects the uterus.

* Because of the oxalate content, this plant should not be consumed regularly.

QUEEN ANNE'S LACE
[WILD CARROT; BIRD'S NEST]
(*Daucus carota*, subspecies *carota*)

[See photos #119 & 120 in rear of book.]

Description

A biennial weed growing to six feet (2-3 ft. being average). It is possessed of a hairy stem and a white, carrot-scented root.

This plant first sees the light of day as a basal rosette of fernlike leaves—compound and finely divided, and sometimes exuding a carrotlike scent when crushed.

A hairy, branching stem appears in the second year, sporting alternate and dissected leaves.

Small, white flowers form in flat-topped clusters (compound umbels), being about 3-5 inches wide. There is often a tiny, solitary flower in the center of the cluster that is colored reddish-brown to dark-purple, which becomes an important identifier distinguishing this plant from some toxic lookalikes. (See "Cautions," below.) The umbels reside on stiff, feathery, three-forked bracts resembling a collar.

Once the flowers fade and the fruits replace them, the umbel's branches begin to curve upward, presenting the appearance of a hollow cup or bird's nest (and thus the plant's alternate common name of "bird's nest").

Range & Habitat

Cosmopolitan weed. Likes dry, open areas—sandy soil, waste places, disturbed ground, roadsides, open woods, fields, and meadows.

Chief or Important Constituents

Volatile oil(inclusive of terpinen-4-ol, limonene, geraniol, carophyllene, carotol, daucol, asarone, pinene), an alkaloid(daucine), flavonoids and misc. glycosides(incl. luteolin, apigenin, chrysin, kaempferol, quercetin), coumarins. Seeds high in myristicin as well as nutrients such as fatty acids(oleic, linolenic, butyric, palmitic) and potassium. Root contains nutrients such as thiamine, riboflavin, choline, and vitamin C.(Beta-carotene—contrary to many authors—may actually occur only in the domesticated carrot.)

Food

The whitish, first-year roots, if gathered in late fall—or even more preferably, in the following spring before the flower stalk emerges—are passable as survival food, but are best mixed into "wild" stews and casseroles instead of being eaten by themselves. (Note that they possess a woody core which must be discarded before they are consumed.)

Peeled, young flower-stalks are also edible. They are crisp and peppery when eaten raw. They can also be boiled in salted water for about ten minutes.

Health/Medicine

Queen Anne's lace is detailed in the *British Herbal Pharmacopeia* and was listed in the *U.S. Pharmacopoeia* from 1820-82 as a **stimulant, menstual excitant**, and **diuretic.** As to the effect upon menstruation, this would seem to accrue primarily from the seeds, which are *warming* and *stimulating.* Such effects might also elucidate the use of the plant by the Lumbee Indians—as a compress applied to the chest of a young child afflicted with a **CHEST COLD** or **PNEUMONIA.***(Mortimer 1983:56)* As to the **diuretic** function, nineteenth-century physician and plant scientist Asahel Clapp wrote: "An infusion of the seeds or the roots and leaves is an excellent diuretic. Thatcher, Chapman, Eberle speak very favorably of the efficacy."*(Clapp 1852:777)* So did the celebrated, seventeenth-century herbalist Nicholas Culpeper, who wrote that wild carrot "provoketh urine...and helpeth to break and expel stones"*(Culpeper 1983:38-39)*

Because such an effect could also tend to promote excretion of uric acid, our wild carrot has historically been utilized to help relieve the pain of **RHEUMATISM** and **GOUT.***(Crellin & Philpott 1990:135; Newall 1996:264-65)* In addition, it has served as a traditional **antiseptic/diuretic** for the treatment of **CYSTITIS** and **PROSTATITIS,** proving especially helpful in relieving the pain of the associated **STRANGURY***(Johnson 1864:152)* and even in alleviating urine stoppage, as the Iroquois Indians discovered.*(Herrick 1995:195)* Queen Anne's lace has also traditionally been used for **DROPSY**, and here it is of significance that cardiovascular effects have been observed for this plant in animal models,*(Gilani et al 1994:150-53)* including even a reduction of heart rate and blood pressure.*(Fetrow & Avila 1999:534)* Finally, Alabama herbalist Tommie Bass heartily utilized the diuretic aspect of wild carrot *tops* (not seeds *per se*) as part of his famous "[weight]-reducing program," but felt that while it "works on the same system [as do] water tablets," it nevertheless "works through the pores."*(Crellin & Philpott 1990:120)* Scientists, however, have suggested instead that the diuretic effect is probably due to the chemical terpinen-4-ol, which stimulates diuresis through renal irritation.*(Newall 1996:264-65;)*

A tea from the seeds has historically been used for **FLATULENCE.** In early America, physician and plant scientist Francis Porcher described the use of the root for "spasmodic vomiting, flatulent colic, and nervous headaches."*(Porcher 1869)* The prime factor here seems to be the **antispasmodic** effect that the seeds have on smooth muscle, owing to their essential oil. Such an effect has been experimentally verified and likened to that of papaverine, though deemed weaker. *(Gambhir et al 1979:225-28; Dobelis 1986:279)*

The blossoms picked at full bloom were steeped as a tea by Mohegan and Oklahoma Delaware Indians for **DIABETES.** *(Tantaquidgeon 1972:118-19, 130-31)* Wild carrot has also traditionally been used for **JAUNDICE** and for other **LIVER DISORDERS**. *Hepatoprotective* properties have been observed in animal studies using the potent liver toxin carbon tetrachloride. *(Bishayee et al 1995:69-74; Handa 1986:307-51)*

A useful vulnerary, Queen Anne's lace has traditionally been employed as a poultice for **BRUISES** and **CUTS**. The Cherokee Indians bathed **SWELLINGS** with a tea made from the herb. *(Hamel & Chiltoskey 1975:51)* A limited *antifungal* activity has been observed and related to the essential oil (it was active against only one organism—*Botrytis cinerea*—out of nine tested). *(Guerin & Reveillere 1985:77-81)* Then, too, the seeds have historically been held to be a valuable *anthelmintic.*

The seeds are usually used by way of infusion: One-third to a full teaspoon (max.) of seeds have traditionally been added to 8-16 ounces of water, and then infused (or simmered gently) for 15-30 minutes, with such a tea being drunk at a dose of ½ - 1 cup, up to three times a day (or simply for flatulence as a single serving after a meal). *(Carse 1989:164, 174; Hutchens 1991:299-300)* The aerial portions (Bass says they should be gathered in June, before they open their blossoms--*Crellin & Philpott 1990:135*) have usually been implemented to the tune of one teaspoon per cup (or, as Bass says, "a big handful makes a quart of tea"-*Ibid*), steeped for 15 minutes, and then used up to three times a day.

Cautions

    * Too high a dose of the seeds can cause nervous-system damage, so do not exceed traditional dosages.

    * Seeds are *estrogenic* and *abortive* *(Farnsworth et al 1975:535-98)* because they inhibit implantation of the embryo. *(Prakash 1984:9)* They should be **STRICTLY AVOIDED DURING PREGNANCY.** They should also be avoided during **LACTATION.**

    * Contact with the leaves may cause *dermatitis* in sensitive individuals.

    * This plant's leaf-and-flower pattern allows it to be *easily confused with toxic relatives* such as water hemlock *(Cicuta spp.),* fool's parsley *(Aethusa cynapium),* and poison hemlock *(Conium maculatum).* (Only the first-named species, however, generally occurs in Minnesota, though the latter creeps into the eastern portion of Wisconsin.) (Our wild carrot is often also confused with the non-toxic wild parsley, *Pastinaca sativa,* which has a hairless and grooved stem.) *Be absolutely certain of identification* before attempting to use any plant you think to be Queen Anne's lace, as a mistake could be deadly. For one thing, one should always look for the solitary, dark-colored flower in the middle of the umbel to identify it. As for the poisonous water hemlock, one should bear in mind that it has lateral leaf veins that end *between* the teeth and not *in* the teeth as with not-so-deadly lookalikes.

RED CLOVER
[TREFOIL]
(Trifolium pratense)

Description

[See photo #121 in rear of book.]

This familiar perennial grows anywhere from six inches to two feet tall and has a hairy, branching stem.

The alternate leaves are pinnately compound, having three leaflets that are dark-green and oval. (The genus name, *Trifolium*, is named after this three-lobed structure.) The leaflets usually have a pale "V" shape, or chevron mark, in their center portions.

Many rosy-red or magenta flowers appear clustered into a round, compact flowerhead. These tiny, individual flowers lack any sort of stalk.

Range & Habitat

Red clover is ubiquitous throughout North America. It flourishes in old fields, waste places, roadsides, trailsides, and lawns.

Chief or Important Constituents

Isoflavones(1 to 2.5%, including genistein, daidzein, formononetin, pratensein, trifoside, and biochanin A), hydroxymethyloxy-flavones(pratol), other flavonoids (trifoliin[isoquercetrin], pectolinarin), saponins(trifolianol), salicylic acid, coumarins, an essential oil(incl.furfural), and carbohydrates(arabinose, rhamnose, xylose, glucose). Nutrients include vitamins C, E, and several of the B vitamins(niacin, thiamine, and riboflavin), as well as substantial amounts of minerals such as calcium, chromium, magnesium, potassium, and phosphorous.

Food

Fresh flowerheads make a nice trail-nibble. (However, see "Cautions," below.) Not only do they taste good, but they are extremely high in water, and this is compounded to the nth degree with a morning dew or after a good rain, since the design of the flower-head is such that it can retain a lot of H_2O. Often, I find that I can sustain myself on a day's foraging by taking advantage of high-water plants like these instead of lugging along a lot of water. And while fresh flower heads should be eaten only in moderation, one can still suck off all the water that one wants from them!

Dried flower heads (see further under "Cautions," below) and seeds can also be ground into a nutritious flour.

Health/Medicine

Red clover blossoms have traditionally been used topically in a salve to heal **SORES, ULCERS,** and **BURNS**. The famous seventeenth-century herbalist Nicholas Culpeper also noted that "boiled in lard and made into an ointment, the herb is good to apply to the bites of venomous beasts"*(Potterton 1983:194)* (Nowadays, however, the **vulnerary** ointment or paste is usually made from a solid extract, available commercially from several sources). Possibly shedding light on these uses, a 1976 lab study found that a related species of clover, *T. arvense*, powerfully inhibited inflammation in the rat-paw edema test.*(Benoit et al 1976:167)*

A reliable **alterative,** red clover is helpful in many **SKIN PROBLEMS,** preeminently **PSORIASIS** and **ECZEMA** (esp. in children, to which the plant seems to have an affinity, as herbalist David Hoffmann points out-*Hoffmann 1986:221*). The plant may also possess **antifungal** properties: An infusion has long been a folk treatment for **VAGINAL** and **ANAL IRRITATION,***(Hutchens 1991:234; Ody 1993:105)* while a poultice of the fresh plant has been traditional for treating **ATHLETE'S FOOT.** *(Lust 1974:395)* Red clover would also seem to harbor **hepatic** properties,*(Krochmal & Krochmal 1973:219; Lust 1974:395)* which would at least partly elucidate the **depurative** aspect of its alterative action. Its bitter principles definitely stimulate bile production and/or flow.*(Pedersen 1998:144)*

The plant's depurative ability includes a strong **lymphatic** component. Consequently, an ointment has been a standard for softening **HARD BREASTS***(Lust 1974:395)* and other **LYMPHATIC SWELLINGS** characterized as "firm and hard."*(Ody 1993:105; Tilgner 1999:99)* As the plant has an affinity to the area of the body running from the bronchioles to the head, it is especially of value for **TONSILLITIS.***(Winston 1999:50)*

The Cherokee Indians had utilized red clover for **BRIGHT'S DISEASE, FEVER,** and **LEUKORRHEA.** *(Hamel & Chiltoskey 1975:29)* Interestingly, across the Atlantic, in jolly 'ol England, Culpeper had likewise used it for leukorrhea.*(Potterton 1983:194)*

Our area's Ojibwe Indians imbibed red clover tea for **PERSISTENT COUGHS.***(Naegele 1980:117)* Most famously, it has been used as a strong infusion for **WHOOPING COUGH,** dosed at a half an ounce at a time.*(Bolyard 1981:73; Millspaugh 1974:188; Wren 1972:255)* For such an illness it is tailor made, due to its **sedative, expectorant,** and **antispasmodic** abilities.*(Wren 1972:255; Bradley 1992)* (The drug company Parke-Davis had even marketed red clover as a sedative in the 1890s.-*Parke-Davis 1890:149*) It is regarded as a specific for **BRONCHITIS**, for which it should be taken in the form of a hot infusion.*(Willard 1993:253)* A 1954 test-tube study showed that it even inhibited a virulent human strain of **TUBERCULOSIS**, though only moderately.*(Fitzpatrick 1954:531)*

Red clover is commonly used today in marketed formulas for *MENOPAUSAL SYMPTOMS*, since studies have shown that plant extracts have improved estrogen levels in menopausal women.*(Duke 1997:325)* However, as its isoflavones can decrease thyroid hormone activity, extended use of red clover should be avoided by those with hypothyroidism or tendencies toward it.

One of red clover's most famous applications has been in the herbal treatment of *CANCER*. The branch of herbal medicine founded by Samuel Thomson in early America (since called Thomsonianism) utilized it thusly, most often in the form of a salve. Writing in 1841, Morris Mattson described this application as follows: "The blossoms boiled in water...and the liquid simmered over a slow fire until it becomes about the consistence of tar, forms the cancer plaster of Dr. Thomson, which has gained so much reputation in the cure of cancers," etc.*(Mattson 1841:242-42)* The flower heads were marketed in the late 1800s and early 1900s as the major ingredient in Merrell's Trifolium Compound, a skin salve suggested for various dermatological problems. In the 1940s/1950s, red clover became a chief ingredient in the nearly identical salve made famous by the controversial Harry Hoxsey for treating skin cancer.

Interestingly, red clover's traditional uses against cancer now have some scientific support. In the late 1980s, a cancer-protective effect was isolated in the plant by means of an *in vitro* experiment.*(Cassady et al 1988:6257-61)* Genistein and daidzein, isoflavones in red clover, have particularly been shown to be anticarcinogenic, even by scientists from the National Cancer Institute, who examined the plant because their own scientist and historian, Jonathan Hartwell, found that it had been used against cancer in 33 different cultures!*(Hartwell 1984)* Genistein, in fact, has been shown to be anti-angiogenic, that is, it chokes out those extra blood vessels that tumors force the body to grow in order to nourish themselves.*(Duke 1997:399)*

Dried flowerheads were official in U.S. drug literature such as the *National Formulary* and *U.S. Dispensatory*, up until the late 1940s. A tincture of the fresh plant is also occasionally used by herbalists, although the infusion has definitely held sway.

Cautions

* Because of *estrogenic* effects, this herb is contraindicated in *PREGNANCY*.

* The plant should *never be used* when only *wilted*, as in such a stage its coumarins can, under the right circumstances, be altered into dicoumarol, a potent anticoagulant that can bring on hemorrhaging. (Unaltered coumarins, however—contrary to popular belief—are not potent blood-thinners.)

SELF-HEAL

[HEAL-ALL; WOUNDWORT; CARPETWEED; CARPENTER-WEED]
(Prunella vulgaris)

Description

[See photos #122 in rear of book]

A perennial mint of varying height—all the way from 3 inches to occasionally 3 feet (6-12 inches being most common). Erect or sometimes matted in appearance, this herb has several square stems forming from a single base.

As is characteristic of the mint family to which the plant belongs, the leaves grow in pairs opposite each other on the stem. They are lance-shaped to egg-shaped, 1-4 inches long, and less than one inch wide. The edges may be either smooth or a bit indented (wavy edged).

Flowers are small, numerous, short-stalked, and snapdragon-like in appearance, with an arched upper lip and a fringed, three-part, lower lip that droops. They vary in color from violet (most common) to blue to pink. They are borne in a dense, spikelike cluster at the tip of the main stem, with smaller flower clusters usually appearing at the tips of any stem branches. Under each blossom are five, hairy, greenish bracts.

Range and Habitat

Self-heal prefers moist places, such as along trails in damp woods or on stream banks. But it is also widely distributed in drier areas, such as in meadows, pastures, and waste areas. It is found throughout most of the US and Canada. It thrives in most of our area except for Minnesota's southwestern counties.

Chief or Important Constituents

In the past, not much work has been done in analysis, but currently this is changing. Self-heal is known to contain flavonoids(rutin, hyperoside, and others), rosmarinic acid, ursolic acid, oleanolic acid, galactonic acid, prunellin, essential oil (containing cineol, pinene, camphor, and other chemicals), a resin, and tannins(up to 50%!). Nutrients include vitamins (A, B, C, and K) and minerals such as calcium, copper, iron, magnesium, manganese, potassium, sodium, and zinc.

Food

Raw leaves may be sprinkled onto a salad or eaten sparingly as a trail-nibble. As with any mint, food use should not be overdone. The plant may also be boiled as a potherb—a popular Appalachian repast. James Duke reports feasting on the herb prepared in this way at a West Virginia home, where the plant was called by a corrupted form: "eel-oil." *(Duke 1992a:158)*

Health/Medicine

Self-heal has a fascinating history of medicinal use among European, Chinese, and Amerindian cultures, having been used so widely and with such good results that it derived the name of "heal all [remedies]."

Various Amerindian tribes utilized our *Prunella* to treat a whole host of afflictions:*(Moerman 1986:1:370)* The Quinault Indians rubbed the juice on *BOILS,(Gunther 1973:45)* while the Blackfeet used it to wash a *BURST BOIL* and also *SORES ON THE NECK.(Hellson & Gadd 1974:78)* The Bella Coola used a weak decoction of the root and aerial portions to support a *WEAK HEART.(Smith 1929:63)* The Cree found it invaluable as a chew or gargle for *SORE THROAT,(Holmes 1884:303)* while our area's Ojibwe (Chippewa) Indians implemented the root in a compound decoction for *"FEMALE PROBLEMS"(Smith 1932:72),* The Iroquois found many uses for self-heal: as a wash for *PILES,* as an astringent for *DIARRHEA,* as a digestive aid for *STOMACHACHE,* and to treat *VENEREAL DISEASES.(Herrick 1977:423-25)* Self-heal was an esteemed *febrifuge* in the materia medica of both the Iroquois*(Herrick 1977:424)* and the Delaware.*(Tantaquidgeon 1972:37).*

The herb boosted in popularity among Europeans in the sixteenth century when German military physicians used it to treat a raging contagion, a type of *QUINSY,* that spread throughout the imperial armies in 1547 and again in 1566. The feverish affliction was called "the browns" because it was characterized by a brown-coated tongue. Our herb derived its species name, "prunella," from the German word for "brown" in consequence of its helpful aid during these times. Several ingredients in the plant may have come to the fore for this affliction, including prunellin, which has been shown to be an active antiviral compound. Interestingly, a standard infusion of self-heal has been demonstrated to be the most effective method of preparation to obtain the antiviral effects,*(Yamasaki et al 1996:65)* and this form is what was primarily used during this contagion.

Modern Western herbalists have been little inclined to use self-heal for the spectrum of afflictions as delineated above. Because of its blatant *astringent* properties, however, they have recommended it for *SORES, WOUNDS* (hence its alternate common name, "woundwort"), *LEUKORRHEA,* and as an ingredient in an ointment for *HEMORRHOIDS.* Internally, it has been employed by way of infusion (tea) as a gargle for *SORE THROAT* and *DIARRHEA* and as a *styptic.* In the latter regard, *INTERNAL BLEEDING* has been treated by steeping one ounce of the herb in one pint of water and drinking two fluid ounces at a time. As to its amazing, overall *vulnerary* properties, the famed European herbalist John Gerard had summarized the situation thusly: "The decoction of Prunell made with wine or water doth joine together and make whole and sound all wounds, both inward and outward."*(Gerard 1975)*

Part of the reluctance on the part of many modern natural healers to use self-heal in a more comprehensive way has been, until recently, the lack of scientific validation as to any significant pharmacological properties possessed by the plant. Yet, new research is uncovering some astounding things about self-heal. For one thing, it is now known that the plant is one of the best-known natural sources of rosmarinic acid, a phenylpropanoid that has powerful *antioxidant* and *antiviral* activities (especially proving antiviral against **HERPES SIMPLEX**; one study even found an ophthalmic form of self-heal to cure, or significantly improve, *Herpes simplex* type 1 keratitis in patients-*Zheng 1990:39-41*) Then, a study conducted in the mid-1970s found *Prunella vulgaris* to possess moderate **anti-inflammatory** activity (as measured by the standard, carrageenan-induced, rat-paw-edema lab test).*(Benoit et al 1976:167; cf. Harborne & Baxter 1993:486).)*

A 1988 study demonstrated that self-heal possesses powerful **antimutagenic** properties. *(Lee&Lin1988:228-34)* It contains, for one thing, ursolic acid, a triterpenoid saponin that has recognized **antitumor** activity.*(Lee et al 1988:308-11; Harborne & Baxter 1993:687)* Interestingly, the Chinese have traditionally used it for **CANCER,** long before this information was known.*(Lee et al 1988:308-11)* And indeed, Traditional Chinese Medicine(TCM) finds self-heal flowers (but not leaves) to be a staple in its repertory. Feeling that it has an affinity for the liver and gall-bladder and understanding it to be a cooling herb, the Chinese employ it to treat **JAUNDICE***(Reid 1992:94)* and various other "heat" conditions in the liver, manifested by indications such as **DIZZINESS, IRRITABILITY, HEADACHE, EYE-BALL PAIN,** and **RINGING EARS.** *(Holmes1997:1:541;Reid1992:94)* Health conditions popularly treated by TCM with *Prunella vulgaris* include **HYPERTENSION, GOUT, EDEMA,** and **SWOLLEN LYMPH NODES**.*(Holmes1997:1:541 Tierra1992: 182; MH 10:4:15;Reid 1992:94)*

Interestingly, a series of Chinese studies have also verified **immune-enhancing** properties in the plant*(Markova et al 1997:58-63)*—so promising, in fact, that it is now being examined in connection with **AIDS**.*(Duke 1992a:159)* Research in the 1980s found that the chemical prunellin in this plant was an active anti-HIV compound.*(Tabba et al 1989:263-73)* Thus, in an important study published in 1992,*(Yao et al 1992:56-62)*, a crude extract of this plant was found to significantly inhibit HIV-1 replication. When the active factor was purified, this was observed to block cell-to-cell transmission of the virus, which it accomplished by preventing the virus from attaching to CD4 receptors. A 1993 study revealed further that a simple *hot*-water (as opposed to cold-water) extract was the preferred method for maximizing the inhibiting effects against HIV replication.*(Yamasaki1993:818-24)* An even more recent study found a simple self-heal hot-water extract to provide protection against HIV-induced cell damage (cytotoxicty) and to enhance protection when used in conjunction with the standard drugs, zidovudine and didanosine.*(John et al 1994:481)*

All in all, this is certainly a plant that deserves to be better known and used!

Caution

Excessive consumption of self-heal over an extended time is not recommended due to the high tannin content.

SHEEP SORREL
(Rumex acetosella)

Description

[See photos #123-25 in rear of book.]

A common, perennial herb that is usually encountered growing in colonies. It possesses a slender, erect, four-angled stem and reaches a height of 4-15 inches.

The plant starts as a basal rosette, with untoothed and halbert-shaped leaves (i.e., having two outward-pointing basal lobes presenting the appearance of a sheep's head, and thus the herb's common name). The center stem shoots up in summer, bearing leaves which are alternate, lanceolate, and squared off at the point of attachment.

Many small flowers appear in slender racemes in the apexes of the upper leaves, often catching the eye when a sheep sorrel colony dominates a field. Furthermore, the plant is dioecious, i.e., it bears male and female flowers on separate plants (the former being greenish and the latter being more reddish-orange).

Range & Habitat

A common, North-American field weed, occurring in the eastern 5/6 of our area. Demanding acidity in the soil and low fertility, it prefers full to partial sun. An inhabitant of old fields, pastures, meadows, abandoned gardens, and untamed lawns.

Seasonal Availability

Look for this plant in practically every month of the year except for those with heavy snow. It can be found, for instance, in the late fall, though died back to basal leaves, in which form it hugs the ground, sheltered by grass. When the snow abates in spring, it can be found in such a low-lying form once again.

Chief or Important Constituents

Organic acids(malic, tannic, tartaric, citric, and oxalic), calcium oxalate, potassium oxalate, catechin, chalcones, rutin, auxin, and high amounts of chlorphyll. Both the aerial portions and the root contain anthraquinones(the plant containing physcion and the root emodin and chrysophanic acid). The root also contains tannins. Nutrients include a broad spectrum of vitamins(B-complex, D, E, K, and very high in A[8-12%, or 11,000 IUs per 100g] and C) and minerals(incl. high amounts of iron).

Food

What can be better on a hot summer day than munching on the refreshing, thirst-quenching leaves of this lemony-tasting plant? It indeed makes a superb trail nibble, and one that I personally treasure very much. (However, see "Cautions" below.)

The leaves can alternately be simmered (or steeped) and then chilled for a delicious sorrel-ade, tasting as good as the very best of lemonades! One can also add the leaves to soups and stews as a thickener, or alternately to fish and potatoes as a seasoning.

This common field weed also makes a much beloved soup. However, due to the oxalate content, the soup can be dangerous if prepared too strong, eaten in excess, or eaten too frequently.(See under "Cautions," below.)

Sheep sorrel can be frozen for future use, too. Simply blanch it in boiling water for one minute, then drain and cool it, and finally stuff it into a plastic bag and freeze it.

Health/Medicine

The revered botanical writer John Gerard discussed the sheep sorrel in his famous herbal of the late sixteenth century, noting that it had *drying* and *cooling* properties.*(Gerard1975:318-21)* In harmony with its drying properties, it has proven its worth throughout history as an **astringent,** having been utilized in the full variety of ways in which plants with such a property have been employed: First, sheep sorrel has been put to use for **DIARRHEA.** Secondly, Cherokee Indians poulticed the leaves and blossoms onto **OLD SORES** that refused to heal.*(Hamel & Chiltoskey 1975:56).* Thirdly, the Russians have traditionally used a compress of sheep sorrel tea for **BLEEDING WOUNDS** and even a decoction of the root orally for **INTERNAL BLEEDING.***(Hutchens 1991:256)* (A decoction for this latter purpose has been used by other cultures as well, and is especially thought to be helpful for **STOMACH HEMORRHAGE**-*Callegari & Durand 1977:37*) Fourthly, a decoction of the root has had a valued place in folk medicine for **MENOR-RHAGIA.***(Callegari & Durand 1977:37)*

Gerard, as we noted, had also mentioned the plant's cooling properties. Here, the **refrigerant***(Wren 1972:278)* nature of the herb, combined with some possible **diaphoretic** ability,*(Grieve 1971:2:754)* has proven useful in **fevers,** even as Gerard also went on to note.*(Gerard 1597:318-21;cf. Potterton 1983:178; Foster & Duke 1990:214)* Moreover, the raw leaves were even eaten by Squaxin Indians of Washington state to inveigh against **TUBERCU-LOSIS.***(Gunther 1973:29)*

This delightful and versatile plant has had a variety of additional applications throughout history as well. Many of these have been with reference to either the gastro-intestinal tract or the urinary tract, for which body systems sheep sorrel has shown an especial affinity.*(Potterton 1983:194)* As to the former, the Oklahoma Delaware Indians chewed leaves of sheep sorrel as a **stomachic.***(Tantaquidgeon 1972:75, 132-33)* Then, too, when an Iroquois Indian had an **UPSET STOMACH,** he would be given a decoction of the root to settle it.*(Herrick 1977:311)* Moreover, a tea of the flowers has been reverenced by all sorts of people as an excellent treatment for **ULCERS** in the GI tract.*(Callegari & Durand 1977:37)* Our herb would also even appear to have **vermifuge** properties.*(Hutchens 1991:255)* And it has proven its merit as a **diuretic,***(Wren 1972:278; Mockle 1955:38)* finding application by various cultures for **KIDNEY/BLADDER STONES***(Hutchens 1991:255)* and sundry other **URINARY CONDITIONS.** *(Wren 1972:278)* (However, see "Cautions," below.)

Sheep sorrel's best known application, however, has been for *CANCER.* Back in medieval times, a European doctor named Odo de Meuse, who lived in the city of Lorraine, had heralded sheep sorrel as an herb to treat this feared condition.*(Moss 2000:29)* Then, too, roasted leaves have been poulticed onto *SKIN CANCERS* by several different national groups.*(Foster & Duke 1990:214)* The expressed juice on bread has been used by native Americans as a plaster on same,*(Erichsen-Brown 1989:420)* which application (along with others) was catalogued by Jonathan Hartwell, a scientist with the National Cancer Institute.*(Hartwell 1984)* America's Eclectic physicians of the late 1800s also heralded the benefits of sheep sorrel "where there is a tendency to degeneration of tissue... cancer."*(Scudder 1870:203)* One of them had written: "Its use in the treatment of cancer has been quite extensive, and if we can believe the reports given, it has proven fully as successful as any other remedy. A full description of this method of treatment will be found in the *Eclectic Medical Journal* for May, 1870, page 142.*"(Scudder 1870:203)*

Then there is the incredible story of Essiac®, an herbal brew containing sheep sorrel and three other herbs (burdock, red clover, and Turkey rhubarb), which merits some detail to do it justice. (The history presented below is a synthesis from Walters 1993:105-19; Moss 1998:108-35; Thomas 1993:*passim*, and Olsen 1998:*passim*) It begins in the 1890s, when an Ojibwe-Indian medicine man supposedly used an herbal tea containing sheep sorrel to help heal a woman with breast cancer. It seems that the woman recovered fully and thirty years later, during an unrelated hospital stay in Ontario, met a nurse named Rene Caisse to whom she told her story. Caisse was intrigued and sought to secure the formula in the event that she herself might come down with cancer. Her efforts to obtain it indeed proved successful.

Several years later, in 1924, when Caisse's own aunt was diagnosed with advanced stomach cancer, Rene asked the attending physician, R. O. Fisher of Toronto, if he would approve her attempting a trial with the formula. When he agreed, she dug it out, prepared it, and used it. After two months of drinking the concoction, the aunt got well and even went on to live another 26 years! Dr. Fisher was so impressed that he agreed to partner with Rene in treating cancer patients with the brew. At the same time, the two humanitarians worked nights and weekends experimenting with the formula on mice inoculated with human cancer, modifying it slightly to achieve optimum results, and eventually naming it "Essiac" (Rene's last name spelled backwards).

As word of their many successes with the brew began to spread, opinions for and against the tea began to crystallize: As to the former, nine physicians who came to be acquainted with its successes petitioned the Canadian federal health department to allow Caisse to test the herbal remedy on a wide scale. Rather than receiving a dignified response, however, two hatchet-men were dispatched by Ottawa's Dept. of Health and Welfare to arrest Caisse or at the least, restrain her from continuing with the treatment. When they realized that she was only seeing terminal cases and not charging for her services, they balked at arresting her.

Caisse next went on to treat her own mother, who at 72 years of age had developed inoperable liver cancer and was given a mere two days to live by a respected physician in Ontario. After a mere ten days of daily injections with Essiac, Mrs. Caisse began to convalesce, finally being restored to health and going on to live eighteen more years.

Eventually 55,000 people petitioned the Ontario parliament to allow Caisse to treat cancer patients freely and without fear of medical reprisals. The bill was presented in 1938 and failed to pass by a paltry three votes. From that time forward, it appears that Caisse began to grow disillusioned by her opposition and decided to close her clinic in 1942, though she continued to treat patients privately out of her home. In 1959, however, Essiac saw new life: Then it was that Caisse was invited by a famous New England physician, Dr. Charles Brusch, to test her formula at his prestigious medical center in Cambridge, Massachusetts, both on cancer patients and in lab mice. Brusch (who was once knighted by Pope Paul VI and who was at the time the personal physician to then-senator John F. Kennedy) and his medical director, Dr. Charles McClure, found that the formula reduced pain and tumor size in the human patients and reduced the mass of tumors in mice as well as changed their cell formation. (Some time later, almost as if by fate, Brusch himself would develop cancer, for which he would make the decision to use Essiac as treatment. On April 6, 1990, he would have a statement notarized which would declare: "I endorse this therapy even today for I have infact cured my own cancer, the original site of which was the lower bowel, through Essiac alone.")

Brusch got Sloan-Kettering involved at this time for some testing, having them submit mice to Caisse for Essiac treatment. Although reportedly impressed with her results, they wound up demanding the formula, which she refused to give. Eventually, therefore, the arrangement was terminated. However, in 1973, Caisse herself approached Sloan-Kettering, asking them to have Essiac re-examined. This time, though, she wound up submitting only one herb from the formula—the one that had had the best historical pedigree and which she believed was the most active: sheep sorrel. After some testing, Sloan-Kettering's Vice-President for Academic Affairs, Dr. Chester Stock, acknowledged in a letter to Caisse (dated 10 June 1975) that the sample had indeed produced regression of sarcoma 180 in mice. Later reports sent, however, were discouraging and confusing. Moreover, Caisse was angered when she discovered that the Institute was freezing the herb, rather than boiling it as she had directed, and that they were using animal carcinoma and not human carcinoma as she had anticipated. Frustrated by what she perceived as their uncooperation, she terminated the testing in 1976.

In 1977, tired and disillusioned and on her deathbed, Caisse sold the formula to a Canadian corporation that eventually began to market it, using her chosen name for the formula, which it also proceeded to trademark. In the 1980s, another source began to market a slightly different formula, claiming to have secured this version from Dr. Brusch. This latter product, which could no longer use the now-trademarked name of Essiac, included red clover, watercress, kelp, and blessed thistle. Countless other versions have appeared since. As, however, there is no consensus among unbiased herbal researchers as to the precise measurements of the herbs in the actual formula used by Caisse, none will be noted here.

My own observation on this alleged Ojibwe-Indian herbal tonic is that I have seen astounding results in some people and abysmal failures with others. The many cases described by the authors cited above are impressive, although at least one of these (Ralph Moss) said that he would not be surprised if the actual percentage of genuine and documentable regressions with the herbal brew as the sole treatment would be only 2.5%, the percentage apparently determined in an unpublished report he had read of 162 patients treated with the tea in Israel.

We seem to be on the verge of some good controlled studies, however. One scientific survey, done under the auspices of researchers from the University of Texas-Austin's Center for Complementary and Alternative Medicine, has just recently been published.*(Richardson et al 2000:40-46)*   In this survey of 1,211 cancer patients who had used the marketed herbal tonic based on Brusch's version of Caisse's formula (containing the four basic herbs plus red clover, watercress, blessed thistle, and kelp), the tonic was reported to have produced very good to excellent benefits in 71% of these people, including a retardation of cancer growth in 40.6% and a purported cure in 16.2%! Moreover, 53% felt better with the tonic, 31.5% had more energy, and 22.3% found that it improved the symptoms of their cancer.   So it would seem that this is a subject that merits further study.

As to preparations of sheep sorrel by itself, an infusion is made from 28g (1 oz) leaves to 1 pint (568ml) of boiling water and is usually administered in doses of 2 fluid ounces (56ml).*(Potterton 1983:178)*   A tincture is also available commercially.

Cautions

* Due to the potential for toxicity with reference to the herb's oxalates, extreme caution must be used in any ingestion of this plant for food. Keep fresh leaves to a minimum, as a trail nibble only. Sheep sorrel prepared as a potherb should always be cooked in more than one water, to reduce toxicity, and quantity should be kept down. (Some 500 grams of garden sorrel prepared as a soup proved to be fatal to one man.*-Farre 1989:1524*) Medicinal doses of the plant are generally regarded as safe as long as dosages are strictly observed.

* Although there are some differing opinions on the matter of whether sheep sorrel should be used by those with a **HISTORY** of **KIDNEY STONES,** most modern botanical researchers would say that it should not.

SHEPHERD'S PURSE
[CASEWEED]
(Capsella bursa-pastoris)

Description

[See photo #126 in rear of book.]

An annual weed, growing 4-20 inches tall, with an erect and branching stem. It starts life under the sun as a basal rosette of long-stalked, dandelion-like leaves, 2-5 inches long. Eventually, the stem forms and with it, a number of alternate, stalkless, arrowhead-shaped leaves which tightly clasp the stalk.

Shortly after the stem starts sprouting, the flowers begin to appear. These are small, white, six-stamened, and with four petals arranged as a cross. They are borne as elongate racemes at the end of the stem's branches. The lower ones open first, leaving green buds at the top of the raceme simultaneously.

Triangular, two-parted seed pods (i.e., fruits) evolve from blossoms appearing at the bottom of the raceme and then also from other flowers as the process moves upward at a rapid pace. Measuring from 5 to 8 millimeters long, these fruits are flattened, notched or indented at their tips, and attached to their stalks at their apex. The plant's common name derives from the fact that these pods resemble the purses that medieval shepherds used to wear.

Range & Habitat

A cosmopolitan weed, thriving in dry, open places—wasteland, old homesteads, old fields, gardens, and ill-kept lawns.

Chief or Important Constituents

An alkaloid(bursine), other unidentified crystalline alkaloids, sinigrin, hyssopin, beta-sitosterol, flavonoids(including rutin, quercetin, diosmetin), choline, acetylcholine, tyramine, histamine, polypeptides, saponins, bursi[ni]c acid, fumaric acid, thiocyanic acid, chlorogenic acid, malic acid, pyruvic acid, tannin. Nutrients include goodly amounts of vitamins A and C (and possibly some K, although authorities are divided on the presence of this nutrient) and minerals such as iron, thiamin, riboflavin, and potassium. This is also one of the higher sources of protein among leafy greens.

Food

Leaves may be consumed raw—sparingly as a trail nibble or in salads with other greens. Best when consumed before the plant flowers, they are peppery in taste like those of the related peppergrass *(Lepidium spp.)*(see under that entry).

The heart-shaped seed pods are also edible and instantly recognizable by taste as something that, when dried and ground, would make a marvelous pepper substitute. Fresh from the plant, they impart pizzazz to other greens when mixed with them in a salad. Fresh or dried, they can be added to the very end of a camp stew as a superb seasoning.

American Indians used to dry and crush the pods to free the seeds, then winnow and parch them, and finally grind them into flour or cook them as mush. That's a lot of work for today's less enterprising soul!

Food can even be procured from the roots of shepherd's purse, which can be dried and ground for use in soups and stews, or even implemented as a ginger substitute.

Health/Medicine

Shepherd's purse is a very unique and interesting plant that has played a major role in the history of herbal therapeutics, but is tragically underused nowadays.

Most well known and utilized has been the plant's role as a *hemostat*. Back in the late 1960s, an animal experiment evinced that shepherd's purse possessed this ability, as well as a *hypotensive* effect, in rats.*(Kuroda & Kaku 1969:151)* A study in the early 1990s expanded on this research, providing further clarification as to the process and the mechanisms involved.*(Vermathen & Glasl 1993:A670)* Certainly the herb's amines are one factor here, yielding powerful *vasoconstrictive* and *hemostatic* effects.*(Stuart 1982:38)*

Historically, then, shepherd's purse has been utilized for **BLEEDING WOUNDS** (it was used thusly by soldiers on the battlefield during WWI), **NOSEBLEEDS** (wadded cotton soaked in the plant's juice was inserted into the nostrils), **FIBROID TUMORS,** and **HEMATURIA.** It is most famous, however, as an application for **UTERINE BLEEDING,** including **MENORRHAGIA** (esp. with heavy bleeding on the first few days, as long as the discharge is constant and bright-red-colored or colorless and not brownish) and **METRORRHAGIA.** Aside from stemming blood flow, however, shepherd's purse has also been implemented to offset the **TENDENCY TO HEMOR-RHAGE** (the German physician and phytotherapist R. F. Weiss cited research demonstrating a stronger effect in this regard than could be had from decoctions of plantain and yarrow—two other classic hemostatic herbs.-*Weiss1988:311*) Astringent properties in the plant have yielded use of it as well for **DIARRHEA** (aside from the tannins, another chemical in the plant has been shown to be a smooth-muscle contractor for the small intestine—*Kuroda&Takagi1969:382-91;Jurison1971:71-79*) and for **ATONIC CATAR-RHALCONDITIONS.** *(Millspaugh1974:95,96;Ellingwood1983:354;EthnobotDB;Carse1989:157;Hoffmann 1986:236)*

This herb seems to have an especial affinity for the genito-urinary system, on which it most directly acts, toning the tissues. Here it has long been successfully used by herbalists for **CHILDREN'S BEDWETTING**, often in conjunction with an even more astringent herb like agrimony *(Agrimonia eupatoria)*.*(Hutchens 1991:248;Carse 1989:60)* Dr. G. F. Parks, an Eclectic physician practicing in the early twentieth century, found it especially called for with those urinary problems where people were passing profuse urine loaded with thick material that was slimy or dusty. He found that it worked best in high-strung women, but that it was also applicable to men who had acute gonorrhea"where they just slobber." It has also traditionally been used in China for **DYSURIA**.*(EthnobotDB)*

Another important study demonstrated that uterine tone is improved by the plant's polypeptides, which were shown to possess *uterine contractile activity* similar to that of oxytocin.*(Kuroda & Tagaki 1968:70)* This was confirmed and further elucidated in a Bulgarian study published in 1981.*(Shipochliev 1981:94-98)* Not suprisingly, then, shepherd's purse has historically been used as a *parturient* to assist the birthing process. (Here its hemostatic properties can also help to manage any unusual bleeding—or even post-partum hemorrhaging—which may occur. However, herbalist Susun Weed cautions that shepherd's purse should only really be used to deal with an existing problem in this regard, not as a mere hemorrhagic preventative, citing several instances when two dropperspersfull of the tincture given by midwives during labor as a preventative resulted in huge blood clots which were painfully hard to excrete and which made it difficult for the uterus to clamp down during the birthing.*-Weed 1986:72*)

Yet another lab study in the late 1960s showed that our herb is possessed of *anti-ulcer* effects.*(Kuroda & Tagaki 1969:392-99)* This study revealed that shepherd's purse, while not directly affecting the secretions of the stomach, yet speeded the recovery of*STRESS-INDUCED ULCERS*. This is interesting in that the Mohegan Indians had used an infusion of the plant's seed pods as a *stomachic,(Tantaquidgeon 1928:265)* while our area's Ojibwe (Chippewa) Indians similarly employed shepherd's purse for **CRAMPS** in the GI tract.*(Densmore 1974:344)* Further as to the plant's affinity for the stomach, the Eclectic physician Finley Ellingwood had noted: "Dr. Heinen of Toledo treats non-malignant abdominal tumors in women with better results by adding five drops of capsella three times a day to the other indicated treatment."*(Ellingwood 1983:354)*

The 1960s lab study referred to immediately above also showed that shepherd's purse was possessed of *anti-inflammatory* and *diuretic* properties.*(Kuroda & Tagaki 1969:392-99)* The former property was avidly appreciated by the Menominee Indians, who used an infusion of the herb as a wash for **POISON IVY**.*(Smith 1923:33)* As to the plant's diuretic capability, which was appreciated by the Eclectics,*(Ellingwood 1983:354)*the study cited above showed that such results accrued from a palpable increase in the glomerular filtration apparatus of the kidneys. This combination of anti-inflammatory and diuretic properties explains why the plant also has a cherished reputation for aiding*GOUT*. Dr. Ellingwood and Charles Millspaugh have both observed as well that the bruised plant has traditionally and effectively been applied to **RHEUMATIC JOINTS**.*(Ellingwood1983:354;Millspaugh 1974:461)*

Various Amerindian tribes and other cultures have insisted that the seed-pods of shepherd's purse are a reliable ***anthelmintic***, with said ability being attributed to their pungency.*(Speck1917:319).* ***Antimicrobial*** activity against certain gram-positive organisms has also been documented.*(Kuroda 1977:207; Moskalenko 1986:231-59)* This may be due in part to the isothiocyanates. It is of interest here that the herb has traditionally been implemented as a ***urinary antiseptic.*** *(Wren 1988:250)*

Finally, shepherd's purse has been used for ages in Poland in the treatment of cancer.*(EhtnobotDB)* Again, modern science has shed some light on this use: Lab studies conducted in the mid-1970s found ***anticarcinogenic*** effects for this plant documented against Ehrlich solid tumors, which effects were originally attributed to the herb's content of fumaric acid.*(Kuroda et al 1976:1900-03)* However, this acid occurs in many plants, very few of which have been linked to anticarcinogenic activity. Current research, therefore, has focused more on the plant's content of bursine and bursinic acid.

Cautions

  * In that shepherd's purse is a member of the mustard family, physical contact with the seeds or leaves for any length of time can cause blistering.

  * As this plant has uterine stimulant properties, do *not* use in **PREGNANCY** unless as an aid to parturition at the end of term or earlier in term as an emergency hemostat during an episode of hemorrhaging (any such use, however, should be done under a qualified practitioner's care). In this latter instance, however, I generally prefer to use another herb, black haw (*Viburnum prunifolium*), especially if the bleeding has been accompanied by cramping and anxiety (its specific indications), since this botanical is not only much safer, but actually even anti-abortive.

SHOOTING STAR
[AMERICAN COWSLIP; PRAIRIE POINTERS]
(Dodecatheon spp.)

Description

[See photo #127 in rear of book.]

An extremely beautiful native perennial! It grows 10-20 inches high and bears a single, smooth stalk.

The leaves are entirely basal, spatula-shaped, smooth-edged, and grow to six inches long. The inch-long flowers are nodding and have five, long, swept-back petals as well as five, yellow, protruding stamens growing to a common point. These flowers, which can be likened to badminton shuttlecocks in appearance, occur long-stalked in clusters at the tip of the stem.

Two species occur in our area, to wit:

Dodecatheon meadia has 6-30 flowers (rose, purple, or white in color) per umbel and leaf bases that are reddish.

D. amethystinum [alt: *D. radicatum*] (amethyst shooting star) has 2-11 rose-purple (occasionally white) flowers per umbel and green leaf bases.

Range & Habitat

The lovely shooting star thrives in open woods (medium to dry), meadows, and prairies. Its range in our area is from southeastern Minnesota eastward into Wisconsin.

Food

Fresh leaves can be added to salads.*(Willard 1992:171)* The texture is tender and the taste is pleasant.

The leaves and roots can also be prepared as a potherb, either by roasting them or boiling them.*(Kirk1975:195)* The Yuki Indians enjoyed both of these plant parts, roasting them in campfire ashes.*(Chestnut 1902:378)*

Health/Medicine

When Blackfeet children would develop **CANKER SORES**, their parents would make an infusion of the Western species, *D. pulchellum ssp. pulchellum*, and have them gargle it to help heal the painful ulcers.*(Hellson & Gadd 1974:76)* They also utilized a luke-warm infusion of the leaves as drops for **SORE EYES**.*(Hellson&Gadd 1974:81)* For the same discomfort, the Okanagan-Colville Indians preferred to use an infusion of the roots.*(Turner et al 1980:117)*

SKULLCAP

[SCULLCAP]

(Scutellaria spp.)

Description

[See photo #128 in rear of book.]

A perennial member of the mint family, with that family's characteristic square stem and paired leaves. The blue, two-lipped flowers are bell-shaped to horn-shaped, with a "hump" on top of the calyx surrounding their base that looks somewhat like the skull cap worn by the Romans (and hence the plant's common name). This "hump" in an important feature distinguishing this plant from lookalikes.

Two major species inhabit our region:

Scutellaria epilobiifolia [S. galericulata] (common skullcap, marsh skullcap), has a slender stem and grows 6-23 inches tall. Its short-stalked or sessile leaves are lanceolate to ovate. The lower lip on its inch-long flowers is flat and three-lobed. These blossoms are sessile (or nearly so) and they appear *singly* in axils of the upper leaves.

S. lateriflora (mad-dog skullcap, bushy skullcap) has oval, pointed leaves that are barely stalked. Its flowers are smaller (only 1/2-inch long) than in the above species and they appear clustered *on one-sided racemes* that spring from the leaf axils.

Range & Habitat

Throughout our area in moist places: marshes, swamps, bogs, shores, wet woods and thickets

Chief or Important Constituents

Iridoid glycosides (incl. catalpol), flavonoids & other glycosides (incl. scutellarin, scutellarein, apigenin, apigenin-7-glucoside, baicalein, baicalin, hispidulin, luteolin, wogonin), volatile oil, bitter principles, tannins. Nutrients include significant amounts of potassium, magnesium, phosphorous, and calcium.

Health/Medicine

Here is an herb that, although being marketed, is incredibly understudied and underutilized as to its full potential. In clarifying its plethora of applications, then, I would like to first discuss its value as an *emmenagogue*, and to do so by pointing out that both the Cherokee and Cree Indians used it for improving the *quality* of menstruation. *(Hamel & Chiltoskey 1975:55; Goodchild 1984:145)* While it is perhaps a bit inferior *in strength* to some other emmenagogues discussed in the present work—plants such as motherwort, blue cohosh, pineapple-weed, smartweed, catnip, wormwood, or horsemint (*see under each of those entries*)—it is nonetheless of great value, especially when applied appropriately. And here it is definitely an herb of choice where the *AMENORRHEA* is accompanied by *nervousness, anxiety*, and *worry*,(*McIntyre 1994:273*) and *the* herb of choice when *exhaustion* accompanies it as well. (Motherwort, catnip, and blue vervain are other choices when anxiety accompanies amenorrhea, but each has specific indications that differ slightly from those of skullcap—*see under each of these entries for specifics*)

In fact, skullcap *is* the remedy *par excellence* for *NERVOUSNESS, REST-LESSNESS, ANXIETY, WORRY,* and *PREMENSTRUAL TENSION* when accompanied by *NEURASTHENIA (NERVOUS EXHAUSTION).* It is often quite helpful, too, in *IRRITABLE BOWEL SYNDROME*, (*Mills & Bone 2000:180, 175*) due both to its nervine effects and to some carminative properties owing to its essential oil. People with*NERVOUS HEART DISORDERS* may also benefit from this herb,(*Smith 1999:51; Moore 1994:5:4; Tilgner 1999:107*) and *S. epilobiifolia* was in fact used by our area's Ojibwe (Chippewa) Indians in this regard.(*Smith 1932:372*) Then, too, skullcap has proven itself to be a real gem for helping people to break an *ADDICTION* to a recreational drug, whether that be a "hard" drug or a "soft" one such as nicotine or alcohol. Of the latter, one of America's Eclectic physicians had written: "In treating heavy drinkers who wish to give up the habit, I know of nothing better than *Scutellaria*. It steadies and sobers the patient and brings on sleep and appetite."(*Felter & Lloyd*) Indeed, in view of its unique combination of minerals, essential oil, and glycosides, skullcap can be said to possess—as herbalist Peter Holmes points out—a *trophorestorative* effect on the nervous system.(*Holmes 1998:2:535*)

Because it also possesesses *antipasmodic*(*Claus 1961:219-20;Wren 1988:247*) and *anticonvulsant*(*Wren 1988:247*) properties, skullcap is specific for what America's Eclectic physicians defined as any "nervous disorder characterized by irregular muscular action, twitching, tremors and restlessness, with or without incoordination."(*Ellingwood 1983:124*) Skullcap has thus been classically and effectively used for *CHOREA, CEREBRAL PALSY, PARKINSON'S DISEASE, DELIRIUM TREMENS, TENSION HEAD-ACHES, STRESS-CAUSED MUSCULAR SPASMS*, and *EPILEPSY* (especially to control *petit-mal* seizures—but also to some extent for *grand-mal* ones).(*BHP 1983:183*)

Skullcap was official as a *tranquilizer* in the *U.S. Pharmacopoeia* from 1863-1916 and thereafter in the *National Formulary* until 1947. It is still official in most European pharmacopoeias for epilepsy, hysteria, and nervous tension. And it goes without saying that this unique and remarkable plant has been a mainstay in my own herbal practice. Here I have probably used it most for its plethora of nervine properties and especially for its *anti-addictive* properties. As for the latter, I have found that a teaspoon of the tincture every hour—combined with one or two teaspoons of Oat (*Avena sativa*) milky-seed tincture—can often pull the most terribly addicted person through withdrawal. As for its neuromuscular-relaxing effects, though it is weaker in this regard than a powerhouse like kava (*Piper methysticum*)—which in my experience seems to possess the strength of diazepam to relax the neuromuscular system—it is safer for long-term use. While I thus often prefer to use kava as an initial treatment for chronic anxiety, tension, and tight muscles—and almost always for acute episodes of either—I much prefer skullcap for sustained use. I value it *highly* for its antispasmodic properties and find that it combines well in this regard with another antispasmodic herb, blue vervain *(Verbena hastata)* *(see under that entry)*.

S. lateriflora was also formerly used for **RABIES**, which manifests symptoms like what we have been discussing (muscular tension and spasms). According to the Natura physician R. Swinburne Clymer, the classic application here had been to use it as a poultice immediately after the bite and also to take it internally as a tea of the *fresh* plant four times a day for seven days, along with a tea of purple coneflower *(Echinacea* spp.), or by way of a tincture of 2-15 drops combined with an *Echinacea* tincture of 15-30 drops, and as frequently as indicated.*(Clymer 1973:110)*

Modern scientists have assumed that this application could not possibly possess any efficacy and so skullcap has never actually been tested for action against rabies. While any actual curative action might seem improbable, it is interesting to note that it was a physician, Lawrence van Derveer, practicing in New Jersey during the 1770s, who is credited by historians as having pioneered just such a rabies-healing campaign with this herb. Van Derveer, who, according to the historical records, is said to have healed both rabid humans and animals, understandably created quite a stir in his area at the time. Unfortunately, one of his patients, a Mr. Daniel Lewis, was so impressed with apparently having been healed of hydrophobia by the doctor, that he endeavored to capitalize on the cure, styling himself "Mad Dog Lewis" and shamelessly attempting to market the skullcap medicine surreptitiously in his own state of New York as "Lewis' Secret Cure." As botanical historian Virgil Vogel has noted, the skullcap treament thus "fell into direpute when it was adopted by quacks who promoted it by advertising."*(Vogel 1970:367)*

However, shortly after Van Derveer's death, in 1819, L. Spalding collected case histories from the good doctor's work and calculated that of 850 rabid patients (some of whom were animals) that this physician had treated with skullcap, only three had died!*(Spalding 1819)* Renowned botanist Constantine Rafinesque, writing somewhat later (in 1830), summarized this matter as follows: "[*Scutellaria lateriflora*] is laterly become famous as a cure and prophylactic against hydrophobia. This property was discovered by Dr. Vandesveer[sic], towards 1772, who has used it with the utmost success, and is said to have till 1815, prevented 400 persons and 1000 cattle from becoming hydrophobus, after being bitten by mad dogs... Many empirics, and some enlightened physicians have employed it also successfully."*(Rafinesque 1830:82-83)*

That is an enthusiastic endorsement, and from the pen of a respected botanical scholar. What prompted such assuredness on Rafinesque's part? Charles Millspaugh informs us that Rafinesque was greatly impressed by the testimony of a Dr. White, who had assured him that he himself had been cured of rabies with skullcap after having been bitten by a mad dog, while others bitten by the same animal had died without the treatment..*(Millspaugh 1974:470)* But, there is more: The Eclectic pharmacist John Uri Lloyd investigated this chapter in medical history in still greater depth and uncovered some additional details that would likewise seem difficult to ignore. For instance, he highlighted a particular case of Van Derveer's where "a man, two hogs, and two cattle, were bitten by a mad dog. *Scutellaria* was given the man and one hog. Both recovered. The other animals died of hydrophobia." *(Lloyd 1929)* Might, then, skullcap actually possess some activity against rabies after all? Lloyd would spend the rest of his life trying to get the scientific community to set up controlled studies to answer that very question, but to no avail. Of course, until such a time as those studies are undertaken, this question cannot be positively answered. But, the evidence we have from the historical records is surely most intriguing.

But, having sung its praises, I would be remiss not to point out that skullcap should almost always be used as a *fresh-plant* extract rather than as a tea or encapsulation of the dried herb. The Eclectic physician John Scudder had elaborated well in this regard: "Here we have another remedy that loses its medicinal properties by drying, until by age they are entirely dissipated. I have seen specimens furnished physicians by the drug trade that were wholly worthless—no wonder they were disappointed in its action.... I value the remedy highly, but only recommend it when prepared [as a tincture] from the fresh [flowering] plant." *(Scudder 1870:213)* This difference seems to be especially pronounced with reference to the plant's antispasmodic properties.

On a final note, our skullcap is quite a different plant from the Chinese baikal skullcap (*S. baicalensis*). The latter plant's constituents are as unlike our own species as apples are from oranges! Its variety of uses also differs greatly from those of our own. What is more, the Chinese skullcap has the advantage of having had numerous scientific studies performed on its behalf, something sorely lacking for our American species. Indeed, the situation has changed little since Rafinesque's days of the early nineteenth century when he lamented concerning our American species: "We lack...a series of scientific and conclusive experiments made by well informed men." *(Rafinesque 1830:83)* Hopefully, then, this great deficit in plant research will soon be remedied.

Cautions:

* Skullcap is **CONTRAINDICATED** in **PREGNANCY** due to its *emmenagogue* and *antiplacental* action.

* For a period running from the late 1980s up until the mid 1990s, some commercially available skullcap purchased from European wildcrafters was adulterated with a similar looking plant, germander, which possesses the potential for liver poisoning and did in fact result in hepatitis for at least one person. This problem seems solved for now, but safety concerns may perhaps still motivate cautious consumers to secure skullcap from reliable American sources if not wildcraft it carefully themselves using identification guidelines delineated in the present work, additional field guides, and the personal aid of an experienced wildcrafter.

SMARTWEED
[WATER PEPPER; ARSSMART]
(Polygonum spp.)

Description

[See photos #129 & 130 in rear of book.]

A group of annual and perennial species having branched stems with knotlike (sheathed) joints, flowers arranged in terminal spikes, an acrid juice, and a preference for damp ground. A number of species exist in our area, including:

Polygonum amphibium (water-smartweed) dwells either on water-soaked ground or on the surface of a body of water (where its stem can reach eight feet long!). It has pink flowers arranged in clusters 1-4 times longer than they are wide.

P. coccineum (common swamp smartweed, bigweed lady's thumb) is a perennial reaching a height of 1-3 feet. It possesses hairy, alternate leaves and pink, rosy, or white flowers.

P. hydropiper (common smartweed, water pepper) grows 8-24 inches high and has wavy-edged, lanceolate leaves with a reddish tinge. Its flowers are green and their spike arrangement is very slender, reaching 1-3 inches in length and sagging at the tip.

P. lapathifolium (dock-leaved smartweed, pale smartweed) is an annual, 3-4 ft tall, with smooth and branched stems. Its leaves are elongate and pointed at both ends. The flowers in this species are white to rose in color and the spike on which they are clustered nods quite blatantly at its tip.

P. pennsylvanicum (pinkweed, or Pennsylvania smartweed) is an annual that may take on either an erect or a sprawling nature, reaching 1-5 ft. It has a branching stem and lanceolate leaves with unfringed sheaths. The flowers are rose, pink, or white and they do not nod, but maintain an upright posture.

Range & Habitat

One or another species grows throughout our area in wet, open places—shores of ponds, streams, lakes and in marshes, swamps, roadside ditches, and wet fields.

Chief or Important Constituents

Tannin(21% in the roots of some species and 18% in their stems!), gallic acid, ellagic acid, acetic acid, malic acid, valerianic acid, flavonoids(including high amounts of rutin), beta-sitosterol-glucoside, essential oil(including borneol, camphor, carvone, p-cymol, fenchone, terpineol, and other chemicals), sesquiterpenes, and polygonolide(an isocoumarin).

Food

The peppery leaves can be added in small quantity to camp stews, but the noted wild-foods authorities Fernald and Kinsey remark that even so cooked and diluted they tend to cause tearing of the eyes owing to their pungency, and should therefore be used with caution. *(Fernald & Kinsey 1958:173)*

Health/Medicine

The 21st edition of the *U.S. Dispensatory* says that *P. hydropiper* possesses "medicinal properties," and that it is "esteemed diuretic, and...used with asserted success in...uterine disorders." It then notes its use by physicians in the treatment of **UTERINE HEMORRHAGES.** *(Wood et al 1926:1228-29)* Concordantly, the Menominee Indians of our own area used *P. pennsylvanicum* for **MOUTH HEMORRHAGE**, implementing a tea of of the dried leaf. *(Smith 1923:47)* The **hemostatic** properties, acknowledged of late in the *PDR for Herbal Medicines,* *(PDRHrbs 1059)* accrue perhaps partly due to the plant's rutin content which strengthens capillary walls, *(Foster & Duke 1990:214)* but undoubtedly mainly because of an **astringency** owing to smartweed's high tannin content. Indeed, the 21st edition of the *U.S. Dispensatory* quoted above cites the research of physician B. Woodward to the effect that dried *P. punctatum* (= *P. acre*) (water smartweed) contains an unusually high amount of tannin (18%) for an herbaceous plant, which content and consequent astringency allowed him to implement a tincture of that species most successfully to treat **DIARRHEA** and **DYSENTERY**.

Smartweed's astringency has lent its use to our area's Mesquaki Indians, enabling them to utilize *P. pennsylvanicum* for **PILES.** *(Smith 1928:236)* It has also allowed *P. coccineum* (common swamp smartweed, or bigweed lady's thumb) to be implemented in Canadian folk medicine—the stems and leaves soaked in cold water for **DIARRHEA** and the root tea as a mouth wash for **CANKER SORES.** *(Angier 1978:160)* Early American botanical writer Manasseh Cutler even informs us that Dr. William Withering, credited with the discovery of digitalis, was adamant that smartweed does indeed relinquish these sores. *(Cutler 1785)* Not incongruously, herbal journalist Will Messenger relates how a gargle of smartweed tea healed—almost overnight—a bad gash in his cheek that had been afflicted by an overbite. *(Messenger 1992:3)* Then, too, children's **FLUX** was treated by the Cherokee Indians with an infusion of *P. hydropiper's* roots. *(Hamel & Chiltoskey 1975:55)* Traditional Chinese Medicine (TCM) also uses *P. hydropiper* for diarrhea and even for **BACTERIAL DYSENTERY,** for which it is regarded as a specific. *(Bareft 1977:927)*

The 21st ed. of the USD also noted that *P. hydropiper* has been of value in treating **AMENORRHEA.** This **emmenagogue** effect, in fact, was a favorite use for the plant by nineteenth-century and early-twentieth-century physicians. *(Ellingwood 1915:482-83; Johnson 1884:237; Clapp 1852:854)* Eclectic physician John Scudder wrote in the 1870s that it is "one of the best emmenagogues, especially when the arrest is from cold." *(Scudder 1870:181)* Dr. John Gunn, in his popular household medical guide of the nineteenth century, recommended a strong tincture for this complaint, dosed at one or two teaspoonfuls, three times a day.*(Gunn 1859-61:858)* Alma Hutchens adds that internal use can be accompanied by a fomentation of the hot tea on the lower back where the pain is particularly present.*(Hutchens 1991:294)* The use for ammenorrhea remains smartweed's most popular application in the United Kingdom today.*(Wren 1998:252-53)*

      Country herbalist Tommie Bass had used smartweed for **KIDNEY PROBLEMS**.*(Crellin & Philpott 1997:171, 176)* The Cherokee Indians found that it stymied **PAINFUL** or **BLOODY URINATION** and also helped to dislodge **URINARY GRAVEL**,*(Hamel & Chiltoskey 1975:55)* an application that Dr. Gunn also endorsed.*(Gunn 1859-61:858)* Indeed, Charles Millspaugh remarked that it was a most "powerful diuretic when fresh."*(Millspaugh 1974: 567)* These **diuretic** properties were undoubtedly also partly responsible for the plant's many successes with **DROPSY** (as practiced by the Malecite/Micmac Indians-*Mechling 1959:244*), and **RHEUMATISM/GOUT**,*(PDR Hrbs 1059)* such as with the Mexicans who bathed rheumatic limbs in a bath into which an infusion of smartweed was poured.*(Millspaugh 1974:567)*. But direct **anti-inflammatory** properties are no doubt contributory as well, since the Cherokees poulticed **SWOLLEN** or **INFLAMED BODY PARTS** with this plant.*(Hamel & Chiltoskey 1975:55)* Houma Indians implemented the boiled roots of *P. punctatum* for **PAIN** and **SWELLING** in the **LEGS** and **JOINTS**.*(Speck 1941:58)* Will Messenger even relates of an 82-year-old acquaintance who swears that smartweed soaked in vodka makes the best rubbing liniment around for **SORE MUSCLES, BRUISES,** and **SPRAINS!***(Messenger 1992:3)*

      In both western and eastern herbology, smartweed is thought to possess *heating* and *drying* energies.*(Grieve 1971:2:743; Gerard 1975:360-62; Bareft 1977:478)* As such, it has traditionally been used for **RESPIRATORY PROBLEMS** of a cold, damp nature. Here the Potawatomi Indians found it most helpful for a **COLD**, but specifically one accompanied by a **FEVER**.*(Smith 1933:72)* Though a warming herb, the use for fever was undoubtedly helpful because smartweed is also, as the Eclectic physician John Scudder explained, "one of our most certain stimulant diaphoretics."*(Scudder 1870:181; cf. Grieve 1971:2:743; Johnson 1884:237)* In other words, as the Iroquois Indians elaborated, it is the presence of "fever when you are continually cold and cannot sweat" that actually indicates the necessity for this herb.*(Herrick 1995:144)*

      The tops of *P. pennsylvanicum* were used as a tea by certain Indian tribes to help deal with **EPILEPSY**.*(Duke 1990:160)* South-Carolinan folk healers have wet the leaves of this species with vinegar and then wrapped them around the head to relieve a **HEADACHE**.*(Morton 1974:115)* Various smartweed species have also been used in folk medicine for treating **CANCER**. *(Hartwell 1970:373)*

      As to the form and dosage best used (see "Cautions," below), David Potterton says to infuse one ounce of leaves in one pint of water.*(Potterton 1983:281)* Hutchens writes of cutting up the herb into small pieces and then infusing one teaspoonful in one cup of warm (not hot) water, imbibing it thereafter in wineglassful amounts.*(Hutchens 1991:294)* Southwest-American herbalist Michael Moore says to use a standard infusion, 2-4 ozs. a dose, as needed.*(Moore 1994b:18)* Maud Grieve and Dr. Gunn both note that any tincture must be made from the fresh plant, but that a cold-water infusion is the best form for **GRAVEL, GOUT, COLDS**, and **COUGHS**. *(Grieve 1971:2:743; Gunn 1859-61:858)*

Cautions

This plant produces very *painful, rubefacient,* and sometimes even *caustic,* effects when the *fresh plant* or *undiluted juice* comes in *contact with mucus membranes!* (It is not called smartweed for nothing!) Because of the severe acridity of this plant and the consequent dangers in handling it, it is undoubtedly best to use it medicinally only by way of a professional preparation and under professional guidance. Then, too, although smartweed can be prepared as a wild food, as we have noted above, this requires great care and should probably not be attempted by the average person. The exception would be if such a person was in the company of an experienced forager or elderly relative who used to prepare it in "the old days," so that one could get "hands-on" training in its proper harvesting, preparation, and consumption.

SOLOMON'S SEAL

(Polygonatum biflorum; P. pubescens)

Description

[See photos #131-32 in rear of book]

A perennial herb growing from 1 to 6 feet (usually 3-4 feet). The stem, elongate and unbranched, is strongly arched so that the plant grows practically horizontally.

Leaves of Solomon's seal are oval to elliptical, smooth and waxy, parallel-veined, and toothless. They are about 4 inches long and 2 inches wide, and are arranged alternately along the stem. Smooth veins on the underside of the leaves characterize *P. biflorum,* while *P. pubescens* is marked by—what else?—pubescence.

Flowers of *P. biflorum* are greenish-white and ½ to 1 inch long. Those of *P. pubescens* are a bit smaller and more yellowish green. *Polygonatum* flowers can dangle singly from the leaf axils (e.g., in *P. officinale* [European species]; sometimes also *P. pubescens*), in pairs (*P. biflorum;* sometimes *P. pubescens*), or in clusters (*P. multiflorum* [Asian species]; sometimes *P. biflorum*). Six short, green-spotted lobes hang from the bottom of each flower, so that the blossom has an overall "bell-like" look.

In late summer, the flowers are replaced by berries. At first green, they finally—after quite a long time—turn blue or blue-black. They remind one of blueberries, but are larger, being three-celled and with one or two seeds in each chamber.

Attached to the roots is a creamy-white, fleshy rhizome, on which can be found one or more rounded scars (the "seals" in the name "Solomon's seal"—they resemble wax seal impressions). One can tell the age of a Solomon's seal plant by counting the number of these "seals" on the rhizome, for it contains one "seal" for each year of life

Range & Habitat

Moist wooded areas, thickets, steambanks, roadsides. Range is eastern and midwestern US and Canada. In our area, *P. pubescens* thrives in the eastern half, while *P. biflorum* can be found throughout, except for the far northeast.

Chief or Important Constituents

Flavonoids, acetidin-2-carboxylic acid, mucilage, gum, and pectin. Steroidal saponins(saponosides A and B)—based upon diosgenin—occur in the rhizomes. Asparagin is a further constituent of the rhizome. Certainly a strong anti-inflammatory compound is present in at least the rhizome, and this probably corresponds to one or more of the saponins and/or flavonoids. In that Solomon's Seal is in the lily family, it is not surprising that the cardioactive glycoside convallarin is also present—even as in its famous relative, lily of the valley (*Convallaria*). As for other plant parts, anthraquinones are known to exist in the berries and saponins in their seeds.

Food

Solomon's seal rhizomes make an excellent cooked repast—boiled or baked. One should be prepared for a strong, unpleasant odor, however. In marked contrast to this repulsive smell, though, the *taste* of the cooked rhizomes is most delicious!

The rhizome is a useful survival food to know because, like many roots and tubers, it provides long-burning fuel, in contrast to the quick and short-lived energy provided by leaves and berries. Edible-wild-plant author Oliver Medsger, in fact, reminds us of Francis Parkman's comment that this plant even helped save certain French colonists in early America from starvation.*(Medsger 1972:163)* Very young shoots can also be prepared as a potherb by cooking them in water for 10 minutes

Health/Medicine

This lovely woodland plant shows significant *anti-inflammatory* properties.*(PDRHrbs)* Thus, a poultice made from the rhizome has been used in many different cultures to treat inflammatory conditions such as *POISON IVY, ARTHRITIS, GOUT,* and *RHEUMATISM.* *(Naegele 1980:206;Hutchens 1973:315)* The juice from the rhizome has been implemented to soothe *SUNBURN,* as well as droppered into ears to treat *EARACHE.*

Probably the plant's most well-known use is as a treatment for *BRUISES.* Immediately coming to mind is sixteenth-century horticulturist John Gerard's famous statement about it proving able to heal—in a mere one or two days—"women's willfulness in stumbling upon their hastie husband's fists."*(Gerard 1975)* Gerard had noted that the fresh rhizome was, in these cases, mashed and applied topically in conjunction with cream or milk. Writing a century and a half later, botanical scholar J. Hill would call Solomon's Seal "a vulnerary of the first rank," but go on to elucidate that "our country uses it *internally* in cases of bruises from blows."*(Hill 1751:654; emphasis added)* A number of Amerindian tribes also used it for bruises, but chiefly by way of topical application. The Rappahannocks, for example, used two methods to make a salve: (1) they boiled down the roots; (2) they stewed the berries and mixed them with elder bark and hog grease.*(Speck 1942:30)* The Cherokees, on the other hand, simply bruised and heated the rhizome and applied it topically.*(Youngken 1924:407)*

Both highly *astringent* and *demulcent* in its effects, Solomon's seal has an *affinity to mucous membranes* and has been used to treat various problems involving these as well as the skin in general. Thus, back in 1884, Laurence Johnson had observed: "In decoction, [Solomon's Seal] is employed as a domestic remedy to allay irritation of mucous surface, and in rhus[poison ivy/sumac] poisoning."*(Johnson 1884:276)* Then, too, the Iroquois Indians used to put a smashed rhizome in a glass of water, soak a cloth in this solution, and squeeze the fluid into *SORE EYES.*(Herrick 1995:244)* The Russians have long used a decoction of the dried rhizome to treat peptic *ULCERS.*(Hutchens1991:254-55)*

Benefits would not seem to be limited to the upper G.I. tract, however. The early-American botanist Constantine Rafinesque, author of a brilliant and valuable tome on the medicinal uses of American plants, made the observation that the powerful mucilaginous effects of powdered *Polygonatum* roots "appear to be equivalent to *Ulmus fulva* [Slippery Elm] and may perhaps be used in bowel complaints."*(Rafinesque 1830:85)* The gastrointestinal applications would become well established in American practice, so much so that the popular nineteenth-century household medical guide by Dr. John Gunn would succinctly note that the rhizome is "good in...irritable conditions of the stomach and bowels."*(Gunn 1859-61:860)* In fact, the most revered application has proven to be for **PILES!** Grieve elucidates the classic method as follows: "4 ozs. Solomon's Seal, 2 pints water, 1 pint molasses. Simmer down to 1 pint, strain, evaporate to the consistency of a thick fluid extract, and mix with it from 1/2 to 1 oz. of powdered resin. Dosage: 1 teasponful several times daily."*(Grieve 1974:2:751)*

Rafinesque had observed that "coughs and pains in the breast" also yielded to the magic of Solomon's Seal.*(Rafinesque1830:85)* Concordantly, our own area's Ojibwe (Chippewa) Indians appreciated the demulcent effects of this plant's rhizomes for ***RES-PIRATORY SORENESS**.(Smith1932:363)* Interestingly, even the quintessential respiratory affliction, **TUBERCULOSIS**, might perhaps find itself somewhat softened by Solomon's Seal, in that a 1954 study found three *Polygonatum* species (including our own *P. biflorum*) to be among several hundred plants that inhibited a virulent human strain of tubercle bacilus (although not until the dilution was at 1:20).*(Fitzpatrick 1954:530)* Another "pain in the breast" benefitted by Solomon's Seal has been **HEART TROUBLE**,*(Wren 1988:254; Holmes 1997:1:453)* with some using it here as a substitute for its famous relative, Lily of the Valley (*Convallaria* spp.). Benefits are explainable by the fact that both species contain the cardioactive glycoside convallarin (although our herb much more sparingly). Since Solomon's Seal is also diuretic,*(Holmes 1997:1:453)* it has also yielded benefits for the sometimes accompanying **EDEMA**.*(Smith 1923:41;Smith 1932:374)*

Solomon's Seal has also been judicially employed for a wide variety of **"FE-MALE PROBLEMS**." Dr. Gunn had put it:"Very useful in female weakness and disease, as in leucorrhea or whites, and excessive and painful menstruation."*(Gunn1859-61:860)* Here, in actuality, he had simply echoed the Cherokee use.*(Hamel & Chiltoskey 1975:56)* Another fascinating use was that employed by our area's Menominee and Mesquaki (Fox) Indians: a smudge of the rhizome, heated on a coal, was implemented to revive an **UNCONSCIOUS** person.*(Smith 1928:230; Smith 1923:41)*

There appears to be a valuable ***tonic*** effect from the plant as well. Dr. Gunn had noted that Solomon's Seal was "good...in general debility," while modern herbalists Michael Tierra and Peter Holmes find it invaluable as a nutritive tonic for **QI DE-FICIENCY** according to the TCM model.*(Gunn1859-61:860; Tierra 1992:322-23; Holmes 1997:453-54)* The Chinese also value it as a ***yin tonic***, in reflection of its moistening effect.

Cautions: The berries have a long tradition of being inedible or toxic—sometimes provoking diarrhea, or vomiting, or both. Research conducted several decades ago uncovered the presence of anthraquinones (strongly cathartic—and sometimes emetic—glycosides) in them.*(North 1967:144; Lewis and Elvin-Lewis 1977; Woodward 1985:38)* There is also a particularly high concentration of irritating saponins in the seeds,*(Cooper & Johnson 1988:60)* which are often lackadaisically consumed when the berries are eaten.

SOW THISTLE
(Sonchus arvensis; S. asper)

Description

[See photos #133-35 in rear of book.]

This is an annual or perennial herb, growing to three feet high. The upright stem is smooth, hollow, and possessed of a milky sap in its walls.

The leaves are alternate, 4-14 inches long, 1½ to 5½ inches wide, spiny-toothed, and with their bases clasping the stem.

The flowering heads are produced in loosely-branched, elongated, erect, terminal clusters. Each head is 1-2 inches across, with lance-shaped bracts on the outside and many small, yellow, ray flowers within.

Several species abound in our area, as well as throughout North America:

Sonchus asper (spiny-leaved sow thistle) is an annual with a smooth, angled stem. It has very spiny leaves, with rounded and curling lobes similar to those possessed by some true thistles. Viewed from the side, the base of these leaves looks like a human ear!

S. arvensis (field sow thistle) is a perennial with weak spines and hairy bracts and stalks. It has a horizontal rootstock and sharp leaf lobes. Its flowerheads are larger than those of the above species. Many, gland-tipped hairs appear on the bracts.

S. oleraceus (common sow thistle) is a hairless annual bearing leaves that have sharp-pointed basal lobes, the terminal one of which looks like a large triangle.

S. uliginosus (marsh sow thistle) is a marsh-loving perennial with sharp-toothed leaves that tightly clasp the stem. Like *S. arvensis*, it has large flowerheads, but its bracts are smooth.

Range & Habitat

Sow thistle flourishes in waste places, fields, and roadsides. It can be found throughout our area, being a common weed throughout most of North America.

Chief or Important Constituents

Flavonol glycosides, sesquiterpene lactones(at least four of them in *S. asper*) Nutrients include goodly amounts of vitamin C.

Food

Sow thistle was a popular food among the Romans, which fact was mentioned by contemporary herbal writers such as Dioscorides and Pliny the Elder. The North-American palate, however, has yet to "get with the program!"

Practically every part of the plant is edible: First, young (spring-gathered) sow-thistle leaves can be collected, carefully trimmed of their thorny spines, and added to salads. Boiled in salted water for 3-5 minutes, then transferred to another pot and boiled for an additional five minutes, they also make a fair potherb. The young leaves can alternately be added to casseroles and soups. Autumn-gathered plants, and especially their uppermost leaves, are likewise not too bad when cooked, but *only after a frost has touched them.* (Warm-weather sow thistle is far too bitter to be palatable, even after boiling in several waters.)

The peeled stalks of young *S. oleraceus* plants—after having reached a foot or so in height and prior to flowering—can also be boiled for a few minutes, then doused with butter and devoured.

Sow-thistle roots are also edible. They can be chopped and added to stews.

Health/Medicine

The Kayenta Navajo Indians found *S. asper* to be helpful with **HEART PALPI-TATIONS,***(Wyman & Harris 1951:50),* as does modern European herbalist Juliette de Bairacli Levy.*(Levy 1974:137)* Success here may be due to *nervine* properties in the herb, since the Cherokees had used *S. arvensis* to arrest **NERVOUSNESS***.(Hamel & Chiltosky 1975:59)* (In my practice as an herbalist, too, I have noted that heart palpitations are often caused by overworked adrenal glands in their interweaving with the autonomic nervous system.) Many herbalists have even discovered that the whitish sap is so potent in this regard that it can even be used to break **ADDICTIONS**! The Chinese also value sow thistle for its calming ability, which is thought by them to accrue most effectively from the herb's yellow flowers.*(Silverman 1990:151)*

The noted European horticulturist John Gerard believed sow thistle to be possessed of powerful *refrigerant* properties and thus useful for "inflammations or hot swellings" and "gnawings of the stomach."*(Gerard 1975:232)* Asian healers have also poulticed the leaves onto various **BODILY INFLAMMATIONS**.*(Foster & Duke 1990:128)* Even **SNAKEBITE** has found successful treatment with this plant, an art perfected by the Navajo Indians.*(Wyman & Harris 1941:125)* Finally, **CAKED BREASTS** were treated by the Potawatomi Indians with a wash made from an infusion of *S. arvensis.*(Smith 1933:54)*

Caution

Be careful in handling this thorny plant! Also, be absolutely sure that you have trimmed off all of the thorns from this plant before consuming it.

SPIDERWORT
[WIDOW'S TEARS]
(Tradescantia spp.)

Description

[See photos #136 & 137 in rear of book]

A pretty summer perennial growing 8 inches to 2½ feet tall, with several succulent stems that are usually erect but are sometimes bent.

Leaves are lanced—long and narrow, onion-like to iris-like in their appearance. Large sheathes can be found at their base. The sprawling stems and bending leaves give, say some, the appearance of a spider on the hunt, which may be the origin of the herb's name. But others say that the name arose because the herbage, when pulled apart, reveals a sticky, juice-like substance that appears in strands quite similar to a spider's webbing.

Largely responsible for the plant's legendary beauty is the three-petaled blue to violet flower, about 1¼ inches long and possessed of six blue stamens with prominent yellow anthers. These flowers, situated singly at the tip of the plant stems, open only in the morning, while in the afternoon a real "sight to see" transpires: the flower head closes up and in its place there is exuded a sticky, gel-like mass. This is due to enzymatic action within the herb, and has given spiderwort its alternate common name, "Widow's Tears."

Range and Habitat

Found in fields, meadows, prairies, open woods, thickets, and roadsides. Various species blanket the entire US except the upper northeastern states. *T. virginiana* (with slightly hairy sepals and stamens) and *T. ohiensis* (with a whitish bloom on leaves and stem) preside in the east (and the latter in the very southeastern tip of Minnesota), and *T. occidentalis* in the west and midwest (in Minnesota, in the east-central portion). Another species, *T. bracteata*, is ubiquitous in the southern two-thirds of Minnesota.

Seasonal Availability

This is a plant of summer—blanketing the areas that it covers in breathtaking beauty!

Chief or Important Constituents

No known chemical work done on this species, but astringent agents (probably tannins) are suspected.

Food

This plant's succulent leaves and stem, not to mention its flower, are quite edible and, in the opinion of many, very tasty. Not surprisingly, then, spiderwort makes an excellent trail snack, especially on a hot summer day in that it is conveniently high in moisture and quickly chews into a mucilaginous mass. I enjoy the taste and texture of this plant very much and eat it extensively while it is available, supplementing my water rations with it (ditto with thistles and a few other such fluid-valuable plants), so that sometimes I don't even need to carry a water canteen. Indeed, spiderwort is a plant that makes me ache for summer!

A fine potherb as well, spiderwort can be added to stews or eaten by itself after being boiled in salted water for about five to ten minutes.

It is said that George Washington Carver, the famed food scientist, raved about the flavor of this succulent herb.*(Fernald & Kinsey 1958:124)* Carver was right—it's downright delicious!

Health/Medicine

A tea made from spiderwort was used by the Cherokee to ease **STOMACH ACHES** (especially from overeating), as a *laxative,* and as part of a compound formula for **KIDNEY TROUBLE** and for **DISORDERS OF THE FEMALE REPRODUCTIVE SYSTEM**.*(Hamel & Chiltoskey 1975:56, 57; Banks 1953:12-13)* These Indians also poulticed leaves for application to **SKIN CANCERS** and rubbed them on **INSECT BITES** for relief.*(Hamel & Chiltoskey 1975:56,57)*

The Mesquaki (Fox) and Navajo-Kayenta Indians also found medicinal uses for this lovely prairie plant: the former utilized it as a **urinary** aid (probably as a diuretic),*(Smith 1928:209)* while the latter prepared a decoction of the root to treat **INTERNAL INJURIES**.*(Vestal 1952:20)* The Navajo-Kayenta also implemented a cold infusion of the root to treat what they called "deer infection" in humans, drinking it and applying it topically in a lotion.*(Vestal 1952:20)* I can't help but wonder: Might this have been what we now know as Lyme Disease?

Utilitarian Uses

Spiderwort has been implemented as a pollution indicator, in that its stamen filaments will transform from blue to pink in a period of 1½ to 2½ weeks in the presence of significant levels of pollution or radiation. It has been used extensively by the EPA.

This herb is also often the plant of choice for cell studies conducted in cytology classrooms in that its chromosomes are unusually large, enabling the nucleus and flowing cytoplasm to be viewed with relative ease.

STARFLOWER
[STAR ANEMONE; CHICKWEED WINTERGREEN]
(Trientalis borealis [alt: T. americana])

Description

[See photo #138 in rear of book]

A lovely woodland plant, growing 2-8 inches tall. Its slender, upright stem arises from a horizontal, underground rhizome.

This plant is appropriately named: The strikingly starlike flowers have pointed, white petals. There are usually seven of them, but they can number anywhere from five to nine. The showy stamens project well above the blossoms on hairlike stalks and are tipped with golden anthers.

Shiny, distinctly-veined leaves are whorled from the stem. They number anywhere from five to nine (but again, are usually seven) and are lanceolate, pointed, and of uneven size. Their arrangement and pointed tips add to the overall starlike appearance of this delicate herb.

A small scale leaf is also present below the whorl and near the middle of the stem.

Range and Habitat

This plant likes to grow in rich woods and at the edges of bogs—especially in shaded areas. In our area, starflower grows from the east-central quadrant of Minnesota to the west-central quadrant of Wisconsin, and then northwards in both states.

Health/Medicine

The Montagnais and Tadoussac Indians of the Quebec region steeped an infusion of this species to allay **GENERAL SICKNESS** and to help manage the dreaded **TUBERCULOSIS.** *(Speck 1917:314;Vogel 1970:271)* In the early 1800s, Indian tribes of the Arkansas territory (including Oklahoma, Missouri, and parts of Kansas) also utilized starflower medicinally, but the details are not known. *(Vogel 1970 :107)*

The Cowlitz Indians of Washington state squeezed the juice from the Western species, *T. latifolia*, into water and used it as an eye wash—whether for **INFECTED EYES** or just **SORE EYES** is not known. *(Gunther 1973:44)*

Herbalist Michael Tierra has found the root of our species to be a **qi tonic** (in the TCM model) and **stimulant**. He thus uses it for **FATIGUE, WEAKNESS**, and **EXHAUSTION**. He says that it affects the heart, spleen, and lung meridians. Energetically, Tierra finds starflower to be acrid, sweet, and warm. *(Tierra 1992:297-98)*

STINGING NETTLE
[ITCHWEED]
(Urtica dioica)

Description

[See photos #139-40 in rear of book.]

A tall (2-6 ft.) perennial with creeping roots and an angled, unbranched stem.

Leaves are opposite, ovate to lanceolate, pointed at the tip, deeply serrated around the edges, and grow to six inches long. Dark green in color and prominently veined, they—like the stem—are covered all over with tiny, stinging hairs. (Note that two similar-looking plants—with which nettle is often confused—lack the stinging hairs; these are false nettle, *Boehmeria cylindrica*, and clearweed, *Pilea pumila* [see under that entry].)

Tiny, cream-colored flowers, lacking petals, droop in loose clusters from the leaf axils—male (with stamens only) and female (with pistils only) often *on different plants* (occasionally both on same). (Contrariwise, flowers of the similar-looking false nettle are *dark green*, while those of clearweed are *medium green*.)

Wood nettle (*Laportea canadensis*) (*see under that entry*), also a perennial, has a more succulent look, with wider—and *alternate*—leaves and ones that are more broadly and conspicuously toothed. However, like its cousin, stinging nettle, it also bears flowers in the leaf axils (although it always has both genders on the same plant) and stinging hairs on leaf and stem.

Range & Habitat

Stinging Nettle makes its appearance in the eastern two-thirds of the USA and in southern Canada. It can be found on disturbed ground—waste places, vacant lots, garbage dumps (but don't forage here, due to possible chemical contamination), woodland edges, streamsides, roadsides, and trailsides. It is ubiquitous in our area in such surroundings.

Seasonal Availability

Leaves available in spring for food use, when plant is under 8 inches tall (see below). For medicinal use as a dried-plant tea or tincture, nettles may be harvested at any time.

Chief or Important Constituents

Above-ground portions contain glucoquinone, caffeic acid, p-coumaric acid, chlorogenic acid, carbonic acid, organic acids(acetic, citric, butyric, malic, oxalic, fumaric), sterols(beta-sitosterol), volatile oil(consisting of ketones, esters, and free alcohols), carotenoids(lycopene, violaxanthin, beta-carotene), flavonoids(quercetin, rutin, kaempferol, isorhamnetinbetaine), lecithin, mucilage, and the enzyme secretin. The seeds contain fatty acids(linoleic, linolenic, oleic, palmitic). The roots contain saponins(lignans, sitosterol), polysaccharides, coumarin(scopoletin), phenolic acids, phenylpropanoid aldehydes, monoterpenes, triterpenes, and tannin. The stinging hairs contain acetylcholine, histamine, 5-hydroxytryptamine, and possibly formic acid. Nettle's vitamins include A, B1, B2, B5, C, D, E, K1, choline, and folic acid. A smorgasboard of minerals includes iron, sodium, phosphorous, sulfur, manganese, calcium, silica, and potassium. Nettle is also extremely high in protein(some studies finding 41% by dry weight and 6.9% when fresh), as well as possessing one of the highest chlorophyll contents known in edible, North-American wild plants.

Food

Nettle's stinging hairs often discourage both man and beast from having anything to do with this plant. But, with the proper precautions, one need not be balked: Gloves and scissors can reap one a bountiful harvest that can be put to good use.

First, the entire plant can be harvested if under eight inches tall. If over that height, only the tender young leaves at the top should be taken. (This is because mature nettles develop a toughness and grittiness to them due to the formation of tiny, hard, calcium crystals called cystoliths, which can irritate the kidneys) To harvest nettles, and many other plants, I like to use a large, plastic, ice-cream pail, through which I have bored two holes and strung some twine in order to loop the bucket around my neck, which leaves my hands free for probing, bending, and clipping the desired plants.

Of course, even when taking reasonable care in harvesting nettles, there are no guarantees that one will not be stung. For example, though I myself have developed collecting techniques whereby I have not been stung by nettles in many years, yet recently, during the spring, while I was lying over what I thought was simply a bed of dead, bent-over grass stalks to photograph a plant from a side angle, I arose with the characteristic feel of nettle stings all over my elbows and forearms! I quickly discovered that the grass had been hiding an extension of a newly emerging nettle colony and that the pressure of my arms had brought me into direct contact with the obscured, stinging plants. Needless to say, my arms became quickly covered with nettle welts.

What to do in such a situation? Foragers have traditionally looked for the leaves of dock, jewelweed, or plantain, with which the sting (usually lasting about seven minutes, from which the plant has garnered the nickname of"Seven Minute Inch") can, it has long been claimed, often successfully be "rubbed out." In my own case as summarized above, these three plants were not to be readily found, so I implemented another age-old remedy—nettle juice! Whether it was the high amount of lecithin contained in the juice, or some other factor, the traditional cure really did work, to my utter delight!

I know of at least one foraging manual that urges the reader to remove nettle leaves from their stems and to rub them with gloved hands so as to rupture the stinging hairs, thereafter using them in a spring salad. The author of another foraging manual says that he has masticated raw leaves and that the stinging stopped by the time the nettles proceeded down his throat. *I consider these statements to be rash and unmerited in a foraging manual, possibly inviting dangerous experimentation*. While some persons might emerge unscathed from such an experience, others might suffer severe reactions. My background in health research has alerted me to the fact that everyone's body chemistry is different, and that one man's food could be another's poison. And this principle is certainly compounded when dealing with the known toxins contained in raw nettle leaves. The possibility of some unruptured, stinging hairs making it through even a thorough glove-rubbing or mastication is quite real, in that they are small, numerous, and quite dexterous. Please, then, do *not* experiment with raw nettles—the results could be truly disastrous. Always remember the true forager's rule: Safety first!

Nettle's sole culinary use, then, should be as a cooked plant. The leaves can be boiled or steamed, which kills the stinging element. A few minutes is all that is required. They can then be eaten as is, tasting somewhat like spinach, or combined with cream-of-mushroom soup or other food items. To avoid a bronze color forming in the leaves while they are being cooked, leave the top off of the kettle or use more water. (This unappealing color change, experienced with bewilderment by foragers when cooking nettles in times past, is now known to be triggered by the high amount of chlorophyll in nettle reacting with natural acids present in the leaves and cooking water, which then further react with carotene pigments in the leaves.-*Rinzler 1991:114*)

A tea made from dried nettle leaves is very good, and the reconstituted pieces in the cup can be chewed and eaten while the tea is being drunk. In fact, dried nettles (at least *U. dioica*) reconstitute readily, freshly, and *nearly fully* in water *(Young 1993:9)*.

The fleshy, pink section of the plant between the stem and the rootstock is especially favored by wild-foods foragers.

Nettle broth can be procured from the roots. A better use of them, however, is to transplant the autumn-dug ones to a cellar or basement in a container of sand, secreting them away in a very dark place and watering them every few days. This procedure, called "forcing," actually fools the plants so that they will put up shoots through the winter, all of which can be clipped for food or tea—and this usually repeatedly throughout the cold months!

Of course, what greens you have collected in the wild can be frozen, too. To accomplish this, simply blanch them for two minutes, then cool, pat dry, and freeze.

Nettles are incredibly healthful, containing a plentitude of vitamins and minerals (see list under "Constituents" above). Two of the more interesting are vitamin D and phosphorous. The former is not available from many plants and is important in that it works synergistically with calcium in the body, making this plant valuable for those requiring a dairy-free diet. The phosphorous content no doubt at least partly accounts for the herb's traditional good effect upon teeth and gums. Another mineral present, silica, contributes to the hardness of hair and nails.

Interestingly, too, Euell Gibbons found nettle to contain more proteinper unit of measurement *than any other, edible, leafy-green plant!(Gibbons 1971:165)* The precise percentages are 6.9% fresh*(Ibid.)* and 42% by dry weight.*(Tull 1987:150)* No wonder that inmates of the notorious Nazi concentration camps found this weed, gathered surreptitiously, to be so nourishing and helpful—it probably provided more sustenance than all the slop combined that the Nazis had tossed their way.

Health/Medicine

The truly amazing nettle has a wide variety of therapeutic uses, often related to problems of a *burning, stinging nature*—similar to the feeling produced by its own stinging needles!*(Smith 1999:47)*

First, this despised weed serves as an excellent *styptic*, both for external **WOUNDS** and for **INTERNAL BLEEDING.** Laurence Johnson, a botanist and physician of the late nineteenth century who was generally critical of herbal remedies of the day, yet noted here: "Internally the drug has been used with asserted benefit in hemorrhages from the nose, lungs, uterus, etc., and in catarrhal affections."*(Johnson 1884:245)* Both the tea and the fresh juice (especially the latter) are useful in this regard. Naturopath Donald Lepore had recommended a teaspoon of the juice every hour,*(Lepore1988:183)* a dosage that might be committed to memory for any serious internal injuries that could occur in the wilderness when hospital help is distant. A scientific article published about a decade ago reported that lectins isolated from this plant were observed to display hemagglutination activity,*(Willer&Wagner 1990:669)* thus providing some scientific testimony for the plant's traditional use as a styptic.

Because nettle contains considerable amounts of iron, it is an excellent treatment for **ANEMIA,** especially in that the vitamin-C content assures excellent absorption by the body. I especially like to use nettle tea as a blood-revitalizing tonic for clients who have been debilitated by a long illness and are thus very pale and weak. Nettle is also a time-honored *galactagogue.(Ellingwood 1983:355; Smith 1999:47; Ody 1993:108)*

British herbalists have long felt that nettle tea has a lowering effect on blood sugar, making it useful for diabetics.*(BHC 1976:1:217; McQuade-Crawford 1996:162)* The experimental evidence, however, is unclear (rabbits fed nettle had *increased* blood-sugar levels; also, the herb aggravated diabetes in mice.*-Roman et al 1992:59-64; Swanston-Flatt et al 1989:69-73*)

Nettle tea has been suggested as a wash for **ITCHY CROTCH**.*(Manning and Jason 1972:79)* In fact, it is often recommended by herbalists to treat **SKIN CONDITIONS** in general, owing to a *depurative* effect.*(Kresanek1989:196)* The *British Herbal Pharmacopoeia* specifically indicates the herb for **NERVOUS** or **PSYCHOGENIC ECZEMA,** as well as for **INFANTILE ECZEMA**.*(BHC 1976:1:217)* Further as to youngsters, Celsus, an herbal scholar of the late first century C.E., found that mashed nettle seeds in water served as a *vermifuge* for **THREADWORMS** in children.(*De Medicina*, 4.24)

This herb has also long been acknowledged as a valuable *diuretic,*(*Kresanek 1989:196; Weiss1988*) which action has been confirmed in a clinical trial,(*Kirchhoff 1983:621-26*) and is pos-sibly attributable to its flavonoids and high potassium content. This property of nettle has allowed herbalists to employ it for **PREMENSTRUAL BLOATING**. Then, too, a number of good studies have found nettle root to be of help in treating urinary retention associated with **BENIGN PROSTATIC HYPERTROPHY (BPH)**, a condition affecting many elderly men.(*Belaiche & Leivoux 1991; Bombardelli & Morazzoni 1997; Krzeski et al 1993; Schneider et al 1995; Sokeland & Albrecht 1997; Vontobel et al 1985; Goetz 1989;Lichius & Muth 1997; Hirano et al 1984; Schonefeld et al 1984; Hirano et al 1994; Hryb et al 1995; BanBer & Spitteler 1995; Lichius et al 1999:666-68; Lichius et al 1999:768-71; Muller et al 2000*).

In Europe, preparations using an extract of nettle root are marketed for this use and it is also gaining popularity in America. There are numerous opinions and argu-ments as to why it works, but it does not seem to have been conclusively proven one way or the other. In fact, as the above studies suggest, various factors seem to be at play. One interesting study found that nettle root interrupted the binding of human sex-hormone binding globulin (SHBG) to its receptor site, thus decreasing the amplification of the androgen signal by estrogen and thereby theoretically slowing down BPH.(*Gansser et al 1995:98-104; Hryb et al 1995:31-32*) Another study suggested that nettle root inhibited prostate Na+/K+ ATPase enzyme, so that prostate cells were prevented from multiplying.(*Hirano et al 1994:30-33*) Then, too, as with the popular, marketed herb saw palmetto (*Serenoa repens*), stinging nettle root may hamper the conversion of testosterone to dihydrotestosterone, as several studies seem to show.(*e.g., Schottner et al 1997:529-32*) If this is the case, it might explain another traditional use of this herb: as a warrior in the battle against baldness, since male pattern baldness has been linked to breakdown products of the *very same testosterone-dihydrotestosterone conversion.* And an infusion of the herb used internally and/or as a hair wash *does* appear to decrease **HAIR LOSS**—not to mention benefitting the scalp in other ways, such as in helping to eliminate **DANDRUFF.**(*Kresanek1989:196; Mabey1988:124*) The plant's high content of quality protein, not to mention its silica, undoubtedly contributes at least partly to the normal growth and health of hair, as well as to its strength.

Nettle's most famous and traditional application has been in the treatment for **HAY FEVER (ALLERGIC RHINITIS).** Interesting here is that a randomized, double-blind study of freeze-dried preparations of the plant found it to be about as effective in the human test group as hay-fever medications of the standard, over-the-counter and prescription variety.(*Mittman 1990:44–47*) The freeze-dried form preserves the histamine content, which many feel is vital in halting the allergic process, either by way of interfering with the body's release of histamine or possibly by stimulating the body to make its own antihistamines. Interesting as this theory goes, however, the clinical experience of numerous herbalists throughout history reveals that nettles do no not need to be freeze-dried to produce relief for allergy sufferers. I myself have been privileged to help many dozens of hayfever sufferers with just plain 'ol nettle tea or tincture! I can't help wonder if this is due at least partly to the plant's renowned strengthening effect upon the adrenal glands, so important in the body's immune defenses. As endocrinologist John Tintera was so fond of communicating to his readers, a lack of potency in these glands allows allergic reactions to proceed unchecked.*(Tintera 1959:28-29, 92)* Interesting, in this regard, is a 1989 study which found four different polysaccharides in *Urtica dioica* that were observed to stimulate the proliferation of human lymphocytes, being thus demonstrative of ***immune-enhancing activity***.*(Wagner et al 1989:452–54)* Follow-up work found five immunologically active polysaccharides.*(Willer & Wagner 1990:669)*

From time immemorial, the nettle plant has been used by sufferers of **ARTHRITIS, RHEUMATISM,** and **GOUT** to whip themselves, for the purpose of a counterirritant, and with claimed results of a fair to an excellent nature. Quinault Indians used the relates species, *U. lyallii*, in the same fashion.*(Gunther 1973:28)* The ancient and learned medical writer Celsus had even urged the use of nettle whips to try to help resolve or ameliorate **PARALYSIS** because of 'stimulating the skin of the torpid limb.'(*De Medicina* 3.27) Orthodox medicine has summarily dismissed nettle-plant flagellation as ignorant and useless. But herbal writer Donald Law cannot agree, as he points out here: "In that mysterious acid of the sting lie some curative properties which both the Romans and the Aztecs have recorded quite independently of one another, and as a young man I experienced myself."*(Law 1976:32)* Interestingly, a recent exploratory study has suggested some clear benefits of this self-flagellation: Eighteen persons undergoing a trial of nettle-sting therapy for joint pain were interviewed by researchers at the Plymouth Postgraduate Medical School in Plymouth, United Kingdom, in 1999, with the result that all but one of the interviewees expressed praise for the therapy and several even considered it a 'cure.' The authors concluded the study with the stated impression that the therapy was safe, useful, and inexpensive and that it should be studied in more detail.*(Meethan et al 1999:126-31)*.

Applied as a poultice to **RHEUMATIC JOINTS,** some have suggested that nettle draws uric acid out of the body.*(Lepore 1988:183; Ody 1993:108)* In one recent British study, a poultice of nettle leaves on osteoarthritic thumb and finger joints greatly reduced the pain in such spots for 27 people after merely a week!*(Randall et al 2000:305-09)* Members of numerous Amerindian tribes experienced relief from rheumatic pains by imbibing the tea.*(Moerman 1986:498)* It is even endorsed for this use by the *German Commission E Monographs,* prepared by German scientists.*(GermCommE)* Moreover, some good scientific evidence exists for ***anti-inflammatory*** functions for the plant. The coumarin scopoletin, occurring in the roots, has been shown to be anti-inflammatory, as have several of the herbage's flavonoids and acids. Moreover, in the 1989 study cited above re: nettle's immune-enhancing constituents, the immunologically active polysaccharides showed antiphlogistic activity in rats.*(Willer & Wagner 1990:669)* Two studies since have even showed that nettle leaf reduced the amount of anti-inflammatory drugs needed to achieve moderation of arthritic pain in humans.*(Ramm & Hansen 1995:3-6; Chrubasik et al 1997:105-08)* Moreover, a very recent study demonstrated that an extract from nettle leaves inhibited the proflammatory transcription factor NF-kappaB.*(Riehemann et al 1999:89-94)*

Southwest-American herbalist Michael Moore cautions that nettle may be irritating to the kidneys if used for an extended period of time without a break.*(Moore 1977:114)*

Utilitarian Uses

The fiber in nettle stalks makes excellent cordage. It can also be woven into fine cloth. The Romans, when moving into cold northern territory to extend their conquests, used to sting their legs with nettles to keep warm. Possibly this was hinted at in the ancient Biblical book of Job as well, in its poetic section (which goes back at least a thousand years before Christ), where we read that outcasts would huddle together under nettles.(Job 30:7) As undoubtedly there was larger and more heat-preserving brush available, one wonders whether these individuals, like the Romans later, had availed themselves of the sting to keep warm. Of course, a person out in the wilderness today usually has available less drastic methods to preserve warmth. But an injured person unable to move might be forced to resort to such a method if the plants were nearby and all else had failed.

Cautions

Wild-food cautions:

* Do not eat fresh, raw nettle plant parts.

* Do not eat whole plants after the height of eight inches is reached (see above, under "Food") or even individual leaves from plants over that height except the uppermost, newly-emerging leaves which can be snipped off and boiled throughout most of the growing season (see also above, under "Food").

* Use gloves—or makeshift gloves of pot-holder-like mullein leaves—to collect nettles so as to avoid the sting. If stung, rub the leaves from dock or plantain on the sting to allay. Or nettle juice itself can be used, if it can be procured without further stinging.

* What puts the "sting" in stinging nettle? The hairs that transmit the sting are actually small capillary tubes, composed of silica at the upper end and calcium at the lower end. As to the exact mechanism of the sting, even the technological wonders of today's phytochemists have not explicated its mysteries to the full, despite tedious work. Many herbals allege that the chemical agent transmitted through the "hair" and responsible for the sting is formic acid (the toxin found in red ants). But I have not been able to locate any studies in verification of this. Numerous published studies I have consulted have indicated, however, that the sting is due to several other compounds: histamine, acetylcholine, 5-hydroxytryptamine (serotonin), and at least one other unidentified substance that is perhaps a histamine liberator. *(Emmelin and Feldberg 1947:440–55; Collier and Chesher 1956:186–89)* Work by Saxena et al in the mid-1960s buttressed the hypothesis of a histamine liberator,*(Saxena et al 1965:869–76)* arguing for an endogenous release of histamine to account for the persistent itching. Some European researchers

have recently opted for a substance with the potential for secondary release of other mediators, or even a nerve toxin, because they also agree that histamine, acetylcholine, and 5-HT could not be responsible for the persistence of the sting, but only for the initial phase of it. *(Oliver et al 1991:1–7; cf. Kulze and Greaves 1988:269–70)* (When I have been stung severely, I have noticed a lingering headache, persisting for up to a day. To me, that indicates some sort of agent, or agents, for sustained ill effects.)

Medicinal-Use Cautions:

* Most herbalists would not contraindicate nettle during **PREGNANCY**. However, if used during this time, it should be *used with caution,* due to possible uterine-contracting properties (these have at least been observed in rabbit experiments). *(Newall et al 1996:201)*

SUNFLOWER
(Helianthus spp.)

Description

[See photos #141 in rear of book.]

Sunflowers are tall, annual or perennial plants with firm and upright stems, composite flowers consisting of a disk flower with 10-25 yellow ray flowers, and dark-green leaves that are serrated.

A large variety of species exists throughout the USA and Canada, with a dozen or more thriving in our area, including the following: (1) *H. annuus* (common sun-flower), an annual growing 3-9 feet tall, with a brownish disk flower some 3-9 inches wide and alternate, oval, or heart-shaped leaves reaching 3-7 inches; (2) *H. giganteus* (giant, great, or tall sunflower), growing 11 feet tall on occasion and having a rough and reddish (or purplish) stem, stalkless or short-stalked leaves that are lanceolate and most-ly alternate (but occasionally opposite), numerous light-yellow flower-heads (1½-3 inches wide) with 10-20 ray flowers, long-pointed bracts, and hairs on the flower stems just below the blossoms; (3) *H. grosseserratus* (saw-toothed sunflower), reaching 4-10 feet, having hairless stems, and possessed of lanceolate and coasely-toothed leaves (like saw blades, and thus the common name) with a whitish down beneath; (4) *H. laetiflorus* [=*H. hirsutus*] (stiff, or showy, sunflower), growing 2-8 feet tall, with a stem covered with stiff hairs; (5) *H. maximiliani* (Maxmilian's sunflower), a three-to-nine-foot prairie perennial possessing a yellow disk flower on a head 2-3 inches wide and having alternate, downy, narrow, and tapering leaves that grow 4-6 inches long and are often folded lengthwise and reflexed downward at the tips; and (6) *H. strumosus* (pale-leaved wood sunflower), reaching 3-7 feet, with a smooth and branching stem often covered with a whitish bloom, a yellow disk flower surrounded by 9-15 ray flowers, mostly opposite leaves growing 3-8 inches long on winged stalks, and basal leaves having downy undersides.

Yet other species in our area include *H. petiolarius*, *H. occidentalis*, and *H. rigidus*. The Jerusalem artichoke *(H. tuberosus)* is also a sunflower, but this is discussed in a separate monograph *(see under that entry)*.

Range & Habitat

Sunflowers can be found throughout most of the USA and Canada. Their preferred habitat is dry fields, prairies, waste places, and dry plains. One or another species is pretty much ubiquitous in our area

Chief or Important Constituents

Quercimetrin(a flavonoid), sesquiterpene lactones, diterpenes, tannins, potassium carbonate, potassium nitrate. Seeds contain lecithin, albumin, betaine(trimethylglycine), and are rich in vitamin E(30-75 mg per 100 g), several B vitamins, iron (7mg per 100g), essential fatty acids(oleic and linoleic), and free-form amino acids(esp. phenylalanine, SAMe, and arginine—apparently the highest known source of the latter!).

Food

Sunflower seeds (available generally from *H. annuus* and *H. maximiliani* only), have been a favorite food of many Amerindian tribes. After gathering them—usually by knocking the head with a stick to loosen the seeds into a collecting implement (some modern foragers prefer simply to cut off the heads and then hang them up to dry, upon which the seeds will spill out)—the Indians would parch them over a fire. Next, they would pound them to loosen the hulls and then winnow them in the breeze. Finally, the kernels were eaten as is or else powdered into meal, prepared as a mush or gruel, or made into cakes and cooked. The explorers Lewis and Clark described eating one of these sunflower-seed repasts in their journal, under date of July 17, 1805:"The Indians of the Missouri...first parch and pound [the sunflower seed] between two stones, until it is reduced to a fine meal.... They add a sufficient portion of marrow-grease to reduce it to the consistency of common dough, and eat it in that manner. This...composition we...thought...at that time a very palatable dish."

The Indians also found that the seed head, or fruit, can be crushed and boiled to release sunflower oil, which can then be skimmed off the surface of the water and used as a salad oil or mixed with other foods as advisable.

Although nearly everyone knows about the edibility of sunflower seeds and oil, few realize that the best investment from the plants in the way of total food volume and accessibility is the unopened flower-buds, which can be prepared as a potherb. As such, however, these must be boiled in several changes of water in order to remove bitter principles. I personally find these buds quite tasty when thus properly prepared, but practically unpalatable otherwise!

Available occasionally in certain species (*H. laetiflorus, H. maximiliani, H. strumosus*) are edible tubers, which can be dug, cleaned, and consumed raw on the spot in a survival situation or else prepared in various ways at home. Another tuberous sunflower is the well-known Jerusalem artichoke (*H. tuberosus*) (*see under that entry*). However, unlike this latter species' tubers, those of the aforementioned species are easier to clean on the spot and less likely to produce flatulence when eaten raw, since they contain less of the indigestible carbohydrate inulin than do Jerusalem artichokes. I have had them frequently as a trail nibble and have always enjoyed them.

Health/Medicine

The common sunflower, *H. annuus*, has been the species most used medicinally. Most of its applications would seem to suggest that it has an affinity to the region of the chest and neck. For example, ***PULMONARY TROUBLES*** were addressed by the Teton Dakota Indians with this plant: they boiled the flower heads (minus bracts) and drank the resultant decoction.*(Gilmore1981:78)* The Houma Indians discovered that sunflower stalks could be made into a tea that was particularly useful for ***WHOOPING COUGH***.*(Speck 1941:60)* White herbalists have also used an infusion or syrup of this plant for lung problems (including the early stages of ***TUBERCULOSIS),*** as well as for related respiratory complaints such as ***BRONCHITIS, COUGHS,*** and ***SORE THROAT.****(Hutchens 1991:269; Wren 1972:297)* Sunflower seeds have also long been regarded as being ***expectorant,****(Wren 1972:296)* an infusion of them having been used as such by America's Eclectic physicians of the late nineteenth and early twentieth centuries.*(Millspaugh 1974:330)*

As noted in the Centennial edition of the *U.S. Dispensatory* and in other sources, this plant has a long tradition of use—and some scientific support—for treating ***MALARIA.****(Wood et al 1937:1405;Millspaugh 1974:330;Carse 1989:162)* Sunflower contains sesquiterpene lactones, the same chemicals found in boneset *(Eupatorium perfoliatum),* another plant traditionally and effectively used for malaria *(see under that entry).* American use here has been to prepare an infusion of the stems,*(Millspaugh 1974:33)* while the British have tended to use an infusion of the leaves,*(Wren 1972:297),* and the Russians a decoction of the seeds, flowers, and leaves.*(Hutchens 1991:271)* Soft, pulpy parts of the stems were also implemented by Russian healers for ***FEVER*** in general.*(Hutchens 1991:271)*

Sunflower has been much appreciated as a ***vulnerary*** by native Americans. Thus, Ojibwe (Chippewa) Indians from our area pounded the roots in order to poultice them onto ***BRUISES*** and ***CONTUSIONS***.*(Hoffmann 1891:199)* The Thompson Indians found that mixing the dried and powdered leaves of *H. annus* together with some animal grease made a healing ointment for ***SORES*** and ***SWELLINGS***.*(Turner et al 1990:46)* The Zuni Indians poulticed the chewed roots of this plant, along with those of three other plants, on ***SNAKE BITES***.*(Youngken1925:17)* Various Indian tribes used a sunflower infusion as a wash for ***RHEUMATISM*** and ***GOUT***.*(Hutchens 1991:268)* None of the abovementioned uses are surprising in view of a 1976 study which showed that an extract from this plant was ***anti-inflammatory*** in the carageenan-induced, rat-paw-edema lab test.*(Benoit et al 1976:164)* Perhaps sunflower's phenylalanine content is a factor in reducing pain as well.

The seeds possess ***diuretic*** properties*(Wren 1972:297)* and the Russians have used a decoction of them for ***HEART PROBLEMS*** and ***KIDNEY*** and ***BLADDER AIL-MENTS***.*(Hutchens 1991:271)* *H. annus* has also shown ***hypoglycemic*** activity.*(Lewis & Elvin-Lewis 1977:218)*

Other sunflower species used medicinally have been our *H. grosseserratus* and *H. occidentalis.* The Mesquaki Indians used a flower poultice of the former for healing ***BURNS,****(Smith 1928:215)* while Ojibwe Indians applied crushed roots of the latter to help take the inflammation and pain out of ***BRUISES*** and ***CONTUSIONS***.*(Hoffman 1891:199)*

SWAMP MILKWEED

[MARSH MILKWEED; ROSE-MILKWEED; SWAMP SILKPLANT]
(Asclepias incarnata)

Description

[See photos #142 in rear of book.]

A lovely native perennial, growing 1-4 ft high, and bearing a smooth and erect stem. Inside the stem flows a milky white juice with a strong odor characteristic of the milkweed genus to which this species belongs.

The top part of this plant is branched and quite leafy. The oblong to lanceolate leaves are stalked, opposite, hairy (esp. below), cordate at their base, and toothless but sharp-edged. They grow 4-7 inches long and 1-2 inches wide

The gorgeous flowers, blooming May to August, are pink to rose-colored and clustered into several, small umbels. In the fall, these are replaced by smooth, erect pods, 3 inches long, which finally burst asunder, releasing many seeds tufted with silky hairs.

The plant's rhizome is oblong, 4-6 inches long, yellowish-brown, knotty and hard, and covered with a thin, strong bark through which grow many, tiny rootlets.

Range & Habitat

Swamp milkweed thrives in sunny, wet ground—bogs, marshes, moist meadows, swamps, borders of streams and ponds. Found throughout our area in such locales.

Chief or Important Constituents

Cardioactive glycosides(including asclepiadin), resins, volatile oil, albumin, pectin.

Health/Medicine

This plant's rhizome was listed in the *U.S. Pharmacopoeia* from 1820-63 and 1873-82 and it was a favorite medicine of the Eclectic physicians, who practiced in America in the late 1800s and early 1900s. Finley Ellingwood, as a representative of such, found it to be *emetic, anthelmintic, stomachic*, and *diuretic*. He especially valued and implemented this species for the latter property, diuresis, finding it to be "speedy and certain" in that regard.*(Ellingwood 1983:451)* The Iroquois Indians likewise appreciated this aspect of swamp milkweed, holding that the plant's rhizome was helpful either where there was too much urine or conversely, too little of it.*(Herrick 1977:418)* They also valued the plant for its beneficial effects on **URINARY STRICTURE**.*(Herrick 1995:198)*

Even ***DROPSY*** was eased by the use of swamp milkweed, Ellingwood found, not simply because of the plant's diuretic properties but because it "affects the heart and arteries...strengthens the heart." *(Ellingwood 1983:451)* (This is no doubt due to the presence of the cardioactive glycosides and perhaps additional factors as well.) Then, too, both the Iroquois and the Mesquaki (Fox) Indians discovered that a wash or bath made from swamp milkweed rhizome was 'strengthening' to the human system, perhaps with reference to some sort of cardiotonic (as per above)—or otherwise stimulating—assist. *(Densmore 1974:364-65; Herrick 1977:418)*

As with many milkweeds (see under the "Milkweed" entry, above), this species has also been deemed ***anthelmintic,*** with Ellingwood advising 10-20 grains of the plant for vanquishing ***WORMS.*** *(Ellingwood 1983:451)* There is an ethnobotanical study that weighs in on this, too: Huron Smith, in studying the Mesquaki (Fox) Indians, related how a Potawatomi Indian healer working with them named John McIntosh had "recovered four long tape worms from a woman by its use. The root is said to drive the worms from a person in one hour's time." *(Smith 1928:205)* If there is no exaggeration involved here, this would indeed mark swamp milkweed as a ***taenifuge*** of the first rank. (Smith relates further that McIntosh had also achieved fame for stabilizing a case of dropsy with swamp milkweed when regular physicians had failed. As we noted above, this would not be out of line with what Ellingwood and other Eclectic physicians had found to be possible for this *Asclepias* species.)

The dreaded ***ERYSIPELAS*** was likewise treated by the Eclectics with swamp milkweed—they using it both internally and externally on the eruptions. *(Ellingwood 1983:451)* The Eclectics also used swamp milkweed for many ***RESPIRATORY AFFLICTIONS***, especially ***CATARRHAL INFLAMMATION.*** *(Ellingwood 1983:451)* Even earlier than the Eclectic use, however, the botanical scholar Constantine Rafinesque had noted how that western Indian tribes had used the roots for ***ASTHMA.*** *(Rafinesque 1828:76)*

As it was held to improve digestion, *Asclepias incarnata* was often prescribed by the Eclectics when either ***CHRONIC GASTRIC CATARRH*** or ***DIARRHEA*** was present. *(Ellingwood 1983:451)* Earlier, too, Rafinesque had noted how that Western Indian tribes had used it for ***DYSENTERY.*** *(Rafinesque 1828:76)* Although swamp milkweed is seldom used by modern herbalists, the recent *PDR For Herbal Medicines* notes that the usage for ***DIGESTIVE DISORDERS*** has persisted in some areas to the present day. *(PDRHrbs)*

Cautions

Owing to this plant's resin and cardioactive glycosides, it can be ***strongly emetic***. In view of this, and because of the very presence of these cardioactive glycosides which can powerfully affect the heart, swamp milkweed ***should not be used by the layman***. Dosage can be crucial with plants containing these glycosides and, as with digitalis, erring on the side of too much could prove disastrous! **Use ONLY under professional guidance** (preferably somebody *well practiced* and *skilled* in its use).

SWEET CICELY
[AMERICAN SWEET CICELY; WILD ANISE; ANISEROOT]
(Osmorhiza spp.)

Description

[See photo #143 in rear of book.]

The American sweet cicely (not to be confused with the European sweet cicely, *Myrrhis odorata,* commonly found in the spice section of the supermarket) is a perennial plant growing 1-4 feet tall. It springs from a fleshy, carrot-like root that can grow to six inches long.

The alternate leaves, spreading out from long leafstalks, are compound and fern-like. There are three leaflets, which are bluntly and irregularly toothed. All in all, the leaves have an appearance similar to those of carrots.

The long-stalked flowers are tiny, white, 5-petaled, and arranged in flat-topped clusters, which are also long stalked. Each umbel is divided into 3-8 rays.

Fruits are linear, oblong, and curved, resembling string beans. They are about an inch long and are possessed of an anise scent and flavor. In the fall, they become more tapered, turn black, and develop stiff, clinging hairs. At this stage, they easily wind up clinging to clothes and fur.

Several species occur in our area, including:

O. longistylis (anise root) has very fleshy roots that are strongly scented and flower styles that are long, slender, and plainly visible while the plant is flowering.

O. claytoni[alt: *claytonii*](scent root) has a root that is less fleshy and scented. Its flower styles are barely visible at first. The fruits end in two points.

Two other species—*O. chilensis* and *O. obtusa*—occur only in the extreme northeastern tip of Minnesota. The latter is pubescent and its fruit styles are bent sharply outwards.

Range & Habitat

Moist woods throughout our area.

Chief or Important Constituents

An essential oil consisting chiefly of anethol.

Food

I enjoy nibbling on this plant's fruit pods and young stems and being rewarded with their rich, anise-like flavor. (These plant parts can also be steeped to make a pleasant tasting tea.) The roots are also edible and can be dug into late fall/early winter (that is, if the ground is not frozen). They can either be eaten raw or they can be grated and cooked. Our area's Menominee Indians relished these roots.*(Black 1973:163)* They looked upon them as being a marvelous fattener for skinny people; however, they felt that they must be consumed slowly—only a section at a time.*(Smith 1923:72)* (Might this have been because of *Osmorhiza*'s strong effects upon blood-sugar levels? See below, under "Health/Medicine")

Both the seeds and the roots can be used as seasoning for baked goods and other dishes. Here the roots are dried and stored, then scraped or powdered when ready to use. (Such a methodology best preserves the potency of the volatile oils, and thereby the full and rich taste.)

Before you use what you think to be sweet cicely, be absolutely certain that you have the right herb, because it very closely resembles a most toxic plant (see under "Cautions," below).

Health/Medicine

Herbalist David Winston, a practitioner since the 1960s, regards this herb as a valuable *adaptogen*.*(Winston 1992:98)* And indeed it was used by Indian tribes for this very purpose—the Pawnee, for instance, regarding it as a staple for **DEBILITY** and **GENERAL WEAKNESS**.*(Gilmore 1991:55)* The Iroquois looked upon the autumn-harvested roots as a general **TONIC**.*(Herrick 1995:196)*

The Omaha Indians mashed the roots as a poultice for **BOILS,***(Gilmore 1991:55)* whereas our area's Winnebago Indians did the same for **WOUNDS**.*(Gilmore 1991:55)* Our area's Ojibwe Indians valued topical use of the plant for **ULCERS** and **RUNNING SORES**.*(Densmore 1974:354)*

Modern herbalists appreciate that the plant tonifies the body's mucous membranes.*(Willard 1992a:157)* Amerindian uses reflected such an understanding as well: For example, sweet cicely was used as a wash for **SORE EYES** by numerous tribes.*(Smith 1928:249; Smith 1933:86)* The Ojibwe used it as a gargle for **SORE THROAT**.*(Smith1932:391)* Potawatomi Indians treasured an infusion of sweet cicely roots as a *stomachic*.*(Smith 1933:86)* The Iroquois valued it for **DIARRHEA,***(Herrick 1995:196)* as did Indian tribes of the Nevada region for a related species.*(Train et al 1957:73)* Here the typical preparation and dosage for a child was to steep two small pieces (approx. 2" long) for five minutes in a cup of water, then give a teaspoonful.*(Herrick 1995:196)* This plant was especially valued by the Indians as a *carminative* for **FLATULENCE** and **COLICKY PAINS**.*(Train et al 1957:73)*

For **COLDS**, the Thompson and Okanagan Indians would chew the root of the western species, *O. occidentalis*.*(Turner et al 1990:45; Turner et al 1980:158)* The Bella Coola Indians of British Columbia found the root to be invaluable for treating **PNEUMONIA**.*(Smith 1928:61)* Iroquois Indians regarded the herb as an important **FEVER** medicine: They would take two medium-sized roots and boil them in three quarts of water until one quart remained. For children, they would instead steep a lone root in 1½ quarts of water until it turned blue and then give the youngsters a little bit at a time.*(Herrick 1995:196)* Winston feels that sweet cicely strengthens what Traditional Chinese Medicine (TCM) refers to as *wei qi*, the body's defensive energy which can help offset pernicious influences such as colds.*(Winston 1992:98)*

Some herbalists find sweet cicely helpful as part of a compound tea (inclusive of licorice and sassafras) to control **SUGAR-METABOLISM PROBLEMS**.*(Willard 1992:157)*

Ojibwe Indians utilized a tea made from the roots for **AMENORRHEA***(Gilmore 1933:137)* and as a **parturient.***(Smith 1932:391)* Blackfoot Indians used *O. occidentalis* for the latter purpose as well.*(Hellson & Gadd 1974:61)*

Kawaiisu Indians discovered than an infusion of a related species, *O. brachypoda*, killed **FLEAS** when used as a hair wash.*(Zigmond 1981:47)*

Caution

Do not confuse this plant with the deadly water hemlock (*Cicuta* species), which also grows in our area and has similar looking leaves and umbelliferous flowers. (*See* pages 103 and 219.) Baneberry (*Actaea* spp.), another toxic plant, looks similar to sweet cicely as well. Hence, do not ingest in any form what you may think to be sweet cicely unless the plant has a distinct *anise* or *licorice scent*. *Even then*, be *absolutely certain of identification* before using.

SWEET EVERLASTING

[CUDWEED; RABBIT TOBACCO; CATFOOT; OLD FIELD BALSAM; INDIAN POSY; SWEET BALSAM]
(Gnaphalium obtusifolium [G. polycephalum])

Description

[See photos #144-45 in rear of book.]

Sweet everlasting is an annual or biennial plant, 12-36 inches tall, with a balsam-like fragrance. It has a white, woolly stem that is erect, slender, and many-branched.

The plant begins as a collection of spatula-like, basal leaves arranged in a rosette. As it sprouts, it gains numerous, alternate, stem leaves that are unstalked, untoothed, linear to lanceolate, 1-4 inches long, and less than 1/2 inch wide. These are whitish underneath due to being covered with dense, woolly hairs.

The tips of this herb's many branches display spreading flower clusters composed of numerous, globular heads of yellowish-white disk flowers surrounded by brown, papery bracts.

Range & Habitat

Thrives in open sunny places: dry fields, disturbed ground, waste places, sandy prairies, open woodlands, roadsides.

Seasonal Availability

Collected while flowering, in August.

Chief or Important Constituents

A volatile oil, gnaphaliin, momomethyl-gnaphaliin, and tannins.

Health/Medicine

Sweet everlasting is primarily known for its *astringent* properties, and thus as a tonic to the mucous membranes. The botanical scholar Charles Millspaugh, writing in the late nineteenth century, drew attention to its usefulness in this regard, noting that it was used as a cold infusion for *LEUKORRHEA, DIARRHEA,* and *BOWEL HEMORRHAGE*.(Millspaugh 1974:351-52) In addition, *Gnaphalium* is held to be *anti-inflammatory* and consequently useful for conditions such as *LARYNGITIS, TONSILLITIS,* and *QUINSY*.(Hoffmann 1986:189) The Cherokees had even used it topically for *DIPHTHERIA,* blowing the warmed liquid into the clogged throat through a stem.(Hamel & Chiltoskey 1975:52)

The Cherokees also chewed the plant for *SORE MOUTH/THROAT* and smoked it for *ASTHMA*.(Hamel & Chiltoskey 1975:52) For the latter condition, famed country herbalist

Tommie Bass recommended putting a large clump of the herb in the sink, running hot water over it, and inhaling the steam (he was also fond of explaining that this procedure could likewise relieve **PLUGGED SINUSES** and **HEADACHES***).(Crellin & Philpott 1990:365)* Bass' biographer, John Crellin, noted that sweet everlasting has an excellent reputation as a *pectoral* as well.*(Crellin & Philpott 1997:140)* Millspaugh had likewise drawn attention to its application in this regard—noting that the dried flowers were used in the pillows of*TB* victims in order to allow them a good night's sleep*(Millspaugh 1973:352)*—as earlier had the medical doctor Asahel Clapp.*(Clapp 1852:802)* Interesting here is that a 1991 study involving the related species *G. viscosum* showed it to be effective against gram-positive bacteria commonly responsible for **RESPIRATORY DISEASES**.*(Caceres et al 1991:193-208)*

Sweet everlasting is especially held to be *anticatarrhal*. David Hoffmann, writing about the related species *G. uligonosum*, says that it is indicated as an*expectorant* for **UPPER RESPIRATORY CATARRH,***(Hoffmann1986:189; cf.Ody1993:191)*with an infusion of the dried herb being the best form for this purpose. Our own species has been employed similarly, and in both species the volatile oil may be the active agent.*(Crellin&Philpott 1990:365-66)* Not surprisingly, in view of the above, this herb—under the common name of rabbit-tobacco—has been a popular remedy in the Carolinas for**COLDS,***(Morton 1974:66)*, while the Cherokees had likewise used it for this purpose.*(Hamel&Chiltoskey 1975:51)*

Millspaugh had also listed the hot decoction as*sudorific* and thus helpful in the early stage of a fever.*(Millspaugh 1974:351-52)* To this day, various*Gnaphalium* species are popular as *febrifuges* in the southeastern USA, having been doggedly and successfully used in this area during the flu epidemic of 1941.*(Morton 1974:66)*

Millspaugh had further remarked that the plant has *anodyne* qualities.*(Millspaugh 1974:351-52)* In this regard, and combined with its anti-inflammatory properties, it is probably not without coincidence that the Cherokees had used sweet everlasting for **MUSCULAR CRAMPS**, **RHEUMATISM**, and **LOCAL PAINS**.*(Hamel & Chiltoskey 1975:51)*. Millspaugh added that a hot fomentation has also been used for*SPRAINS, BRUISES, TUMORS*, and **ULCEROUS SORES**.*(Millspaugh 1974:351-52)*

The plant's astringent, anodyne, and anti-inflammatory properties are undoubtedly key factors in offsetting*STOMACHACHE,* another discomfort historically treated by sweet everlasting (although the plant's bitter qualities may be an important factor here as well.*-Kindscher 1992:250)* The herb's alternate common name, cudweed, evidently arose because farmers discovered that feeding it to their cows allayed the sort of stomach inflammation that was capable of causing them to lose their cud.

Foster and Duke*(Foster & Duke 1990:82)* remind us that the plant has had other uses throughout history, namely as: (1) a*diuretic,* (2) a *nerve sedative*, and (3) an *antispasmodic* (interestingly, the Cherokees had used our plant for*TWITCHING-Hamel & Chiltoskey 1975:51*) Our area's Mesquaki Indians even used to steam the plant to revive a person from **UNCONSCIOUSNESS**.*(Smith 1928:214-15)* The strong, pleasant odor was probably the key factor in any success in this regard. On a personal note, I always relish using this plant owing to that appealing scent.

THISTLE
(Cirsium spp.)

[See photos #146-49 in rear of book.]

Well-known, prickly biennial or annual, with most species possessed of a stout taproot and having basal leaves arranged in a rosette. Flowers are composite and pink, rosy, purple, or white. A number of species flourish in our area, including:

C. arvense (Canada thistle) is a perennial, often found in patches. It reaches a height of 1-4 feet, bears a smooth stem, and has slender, white, creeping roots. The alternate leaves are oblong to lanceolate, 5-8 inches long, and green above but gray below. Thet are deeply cut (lobed) and wavy edged (perhaps best described as being "crinkled"), with the teeth extending into prickly spines. Stalkless, they often clasp the stem. The numerous flowerheads are ½ - ¾ inches across and pink to violet (occasionally white) in color. Male and female flowers occur on separate plants. Hugging the flowerhead are involucral bracts, which are *spineless*—in distinction from other species.

C. discolor (field thistle; pasture thistle) is a native perennial topping off at 3-6 ft. Its alternate, prickly leaves are elongate, grow 6-12 inches long, and are nearly flat. Their undersides are densely woolly. Many (5-20), pale-purple flowerheads occur on the plant, having bracts with long spines that stick out at an angle.

C. vulgare (bull thistle; hog thistle) A biennial forming a large rosette of basal leaves in the first year, and then in the second year shooting forth a powerful, upright, branching stem that is winged and armed with spines. It eventually reaches a height of 2-6 feet (3-4 ft. being average). Its alternate leaves are 3-9 inches long and quite narrow—only 1-1½ inches wide. They are coarsely lobed along their margins and possessed of stiff hairs on their surface. The underside of the leaves displays brownish hairs and a fat, swollen midrib. Each leaf lobe ends in a sharp spine. Blossoms (several per plant, but usually only one per branch) occur as vivid, reddish-purple flower heads, two inches high and 1-2 inches in diameter. These are surrounded by very spiny bracts with yellow tips on the spines.

Other species of note in our area include: (1) *C. altissimum* (tall thistle), a native biennial that grows 3-10 ft tall and has leaves that are toothed but not lobed (except that occasionally the lowermost leaves are lobed) and whose undersides are covered with white hairs; (2) *C. muticum* (swamp thistle), a 3-to-8-foot-tall biennial having bracts which lack spines and mature leaves that are hairless.

Most species occur throughout our area in pastures, roadsides, fields, wasteland, vacant lots, and illkept lawns. They thrive in complete or partial sun.

Chief or Important Constituents

C. arvense contains alkanes, taraxasterol, stigmasterol, and glururonides of apigenin and acacetin. The leaves alone contain flavonoids, other glycosides(linarin, cnicin), and an unidentified alkaloid. The flowers are rich in flavonoids(luteolin, apigenin, kaempferol). The root contains a phototoxic compound.

Food

Carefully trimmed leaves of *C. arvensis* can be eaten raw. (See "Cautions," below.) They are absolutely delicious and chock full of water to help sustain the thirsty forager! I look forward to them on nearly every foraging trek, carrying a round-tipped children's scissors with me to trim them so that I can consume them on the spot. Leaves of most other species are, unfortunately, too hairy or prickly on their tops and bottoms to trim and eat as a trail nibble.

Young stems of most any thistle species in our area can, before having gone to flower or even budding, be trimmed of their leaves, peeled, cut crosswise into pieces, and then boiled for 5-10 minutes in salted water. The taste is reminiscent of cooked artichokes or celery, depending on the species and other factors. These can also be eaten raw and they often make a great addition to a tuna salad. Older stems are too fibrous and tough to prepare as a menu item. But, depending on the species, they may possess an edible pith, which can be separated from the stalks and then consumed—raw or cooked. Often, however, these older stalks are infested with vermin, while the labor of removing the pith is often very time intensive.

The flowers of *C. arvense* make a delicious chewing gum, and one that will not promote tooth decay either. I enjoy the flavor very much, and always pick a few when I am out foraging. These flower heads, and especially those of *C. vulgare*, can even serve like rennet to curdle milk! To do this, simply gather a number of them, then mash them until a chestnutlike liquid is released. Only a tiny amount of this fluid is necessary to curdle a rather large amount of milk.

Did you know that artichokes are really the flower bracts of a cultivated species of thistle? Tis true, and as with artichokes, the bracts which surround our area's thistle burs can be trimmed, cooked, and eaten (never eat them raw, in view of the thorns), as can the base of the flowerhead (corresponding to the artichoke "heart").

The roots of all species, before or after flowering, can be peeled and eaten raw or cooked. Simply cut the peeled roots into disks as if slicing carrots, then boil in two waters for 20 minutes. Alternately, they can be boiled for several hours until soft and mushy and then dried and ground into flour.

All in all, thistles are one of the most utilizable of wild foods and certainly one of the most nutritious, as a thrilling incident in human survival has well evinced: In 1870, during an expedition to the Yellowstones, a man named Truman Everts was separated from his companions, thrown from his horse, and badly injured. Unable to walk, but courageously fending off panic, he survived on thistle roots for more than a month until he was found and rescued.*(Medsger 1966:201; Coffey 1993:254-55)* And yet this life-saving plant is castigated, cursed, and assaulted as a noxious weed by our area's governments! Surely, our officials could do better than bad-mouthing and spraying our area's thistles with what is truly noxious—herbicides! Why not educate the populace on the edible and medicinal benefits of thistles so that they could instead be harvested and utilized?

Health/Medicine

Thistles possess some remarkable healing properties and have appropriately been implemented by many native healers, although unfortunately not so much by contemporary herbalists of developed nations.

The plant's physiological properties well elucidate its healing potentials, however. It is, as widely acknowledged, preeminently an *alterative.(Lewis & Elvin-Lewis 1977:277).* North-Carolinan folk use has also viewed the plant as *tonic, diuretic,* and *hepatic.(Jacobs & Burlage 1958)* The *U.S. Dispensatory* for 1926 even highlighted research findings showing Canada thistle to be *diaphoretic, emetic*, and *tonic* and said that such 'merited attention.' *(Wood et al 1926:1506)* Such tonifying, detoxifying, and normalizing properties tend to elucidate the plant's main use as a *depurative,* so that thistles have been an important tool in the repertory of discriminating herbalists for healing people of *toxic conditions*— including even *PARASITES* and *FUNGAL/BACTERIAL* afflictions. Foremost among such discriminating healers were the American Indians....

For example, *C. discolor* was used by the Iroquois Indians for *BOILS.* The procedure was that seven roots were combined with one root of burdock and one quart of water and then this mixture was boiled down to one pint. Then, one-half wineglassful was taken every hour, followed by salts.*(Herrick 1995:231)* On the other hand, the aerial portions of Canada thistle were appreciated by Delaware Oklahoma Indians as a *pectoral(Tantaquideon 1972:71)* and several tribes (incl. the Mohegan) even employed it in a regimen for *TUBERCULOSIS*.*(Tantaquidgeon1928:269;Speck1917:314)* Our own area's Ojibwe Indians utilized this species as a *bowel tonic.(Smith 1932:364)* Abanaki Indians gave a decoction of the roots to children infected by *INTESTINAL WORMS*.*(Rousseau1947:173)* Delaware-Oklahoma Indians made an infusion of the leaves and then washed out the mouth of infants, probably because they were afflicted with thrush.*(Tantaquidgeon 1972:71)*

According to field research conducted by ethnobotanist Scott Camazine*(Camazine 1986:30)*, Zuni Indians afflicted with *SYPHILIS* drank an infusion of a thistle species (*C. ochrocentrum*) and then ran as heartily as possible for one mile, thereafter bundling themselves with blankets—all in order to raise their body temperature to a point necessary to destroy the dreaded infestation. Camazine cites research showing that the infectious spirochetes causing this disease die in a sustained, two-hour temperature of 105.6° Fahrenheit and that jogging three miles on a hot day could produce a rectal temperature of 106° F. One wonders whether the thistle infusion could somehow have potentiated or accelerated this process—perhaps by powerfully activating the liver?

Then, too, as *LYME DISEASE* is also caused by a spirochete and has even been called "deer syphilis," might there be possible application of the "thistle-brew-and-running-too" to that crippling condition? Of course, most people nowadays, being less conditioned than the Zuni warriors, would probably have a heart attack if they ran a mile, especially struggling with the crippling effects of Lyme disease on top of this. And trying to sustain a high temperature in the process could well invite death from heatstroke or heat exhaustion. However, what about drinking the thistle tea prior to a treatment in a hypothermic chamber (or via some other mechanism or methodology designed to maintain a high body temperature) *that is carefully monitored by a physician*? My thought is that this line of research deserves further exploration.

As Lyme disease also causes what has been called "floating arthritis," it is further of interest that North-Carolinan folk medicine has traditionally viewed the root and leaves as being *anti-inflammatory*. *(Jacobs & Burlage 1958)* Concordantly, a tea of the leaves has had wide and varied use in external application for *SKIN ERUPTIONS, SKIN ULCERS,* and *POISON-IVY RASH*. *(Foster&Duke 1990:166)* A strong decoction of the root of many species of thistle, too, serves ably as a disinfectant and healing wash for *POISON IVY* and *SKIN INFECTIONS*. *(Brown 1985:202)* One of the other species common to our area, *C. vulgare*, was utilized for *STOMACH CRAMPS* by the Ojibwe and for *RHEUMATISM* by the Delaware Oklahoma. *(Shemluck 1982:328)* Interestingly, too, these applications for inflammatory conditions can be viewed as having been scientifically validated by a 1976 lab study which found that Canada thistle was powerfully *anti-inflammatory* in the carageenan-induced, rat-paw edema test. *(Benoit et al 1976:164)*

Lewis & Elvin-Lewis also list Canada thistle as *emetic*. *(Lewis & Elvin-Lewis 1977:277)* East-Indian research has revealed that an extract of the species is *spasmolytic* and *hypotensive*. *(Asalkar 1992:205)* A bevy of other applications was appreciated by native Amerians: *C. vulgare* was put into service for *BLEEDING PILES* and *CANCER* by the Iroquois. *(Herrick 1995:231)* An unidentified species of thistle was used by our area's Ojibwe (Chippewa) Indians for women who had "pains in the back" as well as for "male weakness." *(Densmore 1974:356)* Lumni Indians gave a saltwater decoction of the roots and tips of a *Cirsium* species to their women at childbirth. *(Gunther 1973:49)* Costanoan Indians used the roots of an unidentified species of thistle for *STOMACH PAIN* (chewed raw) and for *ASTHMA* (given as a decoction). They also soothed *SORES* on the face and dried up *INFECTED SORES* with the application of a pulp made from the pounded stems. *(Bocek 1984:254)*

Cautions

* Be *absolutely sure* you have trimmed off *all* thorns from thistles before consuming them. When I feel I am done trimming the leaves of *C. arvense*, I always flip them over to view them from both sides before popping them in my mouth. Invariably, upon doing this, I find a remaining thorn or two!

* Be wary of sun exposure when consuming thistle roots, as some of them contain a phototoxic compound that could conceivably cause severe sunburn.

TURTLEHEAD
[BALMONY]
(Chelone glabra)

[See photo #150 in rear of book.]

Description

A breathtaking perennial with a slender, upright, squarish stem that sometimes branches. Turtlehead grows 1-3 feet tall.

The straplike leaves are opposite, sharply toothed, and have prominent veins. Stalkless—or nearly so—they grow 3-6 inches long and ½-1 inch wide.

Inflorescence is in the form of a number of large, two-lipped, tubular flowers arranged in a tight, spike-like cluster at the tip of the stem. The blossoms are creamy white to purple in color. Altogether, this presents the appearance of a turtle's head, and hence the plant's common name.

Range & Habitat

Thrives in wet areas—ditches, swamps, lakeshores, streamsides, wet meadows. Can be found throughout our area, except for the western one-third of Minnesota.

Chief or Important Constituents

Bitter principles; resins

Health/Medicine

Turtlehead has long been appreciated by herbalists as a gentle *laxative,* working toward this end by improving bile function, and such is how it was implemented by various Amerindian tribes, including the Cherokees.*(Hamel & Chiltoskey 1975:59)* Then, too, early-nineteenth-century botanist Constantine Rafinesque described the use by whites of his time in a similar vein: "It is useful in...jaundice, hepatitis, eruptions of the skin & c. In small doses it is laxative, but in full doses it purges the bile and cleans the system of the morbid or superfluous bile, removing the yellowness of the skin in jaundice and liver diseases. The dose is a drachm of the powdered leaves 3 times daily.... Few plants promise to become more useful in skillful hands."*(Rafinesque 1830:118)* As to the effect upon *JAUNDICE,* the Natura practitioner R. Swinburne Clymer esteemed turtlehead to be one of the best treatments for this in its chronic form, holding the plant to be a stimulating cholagogue.*(Clymer 1973:115)*

The Cherokees also found this herb useful as an ***appetite stimulant.***(Hamel & Chiltoskey 1975:59) Such a use was echoed by early American physicians as well: *Gunn's Domestic Physician*, a common "household remedy" guide of the time, informs us that turtlehead was "employed in costiveness, dyspepsia, loss of appetite, and general languor, or debility."(Gunn 1859-61:740) All in all, it is apparent that turtlehead is an excellent herbal option for the aged, who generally experience not only these problems, but also the previously mentioned constipation for which we have also noted our herb to be of aid.

Anthelmintic properties, too, were culled from this plant by the Cherokees.(Hamel & Chiltoskey 1975:59) Dr. Gunn's household guide borrowed this application as well: "Given to children afflicted with worms, it will generally afford relief.... An even teaspoonful [of the tea] is a dose."(Gunn 1859-61:740) Even some modern herbalists revere turtlehead as one of the best expulsives of parasites around. Such ones often advise a dose of one ounce of the leaves to one pint of boiling water, taken in a wineglassful amount in the evening before retiring and in the morning on an empty stomach, with a purge to be expected (or if not consequently accomplished, to be *induced* at the end of the day after fasting).(Carse1989:167;Clymer 1973:115;Hutchens 1973:22)

Clymer also held turtlehead to be "the itching remedy," generally in the form of a salve composed of two or more ounces of leaves to one-half pound of lard, boiled slowly for 30 minutes. He found it best applicable to ***ITCHING PILES***.(Clymer 1973:163) I can't help but wonder if there a connection here with the fact that many health professionals associate chronic rectal itching with parasites.

Interestingly, the Maritime Indians even used turtlehead in an undisclosed way as a ***contraceptive.***(Mechling 1959) The American Indians, in fact, used so many plants to prevent conception (see the index to physiological functions in the rear of this volume, under "contraceptive," for a small sampling) that this is an area that is assuredly overripe for detailed investigation, given our overflowing population at present.

VIOLET
(Viola spp.)

[See photos #151 & 152 in rear of book.]

There are over 70 species of this delicate and delightful perennial herb growing in the USA and over 20 in our area alone! Although the various species differ a bit in detail, they all share some common features, including the fact that their blossoms bear some distinctive earmarks: five sepals, five stamens, and five petals. The lower petal is terminal and usually larger than the rest, as well as possessing noticeable veins (most often purple in color) and displaying a backward-projecting spur. Leaves of most species (incl. the majority of those native to our area) are basal and have their blossoms on a separate stalk from their leaves.

In our area, the most common violet species is *V. sororia*, known as the common blue violet (alternately as the woolly blue violet or the sister violet), which grows 3-6 inches high. It possesses heart-shaped, basal leaves, about as wide as long, with scalloped margins (i.e., rounded teeth). The long-stalked, blue-violet blossom possesses bilaterally symmetrical petals, with the lateral petals having a "bearded" look to them. The whole plant, and especially the stem and underside of the leaves, is finely pubescent.

A second blue violet inhabiting Minnesota and Wisconsin is the marsh violet, *V. cucullata*, which resembles the former except that its blossoms grow taller than its leaves, the center of its blossom is darker blue than are the exterior portions, and the lower petal is shorter than the others (rather than being the longest, as in other species). Another blue species in our area, but with its blossoms and leaves on the same stalk, is the dog violet, *V. conspersa*. Still another blue-flowered species common to our region is *V. pedata*, called the bird's-foot violet because its leaves are deeply dissected and thus somewhat resemble an avian foot. Its petals lack beards and the upper ones curve backwards.

A yellow-blossomed species common to our area is *V. pubescens*, the common yellow violet, reaching a height of 4-14 inches. It has leaves and blossoms on the same stalk and occasionally one basal leaf besides. Its heart-shaped, scalloped-edged leaves are 2-5 inches wide, and the lateral petals on its flowers are "bearded." The whole plant, but especially the stem and along the leaf veins, is pubescent. A variant yellow-flowered species, *V. pensylvanica*, the smooth yellow violet, has smooth stems and one or more basal leaves.

A white-blossomed species in our area is *V. canadensis*, the Canada violet or tall white violet, which grows larger than our other species—to some 8-16 inches. The purplish stems of this species are hairy and it possesses both basal leaves (2-4 inches long, almost as wide, and pointed) and stem leaves. Its flower has a yellow center and sometimes the backs of the petals are tinged with violet. Often the plant is pleasantly scented with a mint-like odor.

Interestingly, the showy blossoms on violets are not the seed producers for the plants, but are largely sterile. Violets actually bear self-pollinating flowers close to where the stalk emerges from the root. Called "cleistogamous" by botanists, these flowers, which produce numerous seeds, do not open until they release those seeds, and thus are difficult to spot by the untrained eye. And while their wind-blown seeds start new colonies, the existing colonies spread largely by means of subterranean rhizomes.

Range & Habitat

One or another species of violet occurs—often in large colonies—throughout our area and indeed throughout most of USA and Canada. The preferred habitat is open woods, trailsides, pastures, moist meadows, swamps—in short, wherever the combination of shade and moist soil can be found.

Seasonal Availability

Violets should be gathered in the spring, while in flower, for fear of mistaking them for inedible or mildly toxic plants whose leaves resemble theirs.

Chief or Important Constituents

Saponins, mucilage, salicylic acid, methyl salicylate, phenolic glycosides(incl. eugenol), flavonoids(incl. rutin in goodly quantity), and a bitter, purgative principle in the roots which varies from species to species. Nutrients include incredibly high amounts of vitamin A(15,000-20,000 IU's per 100 grams—much more carotene than in spinach, which has the highest content of any marketed vegetable) and C(150-250 mgs. per 100 grams; in other words, in one half-cup of violet leaves there is more vitamin C than in four oranges!)

Food

The raw leaves can be munched as a trail nibble or mixed into salads, and they are often excellent, although at certain stages of their growth they can be too bitter. Nevertheless, I *so* look forward to these greens when they first appear in the spring! The blossoms are also edible in moderation (in not-so-moderate doses they can be laxative!), but are usually not as tasty as the leaves.

The leaves can also be eaten as a potherb. One simply boils them in a small amount of water for 10-15 minutes. Then, too, a tea made from the leaves has a pleasant, mild taste. Chilled, it makes an excellent "iced tea." Euell Gibbons*(Gibbons 1970:64-68)* raved over the taste of violet jam and jelly, but I have yet to try making either.

Health/Medicine

Violet's most acclaimed use, especially that of its blossoms, has been along the respiratory line—to treat *COLDS, BRONCHITIS, CHRONIC COUGHS,* and *SORE THROAT.* The blossoms are prepared by soaking for three minutes in cold water, and then brought to a boil for ten minutes, after which the tea is cooled to a drinkable temperature. Combined with a tincture of blue vervain, violet blossoms have also been used to treat *WHOOPING COUGH.* *(Potterton1983:196)* Violet's mucilage is helpful in these conditions as a soothing, protective agent for the mucous membranes (and thus the tea is gargled for sore throat, as did our Ojibwe Indians-*Hoffman 1891:201*), while the saponins in the herb serve as an *expectorant* because of irritating the respiratory system by means of reflex action from the stomach. In other words, violet serves as a moistening expectorant. Not surprisingly, too, persons with *ASTHMA* have been known to benefit from this herb. The Blackfoot Indians, for instance, gave an infusion of the leaves (as well as a very small portion of the roots) of *V. adunca*—another species scattered throughout the northern part of our area—to asthmatic children. *(Hellson & Gadd 1974:74)*

Violet blossoms are renowned in folk medicine as a safe and effective *laxative* for children. For this purpose, British herbalist Mary Carse recommends pouring one cup of boiled water over one cup of fresh violet flowers, steeping 12 hours, straining, sweetening with one pound of honey, and giving one-half to one full teaspoon of the syrup to constipated youngsters at bedtime so as to produce a morning bowel movement. *(Carse 1989:170)* (Note: *Never* give any product containing honey to an infant under one year of age.)

A much-appreciated *refrigerant*, violet tea has proven useful as a mouth swish for *INFLAMED GUMS* and in compress form for reducing *FEVERS* (as used by French physicians in the early 1900s; a recent Pakistani study found leaf extracts comparable to aspirin in reducing pyrexia in animals-*Khattak et al 1985:45-51*) and *HEADACHES* where the head feels "hot." It is especially effective for hangover headaches, even as the Roman naturalist Pliny had acknowledged, pointing out that the wearing of a garland of violets around the forehead would banish headaches and dizziness from too much drinking! Centuries later, and undoubtedly unaware of Pliny's remarks, Cherokee Indians had likewise poulticed violet leaves on headaches. *(Hamel & Chiltoskey 1975:60)* Certainly here the salicylic acid, eugenol, and methyl salicylate (the latter chemical also saturating the cooling and refreshing wintergreen plant) are factors in relieving headache pain and may also help elucidate why violet (esp. its blossoms) has traditionally been valued for *anodyne* abilities in general.

In accord with this pain-dulling property, our area's Ojibwe (Chippewa) Indians employed a decoction of *V. canadensis* for *BLADDER PAIN* *(Hoffman 1891:201)* Then, too, a poultice of the leaves of various species has traditionally been used to ease the pain of *RHEUMATISM* (the Blackfoot Indians applying a compress of a root-and-leaf infusion to the painful areas-*Hellson 1974:79*) and *GOUT* (Pliny related that a liniment made from violet roots and vinegar was helpful here). Especially significant in this regard is that violet tea is a mild *diuretic* (recently confirmed in animal studies-*Dobelis 1986:316*) and *depurative*, thus enabling it to drive excess uric acids from the body. These physiological activities also lend violets for use with *BOILS, SWOLLEN LYMPH NODES,* and *PIMPLES*.*(de Bairacli Levy 1974:148)* The Iroquois Indians had used it for the latter affliction, taking a decoction internally and as a facial wash.*(Herrick 1977:387)* The Cherokees had appreciated the use for boils, applying a poultice of the crushed roots of one or another species to them.*(Hamel & Chiltoskey 1975:60)* The effect on the lymphatic system can be quite pronounced, and this is the herb of choice in stagnant lymph conditions and chronic lymphatic swelling where, as modern herbalist David Winston points out, such is accompanied by constipation.*(Winston 1999:54)*

Some of violet's other topical uses have included a chewed-leaf poultice for *CRACKED NIPPLES* and *SORE EYES.* Manasseh Cutler, an early American botanical author, also noted that Indians commonly bruised the leaves of a yellow-blossomed species as a topical application to *BOILS* and other *PAINFUL SWELLINGS,* in order to ease the pain and to cause them to suppurate.*(Cutler1785:485)* Then there is a vivid personal account from an American missionary working among the Indians in the early 1800s, revealing that its author "once suffered the most excruciating pain from a felon or whitlow on one of my fingers, which deprived me entirely of sleep. I had recourse to an old Indian woman who in less than half an hour relieved me entirely by the simple application of a poultice made of the root of the common blue violet."*(Heckewelder 1820:229)* (It should be noted that when suppuration is not the goal, mashed violet blossoms should not be left on painful body areas for longer than 2-3 hours, in order to prevent blistering.)

Violets have traditionally been used topically to soften almost any kind of hard lump in the body, especially *SWOLLEN, LUMPY BREASTS,* and even including *TUMORS*. A poultice of the fresh or dried leaves has even long been used as an application to *SKIN CANCERS*. Jonathan Hartwell, an American scientist, collected voluminous accounts of the many uses of violets in anti-cancer regimens from around the world.*(Hartwell 1971; Hartwell 1984)* In fact, in the five-year period from 1901 to 1906, five different articles even appeared in British medical journals on violet treatments for cancers!*(Mockle 1955:50)* And while naysayers claim that the salicylic acid in violet simply acts as a dissolvent to rid the physical structure of these cancerous growths (as indeed a violet ointment has been a favorite for removing *WARTS* and *CORNS*), the incredible range and variety of cultural uses strongly suggests that there is more to it than merely that, and such a conclusion is reinforced by a 1960s lab study which demonstrated that a violet extract (from *V. striata*) brought about genuine damage to cancer in mice.*(Farnsworth et al 1968:237-48)*

Because of violet's above-noted anodyne properties, the herb has been utilized to help ease the *PAIN* of cancer, especially when it is in the throat. For this purpose, a strong tea is sought. A recipe popular in India and Pakistan consists of two ounces of leaves infused in a pint of boiled water, steeped for twelve hours, and then strained and taken in doses of two fluid ounces every two to three hours.*(Hutchens, 1991:288; Potterton 1983:196)*

Then, too, *sedative* abilities (especially where there is **RESTLESSNESS-**de Bairacli Levy *1974:148*) have also been ascribed to our little plant, and thus a violet footbath has been a favored treatment for **INSOMNIA.** Our herb is especially indicated for restlessness in children.

Violet flowers contain a large amount of the bioflavonoid rutin, which has been recognized by medical science for over half a century now as providing support to the walls of blood vessels. In calculating the content of this bioflavonoid in violet, James Duke, a former economic botanist with the USDA, says that a few tablespoons of the plant could provide biologically significant amounts for tonifying **ATONIC BLOOD VESSELS** (e.g. **VARICOSE VEINS**).*(Duke 1997:446)*

Another circulatory affliction traditionally benefited by violets has been **PROB-LEMS WITH THE HEART.** The Potawatomi Indians used the root of our downy yellow violet, *V. pubescens*, as a cardiac medicine,*(Smith 1933:87-88)* while the Flambeau Ojibwe preferred the entire aerial portions of our dog violet, *V. conspersa.(Smith 1932:392).* One species, *V. tricolor*, the garden pansy or Johnny-jump-up, has even been known by the common name of "heart's ease" owing to its traditionally understood cardiotonic properties. This species was at one time official in the USP (the 1883 ed.), as have been our *V. pedata* (1820-70) and Europe's *V. odorata* (our area's *V. cucullata* being the American equivalent to this species).

Utilitarian

Violet juice from crushed blossoms can be used a natural litmus paper—to test a substance as to whether it is primarily acidic or alkaline. The juice turns red when contact with an acid is made, but green when touched by a base (alkaline substance).

Cautions

* *Yellow* violets have been known to cause stomach upset or diarrhea in some persons. Proceed with caution in attempting to use any non-purple-flowered violets.

* No violet *roots*—of any species—should ever be eaten, as they are all purgative and emetic (to one degree or another).

* Collect violets *only when blooming* so as to avoid confusing this species with toxic or inedible plants whose leaf structures resemble that of violets.

*Bear in mind that the common house plant known as African violet is NOT related to wild violets and is strictly inedible. Do *not* eat *African violets*!

VIRGINIA WATERLEAF
[INDIAN SALAD; JOHN'S CABBAGE]
(Hydrophyllum virginianum)

Description

[See photos #153 & 154 in rear of book]

A perennial herb of medium height (1-2½ feet tall), with an erect—though weak—stem.

Leaves are pinnate, with 5-7 leaflets per leaf stalk. Each leaflet is 2-5 inches long, ovate or lanceolate, and irregularly but sharply toothed.

In early spring, the leaves appear to be mottled with striking blotches of gray-green, giving the impression that they have been stained with water droplets. These pseudo watermarks are one excellent identifier of the waterleaf plant.

As summer progresses, the "water spots" appear to fade from view. But visual resplendence is not lost, for this herb has now put forth its dazzling purple (or purplish-white) flowers. These are in the shape of bells, with five petals, and characterized by stamens with long filaments that protrude beyond the rim of the flower. The blossoms grow in dense clusters on long stalks from the leaf axils and arise noticeably above the leaves.

Range and Habitat

In rich, moist woods—often in colonies because of spreading by rootstocks. Flourishes in southern Canada (Quebec to Manitoba) and throughout most of northern two-thirds of the USA east of the Mississippi. A variety of related species grow in the same vicinity and in sandier soils in the West. In our area: Entire, except for the extreme northern strip

Seasonal Availability

Leaves: Spring to early summer; Root: Spring or Autumn

Chief or Important Constituents

It seems that this plant has been largely ignored by phytochemists—at least, I am not aware of any available chemical profile. It is certain, however, that it contains tannins. Known nutrients include vitamins A, C, and E.

Food

Both this species and a near relative (*H. appendiculatum*) make good salad greens when the leaves and/or shoots are taken before flowering. In fact, the Virginia waterleaf of my vicinity is one of my most awaited spring greens, and I have several guarded patches that I return to year after year.

The leaves also make a fine potherb, boiled or steamed for 5-10 minutes. Wild-plant cook Adrienne Crowhurst recommends cooking for three minutes, throwing off the water, and submerging the plant parts in a second pot of boiling—this time salted—water, cooking for five more minutes.*(Crowhurst 1972:47)* According to Fernald and Kinsey, the ethnobotanist Huron H. Smith asserted that it was very important to throw off the initial water.*(Fernald and Kinsey 1958:327)* Some excellent recipes can be found in the foraging manual by March and March entitled *The Wild Plant Companion.*(March & March 1986:121, 143)*

After Virginia waterleaf flowers, the leaves are usually too bitter to eat raw. At this point in their growth, they are not too appealing to me as a trail-nibble anyway, for whenever I have encountered the plant in the flowering stage I have noted that it seems particularly attractive to flies, which are continually landing on its leaves!

In the autumn, various Indian tribes would cook and eat the thick roots.

As mentioned earlier, little study seems to have been done on this herb, although it is known that it contains vitamins A, C, and E. My personal suspicion is that a complete chemical profile would yield some impressive data. All in all, waterleaf is a plant not given nearly enough attention in the foraging manuals.

Health/Medicine

This plant has been sadly ignored by herbalists, which is unfortunate, in that it is a first-rate **astringent**. The entire plant is such, but the roots most powerfully so. Thus, a root decoction—or simply the chewed root—was used by the Iroquois Indians for **MOUTH SORES** and **CRACKED LIPS**.*(Herrick 1977:420; Herrick 1995:203)* Our own area's Menominee and Ojibwe (Chippewa) Indians implemented the root for **FLUX**.*(Smith 1923:37; Smith 1932:371)* As such, it was utilized and deemed safe for men, women, and children.*(Smith 1932:371)*

An analgesic element may be present in the root, since the Menominees used a decoction of it as part of a compound formula for **CHEST PAIN**.*(Densmore 1932:130)* The Ojibwe did likewise, finding it helpful as well for **BACK PAIN**.*(Hoffman 1891:201)*

As noted above, a tea is the most common method of treatment, although a tincture can also be used (the procedure for making such being outlined by Charles Millspaugh—*Millspaugh 1974:478*).

WATERCRESS
(Nasturtium officinale)

Description

[See photo #155 in rear of book.]

A creeping or floating, perennial herb, averaging 4-10 inches long, and possessed of a pungent odor when bruised. The hollow stems are succulent and have white rootlings at their leaf nodes.

The alternate leaves are pinnately compound, with 3-9 leaflets. As with mustard-family relatives like wintercress (*Barbarea* spp.), *the end leaflet is the biggest one.*

Also characteristic of its mustard-family membership is the fact that its tiny, white flowers *have four petals,* as well as six stamens. These blossoms display themselves in clusters, but usually only toward the last half of the growing season.

Range & Habitat

Grows in colonies throughout our area in running—though quiet—waters, such as can be found in brooks and springs.

Chief or Important Constituents

A glycosidal isothiocyanate precursor(gluconasturiin), bioflavonoids, fixed oils. Nutrients include vitamins (A[4900 IU per 100g], B2, folic acid, C[79mg per 100g], D, and E) and minerals (calcium[151mg. per 100g], manganese, copper, iodine, phosphorous[54mg per 100g], sulfur, and zinc).

Food

This plant is one of the most nutritious of all vegetables, as can be ascertained from the nutritional profile provided above. I have savored it since I was a teen and I doubt that I will ever lose my taste for its wonderful tang!

Watercress can be lightly steamed for a few minutes or chopped and added to stir-fried dishes. Ideally, however, it should be eaten raw. This not only best preserves the nutrients, but it tastes much better this way, too! (It's hard to beat fresh watercress on a sandwich, or as an ingredient in a super salad!) The problem, however, is that in nature this herb has a habit of accumulating harmful parasites from infected waters (see "Cautions," below), which can easily be transferred to humans. Thus, in wildcrafting it, unless you have your water tested, you always run some risk, and this could conceivably even be true if you first soak the plant in a water bath to which drops of grapefruit-seed extract, halazone tablets, or other disinfectant agents have been added. All in all, this is one plant that might best be secured from commercial sources, where it is grown in protected beds.

Watercress can also be blanched for two minutes, drained, pat-dried, labeled, and frozen for later use.

Health/Medicine

This plant is a nutritional powerhouse and many of its therapeutic benefits are related to its vast array of nutrients. First, due to its goodly amount of iron, it has long served as a revered treatment for **anemia.***(de bairacli Levy 1974:150)* Second, the high vitamin-C content explains the traditional use of this plant as an **antiscorbutic.***(Dobelis et al 1986:327)* Watercress' huge quantity of this vitamin may also explain why the herb has a tonic effect on the gums when chewed.*(Mabey 1988:53)*

The plant's bioflavonoids and sulfur may also play a role in its aid for the gums; they definitely strengthen both immune and sinus function (which, in fact, are related through the adrenal glands).*(Landis 1997:188)* Then, too, watercress strengthens the lungs and acts as an **expectorant,***(Winston 1999:54)* which fact even Hippocrates had alluded to several centuries before the birth of the Common Era. Macerated in honey, this water-loving plant has even been a traditional **COUGH** remedy.*(Krochmal & Krochmal 156)* Watercress is also used to treat **FLUID IN THE LUNGS**.*(Tierra 1992:225)* There has even been some use of the herb by various communities (including the Chinese working on the railroads in nineteenth-century America) to help ease the ravages of **TUBERCU-LOSIS**.*(Duke 1997:218; Krochmal & Krochmal 1973:156; Grieve 1971:2:845)*

Spanish-speaking communities in New Mexico have long used watercress for **HEART TROUBLE**.*(Krochmal & Krochmal 1973:156)* The herb's aid in this respect may be due to its high content of vitamin E, which increases the body's utilization of oxygen and is helpful for the heart in other respects as well.*(Hutchens 1991:293)*

Watercress's pungency stimulates digestive function,*(Johnson 1884:94)* so that it is often helpful for **INDIGESTION***(Tierra 1992:225)* or even a lack of appetite.*(de Bairacli Levy 1974:150; Grieve 1971:2:845)* The plant is also a **cholagogue** and has thus been used for "biliousness" (poor fat metabolism). Traditional Chinese Medicine (TCM) hails it as a remedy for *stagnant liver-qi*.*(Winston 1999:54)*

Watercress is food for the kidneys*(Landis 1997:188)* and has long been appreciated by Spanish-speaking communities of New Mexico for **KIDNEY TROUBLES**.*(Krochmal & Krochmal 156)* It is a reliable **diuretic,***(Johnson 1884:95; Tierra 1992:225; Stuart 1982:100)* (though it can be too irritating to some—see "Cautions" below), so that it has historically been used for **EDEMA**.*(Tierra 1992:225)* Its diuretic properties may at least partly also explain its historic use for both **GOUT** and **RHEUMATISM**.*(Foster & Duke 1990:34; de Bairacli Levy 1974:150; Lust 1974:390)*

The fresh plant is *laxative*(Krochmal & Krochmal 1973:156), probably due to its content of mustard oil. This oil is also *antibacterial*.(Winston 1999:54) Watercress is also mildly *antipyretic*. (Stuart 1982:100)

Our pungent herb has been used in some lands—such as in various parts of Africa—as an *aphrodisiac*.(Krochmal & Krochmal 1973:156) The Romans even called it *impudica*, meaning "shameless." It is interesting in this regard that another member of the mustard family, peppergrass (*Lepidium* spp.), was used by the Lumbee Indians as an aphrodisiac. (See under "peppergrass.") Could watercress's purported ability in this regard be partly owing to its *stimulant* properties, which even Hippocrates had noted?

Cautions

* Wildcrafting watercress presents some danger in areas where cattle or sheep graze, as it can harbor liver flukes in the waters where it grows due to the excrement of contaminated livestock. *Hence, do not gather watercress from known grazing areas.*

* *Caution in pregnancy*: It is thought that this plant could possibly cause miscarriage if used in quantity during pregnancy.(Krochmal & Krochmal 1973:156) Some Amerindian women even ate it during labor to hasten delivery.(Scully 1970:285)

* Excessive use may cause *bladder irritation*. This is because, as botanist and physician Laurence Johnson had so well put it: "The [plant's] acrid principles...appear, clinically, to be eliminated by the kidneys, and hence, incidentally they produce a decided diuretic effect. The urine is not only increased in quantity, but partakes also of the acrid character of the plant."(Johnson 1884:95)

WILD BERGAMOT

[BEEBALM; SWEETLEAF]
(Monarda fistulosa)

Description

[See photo #156 in rear of book.]

A perennial mint, with a perfumish scent, and a square stem that is hairy at the nodes. This herb reaches a height of 1-4 feet.

The leaves are opposite, lanceolate in shape, and coarely toothed. The grow to three inches long and spring from short stalks.

Many lavender, tubular, inch-long flowers appear clustered in dense, rounded heads at the end of the stems. These flowers are two-lipped—the upper lip being hairy, erect, toothed, and two-lobed, and the lower lip three-lobed and sagging.

Range & Habitat

Thrives in dry, open places: prairies, meadows, thickets, dry fields, open or upland woods. Occurs in the lower half of Minnesota, west of the Mississippi.

Chief or Important Constituents

Essential oil (consisting of alpha-pinene, alpha-terpinene, alpha-terpineol, beta-pinene, camphene, carvacrol, formaldehyde, geraniol, limonene, linalool, myrcene, nerol, p-cymene, pulegone, sabinene, thymol, etc.), rosmarinic acid.

Health/Medicine

This interesting plant has long been appreciated for the *carminative* and *diaphoretic* properties that all mints possess. Like its cousin, horsemint [*M. punctata*] (*see under that entry*), it is also a *stimulant.*

Bergamot's carminative property was greatly appreciated in colonial America, where the oil was extracted and two drops added to a sweetened glass of water to *ALLEVIATE GAS PAINS.(Rafinesque 1830:37-38)* As for the plant's stimulant property, the Iroquois Indians appreciated this for *GENERAL LASSITUDE.(Herrick1995:208)* Lakota Indians even implemented our herb as a nasal stimulant for tribe members who had *FAINTED.(Rogers 1980:50; Munson 1981:236)* Such a property was of value as well with *CONGESTION* due to *COLDS*, for which various Amerindian tribes used the plant, including the Mesquaki (Fox) and the Teton Sioux. *(Densmore 1918:270)*

The latter also utilized an infusion of the flowers as a *febrifuge*,*(Densmore 1918:270)* while the Lakotas appreciated bergamot for both *FEVER* and *COLDS*.*(Rogers 1980:50; Munson 1981:236)* Koasatis Indians even bathed a feverish patient having *CHILLS* with a decoction of this herb's leaves.*(Folson-Dickerson 1965:64)* That enthusiastic botanical writer of the early nineteenth century, Constantine Rafinesque, likewise alluded to its use for fevers, especially intermittents, noting the work of Dr. Johann David Schoepf for such a purpose.*(Rafinesque 1830:37-38)* The Cherokee Indians even used a bergamot infusion to "bring out measles." *(Hamel & Chiltoskey 1975:39)* Navaho Indians found that a cold tea served as a healing wash for *HEADACHES*.*(Elmore 1944:73)*

Undoubtedly the plant's noted diaphoretic property lent a hand to reducing fevers in these instances. But, there seems to be another factor at work here, too—more along an energetic line. As my fellow Minnesota herbalist Matthew Wood explains, in his wondeful *Book of Herbal Wisdom*, Amerindian tribes have always viewed bergamot as a plant that can "draw out fire." They use the principle of "like cures like," believing that the heat of an affliction will go into a "hot" plant like bergamot. And "hot" it is, for if applied directly to the skin, there is a strong *rubefacient* effect. (Among the American colonists, the extracted oil was dissolved in alcohol and used as a rubefacient liniment for *PERIODICAL HEADACHE*, as well as for *CHRONIC RHEUMATISM, DEAFNESS, PARALYSIS,* and *TYPHUS*-*Rafinesque1830:158*) The Indians feel that bergamot is especially indicated for interior heat and where the body surface is cold, so that it is specific for a fever accompanied by chills or a cool, clammy skin. Some tribes even used it for *BURNS,* probably where a cold sweat accompanied such. Our own area's Ojibwe Indians seem to have applied a poultice of the plant's flowers and leaves to such a burn, finding it especially effective for *SCALDS*.*(Densmore 1974:354)*

Other "hot" conditions were treated similarly with bergamot: The Lakotas wrapped bergamot leaves in a soft cloth and positioned such over *SORE EYES* to bring relief.*(Rogers 1980:50; Munson 1981:236)* Then, too, the Crow Indians discovered how to take the "sting" out of *INSECT STINGS* and *BITES*: they simply crumpled bergamot leaves, mixed the powder with spittle, and applied the mixture directly to the painful areas.*(Vestal & Schultes 1939:49)* Blackfeet Indians poulticed bergamot flowerheads over a *BURST BOIL* until it healed, thus drawing the fire out of this "hot" affliction, while the plant's thymol and other chemicals no doubt served to quell any new infection.

Bergamot's high quantity of essential oil also means that the plant is helpful for clearing up respiratory afflictions. Thus, the Flambeau Ojibwe Indians inhaled vapor from the boiled plant for *CATARRH* and *BRONCHIAL AFFECTIONS*.*(Smith 1933:61; Smith 1932:372)* The Menominees used leaves and flowers for *CATARRH*.*(Smith 1923:39)* *COUGHS* were treated with a tea of this plant by the Blackfoot Indians.*(Hellson & Gad 1974:67)* The Crows used the same infusion for various *RESPIRATORY PROBLEMS*.*(Hart 1976:70)*

The Iroquois steeped a handful of roots in a quart of water and had children with *HEADACHE* and *CONSTIPATION* drink 1/2-cup of the tea, three times a day.*(Herrick 1995:208)* South Ojibwe Indians appreciated a decoction of the root for *GASTRO-*

INTESTINAL PAIN*.(Hoffmann 1891:201)* Teton Dakota Indians implemented a decoction of the leaves and flowers for same.*(Gilmore 1991:59)* Mesquaki (Fox) used a lone plant in a compound formula for ***STOMACH CRAMPS****.(Smith 1928:225)*

The Blackfeet imbibed bergamot tea for ***ACHING KIDNEYS****.(Hellson & Gad 1974:67)* Interesting in this regard is the *U.S. Dispensatory's* observation that our herb is "an active diuretic."*(Wood et al 1926:1386)* Thymol is also a ***urinary antiseptic***, as we saw with horsemint, earlier (*see under that entry*). Undoubtedly, too, bergamot's energetics were involved in the Blackfeet's use for painful kidneys, with the herb "drawing" out the heat from what Traditional Chinese Medicine might call a *kidney-fire* situation.

Winnebago Indians splashed a decoction of the leaves on ***PIMPLES*** and other ***SKIN ERUPTIONS****.(Gilmore 1991:59)* Recent scientific studies suggest that bergmot seems to counter ***HERPES-FAMILY INFECTIONS****.(Keville 1991:132)* (Probably this is owing to its content of rosmarinic acid, even as a another plant containing that in abundance, self-heal—*see under that entry*—also inhibits this virus.) The Blackfeet chewed bergamot roots when they had ***SWOLLEN LYMPH NODES****.(Hellson & Gadd 1974:67, 72, 77, 84)* Then, too, the chemical geraniol, found in goodly amounts in this plant, is a decay-preventive compound. Coupled with the hight content of the antibacterial thymol, this suggests— says former USDA scientist James Duke—that bergamot might serve as a good oral rinse.*(Duke 1997:431)*

Like cousin horsemint, wild bergamot is also a ***vermifuge****:* Thus, our area's Ojibwe (Chippewa) Indians utilized a decoction of the flowers and roots to expel ***WORMS*** from the intestinal tract.*(Densmore 1974:346)*

Cautions Because of the high *thymol* and *carvacrol* content, and the classic use as an *emmenagogue*, this herb should *be* ***EXCLUDED FROM USE DURING PREGNANCY.***

WILD GERANIUM

[AMERICAN CRANESBILL; CROWFOOT; SPOTTED GERANIUM; ALUM ROOT]

(Geranium maculatum)

Description

[See photo #157 in rear of book.]

A pretty native perennial, topping off at 1-2 feet, with a hairy and branched stem.

The plant contains both basal leaves and stem leaves. The latter are short-stalked and in pairs, while the former are long-stalked and palmately divided into 5-7, coarsely-toothed segments that are etched with deep veins—the length of each leaf being 4-5 inches. At first light-green with pale spots, the leaves eventually become, as Bradford Angier once so well phrased it, "an unforgettable red before dying."*(Angier 1978:285)*

Two to ten rose to violet-colored flowers appear in a loose corymb at the tip of the stems, just above the uppermost set of paired leaves. Like the leaves, these blossoms are also heavily veined. They are 1-2 inches across, with five petals, ten stamens, and a long pistil that develops into a slender, beaked, five-chambered seed-pod. As the pod dries, its outer surface contracts, gradually building up tension until its segments dramatically curl upwards, explosively releasing the precious seminal cargo.

Range & Habitat

Flourishes in dry, shaded areas, such as open woods and thickets. Throughout our area in such locales, except less common in the westernmost one-fourth and in the Great Lakes region.

Chief or Important Constituents

Tannins (up to 30%!), gallic acid (upon drying, via decomposition of the tannin), pectin, calcium oxalate, gum, resin.

Health/Medicine

This gorgeous woodland plant is possessed of *cooling* and *drying* energies*(Holmes 1989:1:406)* and its root has a cherished place in herbal medicine as an **astringent.** Here it is used by way of a strong decoction to the tune of 1-4 ounces or as a 1:5 tincture of the dried plant (50% alcohol/10% glycerin) at ½-1 teaspoon, either as needed.*(Moore1993:298)* Wild geranium was even an official astringent in American drug literature for *over a century*, having been listed in the *U.S. Pharmacopoeia* from 1820-1916 and subsequently in the *National Formulary* from 1916-1936. The *U.S. Dispensatory* for 1926 even praised our cranesbill heartily, noting: "Geranium is one of our best indigenous astringents, and may be employed for all the purposes to which these medicines are applicable."*(Wood et al 1926:519)*

This, however, was no news to native Americans, who had all along been utilizing the plant or its root for the variety of purposes for which an astringent is typically used. For example, various tribes utilized it for **DIARRHEA**.*(Gilmore 1933:134; Smith1928:222; Smith 1932:370-71)* The Mesquaki (Fox) Indians found that a poultice of the pounded root bound to a **HEMORRHOID** caused it to recede.*(Smith1928:222)* Various tribes also employed a douche of the decoction for **LEUKORRHEA**.*(Winder 1846:10-13;cf. Moore 1994:13.8)* The Cherokee Indians found a swish of cranesbill tea to be quite soothing to *CANKER SORES,(Hamel&Chiltoskey1975:35)* while Pillager Ojibwe Indians used it for **SORE MOUTH**.*(Smith 1932:370-71)* The Iroquois found it helpful for sore mouth as well—even **"TRENCH MOUTH,"** where they implemented two pieces of roots about three inches long in a pint of water and boiled it just a bit, or a small handful of roots in two quarts of water and boiled it for a length of time.*(Herrick 1995:190)* The Mesquaki (Fox) Indians found that a freshly-sliced root portion could be poulticed directly onto an **ACHING** or **INFECTED TOOTH** or **GUMS** with good results.*(Smith1928:222; cf. Moore 1979:69)* (I personally found occasion to use this last-listed remedy to great relief once in the 1980s when bothered by a sore tooth and unable to get to a dentist. So happy was I with the results that I vowed right then never to be without the roots in my cupboards!)

The Eclectic physicans, and later British and American herbalists, refined the astringent applications pioneered by the Indians. With reference to its use for diarrhea, for instance, the Eclectics came to understand that it worked best with *chronic* diarrhea, especially when there were semi-formed feces and frequent evacuations and where the inflammatory stage *had passed* and the mucous membranes *had relaxed*. Here the recommended dose was ten drops of the tincture every two hours.*(Felter & Lloyd; Ellingwood 1983:347)* Laurence Johnson, a physician and a botanist who wrote in the latter part of the nineteenth century, commented glowingly on wild geranium, pointing out regarding its application as an astringent: "Although active and efficient, it is still mild and unirritating and devoid of all unpleasant or offensive properties. It is therefore particularly suited to the later stages of diarrhea and dysentery, especially in children. In such cases a decoction in milk has been found very serviceable."*(Johnson 1884:113)* With reference to the common application of an injection or douche for leukorrhea and other genital discharges, he also made the pointed observation that"the decoction is much more serviceable than a simple injection of tannin, doubtless from the fact that there is present [in cranesbill] mucilaginous material which exerts a soothing influence."*(Johnson 1884:113)*

And as for cranesbill's use for ulcers, the Natura physician R. Swinburne Clymer found that it was especially helpful for the pain and distress characterizing gastric ulcers.*(Clymer 1973:78)* Southwest-American herbalist Michael Moore has likewise found wild-geranium root helpful for gastric ulcers, especially when accompanied by vomiting.*(Moore 1994a:10:24)* Herbalists Simon Mills and Kerry Bone, in their wonderful new textbook on the *Principles and Practice of Phytotherapy*, opine that cranesbill might be even more useful for duodenal ulcers than gastric ulcers.*(Mills & Bone 2000:176)* Indeed, for some time, wild geranium has been an important ingredient in a famed herbal formula developed by noted British herbalist Frank Roberts and which herbalists then and since have traditionally used for sores practically anywhere in the digestive tract, *but especially in the duodenum.* (This formula even successfully knocked out my own duodenal proto-ulcer back in the early 1980s, after lengthy trials of both physician-prescribed cimetidine and ranitidine had failed miserably to even touch it! It has since been adapted and slightly altered by several American supplement companies.)

In that wild geranium possesses **styptic** properties in view of its great astringency, many Indian tribes and herbalists of various cultures have found it helpful as well for **BLEEDING WOUNDS**. In this regard, Dr. William Winder of Montreal said of the usage by Indians of Great Manitoulin Island in Lake Huron that it was "a favorite external styptic, the dried root being powdered and placed on the mouth of the bleeding vessel."*(Winder 1846:11)* In similar accord, the Iroquois Indians used a chew of the root as a poultice for a **SEVERED UMBILICAL CORD**.*(Herrick 1995:190)* Treatment of **MENORRHAGIA** with wild geranium has also been traditional among herbalists of numerous cultures.*(Wren 1988:95; Chevallier 1996:214)* Clymer, referred to above, found that it was "the remedy" for **LUNG HEMORRHAGES**, dosed at 40 to 50 drops of the tincture, and after success obtained repeated in doses of 10-15 drops, up to three times a day.*(Clymer 1973:78)* Its styptic abilities have even been scientifically demonstrated, for in 1938 a lab experiment found that it markedly increased clotting time in the blood of rabbits.*(Spoerke 1990:65)*

Wild geranium's extraordinarily high amount of tannins allows it to serve as a useful **anti-microbial** agent, too. Here it is effective against *Candida* organisms, and thus has been widely used for **ORAL THRUSH**, such as by the Cherokee Indians who used it in combination with chicken grape *(Vitis cordifolia)* for washing out the mouths of infants having this condition.*(Hamel & Chiltoskey 1975:35; cf. Moore 1979:69)* A 1954 study even revealed that it somewhat inhibited a virulent human strain of tuberculosis.*(Fitzpatrick 1954:530)* An even earlier antibiotic test revealed that it inhibited gram+ bacteria.*(Suter 1951)* Canadian research in the 1970s, followed by Bulgarian research since then with the European species *G. sanguineum*, has revealed that wild geranium also possesses powerful **antiviral** activity, especially against **HERPES SIMPLEX INFECTIONS** and **INFLUENZA**.*(Konowalchuk & Speirs 1976:1013-17; Serkedzhieva lu Manolova et al 1986:73-82; idem 1987:66-71 and 1988:16-21)* (The latter is interesting in view of early-American use for **AGUE**-*Vogel 1970:390)* Choctaw Indians considered it to be the most powerful warrior against **venereal diseases***(Campbell 1951:287)* and early-American physician Benjamin Barton specifically extolled it for **gonorrhea** and **gleet**.*(Barton 1900)*

Uses for our gorgeous cranesbill seem almost limitless: Southwest-American herbalist Michael Moore recommends it for **TONSILITIS** and **DIVERTICULITIS**.*(Moore1979:69)* Some have found it helpful in **IRRITABLE BOWEL SYNDROME**.*(Chevallier 1996:21)* Certain Amerindian tribes applied a poultice of the mashed roots over breasts afflicted with **MASTITIS,** greatly relieving the pain. Wild geranium has further shown experimental **hypoglycemic** activity.*(Lewis&Memory-Lewis1977:219)* Canadian herbalist Terry Willard has found that cranesbill can even leech mercury from the body, should poisoning from that toxic element be suspected.*(Willard 1993:266)*

I suspect that herbal birth-control will be the wave of the 21st century, and wild geranium is certainly a plant to keep an eye on in this regard: Nevada Indians used to give the tea to a woman one month after childbirth to keep her from conceiving until the child was at least a year old.*(Scully 1970:121)* Virginia Scully, author of a popular book on Indian ethnobotany, opined that the use of wild-geranium tea was probably the most widely used method of birth control among our continent's Indian tribes.*(Scully 1970:178)*

Finally, in view of cranesbill's cooling and drying properties, herbalist Peter Holmes recommends it for conditions of the urinary system viewed by traditional Chinese medicine as involving "kidney fire"—problems like **ANURIA** and **ACUTE NEPHRITIS**.*(Holmes 1990:2:821)* Interestingly, these applications are in accord with speculation among early-American physicians that cransebill would be helpful "perhaps in nephritis."*(Barton 1900)* A bit later down the stream of time, the Eclectics put cranesbill to the test for "hot" urinary-tract afflictions and were most satisfied.*(Felter & Lloyd)* One of their number, Finley Ellingwood, once remarked in this regard: "[I] treated a case of hematuria for nearly two years with absolutely no permanent impression upon this condition....All of the usual remedies were used. Finally fifteen drops of geranium were given every two hours, and in two weeks the blood was absent and had not returned at the end of three years, except mildly when the patient persistently overworked."*(Ellingwood 1983:347)*

Caution

Wild geranium's extremely high content of tannin contraindicates its use for any length of time, as these substances can be harmful to the liver—and in other ways—if used unjudiciously.

WILD MINT
[CORN MINT; FIELD MINT; CANADA MINT; BROOK MINT]
(Mentha arvensis)

Description

[See photo #158 in rear of book]

A perennial herb, growing 6-24 inches tall, with a square stem that is usually unbranched. The stem may be either hairy or smooth and it often reclines at the base, in that it spreads by runners.

Wild mint has opposite, stalked, freely-toothed leaves possessed of a strong, characteristically "minty" odor, especially when disturbed (such as being brushed by one's leg!). They are lanceolate to ovate, about two inches long, and tapering at their tips. Peculiar, dotted glands adorn their surface.

This herb possesses tiny, bell-shaped flowers that are usually blue, but may have a pinkish or whitish hue. These blossoms are in balled clusters at the nodes of the stem.

This is our area's only native *Mentha* species. The others (including peppermint) are all introduced/garden escapes.

Range and Habitat

Throughout North America, in damp soil such as occurs in ditches, wet fields, moist woods, and on streambanks, lakeshores, and the perimeters of marshes and swamps. In our area, the plant is ubiquitous in such terrain.

Chief or Important Constituents

Essential oil(consisting largely[65-80%] of menthol[strongest source known—up to 26,000ppm], menthone[5-30%], piperitone, cineole, limonene, camphene, pulegone, menthol acetate, carvone, eugenol), tannins, rosmarinic acid, flavonoids(luteolin, hesperidin), resin, and oligosaccharides. Nutrients include good amounts of vitamins(A, C, and K) and the minerals iron, calcium, potassium, and magnesium.

Food

One can use this mint in all recipes in which the domesticated mints are used. I particularly enjoy wild-mint julep (*see* recipes under "Mountain Mint") and tea.

Health/Medicine

Amerindian tribes widely used and esteemed this strongly-scented herb, which powerfully proved its merit relative to infectious illnesses. Here the Mohawk, Tete-de-Boule, Potawatomi, Cherokee, and Pillager Ojibwe Indians implemented wild mint as an

antipyretic. *(Rousseau1945:58;Raymond1945:129;Smith1933:61;Smith1932:371-72; Hamel & Chiltoskey 1975:45; White 1945:562)* Our area's Menominee Indians combined it with catnip *(Nepeta cataria)* *(see under that entry)* and peppermint to treat **PNEUMONIA,** drinking the mixture as a tea and poulticing it on the chest. *(Smith 1923:39)* The Potawatomis used it for **PLEURISY.** *(Smith 1933:61;Vogel 1970:340)* The Cheyenne appreciated its remarkable **anti-emetic** properties. *(Grinnell 1905:39; Hart 1981:27)*

The Flambeau Ojibwe, Cheyenne, and Blackfoot Indians used the infusion as a remedy for **HEART PROBLEMS.** *(Smith1932:371-72;Grinnell1972:186;Hart1981:27)* The cardiac assist is possibly due to the plant's **stimulant** properties. Such activity may also explain wild mint's good reputation as a **libido restorer**, treasured by the Cheyenne. *(Hart 1981:27)*

Like most other mints, *Mentha arvensis* is a powerful **stomachic** and **carminative**. Indian tribes of the Missouri region thus appreciated it for **FLATULENCE,** *(Gilmore 1991:60)* as did our own area's Ojibwe. *(Gilmore 1933:140)* Indians of the Nevada region found it most helpful for **DIARRHEA**. *(Train et al 1957:69)*

Studies published over the last few decades have revealed wild mint's essential oil to be a powerful **bactericide** *(Sanyal&Varma1969:23-24)* and **fungicide.** *(Sarbhoy et al 1978:7-8, 723-25; Kishore et al 1993:211-15)* No wonder the Cree Indians found it to be highly effective for **INFECTED GUMS**—they using the ground flowers of this plant and of yarrow *(Achillea millefolium)* *(see under that heading)* in a moistened cloth as a topical rub. *(Leighton 1985:45)*

This is a **PAIN** herb *par excellence*! Its **anti-inflammatory/analgesic** properties were thus gratefully utilized by numerous Indian tribes: The Lakotas and Goisutes relieved **HEADACHE** with a strong tea made from the roots of this plant. *(Buechel 1983:461,799;Chamberlin1911:351)* The Navajo-Kayenta poulticed the leaves on **SWOLLEN AREAS** of the body, *(Wyman&Harris 1951:40)* as did the Kawaiisu. *(Zigmond 1981:40)* In modern times, too, a sports-medicine study reported at the 24th International Symposium on Essential Oils in Berlin in 1993 found that a gel made from the oil of this herb surpassed the usual salicylate treatment for the care of acute and subacute **SPORTS INJURIES**, based upon the level of **analgesia** experienced. (Altogether, 78% receiving the wild-mint therapy reported good results as contrasted with only 34% in the salicylate group- *MH6:4:12*) Finally, the Woods Cree poulticed ground wild mint on **TOOTHACHES**, to their great relief. *(Leighton 1985:45)* Such a level of analgesia is not surprising in that wild mint contains—as we have seen in the "Constituents" section above—eugenol, the familiar drugstore remedy (formerly marketed as "oil of cloves") long used for tooth pain, as well as the highest concentration of menthol known in the plant kingdom!.

Cautions: *Abortive(Kanjanapothi et al 1981:559-67)* and thus ***should not be used during PREGNANCY. Avoid also in LACTATION,*** due to high menthol and carvone content. *Never apply topically to the head or chest region of infants,* as a member of the mint family containing significant levels of menthol such as this plant can cause *instant respiratory collapse*. *(Leung 1980:232)*

WILD MUSTARD
(Brassica [Sinapsis] spp.)

[See photo #159 in rear of book.]

An anual weed with slender, erect, branching stems covered with stiff hairs

The leaves are alternate and hairy. The lower ones are stalked and deeply divided, while the uppermost blades are stalkless, coarsely toothed, and bristly.

The characteristic, four-petaled, yellow flowers, growing about half an inch wide, are clustered atop the branched stems. As the petals fall, slender seedpods develop with minute, dark-colored, pungent seeds. Our major species include the following:

B. nigra (black mustard) is an annual that grows 2-6 feet and has a round stem. The lower leaves are lobed, with four lateral lobes and a large terminal one. The upper leaves are lanceolate and toothed, but not lobed. The four-sided seedpods are about 3/4-inch long and point upward (i.e., are practically appressed against the stem). Most conspicuously, they are *beaked* at their tip. They are filled with brownish seeds.

B. alba (white mustard) is an annual reaching a height of 2-5 feet. Its leaves are smoother and its flowers larger than in the other species. Unlike *B. nigra*, above, its seed pods are not beaked.

B. kaber (crunchweed, charlock) tops off at 8-30 inches and has green stems that are hairy near the base. The middle and upper leaves are toothed, while the lower leaves are deeply lobed. The fruits lack visible hairs or are very lightly pubescent.

Brassica species should be distingushed from the similar looking wintercress (*Barbarea vulgaris*), which has *hairless* stems and *rounded, glossy* leaves—the lower ones of which are deeply lobed (the terminal lobe being especially prominent).

Range & Habitat

Thrives in fields, meadows, wasteland, and clearings throughout our area.

Chief or Important Constituents

Black mustard contains the alkaloid sinapine, glycosides(sinigrin), an enzyme (myosin), an irritant oil(allyl isothiocyanate, as a result of hydrolysis, which causes myosin to act on sinigrin, thereby producing the oil), fixed oils(up to 37%, including palmitic, oleic, erucic, arachic, and eicosenoic), alkanes, urease, argonase, lipase, and mucilage

White mustard contains a different glycoside, sinalbin, which winds up being enzymatically converted into the irritant oil isothiocyanate, which exudes less of a pungent odor than black mustard's oil.

Mustard greens are rich in vitamins and minerals. A whopping 7,000 IUs of vitamin A occur in a 100-gram portion, as well as 100mg of vitamin C, and goodly amounts of several B vitamins (B1, B2). Minerals occurring in appreciable amounts include: calcium(183mg per 100g), potassium(32mg per 100g), phosphorous(30mg per 100g), and iron(3mg per 100g).

Food

Very young leaves can be added to salads or nibbled on sparingly in the field. Older leaves must be boiled in two waters for about 20-30 minutes. Properly prepared, however, they are quite tasty. (However, if the timing is a bit off, crumpled bacon can be added to tame any overly strong flavor.)

Naturalist and herbalist Tom Brown, Jr. informs us that he relishes the unopened flowerheads, which he cooks like broccoli. He says that they are best when steamed for five minutes, but if accidentally cooked longer they will loose their fine taste and become mushy. Later in the growing season, Brown enjoys the tender young seed pods, added to salads, or as a trail nibble.*(Brown 1985:151, 153)*

Mustard seeds can be collected in late summer and used sparingly on salads or ground in a food mill and then dried and stored for future use.

Health/Medicine

Undoubtedly, the most famous application of mustard for medicinal use is the ages-old mustard plaster. Traditionally, 120g(4oz) of ground, mustard seeds have been mixed with an equal to thrice amount of wheat flour along with some water to make a paste that is sandwiched between two layers of muslin or flannel and then gently placed on the affected area, sometimes being covered with a towel to keep the heat in. This is never applied to children under six years old nor does the application for adults or older children ever exceed *15 minutes*—with 5 or 10 usually being sufficient.*(Jones1991:43; Willard1992b:131;Hoffmann1986:211;Fetrow& Avila 1999:447)* (See under "Cautions," below. Also, irritation to the eyes or nasal passages is possible from the potent fumes that are released, so this should be borne in mind.) After the plaster is removed, and any mustard residue carefully removed, olive oil or egg white is applied to the area or it is sponged with cold water in order to minimize any smarting of the skin. This treatment has been widely used for ***PNEUMONIA, BRONCHITIS, ARTHRITIC PAINS, RHEUMATIC PAINS***, general ***JOINT STIFFNESS***, and many other afflictions.*(de Bairacli Levy 1974:102; Duke 1997:386)* Mesquaki Indians even bound mustard leaves on an***ACHING TOOTH*** or ***HEAD***, being ever mindful of its potential for burning the skin.*(Tantaquidgeon 1928:265)*

But, how does a mustard plaster work? As a***rubefacient*** (counterirritant), the mustard oil induces the peripheral blood vessels to dilate, with the resultant rush of blood drawing the inflammatory chemicals away from the original, deeper sight.*(Dobelis et al 1986:108)* Laurence Johnson, a nineteenth-century physician and botanist often critical of herbal remedies of the day, nevertheless acknowledged of mustard that"as a rubefacient its sphere of usefulness is practically unlimited."*(Johnson 1884:96)*

The ancient Roman medical scholar Celsus even urged that a mustard plaster be ap-plied to *PARALYZED LIMBS* in order to "stimulate the skin of the torpid limb," but he added the customary caution that it be removed "as soon as the skin becomes red." (*De Medicina, 3.27*)   Speaking of the limbs, probably one of mustard's most universally appreciated applications has been the addition of some mustard powder into a tub of water as a footbath for *ACHING FEET, CHILLED FEET*, or even a *CHILLED BODY* in general. (*Carse 1989:136; Jones 1991:43*) Naturalist and herbalist Tom Brown Jr. experienced such a footbath as a lad and later enthused about it in his memoirs, although he was careful to note that he had washed off all of the mustard water afterwards to prevent the skin from being burned. (*Brown 1985:151*) The usual dose for a mustard bath is one tablespoon of bruised seeds per one liter (two pints) of hot water. (*Hoffmann 1986:211*)

Mustard flour can be of great assistance when there are GI problems. It serves ably as a digestive *tonic* for *POOR APPETITE, HALITOSIS,* and *FLATULENCE.* (*de Bairacli Levy 1974:102; Brown 1985:151*) Such uses have a treasured pedigree in American domestic practice: Writing in 1884, Laurence Johnson remarked that "as an aid to digestion it [mustard] is used in every household." (*Johnson 1884:96*) It is of interest in this regard that the seeds serve to improve digestion not only by marshalling the gastric juices but also by stimulating pancreating secretions. (*Lewis & Elvin-Lewis 1977:273, 278*). Then, too, when there are infestations of microorganisms in the GI tract, the use of mustard in foods can lend a helping hand, since its irritant oil is powerfully *antimicrobial*, and that against many kinds of bacteria and fungi. (*Fitzpatrick 1954:531; Abdullin 1962:75*)

Going beyond the traditional culinary use, mustard can serve as a mild *laxative*— the usual dose here being 1/2 tsp. of crushed seeds in a glass of warm water. (*Jones 1991:43*) A bit stronger dose (1-3 drachms in 6-8 ounces of warm water, according to the Eclectics; or 1 tablespoon of mustard flour in a glass of tepid water, according to the herbal of Englishwoman Maud Grieve-*Grieve1971:2:569;cf.Felter & Lloyd*) has long served as a "prompt and efficient emetic" for household poisoning emergencies, (*Johnson 1884:96*) although syrup of ipecac has replaced mustard as an emergency emetic in most American homes.

Cautions

* Those having *VENOUS DISORDERS* should not use mustard medicinally.

* Owing to its irritant oil, mustard is very caustic and should never be ingested or used topically by itself. Nor should it ever be applied to mucous membranes. (Hands should be washed after contact, as thoughtlessly touching the eyes with mustard residue on the fingers could produce severe pain and damage there.)

* *Do NOT use a mustard plaster on a child **under six years old**. Also, do not leave a mustard plaster on an adult longer than 15 minutes, or on a 6+-year-old child longer than 5 minutes!* In fact, in all cases, it should be removed as soon as a burning sensation is felt or the skin is observed to redden. Be sure to wash off any remaining mustard paste from the skin, even if not visible. The subsequent application of olive oil or egg whites may dull any residual smarting, as may dabbing with cold water.

WILD ONION
(Allium stellatum)

Description

[See photo #160 in rear of book.]

As might be expected, this is a *strong-smelling* herb, growing 1-2½ feet tall. It has a long, straight, tubular stem and usually two narrow, flat, grasslike, basal leaves. There are no stem leaves.

Individual flowers are starlike in shape—having three petals, three sepals, and six stamens. They are pinkish to purplish in color.

The faded autumn inflorescence is dotted with small—but showy—black seeds.

A. cernuum (nodding wild onion), thriving in a single county in Minnesota, has a drooping flower cluster because of a bend at the tip of the stem.

Range & Habitat

One species or another of wild onion occurs in every American state except Hawaii and in all of Canada except for the far north. *A. stellatum* thrives pretty much throughout our area. The preferred habitat consists of meadows, prairies, fields, plains, and open and rocky places.

Seasonal Availability

I have dug wild onions as late as the first week of December and even then they were crisp and delicious! Mostly, however, the bulbs start to rot in November

Chief or Important Constituents

Flavonoids(esp. quercetin), sterols, saponins, inulin, and sulfur compounds.

Food

Raw bulbs added to salads make these exceptionally tasty. But the bulbs lend themselves to a variety of recipes for cooking, too. Of course, they can be boiled, but here they need at least one water change in order to reduce their otherwise overly strong flavor. A simpler—and often more appreciated—method of preparing them is to *roast* them, something that is especially apropos in an outdoor setting. One can accomplish this by placing the *unpeeled* bulbs on a bed of medium-hot coals. When tender, poke the skin to allow the steam to be released. The interior can then be spooned out, flavored as

298

desired, and added to hamburgers, sandwiches, or wild dishes. Then, too, the Cheyenne, Blackfeet, and other Amerindian tribes often cooked wild onions with meat in order to flavor it more strongly, even as we do nowadays with domesticated onions.

Wild onions, like their domesticated counterparts, make an excellent soup! They can also be added to stews during approximately the last ten minutes of cooking.

Health/Medicine

A. stellatum was appreciated by our area's Ojibwe (Chippewa) Indians for **RESPIRATORY COMPLAINTS**, especially in children to whom it was given by way of a sweetened decoction.*(Densmore 1974:340-41)*

The chewed, aerial portions of the related *A. cernuum*—found in but one southern Minnesota county—were placed on the chest of those suffering from **PLEURISY** by the Makah and Quinault Indians of Washington state.*(Gunther 1973:24)* The Cherokees fried this species and likewise placed it on the chest, but primarily for **CROUP** (for which they also used the juice internally). They also found nodding wild onion helpful for **COLDS, LIVER COMPLAINTS, HIVES** (for which the juice was particularly used), **URINARY GRAVEL,** and **DROPSY** (after first imbibing a cup of horsemint—*see under that entry*—tea).*(Hamel & Chiltoskey 1975:47)*

The bruised bulbs of various unspecified species of wild onion were used by native Americans (including our own area's Dakotas and Winnebagos) as a poultice for the **STINGS** of **BEES** and **WASPS.** F. Andros, a medical doctor who observed this Indian treatment for years, wrote that it "almost instantly relieves the pain."*(Andros 1883:117)* Eclectic physician Finley Ellingwood offered these similar observations: "The onion poultice stands in high favor with me for all swellings.... 'Roasted[,] the cut surface [can be] applied hot to glandular inflammations and suppurating tumors.'"*(Ellingwood 1983:261)*

Cautions

* Green portions of onions should probably not be eaten, as they readily absorb selenium from the soil and pollutants from the air.

* Persons sensitive to inulin may get indigestion and/or flatulence from eating raw onions.

* Don't collect any plants that may look like wild onions but yet lack the characteristic odor, as several inedible plants—and one very toxic one!—resemble this herb, especially in the feature of its grasslike leaves.

WILD SARSAPARILLA
[FALSE SARSAPARILLIA; SMALL SPIKENARD]
(Aralia nudicaulis)

Description

[See photo #161 in rear of book]

Wild sarsaparilla is a pretty, leafy, woodland plant usually growing about a foot high, but on occasion topping off at around sixteen inches. The stems are thick, square, and lightly prickled.

The three branches of "leaves" are actually but a solitary leaf divided pinnately into three groups of five, toothed leaflets. Each leaflet is 2-4 inches long.

The leaf grows from a strongly-aromatic, underground rhizome, from which also springs the leafless stem upon which this plant's globular flowers grow. As with the leaflets, the greenish-white flower umbels are in threes—each cluster growing on a separate branch of the flower stem.

Round berries, colored bluish-purple (or purplish-black), replace the flowers in July.

Range and Habitat

This species can be found in moist to mesic woods throughout our area, but it definitely predominates in the northern half.

Chief or Important Constituents

Little chemical work has been done of late, but earlier research revealed that the rhizome contains a volatile oil, resin, pectin, and sugars. The berries contain a glycoside.

Food

The dried rhizomes make a very pleasant drink or tea. You need simply to powder three or four of them and stir the results into a cup of boiling water. Then cool, strain, and sweeten. Or decoct the chopped roots until the water becomes a reddish color and then strain, sweeten, and serve as a hot drink. The rhizomes are most famous, however, for the root beer that was made from them in colonial times.

To humans, the plant's berries are toxic when raw as they contain a glycoside that leaves a sting in the throat—as I can well testify! (This principle does not seem to bother animals such as foxes and bears, however, as these critters readily feast upon them!) Some say that the berries are edible when cooked because the heat breaks down the glycoside.

Health/Medicine

This plant has a long history of use by both whites and Indians. Of the former, the renowned botanist Constantine Rafinesque, writing in the early 1800s, noted the plant's then-current physiological applications as follows: *diaphoretic, depurative, stimulant, sudorific, pectoral, cordial,* and *vulnerary*. Regarding the latter, he wrote: "The fresh roots and leaves chewed and applied to wounds, heal them speedily; Dr. Sp. informed me that he was once cured by them alone of a desperate accidental wound by a broad axe." *(Rafinesque 1828-30:1:53-55)* And as to its depurative abilities, Dr. Gunn's famous *Household Guide* of the mid-to-late 1800s explained that wild sarsaparilla was "useful in constitutional disease, such as scrofula, syphilis, skin diseases, and wherever an alterative and purifying medicine is needed." *(Gunn 1859-61:886)* The rhizome of wild sarsaparilla wound up as official medicine in the *U.S. Pharmacopoeia* from 1820 until 1882; it was held to be *diaphoretic, alterative,* and *stimulant.*

As for Indian uses, they were many: Our area's Ojibwe (Chippewa) Indians employed *Aralia nudicaulis* to treat **NOSEBLEEDS** (here the rhizome was dried, powdered and used as a snuff; on other occasions, the fresh root was chewed and inserted into the bleeding nostril). They also found it helpful for **AMENORRHEA** and for **BLOOD DISORDERS.** *(Densmore1974:340-41,350-51,356-57,366-67;Reagan1928:231)*

Vulnerary applications were most frequently utilized by the American Indians, however. Our Ojibwe pounded the rhizome and applied it to **SORES.** *(Densmore 1974:350-511)* Our Mesquaki (Fox) did the same, *(Smith 1928:203)*, while the Potawatomis poulticed it on **SWELLINGS** and **INFECTIONS.** *(Smith 1933:40-41)* The Micmacs boiled the rhizome until soft and then applied it to **WOUNDS,** *(Lacey 1993:64)* while the Woods Cree chewed the rhizome to a pulp and then applied it to wounds to draw out the infection therefrom. *(Leighton 1985:29)* The Flambeau Ojibwe poulticed the pounded, fresh root onto **CARBUNCLES** to heal them; also onto **BOILS** to bring them to a head. *(Smith1932:356)*

Pectoral properties were also utilized by the Indians. The Creeks found wild sarsaparilla helpful for **PLEURISY.** *(Swanton 1924-25:658)* The Penobscots and Kwakiutls made the dried, powdered rhizome into a **COUGH** syrup. *(Vogel 1970:361)* The Iroquois added a decoction of the rhizome to other plant parts to make a medicine for the cough of **TUBERCULOSIS.** *(Herrick 1977:393)* The Woodland Cree used the whole plant in decoction for **CHILDHOOD PNEUMONIA** *(Leighton 1985:29)*

Nephritic/Urinary functions were gratefully implemented by the Creeks, who employed wild sarsaparilla specifically for **DIFFICULT URINATION, BLOODY URINATION,** and **PAIN** in the **BACK/LOWER ABDOMEN.** *(Swanton 1924-25:658)* Numerous tribes, too, valued wild sarsaparilla as a *tonic.* These included the Thompson, *(Steedman1928:471)* Okanagan, *(Perry1952:42)* Mohegan, *(Tantaquidgeon1972:70,12)* Blackfeet, *(EthnobDB)* Montagnais, *(Speck1917:315)* Mesquaki, *(Smith1928:203)* and Delaware. *(Tantaquidgeon1942:74)* It was viewed as a veritable Godsend for **LASSITUDE** and **DEBILITY.** *(Teit 1928:471; Speck 1917:315)*

Caution: *Raw berries* should *not* be consumed. Properly cooked and prepared, they may become edible.

WILD STRAWBERRY
[SCARLET STRAWBERRY; VIRGINIA STRAWBERRY]
(Fragaria virginiana)

Description

[See photos #162 & 163 in rear of book.]

Low (3-6"), perennial herb possessed of roots that send out runners, so that it is often found in colonies.

The dark-green leaves are wholly basal and divided into three, coarsely-toothed, broadly-elliptical leaflets, each growing 1-1½ inches long. The leafstalks are quite hairy.

The pleasantly-scented flowers—each possessing five rounded petals, five sepals, and five bracts—are in stalked clusters. Many yellow stamens create the illusion of a yellow disk center for the blossom.

Fruits are small, dark, hard achenes embedded in a red, fleshy, top-shaped receptacle—a berry. Each berry is cupped by a hull of ten, small, calyx lobes. The thin, filament-like stems on which the berries form can't help but droop under their weight, so that the ruby-red morsels often wind up being somewhat obscured by the leaves.

A related species, wood strawberry (*F. vesca*), has more elongate berries, with achenes standing out a bit from the fleshy receptacle. Its sepals point backward.

Range & Habitat

This plant is found throughout the USA and southern Canada except in arid regions. The preferred habitat is roadsides, open woods, and old, grassy fields

Seasonal Availability

Berries are available in our area for only about three weeks (mid-June to early July).

Chief or Important Constituents

The berries contain glycosides(arbutin), flavonoids(incl. anthocyanins), lutein, pectin, organic acids(citric, malic), salicylates, mucilage, and the amino acid arginine. Nutrients include vitamins (A, B-complex, C, and E) and minerals such as calcium, potassium, magnesium, sulfur, and iron. Leaves contain ellagic acid and are extremely high in vitamin C. Roots contain ellagitannins(including fragarianine).

Food

The berry of this lovely, little, wild plant is universally acknowledged to be tastier than its domesticated counterpart. The famed English writer Izaac Walton (1593-1683) had put it well: "Doubtless God could have made a better berry, but doubtless He never did."

Excellent pies and jam can be made from this fruit, but it is difficult to save enough of the gathered berries to do the job, since one is tempted beyond what one can bear to eat five berries on the spot for every one saved! What berries one manages to save should have their hulls pinched off immediately upon picking, since this is difficult to accomplish later when the weight of many berries in the collecting vessel has somewhat smashed them all.

Steeped leaves—fresh or fully-dried only (see "Cautions," below, on wilted leaves) —make a very mild tea.

Health/Medicine

The berry juice is a powerful *refrigerant*, useful for *FEVERS* and topically for *SUNBURN*. (For the latter, Maud Grieve's herbal recommends rubbing the juice well into the sunburned area and leaving it there for half an hour, afterwards washing it off with warm water into which a few drops of a tincture of benzoin have been added-*Grieve 1971:2:777*) The large amount of iron in the berries also lends their application to iron-deficiency *ANEMIA*, especially in that the high vitamin-C content helps assure the iron's absorption. The high vitamin and mineral content may also at least partly explain the historic use of the fruit and leaves for *NERVOUS DEBILITY*.*(de Bairacli Levy 1974:140)*

The Cherokee Indians also held the fruit in the mouth to remove *TARTAR* from the teeth.*(Hamel & Chiltoskey 1975:57)* This seems to work quite well, especially if the fruits are rubbed on the teeth and the juice allowed to remain there for about five minutes, followed by a tooth brushing of baking soda and water.*(Schofield 1989:183)* Because of their high sulfur content and because they possess the same pH as the skin, the mashed fruits are also often used topically on the face to enhance beauty; most famously, a paste made from mashed strawberries has been used to diminish *FRECKLES*

The leaves are so enormously high in vitamin C that when Euell Gibbons had a sample analyzed for the content of this vitamin at Penn State, the lab tech thought that he had doctored the sample because the amount turned out to be incredulous to her! When she proceeded to re-test with her own sample, she was shocked to find an even greater amount of vitamin C! Not surprisingly, strawberry-leaf tea has been used in folk medicine to prevent *COLDS*. Fortunately, as Gibbons reminds us, strawberry leaves are evergreen and thus continue to exist into winter under the snow where they can be harvested and dried to provide a vitamin-C-rich source during these cold months when other avenues have largely disappeared.

The leaf tea seems to have an affinity for the genito-urinary system, proving helpful for **BLADDER, KIDNEY**, and **UTI PROBLEMS** (esp. **DYSURIA)**, even as the Cherokee and other Indians discovered.*(Hamel & Chiltoskey 1975:57; Scully 1970:267)* It is also somewhat *diuretic.(Wren 1989:260)* **GONORRHEA** and **IRREGULAR MENSTRUATION** were also treated with this herb by the Malecite Indians.*(Mechling 1959:258)* On the other hand, the berry contains arbutin, the same chemical in the renowned urinary herb, uva ursi (bearberry) (*Arctostaphylos uva-ursi*), that hydrolyzes to hydroquinone and then disinfects the urinary tract.

Wild strawberry has a cherished reputation for helping to alleviate **RHEUMA-TISM** and **GOUT**. It is possible that the salicylate content may be mildly helpful here, even as Roger Phillips suggests.*(Phillips1990:65)* But surely of greater importance is the herb's diuretic property, serving to leech out the excess uric acid characterizing these most painful conditions.*(Ody 1993:60)* Modern research would also highlight the importance of the anthocyanin content. But, whatever the exact combination of mechanisms involved, we know that it works. In fact, the famed Swedish botanist and physician Carolus Linnaeus assured us that he had cured himself of gout simply by subsisting for a time on a diet of wild strawberries!*(Scully 1970:267)*

This herb has also traditionally been used for **JAUNDICE.** *(Hamel & Chiltoskey 1975:57; Ody 1993:60)* Probably this is due to the digestive stimulation that the plant supplies, which also explains why it has traditionally been used for **POOR APPETITE** and **DYSPEP-SIA.** *(Dobelis et al 1986:339)* The digestive-stimulating effect is probably also at least partly responsible for wild strawberry's long-appreciated application for **ECZEMA**, which seems to have a connection with sluggishishness of stomach acid, liver function, and pancreatic-enzyme production. The Natura physician R. Swinburne Clymer, in fact, considered it to be the remedy *par excellence* for this irksome condition. Interestingly, he found it to work best in combination with other digestive-stimulating plants such as the roots of dandelion, burdock, and rhubarb.*(Clymer 1973:137)*

The kind and amount of tannins in the leaves and especially the root allows an infusion of either to be implemented for a variety of purposes owing to their powerful *astringency*. Thus, **SORE GUMS** can often be alleviated by swishing a tea of wild strawberry leaves or root in the mouth or even simply by pressing a raw root against the gums. The tea also makes an excellent gargle for **SORE THROAT. DIARRHEA** often yields to its power as well, which application was a favorite one of the Blackfoot, Cherokee, Thompson, and Micmac Indians.*(Hellson 1974:66;Hamel & Chiltoskey 1975:57;Turner et al 1990:259; Lacey 1993:116)* British herbalist Mary Carse discusses an application of the leaf tea to infant diarrhea, noting that 1/2-cup of the carefully strained and unsweetened tea can be put into a baby bottle 3-4 times a day, but that no food should be given for twelve hours.*(Carse 1989:161)* (That last part is easier said than done!)

The Micmacs also appreciated the tea for **STOMACH CRAMPS**, using six teaspoons of fresh or dried leaves to a pint of boiled water and steeping them for fifteen minutes, drinking up to 2-3 cups a day.*(Lacey 1993:116)* Several different Indian tribes, in fact, used the infusion for **STOMACHEACHE** and even for **INFANT COLIC**.*(Smith 1932:384; Turner et al 1990:259)*

Externally, a powder made from the leaf was dusted onto **SORES** by the Okanagan-Colville Indians.*(Turner et al 1980:125).* The Quileute Indians chewed wild strawberry leaves and spit them onto **BURNS**.*(Gunther 1973:36)*

Cautions

* Some people are allergic to domestic strawberries. Bear in mind that these same people *will be allergic to the wild ones as well*!

* When drying the leaves for tea, be sure to dry them fully, as mere *wilted leaves can produce a dangerous mold.*

WOOD LILY

[PRAIRIE LILY; ORANGE-RED LILY; WILD LILY]
(Lilium philadelphicum)

[See photo #164 in rear of book.]

Description

A breathtakingly gorgeous, native perennial, growing 1½-2½ feet tall. The people of Saskatchewan, Canada have so appreciated its beauty that they have made it their floral emblem!

The lanceolate leaves grow 1-4 inches long and are whorled in groups of 3-8 (sometimes the lower leaves are not whorled, but merely scattered). The orange-red, bell-shaped corolla is erect, 2½ inches long and 2-3 inches across. There are three petals and three sepals tapering to a stalked base, which is dotted inside with black or purplish-brown. Six long stamens can be observed. The blossoms are single or in umbels of 2-5.

Range & Habitat

Open and brushy places: meadows, prairies, thickets, clearings, and dry and open woods; most common in pine-forest country. Throughout our area in such environs, but getting scarcer.

Food

The straight, bulbous roots of wood lily were gathered and eaten like potatoes by the Mesquaki Indians. *(Smith 1928:262)* The Woods Cree consumed segments of the fresh bulbs or sometimes dried them for a nibble when traveling; these Indians also relished the plant's seeds. *(Leighton 1985:43)* Blackfoot Indians likewise feasted on the fresh bulbs, although sometimes they saved them for stews. *(Hellson & Gadd 1974:103)*

Health/Medicine

Quebec's Algonquin Indians found that wood-lily roots were good for helping to alleviate **STOMACH DISORDERS.** *(Black 1980:138)* Various other tribes implemented a tea of the roots to treat **COUGH** and **FEVER,** *(Chandler et al 1979:58; Mechling 1959:251)* to **EXPEL AFTERBIRTH,** *(Herrick 1977:282)* and to apply topically to **WOUNDS, SORES, BRUISES**, and **SWELLINGS.** *(Gilmore1933:125; Mechling1959:245; Densmore1932:132;Chandler et al 1979:58)*

Ojibwe Indians of Michigan used this plant topically, too—chiefly for **DOG BITES,** plastering it thereon along with another (unidentified) species. *(Gilmore1933:125)* The chewed or pulverized flowers of *L. umbellatum*, a thin-leaved subspecies of *L. philadelphicum* found in the southern tip of Minnesota and southward along the Missouri-River region, were poulticed onto the **POISONOUS BITES** of a small, brown spider by Dakota Indians, which quickly relieved the painful inflammation. *(Gilmore 1977:19)*

WOOD NETTLE
[CANADA WOOD NETTLE]
(Laportea canadensis)

Description

[See photo #165 in rear of book.]

A large, conspicuous herb growing to four feet and possessing a stem that is covered with stinging hairs.

The alternate, ovate leaves are noticeably long-stalked and coarsely toothed. They grow 3-8 inches long.

The small, greenish flowers occur in loose, branching clusters in the plant's axils: male blossoms (having five sepals and five stamens) in the lower axils and female ones (with four sepals and one pistil) in the upper axils. The flowers of both genders lack petals.

Range & Habitat

Often found growing in dense colonies on streambanks and trailsides in wet woods. This species occurs throughout our area in such environs.

Food

Wood nettle is edible when cooked, and indeed quite sumptious! For preparation and other particulars, the reader is referred to the monograph for "Stinging Nettle," since that information is entirely applicable here.

Health/Medicine

Not nearly as widely used by either white or Amerindian herbal practitioners as is the stinging nettle (*Urtica dioica*) (*see under that heading*), the wood nettle has yet been appreciated by several Indian tribes as a *urinary.* Here the roots were used by both the Pillager Ojibwe and the Meskquakis as a *diuretic.*(Smith 1928:251; Smith 1932:391) The latter found it to be serviceable as well for **URINARY INCONTINENCE.**(Smith 1928:251) Interestingly, down south in the Caribbean region, the natives there use a related species, *L. aestuans*, for **URINARY OBSTRUCTIONS.**(Honeychurch 1990:144) These three widely differing—even seemingly contradictory—uses relative to the urinary system merely give emphasis to what some of today's herbalists believe: that this plant (as well as *Urtica* spp.) is actually a urinary **trophorestorative**, "fixing" the system *regardless of the problem.*

Houma Indians took a decoction of wood nettle to help relieve a **FEVER**.*(Speck 1941:60)* Iroquois Indians celebrated wood nettle as a *parturient*. Again, it was the roots that were implemented: a handful of them were smashed and added to a teacupful of warm water, which was drunk immediately—twice.*(Herrick 1995:132)*

Cautions

   * Do not eat fresh, raw nettle plant parts.

   * Do not eat whole plants after the height of six inches is reached or even individual leaves from plants over that height except uppermost newly-emerging leaves which can be snipped off and boiled throughout most of the growing season.

   * Use gloves, or makeshift gloves of pot-holder-like mullein leaves, to collect so as to avoid the sting. If stung, rub leaves from dock or plantain on the sting to allay. Or nettle juice itself can be used, if it can be procured without be stung further!

WORMWOOD
(Artemisia absinthium)

[See photo #166 in rear of book.]

Description

A pungently fragrant perennial, growing 1½ -3 ft tall. The hoary stems are clumped and sometimes a bit woody at their bases.

The alternate leaves, growing 2-5 inches long, are divided once or twice, and are whitish-gray on both sides, owing to the presence of soft, silky hairs. The leaflet segments are blunted at the tips.

Small, yellowish-gray flowerheads are situated in sagging, elongate, leafy panicles.

The tannish-grey fruits are smooth and flattened achenes, about 1/16 of an inch long.

Range & Habitat

A cosmopolitan weed, occurring throughout our region in old farmsteads, abandoned homesteads, pastures, dry fields, roadsides, waste places, and other open and dry areas.

Seasonal Availability

Should be gathered in late summer, while flowering.

Chief or Important Constituents

Volatile oil(containing terpenes[thujone, pinene, sabinene, phellandrene, beta-carophyllene, myrcene, cadinene, azulenes] and sesquiterpene lactone-glycosides [santonin, absinthin, anabsinthin, artabsin, matricin]), polyacetylenes, phenolic acids(chlorogenic, p-coumaric, syringic, vanillic), flavonoids, lignans(diayangambin, epiyangambin), hydoxycoumarins, tannins. Nutrients include vitamins A & C, silica.

Health/Medicine

Dioscorides, the first-century-A.D. herbalist who may have been a medical officer in the Roman emperor's employ, noted that wormwood had a"warming, binding facultie." He had witnessed the vapor exuding from a decoction of the plant to lessen the pain of both a *TOOTHACHE* and an *EARACHE*. He further outlined its beneficence with *BITES,* including bites of "dragons of ye sea."(*Herbal*, 3.26) Whichever venomous sea creature Dioscorides had in mind is difficult to say, but it is of interest that in Caribbean lands wormwood tincture is venerated as an emergency treatment for *JELLYFISH STINGS*.(*Honeychurch 1980*) European herbalist Juliette de Bairacli Levy has found it useful for easing the pain of stings and bites of just about any kind.(*de Bairacli Levy 1974:159*) The thujone content may be partly responsible, since this chemical is analgesic owing to a narcotic property.(*Rice & Wilson 1976:1054-57*) Another major factor may be its content of azulenes, which are powerful anti-inflammatory agents. At any rate, the *analgesic* effects were also appeciated by our own Ojibwe Indians, who utilized a warm compress made from a decoction of the plant for swollen *SPRAINS*.(*Densmore 1974:362-63*)

Dioscorides also noted the plant's *emmenagogue* effect(*Herbal*, 3.26) and it was used in ancient Pompei for this very purpose,(*Jashemski1999:26*), which application has been perpetuated by herbalists ever since. This effect appears to derive from a combination of the plant's phellandrine and thujone contents.(*Moore1979:162*) Southwest-American herbalist Michael Moore finds wormwood especially of aid for*AMENORRHEA* with *CRAMPS*, particularly when on the tail end of an illness or trauma of some sort.(*Moore 1979:162*) The Cherokee Indians also found wormwood helpful for *DYSMENORRHEA*.(*Hamel & Chiltoksey 1975:62*) It seems to be of aid in a general sense for almost any sort of*PELVIC DIS-COMFORT*, partly owing to a stimulation of blood to this region.(*Kresanek 1989:54*)

A poweful *stomachic,* wormwood was highlighted by Dioscordes as an excellent herb for "taking away ye cholerick matter sticking to ye stomach, and ye belly" as well as for "pains of the belly & of ye stomach."(*Herbal,3.26*) It is most specifically a*diges-tive tonic* for stimulating gastric juices.(*Johnson 1884:183; Kresanek 1989:54; de Bairacli Levy 1974:159; Willard 1993:214*) The German phytotherapist-physician Rudolf Weiss praised it as one of the best, noting its aid in *DYSPEPSIA* and *ATONY* of the *GI TRACT*.(*Weiss 1988: 80-81*) The stomachic effects are of course due to the plant's poignant bitterness, which was recognized and marvelled at from antiquity (even being referrred to in the Bible on several occasions-"bitter as wormwood," Proverbs 5:4 and Revelation 8:11). And how bitter is that? Russian research has revealed that a mere single part of wormwood in 10,000 parts of water still leaves a bitter taste!(*Hutchens 1991:311*)

Pliny, the great naturalist who lived during the time of Dioscorides, outlined wormwood's usefulness for *COLITIS, BOWEL PAIN,* and *FLATULENT COLIC,* while centuries later the Kashaya Pomo Indians implemented it for*CRAMPS* associated with diarrhea.(*Goodrich&Lawson1980:119*) The Eclectics likewise advised it for"obstinate diar-rhea,"(*Felter & Lloyd*) as did the Okanagan and Thompson Indians.(*Perry 1952:36-43*) All of these uses are understandable in light of the plant's *antispasmodic* property (which has also yielded benefits for *EPILEPSY*).(*Millspaugh 1974:349*) Such a property might also shed light on the traditional use of the infusion for *TRAVELLING SICKNESS* (usually dosed at 5-30 drops, 3-4 times a day-*Willard 1993:214*), but here the plant's *disinfectant* property is undoubtedly also an important factor.(*Kresanek1989:54*) Some herbalists regard wormwood to be helpful as well for intestinal permeability (the so-called *"LEAKY-GUT" SYN-DROME*), because of its widespread antimicrobial properties.(*PDRHrbs*)

Wormwood's best-known use is probably as a *vermifuge* for **WORMS** in the intestinal tract. This application has been appreciated since ancient times. Thus, Celsus had urged a wormwood decoction for **THREADWORMS** in children(*De Medicina,* 4:24:2). Later on down the stream of time, Mohegan Indians similarly employed the plant as a *vermifuge*.*(Tantaquidgeon 1972:70, 128-29)* So did the Cherokees, who poulticed the plant on the stomach to rid worms from the GI tract.*(Hamel & Chiltoskey 1975:62)* Our herb is most especially renowned for expelling **THREADWORMS** (as Celsus had noted, above) and **ROUNDWORMS**.*(Felter & Lloyd; Mabey 1988:41; Hutchens 1991:312; Moore 1979:162)* Moore notes that an effective regimen is to imbibe at least two cups a day for a week or two.*(Moore 1979:162)* Hutchens opts for a decoction used as an enema.*(Hutchens 1991:312)* Wormwood's constituents santonin and artemisin are thought to be the chief active **anthelmintic** agents by most researchers, although the thujone content is highlighted by other authorities. Interestingly, the classic worm medicine known as santonine is derived from wormwood's cousin, *A. santonica.* Some feel that wormwood can also rid the body of **FLEAS** and **LICE**.*(de Bairacli Levy 1974:159)*

Weiss considered wormwood to be "a well-proven gallbladder remedy....one of the best remedies for biliary dyskinesia or a gallbladder apt to cause trouble." He advised it especially for a gallbladder that was atonic and he felt that, for this condition, the tea should best be taken some time *after* meals, because otherwise its energies might tend to concentrate on stimulating the stomach.*(Weiss 1988:80)* Celsus regarded wormwood highly for **JAUNDICE**(*De Medicina,* 3.24.2), as did, centuries later, the early-American botanist Constantine Rafinesque*(Rafinesque1830:184)* and also the Eclectic physicians.*(Felter & Lloyd)* This application makes sense in view of the herb's demonstrated *cholagogue* ability,*(Baumann et al 1975:784; PDRHrbs; Kresanek 1989:54; Willard 1993:214)* although, as Weiss pointed out, the herb's quite minor ability in this regard (as compared to other herbs like dandelion and turmeric) is not necessarily to be understood as the chief reason for its value as a biliary tonic.*(Weiss 1988:80)*

Wormwood's *hepatic* effects also help relieve a **FRONTAL HEADACHE** accompanied by indigestion.*(Willard 1992a:203)* The herb even appears to halt lipid peroxidation, thus protecting the liver from the bad effects of rancid fats.*(Willard 1992a:203)* It also holds promise in treating **ACETAMINOPHEN POISONING**, in that a 1995 study revealed that a crude extract ameliorated acetaminophen-induced liver toxicity in mice.*(Gilani&Janbaz1995:309-15)* Wormwood, as Celsus noted, is also *diuretic,*(*De Medicina,* 2.31) which may partly elucidate its time-honored benefits for **GOUT***(Millspaugh 1974:349)* and **DIABETES.** *(Jashemski 1999:26)*

In fact, one may not need acetaminophen at all for fever if one has access to wormwood, as it is a powerful *febrifuge* in itself (confirmed in experiments with rabbits-*PDRHrbs*), at least partly due to its azulenes, as well to its phellandrine content and perhaps as well to its thujone.*(Moore 1979:162; Potterton 1983:203)* In fact, chemicals in the plant directly fight **MALARIA,***(Zafar & Hameed 1992:223-26)* for which affliction the Eclectics sometimes used it prior to the introduction of cinchona.*(Felter&Lloyd)* However, wormwood's cousin, sweet annie *(A. annua),* is the superior **antimalarial.**

As for myself, while I respect the healing power of this plant greatly, and use it when indicated for my clients, I never gather it personally, because a mere whiff of it gives me a *really* bad headache! (I have experienced the same reaction from smelling marijuana smoke, such as back in the 1970s when lowlife punks from my high school would abuse this plant as a recreational drug in the rest rooms. Here such a reaction is of interest in view of scientific research showing that the thujone in wormwood reacts with the same receptor sites in the brain as does the THC in marijuana!-*del Castillo 1975:365-66*)

Wormwood was official in the *U. S. Pharmacopoeia* from 1831-1905 and in the *National Formulary* from 1916-26.

Cautions

* Absolutely contraindicated in both **PREGNANCY** and **LACTATION!** Those with **KIDNEY FAILURE**—or at risk for such—should probably also not use wormwood (especially the oil, since it has been implicated in a case of acute renal failure).

* Wormwood should not be used long-term (Weiss says not to use it continuously for more than 3-4 weeks-*Weiss 1988:81*), as in large and sustained doses its high thujone content could conceivably predispose one to adverse neurological affects. In fact, in nineteenth-century France, a slick, green, alcoholic beverage called absinthe that was largely derived from an extract of wormwood oil became the rage of the day, but eventually brought on the ravages of a condition since designated as "absinthism," imarked by mental confusion, seizures, hallucinations, and other adverse neurological effects. Sadly, many thousands wound up succumbing to absinthe's insiduous, death-dealing ways. (Historians generally feel that artist Vincent van Gogh's slide into mental illness—climaxed by his severing of his ear and his mailing that to a lady friend—was as a result of his addiction to this drink). After (on some occasions, *too* long after) absinthe's toxic effects were scientifically substantiated, it was banned in many European nations (by the early 1900s) and finally in France (in 1915).

YARROW
[MILFOIL]
(Achillea millefolium)

[See photo #167 in rear of book.]

A scented, perennial herb, growing 9-24 inches high, and entirely covered with soft hairs. Its square stem, characteristic of the mint family to which it belongs, is usually unbranched, but occasionally branches near the top.

The alternate leaves are finely divided (twice pinnate) into numerous narrow segments, looking almost like disheveled pipe cleaners. Strongly scented with a bitter, pungent odor (especially noticeable when handled), their divided form and scent are what distinguish this plant from its many lookalikes.

Yarrow's flowers form small heads—all compacting into a terminal, flat-topped cluster. They are almost always white in color, but one can occasionally find pink- or purple-colored flower heads. With the aid of a magnifying glass, one can see that each individual flower is a composite one, made up of a yellow (and female) disk flower and several white/pink (and male) ray flowers.

Range & Habitat

Yarrow is one of those plants well described as "cosmopolitan," occurring throughout North America. The herb likes to grow in loose colonies and its preferred habitat is fields, pastures, roadsides, prairies, meadows, open woods, and any exposed area that is sunny and well-drained.

Chief or Important Constituents

Over 120 compounds, including alkaloids(achilleine[betonicine], trigonelline, stachydrine]), sesquiterpene lactones of the guaianolide and germacranolide types (achillicin, achillin, deacetylmatricarine, millefin), essential oil(containing terpenes such as linalool, carophyllene, eugenol, borneol, camphor, limonene, cineole, menthol, sabinene, pinene, and a trace of thujone), saponins, polyacetylenes, alkanes(tricosane, pentacosane, heptadecane), beta-sitosterol, furanocoumarins(incl. psoralen), flavonoids (rutin, apigenin, luteolin, isorhamnetin, cyanidin), phenolic acids(salicylic, caffeic), isovalerianic acid, butyric acid, fatty acids(linoleic, oleic, palmitic), amino acids, sugars, tannin. Nutrients include vitamins(A, C, folic acid, choline) and minerals(esp. potassium and silica, but also some iron, magnesium, calcium, and phosphorous).

Food

Young leaves can be added sparingly to a salad to give it more of a pungency. Yarrow's distinctive taste is also used to flavor several different alcoholic beverages.

Health/Medicine

Yarrow abounds in our area and has, since my childhood, strongly held my attention, due to its odd scent and appearance. My fascination with it insured that it would become one of the first herbs that I would concentrate on when I initially found myself immersed in an intensive study of herbalism in the 1980s. (Here I was especially impressed when I learned that it had been used by over 58 different tribes of American Indians!) It remains a fascination until the present day.

First and foremost, yarrow is a truly amazing **styptic** and **hemostat.** According to Homer's account in the *Iliad,* the mighty Achilles used poultices of yarrow leaves on the wounds of fellow soldiers during the Trojan War (1200 BCE?). Subsequent peoples have consequently called it "soldier's woundwort" and here it has been observed that it especially excels for *deep* wounds, where the tea is commonly used as a soak.*(Wood 1997:65-67)* But you don't have to be a soldier to benefit from yarrow, in that it has also long been used with success for **INTERNAL BLEEDING** in the **ALIMENTARY CANAL, NOSEBLEEDS, HEMORRHAGES,***(Kresanek 1989:38; Millspaugh 1974:336)* and especially for **MENORRHAGIA,***(Felter&Lloyd; Brooke 1992:151; de Bairacli Levy 1974:160; Carse 1989:176)* which latter application was put to good use by the Cherokee Indians.*(Hamel & Chiltoskey 1975:62)* The great Roman naturalist Pliny the Elder drew attention to its use in this regard and even noted that menorrhagia could be checked by taking a sitz bath in a decoction of yarrow.(HN 7:377) The herb's styptic and hemostatic actions would seem to arise chiefly owing to its alkaloid achilleine, which has been shown to clot rabbit blood more quickly than usual and with such an action persisting for up to 45 minutes!*(Miller & Chow 1954:1353-54)*

The plant's renowned **disinfectant** action—resulting from its essential oil, tannins, and other factors—keeps wounds and sores poulticed with it from getting infected, even as the first-century herbalist Dioscorides had noted.(*Herbal,*4.115) Yarrow is indeed a powerful **antimicrobial**, effective against a wide range of both gram-positive and gram-negative microorganisms.*(Ibragimov & Kazanskaia 1981:108-09; D'amico 1950:77-79; Goranov et al 1983:25-30 Chandler 1982:203-23; PDRHrbs)* An aqueous extract has been found to possess *in vitro*-activity against *Staphylococcus aureus,(Mockle 1955:85)* as well as *Escherichia coli, Bacillus subtilus, Mycobacterium smegmatis, Shigella sonnei,* and *S. flexneri.(Moskalenko 1986:231-59)* Its action against the *Shigella* organisms suggests that it would be a specific for**SHIGELLOSIS**, which it is, but being most effective for that condition when—as herbalist Michael Moore notes—it is combined with echinacea (esp. *E. angustifolia*, I would think) in goodly amounts.*(Moore 1993:274)* Another test-tube study revealed that yarrow had some minor inhibiting effect upon a virulent human strain of the tuberculosis organism,*(Fitzpatrick 1954:530)* which could be due to its psoralen content since this chemical has been shown to be active against *Mycobacteria*.*(Harborne & Baxter 1993:362)* Yarrow also contains a natural peroxide that is a proven **antimalarial**.*(Rucker et al 1991:295)*

Interestingly, irregardless of its styptic properties, a twirled leaf in the nose of a person with a **VASODILATED MIGRAINE** or **SINUS HEADACHE** can *cause* a nosebleed and thus relieve the painful pressure,*(Jones1991:7)* even as the Woods Cree Indians discovered.*(Leighton 1985:23)* Numerous writers have speculated that this is due to the plant's coumarins, which they hail as hemorrhagic. However, coumarin is not normally hemorrhagic when ingested or utilized in moderate amounts, but only so when acted upon in a complex combination of ways involving a wilting of the plant and the introduction of moisture which then transforms it into dicoumarol (the compound upon which rat poison and medical anticoagulant has been patterned). It's difficult to fathom how this is likely to occur from simply plucking a leaf and twirling it in one's nose. Nevertheless, regardless of the mechanism(s) involved, the procedure does often work.

A renowned **astringent**, yarrow has been greatly valued throughout history for **DIARRHEA, UTERINE FIBROIDS,**(*Mills&Bone 2000:243; Brooke 1992:151; Harrar & O'Donnell 1999:80)* **CRACKED NIPPLES,**(*Weed1986:91)***SORE THROAT**(*Hellson&Gadd1974:70)* and **COLITIS**(*Winston 1999:55)* proving helpful as well for the latter condition owing to**anti-inflammatory** properties,*(Goldberg et al 1969:938-41, Goldberg 1972:938-41 Shipochliev&Fournadjiev 1984:99-107, Verzan-Petri&Banh-Nhu 1977:c.24)*, possibly due to its essential oil*(Pedersen1998:175;Mabey1988:40)* and/or to its protein-carbohydrate complexes,*(Goldberg et al 1969:938-41)* and/or to its flavonoids*(Bruneton 1995:290)* and/or to its cyanidin and salicylic acid.*(Mabey 1988:40)*

The anti-inflammatory benefits are legion: Bella Coola Indians poulticed yarrow leaves on **ABSCESSED BREASTS**.*(Turner 1973:201)* Micmac Indians poulticed pulped stems onto **SPRAINS, SWELLINGS,** and **BRUISES**.*(Lacey 1993:95)* Southern Carrier Indians chewed the leaves and applied them to same.*(Smith 1928:65)* Malecite Indians boiled the plant down to a thick paste and used this as a liniment for same.*(Mechling1959:244)* Blackfoot Indians preferred the flowers as a chewed poultice.*(Hellson & Gadd 1974:74)* Winnebago Indians bathed swellings with an infusion of yarrow. They also wadded up a leaf to stick into the ear canal of a person with an**EARACHE** or sometimes simply poured a small bit of the infusion into the affected ear.*(Gilmore 1991:82)* Canada's Thompson Indians implemented a decoction of the plant as a wash for**ARTHRITIC LIMBS**.*(Turner et al 1990:47)* The Flambeau Ojibwe poulticed leaves onto **SPIDER BITES**.*(Smith1932:362)* **ECZEMA** and other **RASHES** were treated topically with fresh yarrow tops by our own area's Menominee Indians.*(Smith 1923:29)*

As a topical pain-killer, yarrow ranks most highly—perhaps even highest—among our area's plants. (However, see "Cautions," below) Thus, Canada's Thompson Indians found that a poultice of the root brought great relief to those afflicted with **SCIATICA**.*(Turner et al 1990:46)* Throughout the ages, too, many persons have chewed yarrow leaves for **TOOTHACHE,***(de Bairacli Levy 1974:160)* even as did the Osage and a few other Amerindian tribes, who successively chewed a dozen or less leaves until a numbing sensation ensued.*(Kavasch 1979:176)* The Costanoan Indians simply held the leaves in their mouths next to the painful tooth,*(Bocek 1984:254)* while Saskatchewan's Woods Cree Indians preferred the root for the same purpose.*(Leighton 1985:23)* Yarrow is no doubt effective here because of its content of eugenol, the famous drugstore remedy (formerly sold as "oil of cloves," but now marketed simply as the distinct chemical eugenol). The **analgesic** effect may also be partly due to the camphor content and perhaps as well to the salicylic acid.

Yarrow flowers are **anti-allergenic** and thus useful for **HAYFEVER,** probably chiefly because of the flavonoid content, but perhaps as well to proazulenes that occur in some suspspecies—which, when the flowers are prepared as an infusion, are converted by the steam into azulenes, which are powerful anti-allergenic substances.*(PDRHrbs; Bruneton 1995:290; Ody 1993:30)*

Yarrow is a treasured **antipyretic** and thus has long been used for **FEVER**, for which it is also often recommended by present-day herbalists.*(BHP 1976:1:145)* It is generally agreed that this herb dilates the blood vessels in the outer surfaces of the body, causing a diaphoretic effect that flushes the skin of excess heat and wastes. Such an effect also makes yarrow useful for **PLEURISY.***(de Bairacli Levy 1974:160)* It is possible that the plant's alkaloids are partly responsible for this effect.*(Pedersen 1998:175)* But, most likely, the sesquiterpene lactones are largely responsible, as they seem to be with yarrow's cousin, boneset *(Eupatorium perfoliatum)* (*see under that entry*), which has a similar diaphoretic effect. Thus, when one of these chemicals, achillin, was administered to rabbits, there was demonstrated a subsequent fall in rectal temperature.*(Falk et al 1975:1838-42)*

Yarrow's application for **FEVER** was appreciated and utilized by numerous Amerindian tribes, including the Woods Cree, who used it by way of a compress on the head.*(Leighton 1985:23)* The Flambeau Ojibwe, on the other hand, singed yarrow blossoms on hot coals and inhaled the steam.*(Smith 1932:362)* The herb's fever-fighting powers were also utilized by the Micmac,*(Lacey 1993:95)* Abenaki,*(Rousseau 1947:154)* Menominee*(Smith 1923:28-29)* Cherokee,*(Hamel & Chiltoskey 1975:62)* and Montagnais,*(Speck 1917:315)* all of whom simply drank an infusion of the herb to deal with the problem. For feverish infants, the Iroquois steeped 3-4 leaves in 1/2-teacupful of cold water for one minute and then gave a little bit of it to the babes throughout the day.*(Herrick 1995:227)*

As for its indications in fever, herbalist Matthew Wood finds it most applicable with a red and flushed face, dry skin accompanied by a full and rapid pulse, and a red and uncoated tongue that is drier in the center and wetter toward the edges.*(Wood 1997:72-73)* Somewhat similarly, the Eclectic physician Finley Ellingwood pointed to its use when there was "hot, dry burning skin, at the beginning of acute asthenic fevers, with suppressed secretion."*(Ellingwood 1983:355)* However, because of its stimulant properties, the herb is felt by some other herbalists to be best applicable to a fever where the skin is cold and the pulse is weak.*(Willard 1993:196)* Personally, I haven't noticed that much difference in its effects with these different types of fever, and I could say the same of that other great febrifuge of the *Compositae* family, boneset (*Eupatorium perfoliatum*) (*see under that entry*). My own experience—in over a decade and a half of use—is that both of these herbs seem to work remarkably well for almost *any* kind of fever, simply because both are so extraordinarily diaphoretic! In fact, as I was writing the above, I remembered a summation that Ellingwood gave of the work of Dr. Cole of Seattle, who "confirmed in a practical matter the action of achillea on the skin....When there is no abnormal temperature, he believes that it has little but a diuretic action. When there is a temperature of 100 or above, *he has never failed* to get profuse diaphoresis without depression. He considers it a *certain remedy."* *(Ellingwood 1983:356; italics added)* I most heartily agree, but also echo Ellingwood's admonition that "it acts best in strong infusion and its use must be persisted in." *(Ellingwood 1983:356)*

Yarrow is a time-honored ***bitter tonic*** for ***APPETITE LOSS*** and ***DYSPEP-SIA***. *(Kresanek 1989:38)* ***WEAK, DISORDERED STOMACHS*** were treated to yarrow tea by the Paiute Indians. *(Palmer 1878:651)* The evidence for these effects is so good that the *German Commission E Monographs* has given its seal of approval to yarrow for appetite loss, dyspepsia, and ***MILD, CRAMPLIKE PAINS IN THE ABDOMEN***. *(GermCommE)* The plant's antimicrobial action (referred to above), along with its content of butyric acid, also lends it for possible use in helping to heal ***"LEAKY-GUT" SYNDROME (INTESTINAL PERMEABILITY)***—much like its cousin, wormwood (*Artemisia absinthium*) (*see under that entry*).

Because of an effect in regulating liver function, yarrow is a genuine ***hepatic*** *(Willard 1993:149)* and even a ***cholagogue*** *(PDRHrbs; Willard 1991:351)* and ***choleretic,*** *(Muller 1955:39; Bisset 1994)* which facts have allowed it to be used by Gosiute Indians for "biliousness." *(Chamberlin 1911:360)* Two animal studies even suggest that it can offer protection against ***LIVER DAMAGE***. *(Castleman 1991:379)* An east-Indian study even showed that the plant may help moderate ***HEPATITIS***. *(Castleman 1991:379)* It is no surprise, then, that yarrow is sometimes considered an herb for the liver above all else! *(Holmes 1989:1:282)*

Yarrow has been long been appreciated for ***CIRCULATORY*** disorders, especially for ***RAYNAUD'S SYNDROME*** and ***THROMBOSIS,*** (the herb's coumarins are ***antithrombotic-*** *Pedersen 1998:175*), and it is the herb of choice for these conditions when ***HYPERTENSION*** is also involved, *(BHC 1976:1:145)* since yarrow has proven ***hypotensive*** properties, *(Leung 1980:327; Newall et al 1996:272; Duke 1985:10),* mainly due to its alkaloids *(Pedersen 1998:175)* and flavonoids. *(Mabey 1988:40)* British herbalists find that it reduces hypertension most effectively, however, when combined with linden and/or hawthorn, two other classic antihypertensives. *(BHC 1976:1:145; McQuade-Crawford 1996:182)* (As a point aside, linden is most specific when *emotional factors* are largely responsible for the hypertension, or at least are aggravating it in some way. I know more people than I could count who have had their hypertension tamed by adding linden flowers to their otherwise failing herbal regimen. Since it is a cooling herb, though, one should not fail to note that it works best in "hot"-bodied types. For "cold"-bodied types with emotionally-aggravated hypertension, valerian is the herb of choice.)

Further as to its circulatory assists, the plant has a special attraction to the venous system of the body *(Felter & Lloyd),* so that it is much appreciated by those suffering from ***HEMORRHOIDS/VARICOSE VEINS,*** in which condition it is employed internally and often as a wash as well. *(Carse 1989:176; Brooke 1992:151; Fishcher-Rizzi 1996:303; Kresanek 1989:38; Moore 1979:164)* Here, the Eclectics deemed yarrow specific for those hemorrhoids which manifested *bloody* or *mucoid* discharges. *(Felter & Lloyd)*

Yarrow has a marked affinity for the pelvic region, which specificity is widely acknowledged.*(Johnson 1884:182; Millspaugh 1974:336; PDRHrbs)* A true trophorestorative for that region, it serves powerfully to allay **PELVIC DISCOMFORT.***(Fisher-Rizzi 1996:303; Brooke 1992:151)* It is even recommended by German Commission E as a sitz bath for **PAINFUL CRAMPS** in the lower quadrant of the pelvis.*(GermCommE)* It is also an **emmenagogue** *par excellence* for **AMENORRHEA,** especially when of an atonic character,*(Scudder 1870:60)* and/or **MENSTRUAL CRAMPS**, mostly when due to cold*(Brooke 1992:151; Harrar & O'Donnell 1999: 80)* or when accompanied by spasms. It is also probably the best single herb to offset the miseries of **ENDOMETRIOSIS**.*(MH 5:3:4)* Yarrow is further one of several plants that modern herbalists find helpful in expelling a **RETAINED PLACENTA,***(Harrar & O'Donnell 1999:80)* even as the Blackfeet Indians had earlier discovered.*(Hellson & Gadd 1974:60)* Then, too, it often proves to be a real dandy for **LEUKORRHEA,***(Scudder 1870:60; Carse 1989:176; Brooke 1982:151)* especially when this discharge is profuse and is accompanied by relaxed vaginal walls, in which case a douche propelled into the area was actually recommended by the Eclectics.*(Ellingwood 1983:356; Felter & Lloyd)*

A **diuretic** and **urinary antiseptic,***(Goldberg 1969:938-41; BHC 1976:1:145; Hellson&Gadd 1974:69)* yarrow can be helpful for **CYSTITIS,***(Brooke 1992:151)* **URETHRAL IRRITATION,***(Winston 1999:55)* **HEMATURIA** with **PAIN,***(Felter & Lloyd; Moore 1994a:9.6)* and as a cold infusion for **URINARY INCONTINENCE***(Felter & Lloyd)* The Eclectics found that it especially fit the bill when there was "irritation of the kidneys and vesical and urethral irritation"*(Scudder 1870:60)* and primarily when such symptoms were indicative of "chronic diseases of the urinary apparatus" as opposed to being acute.*(Felter & Lloyd)* Moore agrees that it especially excels with cystitis and urethritis when these have become low-level and chronic, for which he says that two cups of tea a day for two weeks often proves therapeutic.*(Moore 1993:275)* Holmes find yarrow of aid with urinary conditions when urination is *frequent, scanty,* and *dribbling.(Holmes 1989:1:281)* Not surprisingly, in view of its urinary properties as explicated above, a Chinese species (*A. alpina*) proved marvelously effective in a clinical study with 65 cases of **URINARY-TRACT INFECTION** (UTI's).*(Peng et al 1983:217-18)* (We have seen above that yarrow is active against *E. coli*—often a causative agent in UTI's.) Our herb is ideal for urinary problems because it strengthens the urinary organs without irritating them—unlike some other urinary herbs (e.g., juniper) in popular use.*(Fisher-Rizzi 1996:303; Ellingwood 1983:355)* It may also be appropriate for persons afflicted with **NEPHRITIS/BRIGHT'S DISEASE**.*(Ellingwood 1983:355; Carse 1989:176)*

Yarrow flowers have an **antispasmodic** effect on smooth muscle, due to their flavonoid content.*(Hoerhammer 1961:578-88; PDRHrbs; Pedersen 1998:175; Weiss 1988:92)* The urinary system is lined with smooth muscle, so that yarrow finds ideal application with **SPASMS** of the **BLADDER** and **URINARY TRACT**. Smooth muscle also exists in the GI tract and the uterus, which facts undoubtedly at least partly account for yarrow's **stomachic** and **emmenagogue** effects as noted above. The German phytotherapist-physician Rudolf Weiss drew attention to its application in "spastic conditions in the small pelvis, the parametrium, and neurovegetative disorders in that region."*(Weiss 1988:92)* Its effect on smooth muscle may also explain why it has classically been used for **EPILEPSY/TREMBLING***(Felter & Lloyd; de Bairacli Levy 1974:160)* and for **ANGINA PECTORIS.***(Kresanek 1989:38)* Iroquous Indians even found our herb helpful for **CONVULSIONS** in babies, so that they would boil a whole plant in a pint of water until the fluid turned green and then feed a teaspoonful of the cooled decoction to the infant, while rubbing the rest over its body. *(Herrick 1995:227)*

The plant's sesquiterpene lactones have shown an **anti-tumor** effect in lab studies with mice. *(Tozyo et al 1994:1096-1100)* We have observed in earlier monographs (e.g., boneset, burdock) that these chemicals, widely distributed in the *Compositae* family of plants, have been shown generally to be possessed of this ability. All in all, I would not be surprised should sesquiterpene lactones become the most widely investigated and publicized compounds in cancer-phytotherapy research in the twenty-first century.

Yarrow flowers and leaves were official in the *U.S. Pharmacopoeia* from 1836-1882, listed therein as **tonic, stimulant**, and **emmenagogue.** The herb is still official in the pharmacopoeias of France, Germany, Switzerland, Austria, Romania, Hungary, and the Czech Republic.

Cautions

* Not for **PREGGIES**, as it can be a uterine stimulant in larger doses. Even the trace amounts of thujone present make yarrow a risk for miscarriage.

* Touching bruised leaves can lead to **CONTACT DERMATITIS.** Some studies show that up to half of the population is sensitive in this regard, which problem can be further aggravated—and most severely!—by sunlight.

* People with **EPILEPSY** should avoid using this herb.

* Those on **ANTICOAGULANTS** and **BLOOD-PRESSURE MEDS**. should use yarrow with caution and under supervision of a health-care professional. Persons with **LOW BLOOD PRESSURE** should also use this herb with caution.

YELLOW GOATSBEARD
[MEADOW or YELLOW SALSIFY; JACK-GO-TO-BED-AT-NOON]
(Tragopogon pratensis; T. dubius)

Description

[See photo #168 in rear of book]

A common biennial or perennial herb, 1-3½ feet tall, with a smooth and branching stem. The first-year plant is a rosette, but in the second year, a long stem and herbage appear. At this point, the thick root has transformed into a mere fingerlike form.

Yellow goatsbeard has alternate, untoothed, narrow and grasslike leaves that are parallel-veined.   Blue-green in color, they are 4-10 inches long, pointed at the tip, and contain a milklike juice. Each leaf wraps around the stem at its base.

A solitary flowerhead appears at the tip of the stem—situated on a long, slender, pointed bract. It is 1-2½ inches wide and composed of many, yellow, ray flowers. Quite noticeably, the blossoms "close up shop" by early afternoon (banker's hours!), folding up into their beanpod-shaped bracts.

One of the most noticeable features of this herb is its showy seed-head. Here, when the plant is ready to seed, the flowerheads transform into globular, fuzzy balls similar to to those put forth by the dandelion, but *much* larger (approximately the size of tennis balls!). The dirty-white look of the seeds gives rise to the plant's most oft-used common name: "goatsbeard." The seed-head's structure is loose, enabling the seeds to spread easily on breezy days, facilitated by the fact that each of them possesses a parachute-like structure that allows for a thrill-packed ride along the air currents.

Range and Habitat

Tragopogon is found in sandy waste places, roadsides, railroad rights-of-way, vacant lots, fields, and meadows. It thrives in the eastern two-thirds of the USA, and in Canada from Nova Scotia to Ontario. In our area, the plant grows ubiquitously.

Chief or Important Constituents

Not well determined at present, although the root is known to contain phytosterols, inulin, mannitol, and the following nutrients: B vitamins(especially inositol), vitamin C, potassium, and some smaller amounts of iron, phosphorous, and calcium.

Food

Young leaves are edible raw. (If bitter, a quick blanching in boiling water will take the "nip" out of them.) The leaf crown, located at the bottom of the flower stalk, makes a fine potherb. Being tender, it only needs to be simmered for a few minutes in a little water. (Note that cooked crowns do not store well and should be used right away.)

Master forager and wild-plant author Francois Couplan enjoys the tender, young flowerbuds, finding them to be most sweet. He notes that they can either be used as a trail nibble or lightly steamed to retain their crunchiness. *(personal communication, June 2001)*

The most renowned food use of the plant, however, involves the taproot of first-year (non-flowering) plants, which is good boiled or roasted. Roots from flowering plants can be used, too, contrary to popular belief. They look as if they wouldn't be very palatable—tough and stringy and not having much substance to them. All of this being so, one may find that they taste just fine if enough are gathered and one tries different cooking times until tenderness is achieved.

Health/Medicine

Largely abandoned by modern herbalists, *Tragopogon* yields several important healing properties which those of previous generations were evidently more humbly disposed to learn and utilize. Thus, revered 17th-century herbalist Nicholas Culpeper wrote of this plant: "A large double handful of the plant, roots, flowers, and all bruised and boiled, and then strained, with a little sweet oil, is an excellent clyster in the most desperate cases of *STRANGURY* or *SUPPRESSION OF URINE*, from whatever cause." *(Potterton 1983:86; emphasis added)* Culpeper went on to add that yellow goatsbeard "expels sand and *GRAVEL*, slime and small stones." *(Potterton 1983:86; emphasis added)* And these were not isolated observations: Famed botanical scholar Linnaeus also noted the plant's reputation for strangury, in his *Vegetable Materia Medica* of 1829. *(Linnaeus 1829:147)* Naturopathic writer John Lust commented that the *diuretic* effects are particularly helpful relative to *URINARY PROBLEMS* or *WATER RETENTION*. *(Lust 1974:414)* Probably largely responsible for alleviating the urethral discomforts is the plant's experimentally-verified *anti-inflammatory* effect, established in a 1976 study. *(Benoit et al 1976:165)*

Culpeper observed further: "A decoction of the roots is good for *HEARTBURN*, *LOSS OF APPETITE*, and *LIVER* and *CHEST DISORDERS*." *(Potterton 1983:86; emphasis added)*. The use for heartburn was adopted by the American Indians after the settlers had shared with them the secrets of this European import. *(Dobelis et al 1986:196)* Some such tribes would let the juice coagulate into a gum, which was then chewed to obtain the desired *antidyspepsic* effect. Then, too, the *appetite-stimulant* and *hepatic* properties Culpeper described above have been at least partially evidenced since his time; *(Stuart 1982:146)* it is even thought that *Tragopogon* serves to liquify thickened bile! *(Steinmetz 1957)* Finally, the traditional treatment of *BOILS* with this plant, based upon Indian use, *(Vestal 1952:54)* probably also relates to hepatic toning/cleansing.

Culpeper added: "The roots cooked like parsnips, with butter...strengthen the... weak after a *LONG ILLNESS*." *(Potterton 1983:86)* Today, we would call such use for debilitated persons a *tonic*, and at least partly responsible for such effects, it has long been thought, must be the high vitamin-C content. Now, however, we know that it goes beyond this: A recent Chinese study using mice has discovered powerful *antifatigue* and *anoxia-tolerating* properties in this humble weed (*T. porrifolius* was the species tested). *(Long et al 1990:765)*

YELLOW LOOSESTRIFE

(Lysimachia spp.)

Description

[See photo #169 in rear of book.]

This genus possesses yellow flowers with five (occasionally six) petals and stamens. The leaves are opposite (occasionally whorled) and toothless and the blossoms almost always emerge from the plant's axils. It is represented in our area by several species, among which are:

L. ciliata (fringed loosestrife; tufted loosestrife) is a perennial that grows 1-3 feet tall and has a very weak stem. Its opposite, ovate, light-green leaves reach a length of 2-3 inches and are borne on long leafstalks that are fringed on the upper axils. The flowers *nod* most conspicuously and their petals are *fringed*. They are borne in groups of two or three.

L. quadriflora (prairie loosestrife) reaches a height of 1-2½ feet and possesses opposite, stalkless, lanceolate leaves.. These grow up to three inches long and their edges are smooth and slightly reflexed. The starlike flowers have a reddish center and grow singly on long, skinny stalks.

L. quadrifolia (whorled loosestrife) resembles the previous species, but its leaves are *in whorls*.

L. terrestris (yellow loosestrife; swamp candles) grows 6-20 inches tall and has opposite leaves and starlike blossoms with a circle of red spots. Unlike other species in this genus, its leaves are not borne from the leaf axils but are clustered into a terminal raceme.

L. thrysiflora (tufted loosestrife) is a perennial that tops off at 6-30 inches and has opposite, lanceolate to elongate leaves reaching a length of 2-4 inches. It differs from the other species in having its flowers in balled clusters on short stalks.

Range & Habitat

L. terrestris and *L. thyrsiflora* thrive in swamps and marshes. *L. quadriflora* prefers moist prairies and shores. *L. ciliata* likes moist, shaded areas and can be found in woods and roadside ditches. One or another species can be found throughout our area.

Chief or Important Constituents

Flavonoids

Food

During the period immediately preceding the Revolutionary War, the whorled loosestrife *(L. quadriflora)* was dried and steeped as a tea when the colonists rebelled at drinking the heavily-taxed, British tea. As such, it was known as "Liberty Tea." It is quite mild in taste, however, and often needs to mixed with another herb or two to bring out its flavor.

Health/Medicine

This genus has a long history of use as a medicinal plant. The first-century, Greek herbalist Dioscorides wrote of the **astringent** applications of a European species, *L. vulgaris*, while seventeenth-century herbalist Nicholas Culpeper described such a property in even more detail when he wrote that this plant was implemented widely for **NOSEBLEEDS, MENORRHAGIA, BLOODY DIARRHEA, SORE THROAT,** and **WOUNDS** (for which it was said to "quickly close up the lips").*(Potterton1983:115)* Reflecting on Culpeper's remarks, David Potterton notes that modern European herbalists use this species similarly, in the form of an infusion of the dried herb, prepared from an ounce (28g) of herb to a pint (568ml) of boiling water and dosed to the tune of two fluid ounces (56ml) at a time.*(Potterton 1983:115)*

Traditional Chinese medicine uses yet another species of yellow loosestrife, *L. christinae* (referred to by the Chinese as *jin qian cao*), as a cooling herb to rid the body of pathogenic heat and dampness—factors thought to be connected with conditions such as **JAUNDICE, EDEMA, SNAKEBITE, BURNS, ABSCESSES,** and **CALCULI** of the urinary tract and gallbladder.*(Bareft 1977:719; Yanchi 1995:108)*

As for our own area's species, *L. quadrifolia* was held to be a powerful female remedy by the Cherokee Indians and was also used by them for **BOWEL COMPLAINTS** and **KIDNEY PROBLEMS**.*(Hamel & Chiltoskey 1975:43)* For the latter, they would implement an infusion of the roots.*(Averill 1940:50)* The Iroquois Indians also occasionally employed an infusion of this plant's roots, but as an **emetic** when needed.*(Herrick 1977:411)* *L. terrestris*, another of our local species, has shown significant **anti-inflammatory** properties in the rat-paw-edema lab test.*(Benoit 1976:167)* *L. thyrsiflora* leaves were applied by the Iroquois to women's breasts to halt the flow of milk when that was desired.*(Herrick 1977:411)*

YELLOW POND-LILY
[SPATTERDOCK; COW LILY]
(Nuphar luteum [alt: N. lutea; N. advena] subsp. variegatum)

Description

[See photo #170 in rear of book.]

This is a common, aquatic perennial with several, large (3-15"), heart-shaped leaves having a conspicuous notch between their two basal lobes. One common species, *N. luteum*, has erect leaves appearing a bit above the surface of the water. The other species, *N. variegatum*, has horizontal leaves floating *on* the water.

The leaves are situated at the tip of long, slimy leafstalks arising from a huge (some 1½ - 2½ feet long!) rootstock or rhizome buried in the muck below. (Many have described this rootstock as resembling a green pineapple.)

The flowers, appearing individually on separate stalks resembling the leafstalks, are yellow and cuplike, two inches across, and with five or six prominent sepals resembling petals. The real petals, however, are hidden inside and look more like stamens than they do petals. Most noticeable is the disklike stigma situated in the center of the flower.

In autumn, sepals and petals fall away and green fruit forms with many seeds resembling unpopped popcorn, largely buried out of sight.

Range & Habitat

Yellow pond-lily occurs throughout the northeastern and midcentral states and in our area specifically in the eastern four-fifths. It thrives in lakes, ponds, pools, and slow streams.

Chief or Important Constituents

Steroidal saponins, alkaloids and pseudo-alkaloids[nitrogenic sesquiterpenes] (incl. nupharidine, desoxynupharidine, nupharine, nupachristine, thiobinuphaidine), tannins, mucilage.

Food

The large rhizomes make a passable food if properly prepared, but can be variably toxic if not. The most important element in rendering them palatable and safe is *baking*, as a mid-1980s experiment with rodents—reported in the *Journal of Ethnopharmacology*—demonstrated.*(Airaksinen et al 1986:273-96)* This study revealed that a mere

324

boiling of *Nuphar luteum* poorly eliminated its toxins, while baking it in a 180-200° C [212-232° F] oven rendered it "well tolerated" by the rodents.

One method for human consumption claimed successful by some begins with a boiling of the mass for about twenty minutes. Next, the water is changed and the rhizome is peeled down to its core. Then, the core is either steamed for ten minutes or boiled in two waters. After this, it is *baked in a slow oven*. Another method, popular in a wilderness setting, involves leaving the whole, scrubbed rootstock wrapped in foil in hot campfire ashes for the day. By twilight, it is thought to be sufficiently cooked, enabling it to be peeled and eaten at this time.

There are variables, however, which make safe and effective preparation of this plant's rhizome an inexact science at best, meaning that *there is an element of risk involved*. Hence, I cannot recommend that the neophyte try to prepare yellow pond-lily rhizome as a food. My opinion is that a person wishing to try this repast would be wise to partake of it only from the hand of a cook skilled and practiced in its preparation, until such a time as he or she has carefully learned the technique from his or her teacher. (See further under "Cautions," below.)

The elusive seeds, however, are not viewed as toxic and are frequently prepared by wild-foods foragers as "wild popcorn." To accomplish this successfully takes some skill and practice, however. First, the mature seed pods need to be cut free from the plant. (These fruits are frequently swarming with insects. Scatter them by placing fruits in a container, then beating container with a stick and chanting: "Bugs, bugs, go away....Leave and live another day!" Before you know it, you'll feel like the Pied Piper of the Pond Lilies!) Next, the seed pods should be dried in the sun. Finally, they need to be pounded to loosen the seeds. (If still soft, however, the seeds can sometimes be rubbed free or removed individually by hand.)

The fully dried seeds can be prepared like popcorn *the old-fashioned way*, that is, in a pan and NOT in a modernized popcorn popper! Here's how it is usually done: In a skillet, add vegetable oil and seeds to the tune of one or two layers. Place over high heat on the stovetop. Move (but *do not shake*) pan as necessary in order to avoid burning the popping kernels. When finished, and cooled as necessary, eat the kernels but not the shells. Although most would rank this snack a bit below genuine popcorn in taste, you may still find yourself pleasantly surprised at the flavor. Famed wild-foods connoissuer Euell Gibbons often prepared and served this treat to youngsters and he says that they always wanted more!

The seeds can also be parched, winnowed, and ground into flour. This flour can also be added to soup as a thickener or made into a fair gruel.

Health/Medicine

Yellow pond lily possesses **astringent** properties*(Tierra 1992:341)* and has been used by various peoples to treat **BLEEDING, DIARRHEA, DYSENTERY, URETHRITIS, GLEET,** and **VAGINAL DISCHARGES.** It has also been appreciated for **PAIN** control, as a topical *analgesic*. The Upper Tanana Indians, for instance, implemented a poultice of the sliced and warmed rhizome to control pain,*(Kari 1985:17)* while the Okanagan-Colville Indians applied the stems to an *ACHING TOOTH* to ease the agony of the sufferer.*(Turner et al 1980:110)* The Bella Coola Indians discovered that a decoction of the plant's rhizome soothed the pain of tuberculosis, heart disease, rheumatism, and gonorrhea, as well as "pain in any part of the body."*(Smith 1929:56)*

The analgesic effects may be due, in certain instances, to either an *antispasmodic* element*(Tierra 1992:341)* or an *anti-inflammatory* property possessed by this herb due to its content of steroidal saponins.*(Foster & Duke 1990:88)* Indeed, many inflammatory conditions were treated with yellow pond-lily by a variety of peoples, and most especially an *INFLAMED UTERUS*. The Quinault and Shushwap Indians employed a poultice of the heated rhizomes for *RHEUMATISM*.*(Gunther1973:29;Palmer1975:64)* The Kwakiutl Indians found the rhizomes helpful for *INTERNAL SWELLINGS* and for "sickness in the bones" (arthritis??)*(Turner et al 1973:287)* The Potawatomi Indians pounded the rhizome and used it fresh or dry for "many inflammatory diseases."*(Smith 1933:65)*

The most prominent use to which this plant has been put is as a *vulnerary*. Numerous Amerindian tribes used it as such, in various ways: A poultice of the baked rhizome was applied to *SORES* by the Flathead Indians.*(Hart 1992:33)* The Micmac Indians pounded the rhizome into a mash and applied it to *SWOLLEN LIMBS*.*(Lacey 1993:72)* The Woods Cree Indians grated the rhizome, as well as that of both calamus *(Acorus calamus)* and cow parsnip *(Heracleum lanatum)* *(see under that heading)*, into a mixture with water or grease to use as a poultice for *HEADACHE, SORE JOINTS, SWELLINGS, PAINFUL LIMBS*, and *WORMS IN THE FLESH*. They also poulticed a fresh or rehydrated segment of the rhizome on *INFECTED SKIN LESIONS* to draw out the pus.*(Leighton 1985:46, 47)* Algonquin Indians in Quebec also utilized a poultice of the boiled and mashed rhizomes for *INFECTIONS* and *SWELLINGS*.*(Black 1980:163)*

The application for infections is especially interesting since modern scientific research has substantiated that yellow pond-lily's alkaloids are strongly *antimicrobial*. For example, in a 1954 study of some 300 plants tested to determine any activity against a virulent human strain of tuberculosis, yellow pond-lily was found to be among the very top botanicals active against this organism!*(Fitzpatrick 1954:529,532)* (Interesting in this re-gard, too, is that the Bella Coola Indians have long used yellow pond-lily rhizomes for *TUBERCULOSIS*.-*Smith 1929:56*) Then, too, the Nitinaht Indians discovered that placing large rhizomes of the plant in hot water and taking the resultant liquid acted as an im-penetrable barrier against *EPIDEMIC DISEASES*.*(Turner et al 1983:114)*

Yellow pond-lily evidently has powerful effects on the reproductive system as well. The Gitskan Indians found that they could rely on either an infusion of the toasted rhizome or a decoction of the heart of the rhizome as a *contraceptive*.*(Smith 1929:56)* Some modern-day herbalists prefer to try it as an initial treatment for *menopausal symptoms*. Abnaki Indians even implemented an infusion of the rhizome as an *anaphrodisiac*—especially for men, for whom it was said to curtail the sex drive for a full two months.*(Rousseau 1947:154, 167)* One authority informs us that in some lands medical doctors still use this plant's rhizomes to curtail ***EXCESSIVE SEXUAL EXCITATION*** and ***PREMATURE EJACULATION.****(Stary 1983:140)*

The Iroquois Indians used an infusion of the dried, grated plant for ***HEART PROBLEMS.****(Herrick 1977:319)* They also combined an infusion of the rhizome with those of two other plants to improve the speed of blood circulation in adolescents.*(Rousseau 1945:43)* These vascular functions are no doubt due to the alkaloids, which are ***cardio-active*** as well as ***vasoconstrictive****.(Foster & Duke 1990:88)*

From the standpoint of energetics, yellow pond-lily is bitter and neutral in temperature and affects the stomach, spleen, and liver meridians.*(Tierra 1992:341)*

As to form and dosage usually used, herbalist Michael Moore lists a 1:2 tincture of the fresh root, at a dose of 10-20 drops, or a weak decoction of the rhizome, at a dose of 2-4 ounces—either up to three times a day.*(Moore 1994:17)*

Cautions

* Due to the unpredictable content and nature of yellow pond-lily's alkaloids, this plant is potentially toxic. With the exception of the popcorn-like snack that can be prepared from the seeds as described above, any consumption of the plant should be with caution and is probably best reserved for emergency use only. As for herbal use in health conditions, only professionally prepared extracts of the roots should be used and their intake should best be supervised by an herbal professional.

* Yellow pond-lily should ***not*** be used during **PREGNANCY** or **LAC-TATION.**

YELLOW WOOD-SORREL
[COMMON WOOD-SORREL; SOURGRASS; YELLOW OXALIS]
(Oxalis stricta; O. dillenii)

Description

[See photo #171 in rear of book.]

A delicate, perennial herb, with an upright and freely-branching stem. It grows 4-14 (usually 5-8) inches tall. In *Oxalis stricta*, the stem's hairs are blunt, whereas in *O. dillenii*, they are pointed. Also, the former species spreads from underground stems, whereas the latter lacks these.

Yellow wood-sorrel has alternate, light-green, untoothed leaves that are palmately lobed—that is, they are possessed of three, heart-shaped leaflets notched at their outer edge and sporting a prominent midrib. All in all, the plant looks very much like common white clover. In fact, some feel that the Irish shamrock was based on the *Oxalis* genus rather than on clovers as commonly believed.

Two to five attractive, yellow flowers grow on each plant. They are each possessed of five petals, five sepals, and ten stamens.

The flowers are replaced by pencil-shaped seedpods. These fruits are greenish-tan and five-celled, with two or more seeds in each cell. In *O. stricta*, they are hairless (glabrous) or widespreading, whereas in *O. dillenii*, they are pubescent with appressed hairs (occasionally with some spreading hairs). Also, the seed pods of the former species have straight stalks, whereas those of the latter have bent stalks.

Range and Habitat

Cosmopolitan weed that is especially concentrated in southern Canada and the eastern two-thirds of the USA. In our area, *O. stricta* is ubiquitous, whereas *O. dillenii* is scattered throughout the southern counties.

Chief or Important Constituents

Oxalic acid, potassium oxalate, flavonoids(orientin, isovitexin, vitexin), mucilage. Nutrients include high amounts of vitamin C and phosphorous.

Food

A wonderful trail-nibble, this delicate little plant—full of a zestful, lemony flavor due to its oxalic-acid content—was appreciated as such by the Kiowa, Cherokee, Omaha, Pawnee, and other Indian tribes.

It is also a popular salad ingredient for wild-foods foragers—especially the seed-pods, which some liken to sour pickles. *Oxalis*-soup recipes abound in the foraging manuals, too, but these should be treated with caution (*see under* "Sheep Sorrel").

Woodsmen appreciate it as a stuffing for fish or made into a sauce as a condiment for same. The plant also makes an excellent stuffing for muskrat, beaver, porcupine, and other wild game. The Potawatomi Indians made a dessert out of it. *(Smith 1933:106)*

Health/Medicine

Noted European horticulturist John Gerard highlighted the main uses of this plant in reviewing the European species *(O. acetosella),* which he recommended for "pestilential fevers," "sicke & feeble stomach," and "ulcers of the mouth." *(Gerard 1975:1031)*

Looking at these recommendations in the light of study and experience since, we note the following: The plant, possessed of cooling energies (i.e., a *refrigerant--Johnson 1884:115; Wren 1988:287;de Bairacli Levy 1974:156*), is indeed a useful *antipyretic,(Johnson 1884:115; Stuart 1982:104)* especially for fevers producing a rash or pimples. *(de Bairacli Levy 1974:156)* It was much appreciated as a fever herb by early American healers, *(Crellin&Philpott 1989:459)* as well as by the Iroquois Indians. *(Herrick 1977:365)* As to the second application mentioned by Gerard, yellow wood-sorrel is renowned as a *stomachic,(Grieve 1971:2:752)* proving clinically effective for **HEARTBURN, STOMACH CRAMPS,** and various other **DIGESTIVE DISORDERS.** *(Brown 1985:215)* The Iroquois found it quite useful for digestive afflictions, esp. for **NAUSEA, CRAMPS,** and **DIARRHEA.***(Herrick 1977:365)* And as to Gerard's third recommendation, modern herbalists have indeed found an *Oxalis* mouthwash helpful for various problems in the oral cavity, *(Brown 1985:215)* especially for **CANKERS.** *(Grieve 1971:2:752; Bolyard 1981:102)* The Iroquois likewise appreciated an infusion of this plant as a mouthwash.*(Herrick 1977:366)*

It is probable that an *anti-inflammatory* effect accounts at least partly for the digestive and oral problems described above. *Oxalis* juice has traditionally been used as a compress for **INFLAMMATION,***(Grieve 1971:2:752)* which, in an energetic model, is a heat condition that could be offset by the plant's *refrigerant* properties. Likewise, the Omaha Indians poulticed the fresh herb onto **SWELLINGS**.*(Gilmore 1913:335)* Other probable anti-inflammatory applications are as follows: Naturalist-herbalist Tom Brown Jr. has found a strong tea useful as a wash for **SKIN AFFLICTIONS**.*(Brown 1985:215)* A noted European herbalist recommends it topically for **VARICOSE VEINS**, to be held on by cabbage leaves and an upper cotton bandage and left on overnight.*(de Bairacli Levy1974:156)* Herbalists also use it for "**HANGOVER" HEADACHES**.*(Stuart 1982:104)*

There is evidently an *antimicrobial* factor to this plant as well. It has, for one thing, traditionally been used as an *antiseptic*.*(de Bairacli Levy1974:156)* Lending support to this use, one study found *Oxalis stricta* to strongly inhibit the tuberculosis bacterium *in vitro*.*(Fitzpatrick 1954:530)* (Interesting here is that early-American physicians applied the herb as a poultice to "scrofulous ulcers."*- Crellin&Philpott 1989:459*) In the UK, it is used topically for **SCABIES**.*(Stuart 1982:104)*

Historically prominent has been its celebrated *diuretic* assist (despite the high oxalate content which would contradict extended use that could actually result in kidney damage), an application much appreciated by early-American physicians.*(Bolyard 1981:102; Crellin&Philpott 1989:459)* And application for "urinary disorders" continues up until the present day in many circles.*(Wren1988:287)* Thus, herbalist Peter Holmes notes that yellow wood-sorrel "promotes urination" and is especially helpful for what Traditional Chinese Medicine (TCM) calls *kidney-Qi stagnation,* manifested by fetid stool, a lackluster appetite, and dry skin. He also suggests it for *kidney-fire* conditions such as *ACUTE NEPHRITIS*.*(Holmes 1997:2:651)*

Oxalis was also popular in early America as a topical treatment for *SKIN CANCERS*.*(Crellin&Philpott 1989:459)* This tradition has been perpetuated in Appalachian folk medicine. In this regard, naturalist and botanist Judith Bolyard interviewed Kentucky minister Alvin Boggs, a school superintendent, who relayed his family recipe to her: Boggs said that two bushels of the fresh plant's leaves and stems were boiled in water until a small black lump remained. This was applied daily to the skin cancer for two or three weeks, until it disappeared.*(Bolyard 1981:102)* It is of possible interest here that other plants containing high amounts of oxalic acid have also been used for skin cancers. One of these, sheep sorrel *(Rumex acetosella)* is even a component in the famous Essiac® herbal mix so popular with cancer patients in North America.

Finally, being high in vitamin C, "the fresh plant, eaten raw, is useful in *scurvy."* *(Wood et al 1926:1415)*

Cautions: In view of the high content of soluble oxalates and oxalic acid, consumption of the fresh herb must be restricted to light trail nibbling or sparing use as a salad ingredient. Nor should one use the extracts for medicine for an extended time. Also, for the same reason, this herb should probably not be used if there is *gout, rheumatism,* or a tendency toward *kidney stones of the oxalate variety*

1. **Arrowhead** *(Sagittaria latifolia)*

2. **Arrowhead** (flower close-up)

3. **Aster** *(Aster)*

4. **Aster** *(Aster)*

5. **Bloodroot** *(Sanguinaria canadensis)*

7. **Bluebead Lily** (fruiting stalk)

6. **Bluebead Lily** *(Clintonia borealis)*

8. **Blue Cohosh** *(Caulophyllum thalictroides)*

9. **Blue Cohosh** (fruiting stalk)

10. **Blue-eyed Grass** *(Sisyrinchium)*

11. **Blue Flag** *(Iris versicolor)*

12. **Blue Vervain** *(Verbena hastata)*

13. **Blue Vervain** (flower close-up)

14. **Blue Vervain** (fall-winter seedhead)

15. **Boneset** *(Eupatorium perfoliatum)*   16. **Boneset** (flower close-up)

17. **Bouncing Bet** *(Saponaria officinalis)*

18. **Bugleweed** *(Lycopus virginianus)*

19. **Bugleweed** *(Lycopus uniflorus)*

20. **Bulrush** *(Scirpus)*

21. **Bulrush** *(Scirpus validus)* (close-up)

22. **Bunchberry** *(Cornus canadensis)*   23. **Bunchberry** (fruiting stage)

24. **Burdock** *(Arctium minus)* (immature)   25. **Burdock** (flowering stalk)

26. **Butter & Eggs** *(Linaria vulgaris)*

28. **Butterfly-weed** (flower close-up)

27. **Butterfly-weed** *(Asclepias tuberosa)*

29. **Catnip** *(Nepeta cataria)* (immature)

30. **Catnip** (flowering stage)

31. **Cattail** *(Typha angustifolia)*

32. **Cattail** *(Typha latifolia)*

33. **Chickweed** *(Stellaria aquatica)*

34. **Chickweed** (flower close-up)

35. **Chicory** *(Cichorium intybus)*

36. **Chicory** (flower close-up)

37. **Cinquefoil** *(Potentilla simplex)*

38. **Cinquefoil** (leaf close-up)

39. **Clearweed** *(Pilea pumila)*

40. **Clearweed** (stem/flower)

41. **Cleavers** *(Galium aparine)*

42. **Cleavers** (flower close-up)

43. **Cleavers** (fruit close-up)

44. **Columbine** *(Aquilegia canadensis)*

45. **Columbine** (flower close-up)

46. **Cow parsnip** *(Heracleum lanatum)*

48. **Creeping Charlie** *(Glechoma hederacea)*

47. **Cow parsnip** (leaf close-up)

49. **Culver's Root** *(Veronicastrum virginicum)*   50. **Culver's Root** (flower close-up)

51. **Dandelion** *(Taraxacum officinale)*

52. **Dock** *(Rumex crispus)* (spring rosette)

53. **Dock** (flowering stage)

54. **Dock** (fruiting stage)

55. **Evening Primrose** *(Oenothera biennis)*

56. **Evening Primrose** (flowering stage)

57. **Evening Primrose** (flower stage)

58. **False Solomon's Seal** *(Smilacina racemosa)*

59. **False Solomon's Seal** (fruiting stage)

60. **Star-flowered False Solomon's Seal** *(Smilacina stellata)*

61. **Star-flowered False Solomon's Seal** (fruit)

62. **Fireweed** *(Epilobium angustifolium)*

63. **Fireweed** (flower close-up)

64. **Fireweed** (autumn remnants)

65. **Fleabane** *(Erigeron philadelphicus)*

66. **Fleabane** (flower close-up)

67. **Fragrant Giant Hyssop** *(Agastache foeniculum)*

68. **Fragrant Giant Hyssop** (flower close-up)

69. **Late Goldenrod** *(Solidago gigantea)*

70. **Stiff Goldenrod** *(Solidago rigida)*

71. **Showy Goldenrod** *(Solidago speciosa)*

72. **Gumweed** *(Grindelia squarrosa)*

73. **Hawk's beard** *(Crepis)*

74. **Orange Hawkweed** *(Hieracium aurantiacum)*

75. **Orange Hawkweed** (flower close-up)

76. **Horsemint** *(Monarda punctata)*

77. **Jerusalem Artichoke** *(Helianthus tuberosus)*

78. **Jewelweed** *(Impatiens capensis)* (immature form)

79. **Jewelweed** (mature form)

80. **Jewelweed** (close-up of translucent stem)

81. **Jewelweed** (flower close-up)

82. **Pale Jewelweed** *(I. pallida)* (flower close-up)

83. **Joe-Pye Weed** *(Eupatorium maculatum)*

84. **Joe-Pye Weed** (leaf/stem close-up)

85. **Joe-Pye Weed** (late-flowering stage)

86. **Knotweed** *(Polygonum aviculare)*

87. **Knotweed** (close-up)

88. **Lady's Thumb** *(Polygonum persicaria)*

89. **Lady's Thumb** (leaf close-up)

90. **Lamb's Quarters** *(Chenopodium album)* 91. **Lamb's Quarters** (mature plant)

92. **Marsh Marigold** *(Caltha palustris)*

93. **Milkweed** *(Asclepias syriaca)*

94. **Milkweed** (fruit close-up)

95. **Monkey-flower** *(Mimulus ringens)*

96. **Monkey-flower** (flower close-up)

97. **Motherwort** *(Leonurus cardiaca)*   98. **Mountain mint** *(Pycnanthemum virginianum)*

99. **Mullein** *(Verbascum thapsus)*   100. **Mullein** (2d-year plant)   101. **Mullein** (flower close-up)
(spring rosette)

102. **Nutgrass** *(Cyperus esculentus)*

103. **Pearly Everlasting** *(Anaphalis margaritacea)*

104. **Pennycress** *(Thlaspi arvense)*

105. **Pennycress** (fruiting plant)

106. **Penstemon** *(Penstemon grandiflorus)*

107. **Peppergrass** *(Lepidium)*

108. **Peppergrass** (leaf close-up)

109. **Peppergrass** (fruit close-up)

110. **Pigweed** *(Amaranthus retroflexus)*

112. **Plantain** *(Plantago major)*

111. **Pineapple-weed** *(Matricaria matricarioides)*

113. **Prairie Smoke** *(Geum triflorum)* (flowering colony)

114. **Prairie Smoke** (seed-head close-up)

115. **Prairie Smoke** (leaf close-up)

116. **Puccoon** *(Lithospermum)*

117. **Purple Loosestrife** *(Lythrum salicaria)*

118. **Purslane** *(Portulaca oleracea)*

119. **Queen Anne's Lace** *(Daucus carota)*

120. **Queen Anne's Lace** (flowerhead close-up)

121. **Red Clover** *(Trifolium pratense)*

122. **Self-heal** *(Prunella vulgaris)*

123. **Sheep Sorrel** *(Rumex acetosella)*

124. **Sheep Sorrel** (female flowered plant)

125. **Sheep Sorrel** (leaf close-up)

126. **Shepherd's Purse** *(Capsella bursa-pastoris)*

127. **Shooting Star** *(Dodecatheon)*

128. **Skullcap** *(Scutellaria epilobiifolia)*

129. **Pale Smartweed** *(Polygonum lapathifolium)*

130. **Swamp Smartweed** *(P. coccineum)*

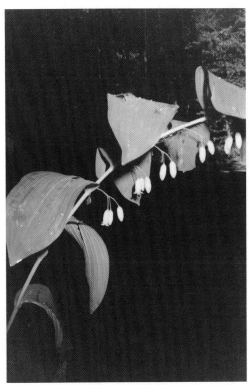

131. **Solomon's Seal** *(Polygonatum)*

132. **Solomon's Seal** (fruiting stage)

133. **Spiny-leaved Sow Thistle** *(Sonchus asper)*

134. **Marsh Sow Thistle** *(S. uliginosus)*

135. **Marsh Sow Thistle** (Leaf close-up)

136. **Spiderwort** *(Tradescantia)*

137. **Spiderwort** (closed-flower stage)

138. **Starflower** *(Trientalis borealis)*

139. **Stinging Nettle** *(Urtica dioica)* (immature)

140. **Stinging Nettle** (flowering plant)

141. **Sunflower** *(Helianthus annuus)*

142. **Swamp Milkweed** *(Asclepias incarnata)*

143. **Sweet Cicely** *(Osmorhiza)*

144. **Sweet Everlasting** *(Gnaphalium)*

145. **Sweet Everlasting** (flower close-up)

146. **Canada Thistle** *(Cirsium arvense)*

147. **Canada Thistle** (flowering stage)

148. **Canada Thistle** (flower closeup)

149. **Bull Thistle** *(Cirsium vulgare)*

150. **Turtlehead** *(Chelone glabra)*

151. **Yellow Violet** *(Viola sp.)*

152. **Common Blue Violet** *(Viola sororia)*

153. **Virginia Waterleaf** *(Hydrophyllum virginianum)*

154. **Virginia Waterleaf** (flowering plant)

155. **Watercress** *(Nasturtium officinale)*

156. **Wild Bergamot** *(Monarda fistulosa)*

157. **Wild Geranium** *(Geranium maculatum)*

158. **Wild Mint** *(Mentha arvensis)*

159. **Wild Mustard** *(Brassica)*

160. **Wild Onion** *(Allium stellatum)*

161. **Wild Sarsaparilla** *(Aralia nudicaulis)*

162. **Wild Strawberry** *(Fragaria virginiana)*

163. **Wild Strawberry** (fruiting plant)

164. **Wood Lily** *(Lilium philadelphicum)*

165. **Wood Nettle** *(Laportea canadensis)*

166. **Wormwood** *(Artemisia absinthium)*

167. **Yarrow** *(Achillea millefolium)*

168. **Yellow Goatsbeard** *(Tragopogon)*

169. **Yellow Loosestrife** *(Lysimachia)*

170. **Yellow Pond-lily** *(Nuphar)*

171. **Yellow Wood-sorrel** *(Oxalis)*

ADAPTOGEN

(Favorably Modifies Stress Response)

Starflower?
Sweet Cicely?

ALTERATIVE

(See *Depurative*)

ANALGESIC/ ANODYNE

(Reduces Pain)

Blue Vervain
Bunchberry
Cattail
Mullein
Penstemon
Violet
Virginia Waterleaf?
Wild Mint
Wild Onion
Yarrow
Yellow Pond-Lily

ANAPHRODISIAC

(Inhibits Sexual Desire)

Yellow Pond-Lily

ANTHELMINTIC

(Expels/Destroys Parasites)

Boneset
Horsemint
Peppergrass
Pineapple-weed
Queen Anne's Lace
Purslane
Self-heal
Shepherd's Purse
Swamp Milkweed

Thistle
Wormwood
Yarrow

ANTI-ANAEMIC

(Maintains Blood-Iron Levels)

Chickweed
Dandelion
Dock
Jerus. Artichoke
Marsh Marigold
Stinging Nettle
Watercress
Wild Strawberry
Yarrow

ANTIBACTERIAL/ ANTIMICROBIAL

(Inhibits/Reduces Microbes)

Bloodroot
Blue-eyed Grass
Blue Flag
Boneset
Bouncing Bet
Burdock
Butter-and-Eggs
Butterfly-weed
Catnip
Cattail
Chickweed
Chicory
Cleavers
Columbine
Creeping Charlie
Dandelion
Dock
False Solomon's Seal
Fireweed
Hawkweed?
Horsemint
Lamb's Quarters

Mullein
Penstemon
Peppergrass
Plantain
Purple Loosestrife
Purslane
Queen Anne's Lace
Red Clover
Self-heal
Shepherd's Purse
Solomon's Seal
Watercress
Wild Geranium
Wild Mint
Wild Mustard
Wormwood
Yarrow
Yellow Pond-lily
Yellow Wood-Sorrel

ANTICATARRHAL

(Reduces Mucosal Inflammation)

Boneset
Creeping Charlie
Knotweed
Milkweed
Plantain
Swamp Milkweed
Sweet Everlasting

ANTICOAGULANT

(Reduces Blood Clotting)

Cattail

ANTICONVULSIVE

(Inhibits Tendency to Convulse/Seizure)

Blue Vervain?
Bunchberry?
Marsh Marigold

Self-heal
Skullcap

ANTIDIARRHEAL

(See *Astringent*)

ANTI-EMETIC

(Inhibits Nausea/Tendency to Vomit)

Cow Parsnip
Horsemint
Wild Mint

ANTIFUNGAL

(Inhibits/Reduces Fungi)

Bloodrrot
Burdock
Clearweed?
Cow Parsnip
Dandelion
Fireweed
Goldenrod
Horsemint
Jewelweed
Milkweed
Plantain
Queen Anne's Lace
Red Clover
Wild Bergamot
Wild Mint
Yarrow

ANTIGALACTIC

(Reduces Breastmilk Production)

Arrowhead

ANTIHISTAMINE

(Inhibits Allergenic Release of Histamine from Mast Cells)

Arrowhead
Plantain
Stinging Nettle
Wild Onion

Yarrow

ANTI-INFLAMMATORY

(Reduces Inflammation)

Bloodroot
Blue Cohosh
Blue Vervain
Boneset
Bouncing Bet?
Bulrush
Bunchberry
Burdock
Butter-and-Eggs
Chickweed
Chicory
Cinquefoil
Clearweed
Cleavers
Cow Parsnip
Creeping Charlie
Dandelion
Evening Primrose
False Solomon's Seal
Fireweed
Fleabane
Goldenrod
Gumweed
Hawkweed
Joe-Pye Weed
Knotweed
Lamb's Quarters
Motherwort
Mullein
Penstemon
Plantain
Self-heal
Shepherd's Purse
Smartweed
Solomon's Seal
Stinging Nettle
Sunflower
Sweet Everlasting
Thistle

Violet
Wild Mint
Wild Onion
Wild Strawberry
Yarrow
Yellow Goatsbeard
Yellow Loosestrife
Yellow Pond-Lily
Yellow Wood-Sorrel

ANTILITHIC

(Prevents or Reduces/Eliminates Calculi[Stones] in the Biliary or Urinary System)

Bluebead Lily?
Blue Vervain
Cleavers
Creeping Charlie
Goldenrod
Joe-Pye Weed
Knotweed
Wild Strawberry
Yellow Goatsbeard

ANTIMALARIAL

(Inhibits Malarial Organisms or Symptoms of Malarial Infection)

Boneset
Sunflower
Wormwood
Yarrow

ANTINEOPLASTIC

(Inhibits Cancerous Cells)

Bloodroot
Boneset
Bouncing Bet
Bunchberry?
Burdock
Cleavers?
Dandelion
Gumweed?
Plantain
Red Clover

Self-Heal
Sheep Sorrel
Shepherd's Purse
Spiderwort?
Violet
Wild Mint?
Yarrow
Yellow Pond-Lily?
Yellow Wood-Sorrel?

ANTIOXIDANT

(Enhances Body's Scavenging of
Free Radicals)

Motherwort
Self-heal

ANTIPARALYTIC

(Reduces/Inhibits Paralysis)

Bunchberry?
Wild Mustard?

ANTIPHLOGISTIC

(See *Anti-Inflammatory*)

ANTIPRURITIC

(Reduces/Inhibits Itching)

Plantain

ANTIPYRETIC

(Reduces Fever)

Arrowhead
Blue Cohosh
Blue Vervain
Boneset
Bunchberry
Butterfly-weed
Catnip
Chickweed
Cinquefoil
Cleavers
Fleabane
Goldenrod

Horsemint
Lamb's Quarters
Mountain Mint
Pineapple-weed
Plantain
Self-heal
Sheep Sorrel
Sunflower
Sweet Everlasting
Violet
Watercress
Wild Mint
Wild Strawberry
Wormwood
Yarrow
Yellow Wood-sorrel

ANTIRHEUMATIC

(Reduces Rheumatic
Inflammation)

Boneset
Burdock
Butterfly-weed
Chickweed
Cow Parsnip
False Solomon's Seal
Gumweed
Joe-Pye Weed
Solomon's Seal
Stinging Nettle
Sunflower
Wild Strawberry
Yarrow
Yellow Pond-Lily

ANTISCORBUTIC

(Prevents Vitamin-C Deficiency)

Chickweed
Cleavers
Columbine
Creeping Charlie
Lamb's Quarters
Peppergrass
Sheep Sorrel

Watercress
Wild Strawberry
Yellow Goatsbeard
Yellow Wood-Sorrel

ANTISPASMODIC

(Inhibits/Reduces Spasms)

Blue Cohosh
Blue Vervain
Butterfly-weed
Catnip
Cinquefoil
Cow Parsnip
Evening Primrose
False Solomon's Seal
Fireweed
Fleabane
Goldenrod
Gumweed
Marsh Marigold
Queen Anne's Lace
Red Clover
Self-Heal
Skullcap
Sweet Everlasting?
Thistle
Wormwood
Yarrow
Yellow Pond-Lily

ANTITUSSIVE

(Inhibits/Reduces Coughing)

Blue Vervain
Burdock
Chickweed
Gumweed
Peppergrass
Red Clover
Violet
Wild Mint
Yarrow

ANTIVIRAL

(Inhibits/Reduces Viral Infections)

Boneset
Burdock
Fireweed
Fragrant Giant Hyssop
Self-heal
Wild Geranium
Yarrow

APERIENT

(See *Laxative*)

APHRODISIAC

(Stimulates Sexual Desire)

Cow Parsnip
Peppergrass
Watercress
Wild Mint

APPETITE STIMULANT

(Stimulates Hunger)

Bunchberry
Turtlehead
Yellow Goatsbeard

ASTRINGENT

(Contracts/Tightens Mucous
Membrane or other Tissue)

Aster
Bugleweed
Bulrush
Butter-and-Eggs
Catnip
Cattail
Cinquefoil
Cleavers
Columbine
Creeping Charlie
Dock
Evening Primrose
Fireweed

Fleabane
Goldenrod
Hawkweed
Knotweed
Lady's Thumb
Lamb's Quarters
Monkey-flower
Motherwort
Mullein
Pearly Everlasting
Pigweed
Plantain
Prairie Smoke
Puccoon
Purple Loosestrife
Purslane
Self-heal
Sheep Sorrel
Shepherd's Purse
Smartweed
Sweet Cicely
Sweet Everlasting
Thistle
Virginia Waterleaf
Wild Geranium
Wild Mint
Wild Strawberry
Yarrow
Yellow Loosestrife
Yellow Pond-Lily
Yellow Wood-Sorrel

CARDIOTONIC

(Enhances cadiac tone/function)

Bloodroot
Bluebead Lily
Bugleweed
Chicory
False Solomon's Seal
Gumweed
Milkweed
Motherwort
Solomon's Seal
Sunflower

Violet
Watercress?
Wild Mint
Yellow Pond-Lily

CARMINATIVE

(Inhibits/Reduces the Spasms that
Cause Intestinal Gas)

Bunchberry
Catnip
Cow Parsnip
Creeping Charlie
Goldenrod
Gumweed
Horsemint
Motherwort
Mountain Mint
Pineapple-weed
Self-heal
Wild Mint
Wild Strawberry
Wild Bergamot
Yarrow

CATHARTIC

(See *Laxative*)

CHOLAGOGUE

(Stimulates the Flow of Bile)

Blue Flag
Burdock
Chicory
Dandelion
Watercress
Wormwood
Yarrow
Yellow Goatsbeard

CHOLERETIC

(Stimulates the
Production/Excretion of Bile)

Culver's Root
Dandelion
Yarrow

CONTRACEPTIVE

(Inhibits Conception)

False Solomon's Seal
Puccoon
Turtlehead
Wild Geranium
Yellow Pond-Lily

DECONGESTANT

(Relieves or Reduces Congestion)

Catnip
Creeping Charlie
Horsemint
Mountain Mint
Mullein
Wild Bergamot
Wild Mint

DEMULCENT

(Locally Soothes Sore/Irritated Mucous Membranes)

Chickweed
Evening Primrose
Gumweed
Joe-Pye Weed
Mullein
Plantain
Solomon's Seal
Violet

DEOBSTRUENT

(Removes Obstruction)

Butter-and-Eggs
Peppergrass

DEODERANT

(Inhibits/Reduces Body Odor)

Cleavers
Wild Mint

DEPURATIVE

(Cleanses the Blood/Lymph)

Blue Flag
Burdock
Chickweed
Cleavers
Dandelion
Dock
Evening Primrose
Pigweed
Plantain
Red Clover
Stinging Nettle
Thistle
Violet
Wild Sarsaparilla

DIAPHORETIC

(Stimulates Sweating)

Aster
Blue Vervain
Boneset
Burdock
Butterfly-weed
Catnip
Columbine
Fleabane
Goldenrod
Horsemint
Joe-Pye Weed
Milkweed
Motherwort
Pineapple-weed
Sheep Sorrel
Sweet Everlasting
Wild Bergamot
Wild Sarsaparilla
Yarrow

DISCUTIENT

(Scatters Tumor or Coagulation)

Bloodroot
Burdock
Cleavers
Dock
Red Clover

Self-heal
Sheep Sorrel
Shepherd's Purse
Violet
Yellow Wood-Sorrel

DISINFECTANT

(Inhibits/Reduces Infection)

Burdock
Cattail
False Solomon's Seal
Pigweed
Plantain
Wild Mint
Wormwood
Yarrow
Yellow Wood-Sorrel

DIURETIC

(Stimulates Reduction of Water from Body Tissues by Enhancing Elimination through the Kidneys)

Arrowhead
Blue Cohosh
Blue Flag
Blue Vervain
Bouncing Bet
Burdock
Butter-and-Eggs
Cattail
Chickweed
Chicory
Columbine
Creeping Charlie
Dandelion
Fleabane
Goldenrod
Hawkweed
Jerusalem Artichoke
Joe-Pye Weed
Knotweed
Lady's Thumb
Milkweed
Peppergrass

Queen Anne's Lace
Plantain
Purslane
Sheep Sorrel
Shepherd's Purse
Smartweed
Swamp Milkweed
Stinging Nettle
Sunflower
Thistle
Watercress
Wild Bergamot
Wild Strawberry
Wood Nettle
Wormwood
Yarrow
Yellow Wood-Sorrel

EMETIC

(Stimulates Vomiting)

Bloodroot
Blue Cohosh
Blue Flag
Blue Vervain
Boneset
Butterfly-weed
Peppergrass
Solomon's Seal
Swamp Milkweed
Thistle

EMMENAGOGUE

(Stimulates Menstruation)

Bloodroot
Blue Cohosh
Blue Vervain
Catnip
Fleabane
Goldenrod
Horsemint
Joe-Pye Weed
Marsh Marigold
Motherwort
Mountain Mint

Pineapple-weed
Queen Anne's Lace
Skullcap
Smartweed
Wild Strawberry
Wormwood
Yarrow
Yellow Wood-Sorrel

ESCHAROTIC

(Locally Dissolves Tissue)

Bloodroot
Violet

EXPECTORANT

(Stimulates Coughing/Sneezing to
Eliminate Excess Respiratory
Mucus)

Bloodroot
Blue Vervain
Boneset
Butterfly-weed
Chickweed
False Solomon's Seal
Gumweed
Marsh Marigold
Milkweed
Mullein
Peppergrass
Plantain
Red Clover
Sunflower
Violet
Watercress
Wild Onion

FEBRIFUGE

(See *Antipyretic*)

GALACTAGOGUE

(Stimulates Breastmilk)

Blue Vervain
Dandelion
Pineapple-weed

Stinging Nettle

HEMOSTATIC/ STYPTIC

(Inhibits Bleeding)

Blue Vervain
Bugleweed
Bulrush
Cattail
Cleavers
False Solomon's Seal
Fleabane
Goldenrod
Knotweed
Pigweed
Plantain
Purple Loosestrife
Self-heal
Sheep Sorrel
Shepherd's Purse
Stinging Nettle
Yarrow
Yellow Pond-lily

HEPATIC

(Enhances Liver Function)

Blue Flag
Blue Vervain
Boneset
Burdock
Butter-and-Eggs
Chickweed
Chicory
Cleavers
Culver's Root
Dandelion
Dock
Lamb's Quarters
Plantain
Self-heal
Thistle
Wormwood
Yarrow
Yellow Goatsbeard

PHYSIOLOGICAL ACTIONS OF THE FLORA OF MINNESOTA & WISCONSIN

HEPATO-PROTECTIVE

(Protects the Liver from Toxins)

Boneset?
Plantain
Queen Anne's Lace

HERPATIC
[alt: HERPETIC]

(Inhibits/Reduces Skin Lesions)

Bloodroot
Bluebead Lily
Burdock
Cattail
Chickweed
Cleavers
Dock
Evening Primrose
Gumweed
Lamb's Quarters
Peppergrass
Red Clover
Stinging Nettle
Thistle
Wild Bergamot
Yellow Wood-Sorrel

HORMONAGOGUE

(Affects Hormone Production, Action, or Levels)

Bugleweed
Chickweed

HYPOGLYCEMIC

(Reduces Levels of Blood Glucose)

Bluebead Lily?
Burdock
Dandelion
Fleabane
Jerusalem Artichoke
Joe-Pye Weed
Purple Loosestrife
Solomon's Seal

Sunflower
Stinging Nettle
Wild Geranium

HYPOTENSIVE

(Reduces Arterial Blood Pressure)

Cleavers
Joe-Pye Weed
Purslane
Self-heal
Shepherd's Purse
Thistle
Wild Onion
Yarrow

IMMUNO-STIMULANT

(Stimulates or Enhances Immune-system Function)

Boneset
Bouncing Bet
Self-heal
Stinging Nettle

LAXATIVE

(Stimulates Defecation or Maintains Regular Bowel Movement)

Blue Flag
Blue Vervain
Boneset
Burdock
Chickweed
Chicory
Cow Parsnip
Culver's Root
Dandelion
Dock
False Solomon's Seal
Plantain (Seeds)
Spiderwort
Turtlehead
Violet
Watercress

Wild Mustard

LITHOTRIPTIC

(See *Antilithic*)

LYMPHATIC

(Stimulates or Maintains Healthy Functioning of the Lymphatic System)

Blue Flag
Cleavers
Mullein
Red Clover
Stinging Nettle

MUSCLE RELAXANT

(Relaxes Tight/Spastic Muscles)

Blue Vervain
Purslane
Skullcap

NEPHRITIC

(Enhances Function of the Kidneys)

Bunchberry
Chickweed
Dandelion
Milkweed
Peppergrass
Shepherd's Purse
Spiderwort
Stinging Nettle
Thistle
Watercress
Wild Strawberry
Yellow Wood-Sorrel

NERVINE

(Soothes the Nerves)

Aster
Blue Vervain
Bugleweed

Catnip
Chicory
Cinquefoil
Gumweed
Joe-Pye Weed
Motherwort
Pineapple-weed
Skullcap
Sow Thistle?
Wild Strawberry

OPHTHALMIC

(Enhances/Preserves Eye Health)

Chickweed
False Solomon' s Seal
Lamb's Quarters
Plantain
Purple Loosestrife

OXYTOCIC

(Accelerates Uterine Contraction/Childbirth)

Blue Cohosh

PARTURIENT

(Aids Childbirth)

Blue Cohosh
Bluebead Lily
Motherwort
Shepherd's Purse

PECTORAL

(Relieves Conditions in the Region of the Chest)

Blue Cohosh
Blue Vervain
Butterfly-weed
Gumweed
Jerusalem Artichoke
Marsh Marigold
Mullein
Plantain

Red Clover
Sunflower
Sweet Everlasting
Thistle
Violet
Wild Mint
Wild Mustard
Wild Onion
Wild Sarsaparilla
Yarrow
Yellow Pond-lily

REDUCENT

(Reduces Body Fat and Weight)

Chickweed
Cleavers
Peppergrass

REFRIGERANT

(Cools the System)

Chickweed
Chicory
Cleavers
Fragrant Giant Hyssop
Lamb's Quarters
Purslane
Sheep Sorrel
Violet
Wild Strawberry
Yellow Wood-Sorrel

REPELLENT

(Repels Insects or Vermin)

Bluebead Lily
Mountain Mint

RUBEFACIENT

(Irritates the Skin by Enhancing the Flow of Blood)

Horsemint
Wild Bergamot
Wild Mustard

SEDATIVE

(Tranquilizes the Nervous System)

Blue Vervain
Catnip (Infants)
Chicory
Mothewort
Mullein
Red Clover
Violet

SIALAGOGUE

(Stimulates Salivary Secretion)

Blue Flag

STIMULANT

(Increases Functional Activity, Often of a Particular System, e.g., Circulatory)

Boneset
Cow Parsnip
Gumweed
Horsemint
Mountain Mint
Peppergrass
Queen Anne's Lace
Watercress
Wild Bergamot
Wild Mint
Wild Sarsaparilla
Yarrow

STOMACHIC

(Stimulates or Improves Stomach Function)

Arrowhead
Blue Vervain
Cow Parsnip
Creeping Charlie
Dandelion
Evening Primrose
Fireweed
Gumweed

Lamb's Quarters
Nutgrass
Pineapple-weed
Plantain
Sheep Sorrel
Shepherd's Purse
Solomon's Seal
Spiderwort
Swamp Milkweed
Sweet Everlasting
Wild Mint
Wild Strawberry
Wormwood
Yarrow
Yellow Goatsbeard
Yellow Wood-Sorrel

SUDORIFIC

(See *Diaphoretic*)

TONIC

(Restores Tone to a Particular Body System; Especially Used of the Gastrointestinal System)

Arrowhead
Blue Vervain
Boneset
Chicory
Creeping Charlie
Culver's Root
Dandelion
Fireweed
Goldenrod
Gumweed
Jerusalem Artichoke
Plantain
Red Clover
Solomon's Seal
Starflower
Stinging Nettle
Thistle
Watercress
Wild Mustard
Wild Sarsaparilla

Wormwood
Yarrow
Yellow Goatsbeard

URINARY

(Enhances the Tone or Function of the Urinary System)

Chickweed
Cleavers
Fireweed
Horsemint
Joe-Pye Weed
Knotweed
Lamb's Quarters
Marsh Marigold
Mullein
Plantain
Shepherd's Purse
Spiderwort
Wild Onion
Wild Sarsaparilla
Wild Strawberry
Wood Nettle
Yarrow
Yellow Goatsbeard
Yellow Wood-Sorrel

VASOCONSTRICTIVE

(Constricts the Blood Vessels)

Shepherd's Purse
Yellow Pond-Lily

VASOTONIC

(Improves the Tone of the Vascular System)

Dandelion
Purple Loosestrife
Violet
Yarrow

VERMIFUGE

(See *Anthelmintic*)

VULNERARY

(Helps Heal Wounds)

Bluebead Lily
Blue Flag
Butterfly-weed
Cattail
Chickweed
Cleavers
Cow Parsnip
Dock
Evening Primrose
False Solomon's Seal
Fireweed
Goldenrod
Gumweed
Joe-Pye Weed
Lady's Thumb
Lamb's Quarters
Monkey-flower
Mullein
Penstemon
Pineapple-weed
Plantain
Puccoon
Purple Loosestrife
Purslane
Queen Anne's Lace
Red Clover
Self-heal
Shepherd's Purse
Solomon's Seal
Spiderwort
Stinging Nettle
Sunflower
Wild Bergamot
Wild Onion
Wild Sarasaparilla
Wormwood
Yarrow
Yellow Pond-lily
Yellow Wood-Sorrel

IN VITRO ANTIMICROBIAL ACTIVITY DEMONSTRATED
FOR THE PLANTS DESCRIBED IN THE TEXT
(Bacteria, Fungi, and *Viruses*—Alphabetized by Infectious Organism)

Bacillus subtilis
Burdock
Plantain
Yarrow

Candida albicans
Bloodroot
Dandelion
Fireweed
Goldenrod

Escherichia coli
Burdock
Catnip
Dock
Yarrow

Herpes simplex
Mullein
Self-Heal
Wild Geranium

Human Immunodeficiency Virus
Bloodroot
Burdock
Dock
Self-Heal

Influenza
Mullein
Wild Geranium

Mycobacterium smegmatis
Burdock
Dock
Yarrow

Mycobacterium tuberculosis
Bloodroot
Blue Flag
Blue-Eyed Grass
Bouncing Bet
Butter-and-Eggs
Butterfly-weed

Cattail
Chicory
Cleavers
Columbine
Creeping Charlie
Dock
False Solomon's Seal
Hawkweed
Lamb's Quarters
Milkweed
Mullein
Penstemon
Plantain
Queen Anne's Lace
Red Clover
Solomon's Seal
Wild Geranium
Wild Mustard
Yarrow
Yellow Pond-lily
Yellow Wood-Sorrel

Pseudomonas pyocyanea
Fireweed

Shigella flexneri
Burdock
Dock
Yarrow

Shigella sonneii
Burdock
Dock
Yarrow

Staphylococcus albus
Fireweed

Staphylococcus aureus
Boneset
Burdock
Dock
Fireweed
Plantain
Yarrow

341

HOW HERBAL EXTRACTS MAY BE MADE AT HOME

Below presented are some methods popularly used by herbalists (including myself) to prepare herbal extracts for human use. As per the disclaimer printed on the copyright page of this book, this material is presented for informational purposes only. Any use made of it by the reader, or anyone else, is done at his/her own risk and responsibility.

Infusions

These are aqueous extracts made by steeping the *soft* parts of an herb or herbs (i.e., leaves, flowers) in wat-er that has been heated to the point of boiling, but is not actively boiling.

Supplies Required:
Herb
Mortar and pestle
Pure water
Tablespoon
Teacup
Creamer cup (teacup with a
 pour spout)
Paper coffee filter or cheese-
 cloth
Stoneware or porcelain
 saucer plate
Non-aluminum tea kettle

Measurement Guidelines:
 Usually just one tablespoon of herb to one cup (8oz) of water.

Steps Commonly Taken:

1.) Approximate amount of herb needed is lightly crushed with mortar and pestle.

2.) Required amount of herb is measured and placed into creamer cup (cup with pour spout)

3.) Required amount of water is put into tea-kettle and heated on stove just to point of boiling.

4.) At this point, water is removed from stove and poured into creamer cup un-til a bit below the rim.

5.) Cup is immediately cov-ered with non-plastic, non-metallic saucer plate.

6.) Herb is steeped for 15 minutes to 12 hours, de-pending on kind of herb and preparation needed.

7.) Straining device (coffee filter, cheesecloth, etc.) is fitted over second cup (nor-mal teacup) and secured with rubber band. Center of straining material is slightly depressed with thumb to create a slight indentation without loosening rubber band from edges.

8.) Steeped fluid is carefully poured through filter into cup.

9.) Rubber-banded filter is removed with care (so as not to drop it, or rubber band, into the freshly made tea).

10.) Beverage (cooled to drinkable temp) is imbibed per individual need.

Decoctions

These are aqueous extracts of plants made by *actively boiling* their *hard, tough* parts (root, seeds, bark, fibrous stems) in water.

Supplies Required:
Plant part
Mortar and pestle
Kitchen scale
Pure water
Medium-size saucepan
Tablespoon
Teacup
Creamer cup (teacup with a
 pour spout)
Paper coffee filter or cheese-
 cloth
Stoneware or porcelain
 saucer plate
Non-aluminum tea kettle

Measurement Guidelines
 Usually, one ounce (by weight) of dried herb if using a pint of water, or two ounces by weight of fresh herb to the same amount of water; to make only a cup's worth, generally what is re-quired is one rounded tea-spoon of dried plant part to one cup of water or three lev-el teaspoons of fresh plant part to the same amount of water

Steps Commonly Taken:

1.) Plant part is crushed, or broken up, with mortar and pestle.

2.) Plant part is measured out according to specs and added to saucepan.

3.) Appropriate amount of water is measured out and added to saucepan.

4.) Contents are heated to a boil, then lightly simmered for 15 minutes or more, or until mixture is reduced by 1/3 to 3/4 (depending on de-sired strength).

5.) Results are strained as per procedure described under "Infusions," above.

6.) Tea is allowed to cool to desired temperature and then imbibed as necessary.

Tinctures

These are alcoholic extracts of plants, either fresh or dried as may be required.

Supplies required:
Herb
Alcohol (100-proof vodka)
Glass Jar with Lid
Measuring Cups
Cheesecloth or Tincture
 Press
Large Beaker
Small Amber Glass Bottles
 with Glass Droppers
Stick-on Label and Felt Pen

Measurement Guidelines:
Dried-herb Tincture(1:5ratio = 1 part herb by weight to 5 parts vodka by volume)

This works out on a gram-to-millileter basis as well as correspondingly on an ounce (by weight)-to-ounce(by vol-ume) basis.

Examples (note below how multiplying the grams by 5 equals the millileters and also how multiplying the weight in ounces by 5 equals the volume in ounces):

2 oz (60g) herb to 300ml (10oz) vodka

4 oz (120g) herb to 1 pint (600 ml=20 oz) vodka

8 oz (240g) herb to 1200ml (40oz) vodka

Fresh-herb Tincture(1:2ratio = I part herb by weight to 2 parts vodka by volume)

This works out on a gram-to-millileter basis as well as correspondingly on an ounce (by weight)-to-ounce(by vol-ume) basis.

Examples (note below how multiplying the grams by 2 equals the millileters and also how multiplying the weight in ounces by 2 equals the vol-ume in ounces):

2 oz (60g) herb to 120ml (4oz) vodka

4 oz (120g) herb to 1 cup (240ml=8 oz) vodka

8 oz (240g) herb to 1 pint (480ml =16oz) vodka

Steps Commonly Taken:

1.) Herb is weighed out and placed in bottom of glass jar.

2.) Herb is covered with appropriate amount of vod-ka. (If fresh herbs are used, contents are mixed in blen-der and then returned to jar.)

3.) Marking pen is used to write the current date on the label, which is then affixed to jar.

4.) Jar is sealed by screwing the lid on tightly.

5.) Jar is stored in a dark place. Dried-herb tinctures are shaken 1-2 times a day for 2-4 weeks. Fresh-herb tinctures are simply allowed to sit for 10-14 days. If any evaporation occurs, level is often topped off with more vodka.

6.) When steeping time has expired, fluid is strained through chosen filter (folded cheesecloth or whatever) into beaker. When filtering is done, as much fluid as possible is squeezed out from filter into the beaker as well.

7.) Strained liquid is care-fully poured into tincture bottle and dropper is tightly screwed on.

8.) Contents are described via use of marking pen on stick-on label and this is affixed to tincture bottle.

9.) Tincture is stored in a dark place that will not rise above room temperature and used when necessary.

Compresses

A compress is a cloth or gauze soaked in an herbal in-fusion or decoction (see methods for preparation of these above) which is wrung out a bit and then applied to a body part.

MEASUREMENTS & EQUIVALENTS

Volume Equivalents

1 minim =	1 drop
1 ml =	.0338 fluid ounce
3.7 ml (60 min) =	1 fluid drachm(just under a teaspoonful = 1 & 1/3 fl dr)
15ml =	1/2 fluid oz = 4 fl dr(a tablespoonful)
29.58ml	1 fluid oz = 8 fl dr
118.291ml =	4 fluid oz = 1 gill(a wineglassful)
147.9ml =	5 fluid oz
295.8ml =	10 fluid oz
473.3ml =	16 fluid oz (1 pint, or 4 gl)
946.6ml=	32 fluid oz (1 quart)

Weight Equivalents

1 grain =	.065mg
10 grains =	650mg (= 1 "00"-size capsule)
15.43 grains =	1 gram
20 grains =	1 scruple
60 grains=	1 drachm (3 scruples)
480 grains(1 oz) =	28.35g = 8dr
16oz (7000grains)=	454g

GLOSSARY

ACHENE

A small, hard, one-seeded fruit not splitting open at maturity.

ADAPTOGEN

Favorably modifying the stress response, or an agent for such.

ALKALOID

A bitter, alkaline, nitrogenous, organic substance found in certain higher plants and often possessed of toxic and/or medicinal properties.

ALTERATIVE

The 'altering' of a body's condition so as to gradually restore health, or an agent for such.

AMENORRHEA

Where normal menstruation is absent or restricted.

ANALGESIC

Relieving pain, or an agent for such.

ANAPHRODISIAC

Inhibiting sexual desire, or an agent for such.

ANNUAL

A plant that sprouts, lives, and dies all within one growing season.

ANODYNE

Reducing pain, or an agent for such.

ANTHELMINTIC

Serving to expel or destroy internally resident parasitical creatures, or an agent for such.

ANTHER

Clublike tip of a plant stamen and the site where the pollen is held.

ANTHOCYANINS

Pigments in plants often responsible for the blue, violet, and red colors occurring in their flowers or other parts. Usually grouped as a subdivision of flavonoids (*see below under that term*).

ANTI-ANAEMIC

Maintaining healthy blood-iron levels, or an agent for such.

ANTICATARRHAL

Reducing mucosal inflammation, or an agent for such.

ANTICOAGULANT

Reducing blood clotting, or an agent for such.

ANTICONVULSIVE

Inhibiting the tendency to convulse or seizure, or an agent for such.

ANTI-EMETIC

Inhibiting nausea or a tendency to vomit, or an agent for such.

ANTIFUNGAL

Inhibiting fungi, or an agent for such.

ANTIGALACTIC

Inhibiting the production, or flow, of breast-milk, or an agent for such.

ANTIHISTAMINE

Inhibiting the allergenic release of histamine from the mast cells of the body, or an agent for such.

ANTILITHIC

Preventing, reducing, or eliminating calculi (stones) in the biliary or urinary system, or an agent for such.

ANTINEOPLASTIC

Inhibiting cancerous cells, or an agent for such.

ANTIOXIDANT

Enhancing the body system's scavenging of free radicals, or an agent for such.

ANTIPHLOGISTIC

Reducing inflammation, or an agent for such.

ANTIPRURITIC

Reducing/inhibiting itching, or an agent for such.

ANTIPYRETIC

Tending to reduce fever, or an agent for such. (*See also under* Febrifuge.)

ANTISPASMODIC

Alleviating cramps or spasms, or an agent for such.

ANTITUSSIVE

Helping to settle, or terminate, a cough, or an agent for such.

APHRODISIAC

Stimulating sexual desire, or an agent for such.

ASTRINGENT

Causing tissues to contract and tighten, or an agent for such.

AXIL

The upper angle of a leaf stalk and its stem.

BASAL

Referring to leaves emanating directly from the crown of a plant and thus at, or near, ground level.

BERRY

A pulpy fruit evolving from a single ovary and having a thin skin and several, or many, tiny seeds which are often distributed throughout the pulp. The term is often misapplied to other kinds of fruits by modern herbalists and foragers who should know better. (For example, fruits produced by hawthorn shrubs are really pomes—*see under that entry*—and are properly termed "haws" or "hawthorn fruits," *not* "hawthorn *berries*" as pop herbalists generally call them).

BIENNIAL

A plant that lives for only two seasons—growing only vegetatively in the first season, and then flowering, seeding, and dying in the second season.

BIPINNATE

Twice pinnate. (*See* Pinnate.)

BRACT

A reduced, or imperfectly developed, leaf growing directly beneath a flower.

BULB

An underground food-storage structure consisting of leaves that are very tightly wrapped around the stem.

CALYX

The green, outer, floral envelope consisting of sepals, the function of which seems to be to protect the flowers (especially when unopened).

CARDENOLIDE

A cardioactive glycoside.

CARDIOACTIVE

Any agent affecting the heart.

CARDIOTONIC

Enhancing cardiac tone/function, or an agent for such.

CARMINATIVE

Relieving or expelling intestinal gas, or an agent for such.

CAROTENOID

A type of terpene (namely, a tetraterpene) found in plants and possessed of bioactive capabilities; beta-carotene, responsible for the orange-yellow coloring in some plants, is one example, being a precursor to vitamin A.

CATHARTIC

Evacuating the intestines quickly and powerfully, or an agent for such.

CHOLAGOGUE

Improving the flow of bile in the digestive process, or an agent for such.

CHOLERETIC

Stimulating the production, or excretion, of bile, or an agent for such.

COMPOSITE FLOWER

A flower in the *Compositae* family consisting of a central disk with many ray flowers extending therefrom.

COMPRESS

A cloth or bandage that has been soaked in a solution and applied externally to a body part to achieve relief of pain, infection, malady, or symptoms.

COMPOUND LEAF

A leaf whose blade is divided into two or more parts called "leaflets."

CORM

An underground enlargement (especially a vertical growth) of a plant's stem, resembling a bulb, but without layers.

COROLLA

The showy parts of a flower, that is, the flower petals considered collectively.

CORYMB

A flattened, or convex, flower cluster in which the outer flowers bloom before the central ones.

CROWN

The base of a perennial plant from which the shoots grow upward and the roots downward.

CYME

A flattened, or convex, flower cluster wherein the central flowers bloom before the outer ones.

CYTOTOXIC

Toxic to organs, tissues, or cells, or an agent for such. Cytotoxic agents are studied by scientists for any possible selective role that they might possess (e.g., destroying cancerous cells while leaving healthy cells intact).

DECIDUOUS

The nature of those plants which spurt through a cycle of growth and then shed parts and/or die back at the end of the growing season; in contrast to "evergreen."

DECOCTION

An herbal extract wherein a hard plant part (a root, stem, seed, or bark) has been actively boiled in water.

DECONGESTANT

Breaking up excessive mucus in the respiratory tract that is making breathing difficult (esp. in the nasal/sinus region), or an agent for such.

DEMULCENT

Locally soothing to sore mucus membranes, or an agent for such.

DEOBSTRUENT

Removing obstruction, or an agent for such.

DEPURATIVE

Cleansing impurities from the system (usually with reference to the blood/lymph), or an agent for such.

DIAPHORETIC

Promoting perspiration, or an agent for such.

DIOECIOUS

Possessing either male or female flowers only.

DISCUTIENT

Scattering tumors or coagulation, or an agent for such.

DISK FLOWER

The central, rounded part of a composite flower.

DIURETIC

Increasing urination, or an agent for such.

DRUPE

A fleshy fruit with a single seed encased in a hard, bony substance.

DYSENTERY

Diarrhea accompanied by painful inflammation of the colon, and thus often marked by the passing of mucous and blood.

DYSMENORRHEA

Painful menstruation.

DYSPEPSIA

A bygone term for *indigestion*.

EDEMA

A condition of swelling resulting from retained fluids in the connective tissues of the body.

EMETIC

Inducing vomiting, or an agent for such.

EMMENAGOGUE

Tending to promote menstruation, or an agent for such.

EMOLLIENT

Tending to soften (the skin, etc.), or an agent for such.

ESCHAROTIC

Locally dissolving tissue, or an agent for such.

ESSENTIAL OIL

Aromatic oil contained in many plants, e.g., the mint family.

ETHNOBOTANIST

A person who studies the use of plants by native peoples.

EXPECTORANT

Promoting the explusion of mucus from the respiratory tract, or an agent for such.

FEBRIFUGE

Reducing fever, or an agent for such.

FILAMENT

In a plant's stamen, this is the hairy stalk on which rests the anther.

FLAVONOIDS

Polyphenolic plant pigments possessed of a chemical structure based on a C15 skeleton, specifically:C6-C3-C6. These compounds are frequently responsible for the yellow and white colors in flowers or in other plant parts. (A sub-category, anthocyanins, often produce red, blue, and purple coloration in plants.*See above under this term*.) They are all to one extent bioactive, often being anti-inflammatory, antiviral, vaso-tonic, or otherwise helpful to the organism.

FOMENTATION

A warm dressing, ointment, or poultice.

GALACTAGOGUE

Stimulating the production, or flow, of breast-milk, or an agent for such.

GLAUCOUS

Covered with a fine, waxy, whitish film.

GLYCOSIDE

A plant compound decomposable into a sugar molecule and another molecule, often bioactive, known as an aglycone.

HEMOSTATIC

Checking or arresting the flow of blood, or an agent for such.

HEPATIC

Benefitting the liver, or an agent for such.

HEPATOPROTECTIVE

Protecting the liver from toxins, or an agent for such.

HERB

A tender, succulent plant having aboveground portions that are entirely non-woody (and which thus usually die back at the end of the growing season).

HERBOLOGY

The study, or science, of herbs as medicinal plants.

HERPATIC [HERPETIC]

Inhibiting/reducing lesions of the skin, or an agent for such.

HORMONAGOGUE

Affecting hormone production, action, or levels, or an agent for such.

HYPOTENSIVE

Lowering blood pressure, or an agent for such.

HYPOGLYCEMIC

Reducing levels of blood glucose, or an agent for such.

INFLORESCENCE

The flowering portion(s) of a plant, applicable to either a simple blossom or a flower cluster; also, the particular arrangement of the flower cluster, whether as a cyme, raceme, panicle, etc.

INFUSION

An extract made from steeping a soft plant part, or parts, in hot (but not actively boiling) water.

INVOLUCRE

An arrangement of whorled bracts occurring in many plants, functioning as a shield for the plant's reproductive elements.

LANCEOLATE

Lance-shaped, i.e., where leaves are much longer than they are wide and where they taper to a point.

LEAFLET

The main subdivision of a compound leaf, often looking like a small and entire leaf.

LITHOTRIPTIC [LITHONTRIPTIC]

Breaking up, or dissolving, calculi (stones) in the urinary or biliary systems, or an agent for such.

LOBE

A rounded projection on a leaf formed by indentation of its margin.

LYMPHATIC

Stimulating, or maintaining, healthy functioning of the lymphatic system.

MACERATE

To soften, or separate, a plant part by steeping it in water (or any other fluid) for extraction.

MATERIA MEDICA

Latin for *medical material* and thus relating to the systematized corpus of healing agents possessed by a particular healer or group of healers.

MENORRHAGIA

Excessive menstrual flow.

METRORRHAGIA

Uterine blood flow occurring midcycle or at any time in the cycle when it should not normally be occurring.

MUCILAGE

Gelatinous substance occurring in certain plants; mucilage has healing potential for irritated mucous membranes, which it can protect and soothe.

NEPHRITIC

Enhancing function of the kidneys, or an agent for such.

NERVINE

Tonifying, or soothing, the nerves or nervous system, or an agent for such.

NODE

The marked joint of a plant stem where the branches or leaves attach.

OPHTHALMIC

Enhancing, or preserving, eye health, or an agent for such.

OXYTOCIC

Accelerating uterine contraction/childbirth, or an agent for such.

PALLIATIVE

Providing relief without effectively curing, or an agent for such.

PALMATE

A leaf shaped like a human hand.

PANICLE

A loosely branched flower cluster, often pyramidical in shape, and with each branch consisting of a raceme.

PARBOIL

To boil briefly.

PARTURIENT

Aiding childbirth, or an agent for such.

PECTORAL

Relieving conditions in the region of the chest, or an agent for such.

PEDICEL

In a flower cluster, this is the stalk of a solitary flower.

PEDUNCLE

The primary stalk of a flower cluster or of a solitary flower when it is the only one on the flowering stem.

PERENNIAL

A plant that lives for several or more seasons, growing from the same root year after year.

PERFOLIATE

A leaf that looks like it has been perforated by the plant's stem.

PERIANTH

Where a plant's calyx and corolla are considered together, especially when they are visually similar.

PETIOLE

A leaf's stalk.

PINNATE

A condition wherein a leaf is divided into leaflets resembling feathers arranged on a bird's wing.

PISTIL

The central organ of a flower, consisting of the ovary, style, and stigma.

POME

A fleshy fruit having a central core consisting of seeds enclosed in a hard, membranaceous capsule.

POTHERB

An edible plant that is cooked before being consumed.

POULTICE

A moistened mass of herb, or herbs, wrapped in a cloth and applied externally to a body part.

PROPHYLACTIC

Preventing, or protecting against, disease, injury, etc., or an agent for such.

PUBESCENT

Hairy and soft.

PURGATIVE

Forcefully emptying the colon, or an agent for such.

RACEME

An inflorescence consisting of an elongate flower cluster of short-stalked flowers arising from a single axis.

RAY FLOWER

In a composite flower, one of the petal-like flowers surrounding the central, disk-shaped flower.

RECEPTACLE

The base of a flower, from which its parts extend; in some plants, the fruit wall is produced therefrom.

RENAL

Pertaining to the kidneys.

RHIZOME

A horizontal, and often elongated, underground stem of certain perennial plants. It is usually situated just below the surface of the soil and it possesses nodes (by which it is distinguished from a root).

ROOTSTOCK

The crown-and-root system of a perennial plant; some writers use it interchangably for the plant's rhizome (*see above under that term*), but such usage is arguable at best.

ROSETTE

A circular cluster of basal leaves, often looking as if flattened against the ground.

RUBEFACIENT

Irritating the skin by enhancing the flow of blood to the area, or an agent for such.

SAPONIN

A chemical found in many plants that foams when shaken with water. Many saponins are bioactive, often serving as expectorants, anti-inflammatories, etc.

SEPAL

A sort of floral "leaf," this is the outermost flower part, usually green in color and petal-like. It serves to protect the flower petals (especially when unopened) and is one of several like members of the calyx.

SERRATE

When the edge, or margin, of a leaf is saw-toothed (as opposed to being round-toothed, lobed, or entire).

SESSILE

Lacking a stalk, i.e., where a leaf clasps the stem.

SHOOT

The newly sprouted portions of a plant.

SIALAGOGUE

Stimulating salivary secretion, or an agent for such.

STAMEN

The male (pollen-bearing) part of a plant, consisting of an anther and a filament.

STIGMA

The pollen-receiving tip of a plant's pistil.

STIMULANT

Increasing functional activity, often of a particular system (e.g., circulatory)

STIPULE

A small appendage at the base of the leafstalk of certain plants.

STOMACHIC

Beneficially affecting the stomach, digestion, or appetite, or an agent for such.

STYLE

In a plant's pistil, this is the middle portion connecting the stigma with the ovary.

STYPTIC

Serving to check blood flow by contricting the vessels, or an agent for such.

SUDORIFIC

Promoting perspiration, or an agent for such.

TANNIN

An astringent, phenolic plant compound that possesses the ability to precipitate protein.

TAPROOT

The stout, main, vertical root of a plant, which in turn gives rise to smaller roots.

TENDRIL

The thin, coiling part of a vine that grasps onto other plants or structures for support.

TINCTURE

An alchoholic extract of a plant or plant part.

TISANE

A bygone term for an herbal infusion.

TONIC

Strengthening and restoring tone (originally used with reference to the digestive system).

TRIFOLIATE

A compound leaf possessed of three leaflets.

TUBER

A subterranean swelling on a root or stem—a sort of modifed branch, usually short and thick, and with buds or "eyes." It is used by the plant for storing food and sometimes occurs quite a distance from the aboveground portions.

TUFTED

Tightly clustered.

UMBEL

A type of flower cluster wherein the individual flower stalks all rise from a common point, giving the appearance of an umbrella.

URINARY

Enhancing the tone or function of the urinary system, or an agent for such.

VARIEGATED

Leaves having the markings of a second color.

VASOCONSTRICTIVE

Constricting the blood vessels, or an agent for such.

VASOTONIC

Tonifying the blood vessels, or an agent for such.

VERMIFUGE

Serving to kill parasitical worms, or an agent for such.

VOLATILE OIL

An alternate term for essential oil (*see above*).

VULNERARY

Promoting the healing of wounds, or an agent for such.

WEED

A small, non-woody plant growing wild or uncultivated.

WHORL

A cluster of three or more leaves radiating from a single node.

REFERENCES

Prefatory Note: The works of the following ancient authors are referenced in the text solely by their standard abbreviations, as follows:

(*De Medicina*) Celsus. *De Medicina.*
(HN) Pliny the Elder. *Natural History.*
(*Herbal*) Dioscorides. *The Greek Herbal of Dioscorides.*

To locate details concerning these works, see under *the abbreviations* in the bibliographic references, below—NOT under the author's name.

anonymous. 1977. *The Barefoot Doctor's Manual: The American Translation of the Official Chinese Paramedical Manual.* Philadelphia: Running Press, 942pp.

anonymous. 1990. "Debugging Precautions," *Prevention* 42(6):22.

anonymous. 1990. "Vorinformation Pyrrolizidinalkaloid-haltige Humanarzneimittel," *Pharmaceutische Zeitung* 135 (1990):2532-33; 2623-24.

anonymous. 1998. "Vaccine Ready to Fight Lyme Disease," *Saint Paul Pioneer Press*, 22 Dec. 1998:1A.

Abad, M.J. et al. 1999. "Antiviral Activity of Some South American Plants," *Phytotherapy Research* 13(2):142-46.

Abdullin, K. K. 1962. *Zap Dazansk, Vet Inst.* 84:75; through *Chem Abstr*, 1964, 11843b.

Aggarwal, M., et al. 1986. "Effect of *Rumex nepalensis* Extracts on Histamine, Acetylcholine, Carbachol, Bradykinin, and PGs-evoked Skin Reactions in Rabbits," *Ann Allergy* 56:177-82.

Ahsan, S. K. et al. 1989. *International Journal of Crude Drug Research* 27(4):235-39.

Airaksinen, M. M. et al. 1986. "Toxicity of Plant Materials Used as Emergency Food During Famines in Finland," *Journal of Ethnopharmacology* 18(3):273-96.

Akhtar, M. S. et al. 1985. "Effects of *Portulaca oleracea* (Kulfa) and *Taraxacum officinale* (Dhudhal) in Normoglycemic and Alloxan-treated Hyperglycemic Rabbits," *J Pak Med Assn* 35(7):207-10.

Akihisa, T. et al. 1996. "Triterpene Alcohols from the Flowers of Compositae and their Anti-inflammatory Effects," *Phytochemistry* 43(6):1255-60.

Alfs, Matthew. 1999. "Boneset," *Herb Quarterly* (Winter)84:40-43.

Alfs, Matthew. 2000. "Euell Gibbons," *American Survival Guide* 22(9):30-33, 64.

Alfs, Matthew. 1996. "Spiderwort: The Forager's Friend," *Wild Foods Forum* 7(3):8-9.

Alfs, Matthew. 1999. "Wild Plants to the Rescue: Use of Local Flora for Wilderness Emergencies," *American Survival Guide* 21(8):58, 60-62, 100-02, 104.

357

Allport, Noel L. 1944. *The Chemistry and Pharmacy of Vegetable Drugs.* Brooklyn: Chemical Publ. Co., 252pp.

Andros, F. 1883. "The Medicine and Surgery of the Winnebago and Dakota Indians," *American Medical Association Journal* 1:116-18.

Angier, Bradford. 1970. *Gourmet Cooking for Free.* Harrisburg, PA: Stackpole Bks, 190pp.

Angier, Bradford. 1972. *Feasting Free on Wild Edibles.* Harrisburg, PA: Stackpole Books, 288pp.

Angier, Bradford. 1974. *Field Guide to Edible Wild Plants.* Harrisburg, PA: Stackpole Bks., 256pp.

Angier, Bradford. 1978. *Field Guide to Medicinal Wild Plants.* Harrisburg, PA: Stackpole Bks, 320pp.

Anisimov, M. M. et al. 1972. "The Antimicrobial Activity of the Triterpene Glycosides of *Caulophyllum thalictroides* [ET]," *Antibiotiki* 17(9):834-37.

Anton, R. and M. Haag-Berrurier. 1980. "Therapeutic Use of of Natural Anthraquinone for Other than Laxative Actions," *Pharmacology* 20. Suppl. 1, 104-12.

Arustamova, F. A. 1963. "Hypotensive Effect of *Leonurus cardiaca* on Animals in Experimental Chronic Hypertension," *Izvestiya Akademii Nauk Armyanski SSR, Biologicheski Nauki* 16(7):47-52.

Asolkar, L. V. et al. 1992. *Second Supplement to Glossary of Indian Medicinal Plants with Active Principles: Part 1 (A–K) (1965–1981).* New Delhi, IN: Pubn's & Information Directorate, 414pp.

Atlee, E.A. 1819. "On the Medicinal Properties of *Monarda punctata*," *Amer. Med. Rec.* 2:496-501.

Auf'Mkolk, M. et al. 1985. "Extracts and Auto-oxidized Constituents of Certain Plants Inhibit the Receptor-Binding and the Biological Activity of Graves' Immunoglobulins," *Endocrinology* 116:1687-93.

Ausubel, Kenny. 2000. "Tempest in a Tonic Bottle: A Bunch of Weeds?" *HerbalGram* 49:32-43.

(BHC) 1976-81. *British Herbal Pharmacopoeia.* Keighley, UK: 229 + 248 + 110pp.

Baba, K. S. et al. 1981. "Antitumor Activity of Hot Water Extract of Dandelion, *Taraxacum officinale--* correlation between Antitumor Activity and Timing of Administration[E.T.]," *Yakugaku Zasshi* 101(6):538-43.

Bader, G. et al. 1987. "The Antifungal Action of Triterpene Saponins of *Solidago virgaurea* L. [E. T.]," *Pharmazie* 42(2):140.

Baille, N., and P. Rasmussen. 1997. "Black and Blue Cohosh in Labour," *N Z Med J* (24 Jan) 110(1036):20-21.

Balboa, S. I. et al. 1973. "Preliminary Phytochemical and Pharmacological Investigations into the Roots of Different Varieties of *Cichorium intybus*," *Planta Medica* 24:133-44.

Banks, W. H. 1953. "Ethnobotany of the Cherokee Indians," Master's Thesis, University of Tennessee.

Barton, Benjamin Smith. 1900. *Collections for an Essay towards A Materia Medica of the United States.* 2d ed. Philadelphia: A. & G. Way, 1801-1804; reprint. Bulletin of the Lloyd Library, No. 1, Series No. 1, Cincinnati: Lloyd Library, 150pp.

Barton, William P.C. 1817-18. *Vegetable Materia Medica of the United States, or Medical Botany*, 2 vols. Philadelphia: M. Carey & Son, 273 + 239pp.

Bartram, William. 1958. *The Travels of William Bartram*, ed. F. Harper, Naturalist's edition. New Haven: Yale University Press.

Basu, N., and R. P. Rastogi. 1967. "Triterpenoid Saponins and Sapogenins," *Phytochemistry* 6(9):1249–70.

Baumann, I. C. et al. 1975. "Studies on the Effects of Wormwood (*Artemisia absinthium* L.) on Bile & Pancreatic-juice Secretion in Man[E.T.]," *Z Allgemeinmed* 51(17):784-91.

Beatty, Bill and Bev. *Bill & Bev Beatty's Wild Plant Cookbook*. Happy Camp, CA: Naturegraph Pubs., 174pp.

Beardsley, G. 1941. "Notes on Cree Medicines, Based on a Collection Made by I. Cowie in 1892," *Michigan Academy of Arts, Sciences, & Letters* 27:483-96.

Béládi, I. et al. 1977. "Activity of Some Flavonoids Against Viruses," *Annals of the New York Academy of Sciences* 284(2):358–64.

Belaiche, P. and O. Lievoux. 1991. "Clinical Studies on the Palliative Treatment of Prostatic Adenoma with Extract of Urtica Root," *Phytotherapy Research* 5:267-69

Belanger, Charles. 1997. *The Chinese Herb Selection Guide*. Richmond, CA: Phytotech, 892pp.

Benigni, R. et al. 1964. *Plante medicinali Intervnie Della Betta* 11:1593.

Benoit, P. S. et al. 1976. "Biological and Phytochemical Evaluation of Plants, Part XIV:Antiinflammatory Evaluation of 163 Species of Plants," *Lloydia* 39(2–3):160–71

Bentley, R. and H. Trimen. 1880. *Medical Plants*, 4 vols. London: J & A Churchill, 2000pp.

Berglund, Berndt, and Clare E. Bolsby. 1971. *The Edible Wild: A Complete Cookbook and Guide to Edible Wild Plants in Canada and Eastern North America*. Toronto/New York: Pagurian Press/Charles Scribner's Sons, 188pp.

Berglund, Berndt, and Clare E. Bolsby. 1977. *The Complete Outdoorsman's Guide to Edible Wild Plants*. New York: Charles Scribner's Sons, 189pp.

Best, William P. 1928. "Eupatorium—Boneset," *Eclectic Medical Journal* 88(2):93-94.

Bever, B.O., and G. R. Zahnd. 1979. "Plants with Oral Hypoglycaemic Action," *Quarterly Journal of Drug Research* 17:139-96.

Bhandari, Prabha et al. 1987. "Triterpenoid Saponins from *Caltha palustris*," *Planta Medica* 53(1):98–100.

Bigelow, Jacob. 1817-20. *American Medical Botany, Being A Collection of the Native Medicinal Plants of the United States*. 3 vols. Boston: Cummings & Hilliard.

Bishayee, A. et al. 1995. "Hepatoprotective Activity of Carrot *(Daucus carota L.)* Against Carbon Tetrachloride Intoxication in Mouse Liver," *Journal of Ethnopharmacology* 47:69-74.

Bissett, N. G. 1994. *Herbal Drugs and Phytopharmicals*. Stuttgart: MedPharm GmbH Scientific Publishers, 568pp.

Black, Meredith Jean. 1973. *Algonquin Ethnobotany: An Interpretation of Aboriginal Adaptations in Southwestern Quebec.* Mercury Series, No. 65. Ottawa: National Museums of Canada, 174pp.

Black, Meredith Jean. 1980. *Algonquin Ethnobotany: An Interpretation of Aboriginal Adaptations in Southwestern Quebec*, Rev. ed. Mercury Series, No. 65. Ottawa: National Museums of Canada, 266pp.

Blair, Thomas S. 1917. *Botanic Drugs.* Cincinnati: Therapeutic Digest Publ. Co., 1917, 394pp.

Bloyer, W. E. 1899. "Epilobium," *Eclectic Medical Journal* 59:276.

Bloyer, W.E. 1901. "Eupatorium," *Eclectic Medical Journal* 61(6):336-37.

Blumenthal, Mark et al., eds. 2000. *Herbal Medicine: Expanded Commission E Monographs.* Newton, MA: Integrative Medicine Communications, 519pp.

Bocek, Barbara R 1984. "Ethnobotany of the Costanoan Indians, California, Based on Collections by John P. Harrington" *Economic Botany* 38(2):240-55.

Bogoiavlenskii, A. P. et al. 1999. "Immunostimulating Activity of a Saponin-containing Extract of *Saponaria officinalis*[E.T.]," *Vopr Virusol* 44(5):229-32.

Bohlmann, Ferdinand et al. 1977. "Neue sesquiterpenlactone und andere inhaltsstoffe aus vertretern der *Eupatorium*-gruppe," *Phytochemistry* 16:1973-81.

Bohm, K. 1959. "Choleretic Action of Some Medicinal Plants,"*Arzneimittel-Forschung* 9:376-78.

Bol'shakova, I. V. 1997. "Antioxidant Properties of a Series of Extracts from Medicinal Plants," *Biofizika* 42(2):480-83.

Bol'shavoka, I. V. 1998. "Antioxidant Properties of Plant Extracts[E.T.]," *Biofizika* 43(2):186-88.

Bolyard, Judith L. 1981. *Medicinal Plants and Home Remedies of Appalachia.* Springfield, IL: Chas. C. Thomas, 187pp.

Bombardelli, E., and P. Morazzoni. 1997. "*Urtica dioica* L.—Review," *Fitoterapia* 68(5):387-401.

Boyd, E.M., and Mary E. Palmer. 1946. "The Effect of *Quillaia, Senega, Squill, Grindelia, Sanguinaria, Chionanthis,* and *Dioscorea* upon the Output of Respiratory-tract Fluid," *Acta Pharmacol et Toxicol* 2(2):235-46.

Brackett, Babette, and Maryann Lash. 1975. *The Wild Gourmet: A Forager's Cookbook.* Boston: David R. Godine, 160pp.

Bradley, P. R., ed. 1992. *British Herbal Compendium*, Vol. 1, London: British Herbal Medicine Assn.

Bramwell, W. 1915. "Scutellaria in Epilepsy," *British Medical Journal* 2:880.

Brenerman, W. R. et al. 1996. "*In vivo* Inhibition of Gonadotropins and Thyrotropin in the Chick by Extracts of *Lithospermum ruderale*," *Gen Comp Endrocrinol* 28(1):24-32.

Brill, Steve, and E. Dean. 1994. *Identifying and Harvesting Edible & Medicinal Plants in Wild (and Not So Wild) Places.* New York: Hearst Books, 317pp.

REFERENCES

Brinker, Francis. 1997. *Herb Contraindications and Drug Interactions*. Sandy, OR: Eclectic Institute.

Brinker, F. 1990. "Inhibition of Endocrine Function by Botanical Agents," *Journal of Naturopathic Medicine* 1:1-14.

Brinker, F. 2000. *The Toxicology of Botanical Medicines*, 3d ed. Sandy, OR: Eclectic Medical Pubns., 296pp.

Britton, Nathaniel Lord, and Hon. Addison Brown. 1970. *An Illustrated Flora of the Northern United States and Canada*. 3 vols., 2d ed. New York: Dover Pubns., 2052pp.

Brooke, Elisabeth. 1992. *Herbal Therapy for Women*. San Francisco: Thorsons, 160pp.

Brown, Lauren. 1986. *Weeds in Winter*. New York: W. W. Norton and Co., 252pp.

Brown, Tom, Jr. 1985. *Tom Brown's Guide to Wild Edible and Medicinal Plants*. New York: Berkeley Bks., 241pp.

Bruneton, Jean. 1995. *Pharmacognosy, Phytochemistry, Medicinal Plants*, Trans. C. A. Hatton New York: Lavoisier Publ. Inc., 915pp.

Bruni, A. et al. 1986. "Protoanemonin Detection in *Caltha palustris*," *Journal of Natural Products* 49(6):1172-73.

Buechel, Eugene. 1983. *A Dictionary of Teton Sioux Lakota-English: English-Lakota*. Pine Ridge, SD: Red Cloud Indian School, 853pp.

Buesemaker. 1936. "Concerning the Choleretic Activity of Dandelion," *Naunyn-Schmiederbergs Archiv fuer Experimentelle Pharmakology und Pathologie.*

Burlage, Henry M. 1968. *Index of Plants of Texas with Reputed Medicinal and Poisonous Properties*. Austin: by the author, 272pp.

Caceres, A. et al 1995. "Antigonorrhoeal Activity of Plants Used in Guatemala for the Treatment of Sexually Transmitted Diseases," *Journal of Ethnopharmacology* 48(2):85-88.

Caceres, A. et al. 1991. "Plants Used in Guatemala for the Treatment of Respiratory Diseases: Screening of 68 Plants Against Gram-Positive Bacteria," *Journal of Ethnopharmacology* 31:193-208.

Cai, D.G. 1983. "Expectorant Constituents of *Eupatorium fortunei*[E.T.]," *Zhong Yao Tong Bao* 8(6):30-31.

Callegari, Jeff, and Keith Durand. 1977. *Wild Edible & Medicinal Plants of California*. El Cerrito, CA: by the authors, 96pp.

Camazine, Scott, and Robert A. Bye. 1980. "A Study of the Medical Ethnobotany of the Zuñi Indians of New Mexico," *Journal of Ethnopharmacology* 2:365-88.

Campbell, A. C., and G. C. MacEwan. 1982. *Systematic Treatment of Sjogren's Syndrome and Sicca Syndrome with Efamol (EPO), Vitamin C, and Pyridoxine*. Eden Press.

Campbell, T. N., and Frank H. H. Roberts, Jr. 1951. "Medicinal Plants Used by Choctaw, Chickasaw, and Creek Indians in the Early Nineteenth Century, *Journal of the Washington Academy of Sciences* 41(9):285-90.

Carlson, Gustave and Volney H. Jones. 1940. "Some Notes on Uses of Plants by the Commanche Indians," *Papers of the Michigan Academy of Science, Arts, & Letters* 25:517-42.

Carr, Lloyd G. and Carlos Westey. 1945. "Surviving Folktales and Herbal Lore Among the Shinnecock Indians," *Journal of American Folklore* 58:113-23.

Carse, Mary. 1989. *Herbs of the Earth: A Self-Teaching Guide to Healing Remedies, Using Common North American Plants and Trees.* Hinesburg, VT: Upper Access Pubs., 238pps.

Carvalho, L.H. et al. 1991. "Antimalarial Activity of Crude Extracts from Brazillian Plants Studied *in vivo* in Plasmodium-berghei-infected Mice and *in vitro* against *Plasmodium falciparum* in Culture," *Braz J Med Biol Res* 24(11):1113-23.

Carvalho, L.H. and A.U. Krettli. 1991. "Antimalarial Chemotherapy with Natural Products and Chemically-defined Molecules," *Mem. Inst. Oswaldo Cruz* 86 Suppl 2:181-84.

Cassady, J. M. et al. 1988. "Use of A Mammalian Cell Culture Benzo(a)pyrene Metabolism Assay for the Detection of Potential Anticarcinogens from Natural Products: Inhibition of Metabolism by Biochanin A, An Isoflavone from *Trifolium pratense* L." *Cancer Research* 48(22):6257-61

Cassady, J.M. et al. 1969. "Terpene Constituents from *Eupatorium* Species,"*Lloydia* 32:522.

Castleman, Michael. 1991.*The Healing Herbs: The Ultimate Guide to the Curative Power of Nature's Medicines.* Emmaus, NJ: Rodale Press, 436pp.

Chabrol, Etienne and R. Charonnat. 1935. "Therapeutic Agents in Bile Secretion[E.T.]," *Annales de Medecine* 37:131-42.

Chabrol, Etienne et al. 1931. "L'action choleretique des Composee," *CR Soc Biol* 108:1100-02.

Chai, Mary Ann P. 1978. *Herbwalk Medicinal Guide.* Provo, UT: Gluten Co., 127pp.

Chakurski, I., et al. 1988. "Treatment of Chronic Colitis with an Herbal Combination *of Taraxacum officinale Hypericum perforatum, Melissa officinalis, Calendula officinalis, and Foeniculum vulgare*[E.T.]," *Vutr Boles* 20(6):51-54.

Chamberlin, Ralph V. 1911. "The Ethno-Botany of the Gosiute Indians of Utah," *Memoirs of the American Anthropological Association* 2(5):329-405.

Chandler, R. Frank et al. 1982. "Ethnobotany and Phytochemistry of Yarrow, *Achillea millefolium*, Compositae," *Economic Botany* 36:203-23.

Chandler, R. Frank et al. 1979. "Herbal Remedies of the Maritime Indians," *Journal of Ethnopharmacology* 1(1):49-68.

Chardenon, Ludo. 1984. *In Praise of Wild Herbs: Remedies & Recipes from Old Provence.* Santa Barbara: Capra Press, 111pp.

Chang, I-M., and H. S. Yun (Choi). 1985. "Plants with Liver-protective Activities: Pharmacology and Toxicology of Aucubin," in H. M. Chang et al., eds., *Adv Chin Med Mat Res.* Singapore: World Scientific.

(ChemAbstr) 1963. *Chemical Abstracts* 59:5535c

REFERENCES

Chestnut, V. K. 1902. "Plants Used by the Indians of Mendocino County, California," *Contributions from the U.S. National Herbarium* 7:295-408.

Chevallier, Andrew. 1996. *The Encyclopedia of Medicinal Plants*. London: Dorling Kindersley, 336pp.

Chodera, A. et al. 1985. "Diuretic Effect of the Glycoside from a Plant of the *Solidago* L. Genus*,"* *Acta Pol Pharm* 42(2):199-204.

Chodera, A. et al. 1991. "Effect of Flavonoid Fractions of *Solidago virgaurea* L. on Diuresis and Levels of Electrolytes," *Acta Pol Pharm* 48(5-6):35-37.

Chopra, R. N. et al. 1956. *Glossary of Indian Medicinal Plants*. New Delhi, India: Council of Scientific & Industrial Research, 330pp.

Christianson, E. H. 1987. "The Search for Diagnostic and Therapeutic Authority in the Early American Healer's Encounter with 'The Animals which Inhabit the Human Stomach and Intestines,'" in: J. Scarborough, ed. *Folklore and Folk Medicines*. Madison: American Institute of History of Pharmacy, 69-85.

Christopher, John R. 1976. *School of Natural Healing*. Springville, UT: Christopher Pubs., 653pp.

Chrubasik, S. et al. 1997. "Evidence for Antirheumatic Effectiveness of Herba *Urtica dioica* in Acute Arthritis: A Pilot Study," *Phytomedicine* 4(2):105-08.

Clapp, Asahel. 1852. "A Synopsis, Or Systematic Catalogue of the Indigenous and Naturalized, Flowering, and Filicoid,...Medicinal Plants of the United States,. Being A Report of the Committee on Indigenous Medical Botany and Materia Medica for 1850-51*" Transactions of the American Medical Association,* 5:689-906.

Clarke, C. B. 1977. *Edible and Useful Plants of California*. Berkeley, CA: Univ. of Calif. Press, 1977, 280pp.

Claus, Edward.P. and Varro Tyler. 1965. *Pharmacognosy*. 5th ed. Philadelphia: Lea & Febiger.

Claus, Edward P. 1961. *Pharmacognosy*. 4th ed. Philadelphia: Lea & Febiger.

Claus, Edward. P. et al. 1956. *Pharmacognosy*. 3rd ed, rev. Philadelphia: Lea & Febiger.

Clymer, R. Swinburne. 1973. *Nature's Healing Agents*. Quakertown, PA: Humanitarian Soc., 230pp.

Coffey, Timothy. 1993. *The History and Folklore of North American Wildflowers*. New York: Facts on File, 356pp.

Colby, Benjamin. 1965. *Guide to Health, Thomsonian System of Practice*. Mokelumne Hill, CA: Health Research, 181pp.

Collier, H. O. J., and G. B. Chesher. 1956. "Identification of 5-Hydroxytryptamine in the Sting of the Nettle (*Urtica dioica*)," *British Journal of Pharmacology* 11(2):186–89.

Colton, Harold S. 1974. "Hopi History and Ethnobotany," in D. A. Horr, ed., *The Hopi Indians*. New York: Garland Press.

Cook, W.H 1985. *The Physio-Medical Dispensatory, 1869;* reprint, Portland: Eclectic Medical Publications, 832pp

Coon, Nelson. 1969. *Using Wayside Plants*. 4th rev. ed. New York: Hearthside Press, 188pp.

Coon, Nelson. 1974. *The Dictionary of Useful Plants*. Emmaus, PA: Rodale Press, 290pp.

Cooper, Marion R., and Anthony W. Johnson. 1988. *Poisonous Plants & Fungi: An Illustrated Guide*. London: Her Majesty's Stationery Office, 134pp.

Costello, C.H. and C.L. Butler. 1950. "The Estrogenic and Uterine-stimulating Activity *of Asclepias tuberosa*: A Preliminary Investigation," *Jour Amer Pharm Assoc Sci Ed* 39:233-37.

Couplan, Francois. 1998. *Encyclopedia of the Edible Plants of North America*. New Canaan, CT: Keat Publ. Co., 584pp.

Court, W.E. 1986. "A History of Mustard in Pharmacy and Medicine," *Pharm Hist*. 6:9-11.

Courtenay, Booth, and James H. Zimmerman. 1972. *Wildflowers and Weeds: A Guide in Full Color*. New York: Van Nostrand Reinhold Co., 144pp.

Crellin, John K., and Jane Philpott. 1990b. *Herbal Medicine Past and Present*, Vol.2: *A Reference Guide to Medicinal Plants*. Durham: Duke Univ. Press, 549pp.

Crellin, John K., and Jane Philpott. 1990a. *Trying to Give Ease: Tommie Bass and the Story of Herbal Medicine*. Durham: Duke Univ. Press, 335pp.

Croom, Edward Mortimer. 1983. *Medicinal Plants of the Lumbee Indians*. Ph.D. dissertation, North Carolina State University at Raleigh, 1982; reprint, Ann Arbor: University Microfilms, 183pp.

Crowhurst, Adrienne. 1973. *The Flower Cookbook*. New York: Lancer Bks., 1973, 198pp

Crowhurst, Adrienne. 1972. *The Weed Cookbook*. New York: Lancer Bks., 190pp.

Culbreth, David M.R. 1927. *A Manual of Materia Medica and Pharmacology*. 7th ed. Philadelphia: Lea & Febiger, 627pp. + Index. [Web reprint at: http://chili.rt66.com/hrbmoore/ManualsOther/Culbrth5.pdf]

Culpeper, Nicolas. 1983. *Culpeper's Color Herbal*. London: 1649; reprint, New York: Sterling Publ. Co., 224pp.

Curtis, John T. 1959. *The Vegetation of Wisconsin: An Ordination of Plant Communities*. Madison: Univ. of Wisconsin Press, 657pp.

Cuthbert, Mabel Jaques. 1949. *The Spring Flowers*. Rev. ed. Dubuque, IA: Wm. C. Brown Co., 194pp.

Cutler, Manasseh. 1903. *An Account of Some of the Vegetable Productions Naturally Growing in this Part of America, Botanically Arranged*. Boston: by the author, 1785; reprint (Bulletin of the Lloyd Library, No. 7, Reproduction Series No. 4), Cincinnati: Lloyd Library.

D'Adamo, P. 1992. "*Chelidonium* and *Sanguinaria* Alkaloids as Anti-HIV Therapy," *Journal of Natural Medicine* 3:31-24.

D'Amico, L. 1950. "Ricerche sulla presenza di sostanze ad azione antabiotica nelle piante superiori," *Fitoterapia* 21(1):77-79.

REFERENCES

Dana, Mrs. William Starr. 1962. *How to Know the Wild Flowers*: *A Guide to the Names, Haunts, and Habits of Our Common Wild Flowers*, Rev. ed., ed. Clarence J. Hylander., 1900; reprint, New York: Dover Pubns., 418pp.

Das, P. K. et al. 1976. "Pharmacology of Kutkin and its Two Organic Constituents, Cinnamic Acid and Vanillic Acid," *Indian J Expl Biol* 14:456-58.

de Bairacly Levy, Juliette. 1974. *Common Herbs for Natural Health*. New York: Schocken Books, 200pp.

de Bray, Lys. 1978. *The Wild Garden: An Illustrated Guide to Weeds*. New York: Mayflower Bks., 191pp.

de Laszlo, Henry, and Paul S. Henshaw. 1954. "Plant Materials Used by Primitive Peoples to Affect Fertility," *Science* 119(3097, 7th May):626-31.

(De Medicina) Celsus. 1938. *De Medicina, with an English Translation by* W. G. Spencer. [Reprint Ed, Birmingham: Classics of Medicine Library, 1989], 499 + 649 + Indices.

Delas, R. et al. 1947. *Toulouse Medicale* 49:57, cited in Knott & McCutcheon, 1961, below.

del Castillo, J. et al. 1975. "Marijuana, Absinthe, and the Central Nervous System," *Nature* 253:365-66.

Densmore, Frances. 1974. *How Indians Use Wild Plants for Food, Medicine, & Crafts* [*Uses of Plants by the Chippewa Indians,* Forty-fourth Annual Report of the Bureau of American Ethnology, 1926-1927]. Washington DC: Government Printing Office, 1928; reprint ed., New York: Dover Pubns, 277-397.

Densmore, Frances. 1932. *Menominee Music*, SI-BAE Bulletin #102. Washington DC: Government Printing Office, 230pp.

Deschauer, Thomas. 1940. *Deschauer's Complete Course in Herbalism, Describing the Great Curative Properties of Herbs*, 2 vols. Publ. by the author, 432pp. This very rare set is one of my prized possessions!

Diel, J. F. 1990. *Safety of Irradiated Foods*. NY & Basel: Marcel Dekker Co., 464pp.

Doan, D. D. et al. 1992. "Studies on the Individual and Combined Diuretic Effects of Four Vietnamese Traditional Herbal Remedies (*Zea mays, Imperata cylindrica, Plantago major* and *Orthosiphon stamineus*)," *Journal of Ethnopharmacology* 36(3):225-31.

Dobelis, Inge N. et al, eds. 1986. *Magic and Medicine of Plants*. Pleasantville, NY: Reader's Digest Assn., 464pp.

Dombradi,. C. A, and S. Foldeak. 1966. "Screening Report on the Antitumor Activity of Purified *A. lappa* Extracts" *Tumori* 52(3):173-75.

Dombradi, C. A. 1970. "Tumor-Growth Inhibiting Substances of Plant Origin. II. The Experimental Animal Tumour-Pharmacology of Arctigenin-Mustard,"*Chemotherapy* 15:250.

Donson, Alexandra. 1982. *Healing Herbs for Arthritis and Rheumatism*. New York: Sterling Publ. Co., Inc., 143pp.

Duckett, S. 1980. "Plantain Leaf for Poison Ivy," *New England Journal of Medicine* 303(10):583.

Ducrey, B. et al. 1997. "Inhibition of 5a-Reductase and Aromatase by the Ellagitannins Oenothein A and Oenothein B from *Epilobium* Species," *Planta Medica* 63:111-114.

Duke, (James A.), Jim. 1994. "Balm *Monarda punctata*, for the Balmy," *Coltsfoot* 15(3):12-14.

Duke, James A. 1985. *CRC Handbook of Medicinal Herbs*. Boca Raton, FL: CRC Press, 704pp.

Duke, James. 1997. *The Green Pharmacy*. Rodale Press, 507pp.

Duke, James. 1992. *Handbook of Edible Weeds*. Boca Raton: CRC Press, 246pp.

Edmunds, J. 1999. "Blue Cohosh and Newborn Myocardial Infarction?" *Midwifery Today/Int Midwife* (Winter) 52:34-35.

Edsall, Marian S. 1985. *Roadside Plants And Flowers: A Traveler's Guide To The Midwest and Great Lakes Area, With a Few Familiar Off-Road Wildflowers*. N.p.: Univ. of Wisconsin Press, 144pp.

Eisner, Thomas. 1965. "Catnip: its *raison d'etre*," *Science* 146(3649):1318–20.

el-Ghazaly, M. et al. 1992. "Study of the Anti-inflammatory Activity of *Populus tremula, Solidago virgaurea* and *Fraxinus excelsior*," *Arzneimittelforschung* 42(3):333-36.

Elias, Thomas S., and Peter A. Dykeman. 1982. *Field Guide to North American Edible Wild Plants*. New York: Outdoor Life Bks., 286pp.

Ellingwood, Finley. 1983. *American Materia Medica, Therapeutics, and Pharmacognosy*. Evanston: Ellingwood's Therapeutist, 1915; reprint, Portland: Eclectic Medical Pubns., 564pp.

Elliott, Douglas B. 1976. *Roots: An Underground Botany and Forager's Guide*. Old Greenwich, CT: Chatham Press, 128pp

Elliott, S. 1821-24. *A Sketch of the Botany of South-Carolina and Georgia*, 2 vols. Charleston: Schenk, Hoh, 635pp.

el-mekkawy, S. et al. 1995. "Inhibitory Effects of Egyptian Folk Medicines on Human Immunodeficiency Virus (HIV) Reverse Transcriptase," *Chem Pharm Bull* (Tokyo) 43:641-48, Abstract.

Elmore, Francis H. 1944. *Ethnobotany of the Navajo*, University of New Mexico Monographs of the School of American Research, 8. Albuquerque:University of New Mexico, 136pp.

Emmelin, N., and W. Feldberg. 1947. "The Mechanism of the Sting of the Common Nettle (*Urtica urens*)," *Journal of Physiology* 106(4):440–55.

Erichsen-Brown, Charlotte. 1989. *Medicinal and Other Uses of North American Plants: A Historical Survey with Special Reference to the Eastern Indian Tribes*. Aurora: Breezy Creeks Press, 1979; reprint, New York: Dover Pubns., 512pp.

Erickson, David W., and James S. Lindzey. 1983. "Lead and Cadmium in Muskrat and Cattail Tissues," *Journal of Wildlife Management* 47(2):550–55.

(EthnobDB) Ethnobotanical Database. Web resource at: http://probe.nal.usda.gov:83

Faber, K. 1958. "The Dandelion, *Taraxacum officinale*[ET]," *Pharmazie* 13:423-36.

Falk, A. J. et al. 1975. "Isolation and Identification of Three New Flavones from *Achillea millefolium* L.," *J Pharm Sci* 64(11):1838-42.

Farnsworth, Norman R. 1966. "Biological and Phytochemical Screening of Plants," *Journal of Pharmaceutical Sciences* 55(3):225–76.

Farnsworth, Norman R. 1968. "Biological and Phytochemical Screening of Plants," *Lloydia* 31:225-76.

Farnsworth, Norman R. and A.B. Segelman. 1971. "Hypoglycemic Plants," *Till and Tile* 57:52-56.

Farnsworth, Norman R. 1985. "Medicinal Plants in Therapy," *Bulletin of the World Health Organization* 63(6):968-81.

Farnsworth, Norman R. et al. 1975. "Potential Value of Plants as Sources of New Antifertility Agents," *Journal of Pharmaceutical Sciences* 64:535-98, 717-54.

Farre, M. 1989. "Fatal Oxalic Acid Poisoning from Sorrel Soup," *Lancet* Dec 23-30, 2(8668-70):1524.

Fassett, Norman C. 1976. *Spring Flora of Wisconsin*, rev. O. S. Thompson. Madison: Univ. of Wisconsin Press, 413pp.

Favel, A. et al. 1992. "Screening of Triterpenoid Saponins for Antifungal Activity," *Planta Medica* 58(Suppl. 1):A635–36.

Fearn. 1923. "Fearn's Eclectic Therapeutics: Epilobium," *Eclectic Medical Journal* 83(12):582-83

Felklova, M. 1958. "Antibacterial Properties of Extracts from *Plantago lanceolata* Extracts," *Pharm Zentralhalle Deutsch* 97:61-65.

Fell, J.W. 1857. *A Treatise on Cancer, and its Treatment.* London: J. Churchill, 95pp.

Felter, Harvey Wickes. 1922. *The Eclectic Materia Medica, Pharmacology, and Therapeutics.* Cincinnati: John K. Scudder, 743pp. Web reprint at: http://chili.rt66.com/hrbmoore/FelterMM.pdf

Felter, Harvey Wickes. 1924. "Eupatorium (Boneset)," *Eclectic Medical Journal* 84(4):200-02.

(Felter & Lloyd) Felter, Harvey Wickes Felter and John Uri Lloyd. 1983. *King's American Dispensatory.* 18th ed., 2 vols, 1906; reprint, Portland: Eclectic Medical Publications. As of this writing, there is a web reprint at: http://metalab.unc.edu/herbmed/eclectic/kings.html

Fenton, William N. 1941. "Contacts between Iroquois Herbalism and Colonial Medicine," *Annual Report of the Smithsonian Institution for 1941.* Washington DC: Government Printing Office, 503-26.

Fernald, M. L., and A. C. Kinsey. 1958. *Edible Wild Plants of Eastern North America*, Revised by Reed Rollins. New York: Harper & Row, 452pp.

Fetrow, Charles W. and Juan R. Avila. 1999. *Professional's Handbook of Complementary & Alternative Medicines.* Springhouse, PA: Springhouse Pubns, 762pp.

Fewkes, J. Walter. 1896. "A Contribution to Ethnobotany," *The American Anthropologist* 9:14-21.

Fielder, Mildred. *Plant Medicine and Folklore.* New York: Winchester Press, 1975, 268pp.

Fischer-Rizzi, Susanne. 1996. *Medicine of the Earth: Legends, Recipes, Remedies, and Cultivation of Healing Plants.* Portland: Rudra Press, 320pp.

Fisher, C. N. 1997. "Nettles—An Aid to the Treatment of Allergic Rhinitis," *European Journal of Herbal Medicine* 2(2):34-35.

Fitzpatrick, Florence K. 1954. "Plant Substances Active Against *Mycobacterium tuberculosis*," *Antibiotics and Chemistry* 4(5):528–36.

Fokina, G. I. et al. 1991. "Experimental Phytotherapy of Tick-borne Encephalitis[E.T.]," *Vopr Virusol* 36(1):18-21.

Foldeak, S. & C. A. Dombradi. 1964. "Tumor-growth Inhibiting Substances of Plant Origin. 1. Isolation of Active Principle of *Arctium lappa*," *Acta Physiology & Chemistry (Szeged)* 10:91-93.

Folsom-Dickerson, W. E. S. 1965. *The White Path*. San Antonio: The Naylor Co., 148pp.

Forst, A. W. 1940. *Naunyn-Schmiedebergs Archiv fur experimentelle Pathologie und Pharmakologie* 195:1-25.

Foster, Steven, and James A. Duke. 1990. *A Field Guide to Medicinal Plants: Eastern and North Central North America*. Boston: Houghton Mifflin Co., 366pp.

Foster, Steven. 1993. *Herbal Rennaisance*. Salt Lake City: Gibbs Smith Publisher, 234pp.

Freitus, Joe. 1980. *The Natural World Cookbook: Complete Gourmet Meals from Wild Edibles*. Washington D. C.: Stone Wall Press, 283pp. + Index.

Fukuchi, K. et al. 1989. "Inhibition of Herpes Simplex Virus Infection by Tannins and Related Compounds," *Antiviral Research* 11(56):285-98.

Fuller, Thomas C., and Elizabeth McClintock. 1986. *Poisonous Plants of California*. Berkeley: Univ. Calif. Press, 433pp.

Gaertner, Erika E. 1979. "The History and Use of Milkweed," *Economic Botany* 33(2):119-23.

Galecka, H. 1969. "Choleretic and Cholagogic Effects of Certain Hydroxy Acids and their Derivatives in Guinea Pigs," *Acta Pol Pharm* 26:479-84.

Gambhir, S. S. et al. 1979. "Antispasmodic Activity of the Tertiary Base of *Daucus carota*, Linn. Seeds." *Indian Journal of Physiol. Pharmacology* 23:225-28.

Gansser, D. and G. Spiteller. 1995. "Aromatase Inhibitors from *Urtica dioica* Roots," *Planta Medica* 61:138-40.

Gansser, D. and G. Spiteller. 1995b. "Plant Constituents Interfering with Sex Hormone-binding Globulin: Evaluation of a Test Method and its Application to *Urtica dioica* Plant Extracts[E.T.]," *Z Naturforsch* [C]:50(1-2):98-104.

Garrett, Blanche Pownall. 1975. *A Taste of the Wild*. Toronto: James Lorimer & Co., 132pp.

Gasperi-Campani, A., et al. 1991. "Inhibition of Growth of Breast Cancer Cells *in Vitro* by the Ribosome-Inactivating Protein Saporin 6," *Anticancer Res* 11:1007-11.

Gassinger, C.A. et al. 1981. "Klinische Prufung zum Nachweis der therapeutishen Wirksamkeit des homoopathischen Arzneimittels Eupatorium perfoliatum D 2 (Wasserhanf composite) bei der Diagnose 'Grippaler Infekt.'" *Arzneim.-Forsch./Drug Res.* 31(1):732-36.

REFERENCES

Gathercoal, Edmund N. and Elmer H. Wirth. 1936. *Pharmacognosy*. Philadelphia: Lea & Febiger, 852pp.

Gathercoal, Edmund N. and Elmer H. Wirth. 1947. *Pharmacognosy*. 2d ed. Philadelphia: Lea & Febiger, 756pp.

Genders, Roy. 1988. *Edible Wild Plants*. New York: van der Marck Eds., 208pp.

Gerard, John. 1975. *The Herball, or Generall Historie of Plantes, The Complete 1633 Edition as Revised & Enlarged by Thomas Johnson.* London: 1633, reprint ed, New York: Dover Publications.

(Germ Comm E) 1998. *The Complete German Commission E Monographs: Therapeutic Guide* E (1998) *to Herbal Medicines*. Integrative Medicine Communications, 685pp.

Gibbons, Euell, and Gordon Tucker. 1979. *Euell Gibbons' Handbook of Edible Wild Plants*. Virginia Beach, VA: Donning Co., 319pp.

Gibbons, Euell. 1973. *Stalking the Faraway Places*. New York: David McKay Co., 279pp.

Gibbons, Euell. 1971. *Stalking the Good Life*. New York: David McKay Co., 247pp.

Gibbons, Euell. 1970. *Stalking the Healthful Herbs*. Field Guide Ed. New York: David McKay Co., 303pp.

Gilani, A.H. et al. 1994. "Cardiovascular Actions of *Daucus carota*," *Archives of Pharmacological Research* 17:150-53.

Gilani, A.H. and K.H. Janbaz. 1995. "Preventative and Curative Effects of *Artemisia absinthium* on Acetaminophen and CC14-Induced Hepatotoxicity," *General Pharmacology* 26(2):309-15.

Gilmore, Melvin. 1913. "A Study on the Ethnobotany of the Omaha Indians*," Nebraska State Historical Society* 17:314-57.

Gilmore, Melvin. 1933. "Some Chippewa Uses of Plants," *Papers of the Michigan Academy of Science, Arts, & Letters, 17*. Ann Arbor: Univ Michigan Press:119-43.

Gilmore, Melvin. 1991. *Uses of Plants by the Indians of the Missouri River Region, 33rd Annual Report of the Bureau of American Ethnology, 1911-12,* Washington DC: Government Printing Office, 1919:43-154; enlarged ed., Lincoln: Univ. Nebraska , 125pp.

Gilmore, Melvin. 1977. *Uses of Plants by the Indians of the Missouri River Region, 33rd Annual Report of the Bureau of American Ethnology, 1911-12,* Washington DC: Government Printing Office, 1919:43-154; Lincoln: Univ. Nebraska Press., 109pp.

Gladstar, Rosemary. 1993. *Herbal Healing for Women*. New York: Simon & Schuster, 303pp.

Goetz, P. 1989. "Die behandlung der benignen prostatahyperplasie mit Brennesselwurzeln." *Z Phytother* 10:175-78.

Goldberg, A. S. et al. 1969. "Isolation of the Anti-inflammatory Principles from *Achillea millefolium* (Compositae)," *Journal of Pharmaceutical Sciences* 58(8):938-41.

Goldberg, A. S. 1972. *Journal of Pharmaceutical Sciences* 58 (1972):938-41.

Goodchild, Peter. 1984. *Survival Skills of North American Indians*. Chicago: Chicago Review Press, 234pp.

Goodrich, Jennie, et al. 1980. *Kashaya Pomo Plants*. Borgo Press, 176pp.

Goranov, K. et al. 1983. "Clinical Results from the Treatment of Hemorrhagic Form of Periodontosis with a Complex Herb Extract and 15% DMSO," *Stomatologia* 65(6):25-30.25-30.

Gosse, P. H. 1971. *The Canadian Naturalist: The Natural History of Lower Canada*. London: John van Voorst, 1840; facsimile ed., Toronto: Coles, 372pp.

Graber, J.D. 1907. "Sanguinaria in Eczema," *Journal of the American Medical Association* 49:705.

Grases, F. et al. 1994. "Urolithiasis and Phytotherapy," *Int Urol Nephrol* 26:507-11

Green, James. 1991. *The Male Herbal: Health Care for Men and Boys*. Freedom, CA: Crossing Press, 267pp. + Index.

Grieve, M. 1971. *A Modern Herbal*, ed. Mrs. C. F. Leyel. 2 vols. London: Jonathan Cape, 1931; reprint, New York: Dover Pubns., 902pp.

Griffith, R. E. 1847. *Medical Botany*. Philadelphia: Lea & Blanchard 704pp.

Griggs, Barbara. 1981. *Green Pharmacy: A History of Herbal Medicine*. London: Jill Norman & Hobhouse, 379pp.

Grimes, William, ed. 1973. *Ethno-Botany of the Black Americans*. Michigan: Reference Pubns.

Grinnell, George Bird. 1905. "Some Cheyenne Plant Medicines," *American Anthropologist* 7:37-43.

Grinnell, George Bird. 1972. *The Cheyenne Indians--Their History and Ways of Life*, Vol. 2. Lincoln: Univ.of Nebraska, 402pp.

Guerin, J. C., and H. P. Reveillere. 1985. "Antifungal Activity of Plant Extracts Used Therapeutically. II. Study of 40 Extracts on 9 Fungal Strains," *Ann Pharm Fr* 43(1):77-81.

Gui, CH. 1985. "Antitussive Components of Verbena officinalis[E.T.]," *Chung Yao Tung Pao* 10(10):35.

Guin, J.D. and R. Reynolds. 1980. "Jewelweed Treatment of Poison Ivy Dermatitis," *Contact Dermatitis* 6:287-88.

Gunn, John. 1859-61. *New Domestic Physician, or Home Book of Health*, 2nd rev. ed. Cincinnati: Moore, Wilstach, & Keys, 1046pp.

Gunn, T. R. and I. M. Wright. 1996. "The Use of Black and Blue Cohosh in Labour," *N Z Med J* (25 Oct) 109(1032):410-11..

Gunther, E. 1973. *Ethonobotany of Western Washington*, Rev. Ed. Seattle:Univ. Washington Press, 71pp.

(HN) Pliny the Elder. 1938. *Natural History,* 10 vols., Trans. W. H. S. Jones (Loeb Classical Library) Cambridge, MA: Harvard Univ. Press. Another good English translation is: Pliny the Elder. 1855. *The Natural History of Pliny*, 5 vols., Trans. J. Bostock & H. T. Riley. London: Henry H. Bohn. A good, one-volume condensation is: Pliny the Elder [C. Plinius Secundus]. 1962. *Selections from The History of the World, Commonly Called the Natural History of C. Plinius Secundus*, Trans. Paul Turner. Carbondale, IL: Southern Illinois Univ. Press, 1962, 496pp.

REFERENCES

Habtemariam, S. and A. M. MacPherson. 2000. "Cytotoxicity and Antibacterial Activity of Ethanol Extract from Leaves of a Herbal Drug, Boneset *(Eupatorium perfoliatum)*," *Phytotherapy Research* 14(7):575-77.

Habtemariam, S. 1998. "Cistifolin, an Integrin-dependent Cell-adhesion Blocker from the Anti-Rheumatic Herbal Drug, Gravel Root (Rhizome of *Eupatorium purpureum*)," *Planta Medica* 64(8):683-85.

Hall, Alan. *The Wild Food Trail Guide*. New York: Henry Holt & Co., 230pp.

Hall, Thomas B. 1974. "*Eupatorium perfoliatum*: A Plant with A History," *Missouri Medicine* 71(9):527-28.

Hall, Walter, and Nancy Hall. 1980. *The Wild Palate: A Serious Wild Foods Cookbook*. Emmaus, PA: Rodale Press, 374pp.

Hamel, P.B., and M. U. Chiltoskey. 1975. *Cherokee Plants:Their Uses--A 400-Year History*. Sylva, NC: Herald Publ. Co., 65pp.

Hamerstrom, Frances. 1989. *Wild Food Cookbook*. Ames, IA: Iowa State Univ. Press, 126pp.

Handa, Sharma, and Chakraborti. 1986. "Natural Products and Plants as Liver Protecting Drugs," *Fitotherapia* 57:305-51.

Harborne, Jeffrey B., & Herbert Baxter, eds. 1993. *Phytochemical Dictionary: A Handbook of Bioactive Compounds from Plants.* Washington DC: Taylor & Francis, 791pp.

Harlan, Richard. 1830. "Experiments Made on the Poison of the Rattle-snake; in which the Powers of the *Hieracium venosum*, as A Specific, Were Tested," Transactions of the American Philosophical Society, New Series, 3:300-14.

Harney, John W. et al. 1978. "Behavioral and Toxicological Studies of Cyclopentanoid Monoterpenes from *Nepeta cataria*," *Lloydia* 41(4):367–74.

Harrar, Sari and Sara Altshul O'Donnell, eds. 1999. *The Woman's Book of Healing Herbs: Healing Teas, Tonics, Supplements, and Formulas*. Emmaus: Rodale Press, 495pp.

Harrington, H. D. 1967. *Edible Native Plants of the Rocky Mountains*. Albuquerque: Univ. New Mexico Press, 392pp.

Harrington, H.D. 1972. *Western Edible Wild Plants*. Albuquerque: Univ. of New Mexico Press, 156pp.

Harris, Ben Charles. 1973. *Eat The Weeds*. New Canaan, CT: Keats Publ. Co., 253pp.

Hart, Jeffrey A.. 1981. "The Ethnobotany of the Northern Cheyenne Indians of Montana," *Journal of Ethnopharmacology* 4:1-55.

Hartmann, R. W. et al. 1996. "Inhibition of 5 *a*-reductase and aromatase by PHL-00801 (Prostatonin) a combination of PY 102 (*Pygeum africanum*) and UR 102 (*Urtica dioica*) Extracts." *Phytomedicine* 3(2):121-28.

Hartwell, Jonathan L. 1967. "Plants Used Against Cancer: A Survey," *Lloydia* 30:379–436.

Hartwell, Jonathan L. 1968. "Plants Used Against Cancer: A Survey," *Lloydia* 31:71–170.

Hartwell, Jonathan L. 1969. "Plants Used Against Cancer: A Survey," *Lloydia* 32:79-107, 153-205, 247-96

Hartwell, Jonathan L. 1970. "Plants Used Against Cancer: A Survey," *Lloydia* 33:97-124, 288-392.

Hartwell, Jonathan L. 1971. "Plants Used Against Cancer: A Survey," *Lloydia* 34:103-60, 204-55, 310-60, 386-425.

Hartwell, Jonathan. 1984. *Plants Used Against Cancer*. Lawrence: Quarterman Pubns., 710pp., reprinting his lengthy series of articles referred to immediately above.

Hatfield, Audrey. 1973. *How to Enjoy Your Weeds*. New York: Collier Bks., 192pp.

Hausen, B. M. 1992. "Sesquiterpene Lactones—General Discussion," in P. A. G. M. De Smet et al, eds. *Adverse Effects of Herbal Drugs*. 2 vols. Berlin: Springer-Verlag, 1:227–36.

Heatherly, Ana Nez. 1998. *Healing Plants: A Medicinal Guide to Native North American Plants and Herbs*. New York: Lyons Press, 252pp.

Heckewelder, J. G. E. 1820. *A Narrative of the Mission of the United Brethren Among the Delaware and Mohegan Indians*. Philadelphia: McCarty & Davis, 429pp.

Heinerman, John. 1979. *Science of Herbal Medicine: Pharmacological, Medical, Historical, Anthropological*. n.p.:Bi-World Pubns, 318pp.

Heller, Christine A. 1953. *Edible and Poisonous Plants of Alaska*. College, AK: Univ. of Alaska and U.S. Dept. of Agriculture, 167pp.

Hellson, J. C. and M. Gadd. 1974. *Ethnobotany of the Blackfoot Indians*. Canadian Ethnology Service Paper No. 19, Ottawa: National Museums of Canada, 138pp.

(Herbal) Dioscorides. 1996. *The Greek Herbal of Dioscorides, Illustrated by A Byzantine A.D. 512, Englished by John Goodyear A.D. 1655, Edited and First Printed A.D. 1933*, ed. Robert Gunther. Oxford: John Johnson, 1959; reprint ed., New York: Classics of Medicine Library, 701pp.

Herrick, James William. 1977. *Iroquois Medical Botany*. Ann Arbor: Univ. Microfilms Int'l, 518pp.

Herrick, James W. 1995. *Iroquois Medical Botany*. Syracuse: Syracuse University Press, 278pp.

Herz, Werner and Ram P. Sharma. 1976. "Sesquiterpene Lactones of *Eupatorium hyssopifolium*: A Germacranolide with an Unusual Lipid Ester Side Chain," *Journal of Organic Chemistry* 1976:1015-20.

Herz, W. et al. 1977. "Sesquiterpene Lactones of *Eupatorium perfoliatum*," *Journal of Organic Chemistry* 42:2264-71

Heywood, V.H. and J. B. Harborne et al, eds. 1977. *The Biology and Chemistry of the Compositae*, Vol. 1 NY: Academic Press.

Hiermann, A. et al. 1991. "Isolation of the Antiphogistic Principle from *Epilobium angustifolium*[E.T.]," *Planta Medica* 57(4):357-60.

Hill, John A. 1751. *A History of the Materia Medica*. London: Longman, 895pp.

REFERENCES

Hiller, K. 1987. "New Results on the Structure and Biological Activity of Triterpenoid Saponins," in Hostettmann, K., and P. J. Lea, eds. *Biologically Active Natural Products*. Oxford: Clarendon Press, pp. 175–84.

Hippocrates. n.d. *The Genuine Works of Hippocrates, Translated from the Greek, with A Preliminary Discourse and Annotations* by Francis Adams. New York: William Wood & Co., 2 vols. in 1, 390 + 366pp.

Hirano, T. et al. 1994. "Effects of Stinging Nettle Root Extracts and their Steroidal Components on the Na+, K+-ATPase of the Benign Prostatic Hyperplasia," *Planta Medica* 60(1):30-33.

Hirose, M. et al. 2000. "Effect of Arctiin on PhIP-induced Mammary, Colon and Pancreatic Carcinogenesis in Female Sprague-Dawley Rats and MeIQx-induced Hepatocarcinogenesis in Male F344 Rats," *Cancer Letter* 155(1):79-88.

Hitchcock, Susan Tyler. 1980. *Gather Ye Wild Things: A Forager's Year*. New York: Harper and Row, 182pp.

Hnatyszyn, O. et al. 1999. "Argentine Plant Extracts Active Against Polymerase and Ribonuclease H Activities of HIV-1 Reverse Transcriptase," *Phytotherapy Research* 13(3):206-09.

Hocking, George Macdonald Hocking. 1997. *A Dictionary of Natural Products,* 2nd ed. Medford, NJ: Plexus Pubs., 1024pp.

Hocking, G. M. 1956. "Some Plant Material Used Medicinally and Otherwise by the Navaho Indians in the Chaco Canyon, New Mexico," *El Palacio* 63:161.

Hodgson, Michael. 1992. "Much Ado About DEET," *Backpacker* 20(118):20–21.

Hoerhammer, L.1961. "Flavone Concentration of Medical Plants with Regard to their Spasmolytic Action," *Congr Sci Farm Conf Commun 21st Pisa*: 578-81.

Hoffmann, David. 1986. *The Holistic Herbal: A Herbal Celebrating the Wholeness of Life.* 2nd ed. The Park, Forres, Scotland: Findhorn Press, 281pp.

Hoffman, W. J. 1891. "The Mide'wiwin or 'Grand Medicine Society' of the Ojibwa," *Seventh Annual Report of the Bureau of American Ethnolog,y,* 1885-86. Washington DC: Government Printing Office, 149-300.

Holmes, E. M. 1884. "Medicinal Plants Used by the Cree Indians, Hudson's Bay Territory," *The Pharmaceutical Journal and Transactions* 15:302-04.

Holmes, Peter. 1989–90. *The Energetics of Western Herbs: Integrating Western and Oriental Herbal Medicine Traditions*. 2 vols. Boulder, CO: Artemis Press, 421pp. + Index & 416pp. + Index.

Holmes, Peter. 1997-99. *The Energetics of Western Herbs: Treatment Strategies Integrating Western and Oriental Herbal Medicine.* 2 vols., Rev. 3rd ed. Boulder, CO: Snow Lotus Press, 442 + 442pp.

Honeychurch, Penelope. 1980. *Caribbean Wild Plants & their Uses: An Illustrated Guide to Some Medicinal and Wild Ornamental Plants of the West Indies*. 166pp.

Hook, I. et al. 1993. "Evaluation of Dandelion for Diuretic Activity and Variation in Potassium Content," *International Journal of Pharmacognosy* 31:29-34.

Hooper, S.N. and R.F. Chandler. 1984. "Herbal Remedies of the Maritime Indians: Phytosterols and Triterpenes of 67 Plants," *Journal of Ethnopharmacology* 10:181-94.

Horrobin, D.F. 1990b. "Evening Primrose Oil and Atopic Eczema," *Lancet* 7 July; 336(87-6):50.

Horrobin, D.F. 1990. "Evening Primrose Oil and Premenstrual Syndrome," *Med J Aust* 153(10):630-31.

Horrobin, D.F. 1984. "Placebo-controlled Trials of Evening Primrose Oil," *Swedish Journal of Biological Medicine*, 3:13-17.

Hostettmann, K., and P. J. Lea. 1987. *Biologically Active Natural Products*. Oxford: Clarendon Press, 283pp.

Howorth, Peter. *Foraging Along the Pacific Coast: From Mexico to Puget Sound*. Santa Barbara: Capra Press, 213pp.

Hriscu, A. et al. 1990. "A Pharmacodynamic Investigation of the Effect of Polyholozidic Substances Extracted from *Plantago* sp. on the Digestive Tract," *Rev Med Chir Soc Med Nat Iasi* 94(1):165-70.

Hryb, D. J. et al. 1995. "The Effect of Extracts of the Roots of the Stinging Nettle *(Urtica dioica)* on the Interaction of SHBG with its Receptor on Human Prostatic Membranes," *Planta Medica* 61(1):31-32.

Hudson, James B. 1990. *Antiviral Compounds from Plants*. Boca Raton, FL: CRC Press, 200pp.

Hutchens, Alma. 1991. *Indian Herbalogy of North America*. Boston: Shambala, 382pp.

IRCS Medical Science. 1986. Ab. 14, 212.

Ibragimov, D. I., and G. B. Kazanskaia. 1981. "Antimicrobial Action of Cranberry Bush, Common Yarrow, and *Achillea biebersteinii*," *Antibiotiki* 26(2):108-09.

Ikram, M. 1980. "Medicinal Plants as Hypocholesterolemic Agents," *J Pak Med Assn* 30(12):278-81.

Inouye, H. et al. 1974. "Purgative Activities of Iridoid Glycosides," *Planta Medica* 25:285-88.

Isaev, I., and M. Bojadzieva. 1960. "Obtaining Galenic and Neogalenic Preparations and Experiments for the Isolation of an Active Substance from *Leonurus cardiaca*," *Nauchnye Trudy Visshiia Meditsinski Institut* (Sofia) 37(5):145-52.

Ito, W. et al. 1986. "Suppression of 7,12 dimethylbenz(a)anthracene-induced Chromosome Aberrations in Rate Bone Marrow Cells by Vegetable Juices," *Mutation Research* 182:55-60.

Jackson, John R. 1876. "Notes on Some Medicinal Plants of the Compositae," *Canadian Pharm. Journal* 9:231-26.

Jacobs, Marion Lee,and Henry M. Burlage. 1958. *Index of Plants of North Carolina with Reputed Medicinal Uses*. Chapel Hill: Henry M. Burlage, 332pp.

Jaeger, Ellsworth. 1945. *Wildwood Wisdom*. New York: Macmillan Co., 491pp.

Jashemski, Wilhelmina Feenster. 1999. *A Pompeian Herbal: Ancient and Modern Medicinal Plants*. Austin: Univ. Texas Press, 107pp.

Jensen, Bernard. 1977. *Health Magic Through Chlorophyll from Living Plant Life*, ed. Leslie Goldman. Rev. ed. Provo, UT: BiWorld Pubs., 154pp.

REFERENCES

John, J. F. et al. 1994. "Synergestic Antiretroviral Activities of the Herb, *Prunella vulgaris*, with AZT, ddI and ddC," *Abst Gen Meet Am Soc Microbiol* 94:481, 484. Abstract S-27.

Johnson, Cathy. 1989. *The Wild Foods Cookbook*. New York: Stephen Greene Press/Pelham Bks, 236pp.

Johnson, Laurence. 1884. *A Manual of the Medical Botany of North America*. New York: Wood, 292pp.

Johnston, A. 1970. "Blackfoot Indian Utilixation of the Flora of the Northwestern Great Plains," *Economic Botany* 24:301-24.

Johnston, Alex. 1987. *Plants and the Blackfoot*. Lethbridge: Lethbridge Historical Soc., 56pp.

Jones, Pamela. 1991. *Just Weeds: History, Myths, and Uses*. New York: Prentice-Hall Press, 303pp.

Jones, T.K. and B. M. Lawson. 1998. "Profound Neonatal Congestive Heart Failure Caused by Maternal Consumption of Blue Cohosh Herbal Medication," *Journal of Pediatrics* 132:550-52.

Jones, Volney H. 1931. *The Ethnobotany of the Isleta Indians*. Univ. of New Mexico, M.A. thesis.

Jordan, Michael. 1976. *A Guide to Wild Plants: The Edible and Poisonous Species of the Norther Hemisphere*. London: Milligan Books, 240pp.

Jordan, Philip D. 1943. "The Secret Six: An Inquiry into the Basic Materia Medica of the Thomsonian System of Botanic Medicine," *Ohio State Archaeological and Historical Quarterly* 52:347-55.

Joselyn, John. 1860. "New-England's Rarities Discovered," *Archaeologica Americana: Transactions and Collections of the American Antiquarian Society*, Boston, 4:105-238.

Jurrison, S. 1971. "Determination of Active Substances of *Capsella bursa-pastoris*," *Tartu Riiliku Ulikooli Toim* 270:71-79.

Kakiuchi, N. et al. 1987. "Effect of Benzo[c]phenanthridine Alkaloids on Reverse Transcriptase and their Binding Properties to Nucleic Acids," *Planta Medica* 53(1):22-27.

Kalm, Peter. 1937. *Peter Kalm's Travels in North America*, English version of 1770, 2 vols., rev. & ed. Adolph B. Benson. New York: Wilson-Erickson Inc.

Kanjanapothi, D. et al. 1981. "Postcoital Antifertility Effect of *Mentha arvensis*," *Contraception* 24(5):559-67.

Kari, Priscilla R. 1977. *Dena' ina K'et'una, Tanaina Plantlore*. Anchorage, Alaska: Adult Literacy Laboratory.

Kari, Priscilla R. 1985. *Upper Tanana Ethnobotany*. Anchorage: Alaska Historical Commission.

Karlowski, J. A. 1991. "Bloodroot: *Sanguinaria canadensis* L.," *Canadian Pharmaceutical Journal* 124(5):260-67.

Karpilovskaia, E. D. et al. 1989. "Inhibiting Effect of the Polyphenolic Complex from *Plantago major* (Plantastine) on the Carcinogenic Effect of Endogenously Synthesized Nitrosodimethylamine," *Farmakol Toksikol* 52(4):64-67.

Katoch, R. et al. 2000. "Hepatoxicity of *Eupatorium adenophorum* to Rats," *Toxicon* 38(2):309-14.

Kaushal, V. et al. 2001. "Hepatoxicity in Rat Induced by Partially Purified Toxins from *Eupatorium adenophorum (Ageratina adenophora)," Toxicon* 39(5):615-19.

Kavasch, Barrie. 1979. *Native Harvests: Botanicals and Recipes of the American Indian.* Washington, CT: American Indian Archaeological Institute, 73pp.

Kay, Margarita A. 1996. *Healing with Plants in the American and Mexican West.* Albuquerque: Univ. Arizona Press, 315pp.

Kenton, Leslie, and Susannah Kenton. 1984. *Raw Energy.* New York: Warner Bks., 315pp.

Keville, Kathi. 1996. *Herbs for Health and Healing.* Emmaus: Rodale Press, 364pp.

Keville, Kathi. 1991. *The Illustrated Herb Encyclopedia: A Complete Culinary, Cosmetic, Medicinal, and Ornamental Guide to Herbs.* New York: Mallard Press, 224pp.

Khattak, S. G. et al. 1985. "Antipyretic Studies on Some Indigenous Pakistani Medicinal Plants," *Journal of Ethnopharmacology* 14:45-51.

Khuroo, M. A. et al. 1988. "Sterones, Iridoids and a Sesquiterpene from *Verbascum thapsus," Phytochemistry* 27(11):3541–44.

Kim, H.K. et al. 1999. "HIV Integrase Inhibitory Activity of *Agastache rugosa," Arch Pharm Res* 22(5):520-23.

Kim, et al. 1998. "*Taraxacum officinale* Restores Inhibition of Nitric Oxide Production by Cadmium in Mouse Peritoneal Macrophages," *Immunopharmacol Immunotoxicol* 20(2):283-97.

Kimball, Yeffe, and Jean Anderson. 1965. *The Art of American Indian Cooking.* Garden City, NY: Doubleday & Co., 215pp.

Kindscher, Kelly. 1987. *Edible Wild Plants of the Prairie: An Ethnobotanical Guide.* Lawrence, KS: Univ. Press of Kansas, 278pp.

Kindscher, Kelly. 1992. *Medicinal Wild Plants of the Prairie: An Ethnobotanical Guide.* Univ. Press of Kansas, 340pp.

Kirby, E.D. 1902. "Violets and Cancer," *British Medical Journal* 1:55.

Kirchhoff, H. W. 1983. "Brennesselsaft als Diuretikum," *Z. Phytother* 4:621-26.

Kirk, Donald. R. 1975. *Wild Edible Plants of Western North America.* Happy Camp, CA: Naturegraph Publishers, 307pp. + index.

Kishore, N. et al. 1993. "Fungitoxicity of Essential Oils Against Dermatophytes," *Mycoses* 36(5-6):211-15.

Klimas, John E., and James A. Cunningham. 1974. *Wildflowers of Eastern America.* New York: Alfred A. Knopf, 273pp.

Kluger, Marilyn. *The Wild Flavor.* New York: Henry Holt & Co., 285pp.

Knap, Alyson Hart. 1975. *Wild Harvest: An Outdoorsman's Guide to Edible Wild Plants in North America.* New York/Toronto: Arco Press/Pagurian Press, 192pp.

REFERENCES

Knott, Robert P., and Rob S. McCutcheon. 1961. "Phytochemical Investigation of a *Rubiaceae, Galium triflorum,*" *Journal of Pharmaceutical Sciences* 50(11):963–65.

Knutsen, Karl. 1975. *Wild Plants You Can Eat: A Guide to Identification and Preparation.* Garden City, NY: Dolphin Bks., 89pp. + Appendix and Bibliography.

Koichev, A. 1983. "Complex Evaluation of the Therapeutic Effect of A Preparation from *Plantago major* in Chronic Bronchitis," *Probl Vatr Med* 11:61-69.

Kong, Y. C. et al. 1976. "Isolation of the Uterotonic Principle from *Leonurus cardiaca,* the Chinese Motherwort," American Journal of Chinese Medicine 4(4):373-82.

Konoshima, T. et al. 1993. "Studies on Inhibitors of Skin Tumor Promotion, XII. Rotenoids from *Amorpha fructiosa,*" *Journal of Natural Products* 56:843-48.

Konowalchuk, Jack, and Joan I. Speirs. 1976. "Antiviral Activity of Fruit Extracts," *Journal of Food Science* 41:1013-17.

Kotobuki Seiyaku, K.K. 1979. "*Taraxacum* Extracts as Antitumor Agents," *Chem Abstr* 94:14530m.

Kourennoff, Paul M. 1971. *Russian Folk Medicine,* Ed. & Trans. George St. George. London: W. H. Allen, 287pp.

Kresanek, Jaroslav. 1989. *Healing Plants.* New York: Dorset Press, 223pp.

Krochmal, Arnold & Connie. 1973. *A Guide to the Medicinal Plants of the United States.* New York: Quadrangle, 259pp.

Kroeber, L. 1950. "Pharmacology of Inulin Drugs and their Therapeutic Use. II. *Cichorium intybus; Taraxacum officinale,*" *Pharmazie* 5:122-27.

Krzeski, T. et al. 1993. "Combined Extracts of *Urtica dioica* and *Pygeum africanum* in the Treatment of Benign Prostatic Hyperplasia: Double-blind Comparison of Two Doses," *Clin Ther* 15(6):1011-20.

Kuang, P. G. et al. 1988. "Motherwort and Cerebral Ischemia," *Journal of Traditional Chinese Medicine* 8(1):37-40.

Kuhnlein, Harriet V., and Nancy J. Turner. 1991. *Traditional Food Plants of Canadian Indigenous Peoples.* New York: Gordon & Breach, 633pp.

Kulze, Anne, and M. Greaves. 1988. "Contact Urticaria Caused by Stinging Nettles," *British Journal of Dermatology* 119(2):269–70.

Kupchan, S.M. 1972. "Recent Advances in the Chemistry of Tumor Inhibitors of Plant Origin," *in Plants in the Development of Modern Medicine,* ed. Tony Swan. Cambridge: Harvard Univ. Press, 261-78.

Kupchan, S.M. et al. 1973. "Structural Elucidation of Novel Tumor-Inhibitory Sesquiterpene Lactones from *Eupatorium cuneifolium,*" *Journal of Organic Chemistry* 38(12):2189-96.

Kupchan, S.M. et al. 1969. "Tumor Inhibitors. 33. Cytotoxic Flavones from *Eupatorium* Species," *Tetrahedron* 25(8):1603-15.

Kupchan, S.M. et al. 1965. "Tumor Inhibitors VIII. Eupatorin, New CytotoxicFlavone from *Eupatorium semiserratum,*"*Journal of Pharmaceutical Sciences* 54 (6):929.

Kuroda, K. et al. 1976. "Inhibitory Effect of *Capsella bursa-pastoris* Extract on Growth of Ehrlich Solid Tumor in Mice," *Cancer Research* 36(6):1900-03.

Kuroda, K. 1977. "Neoplasm Inhibitor from *Capsella bursa-pastoris* ," *Japan Kokai* 41:207.

Kuroda, K. and T. Kaku. 1969. "Pharmacological and Chemical Studies on the Alcohol Extract of *Capsella bursa-pastoris*," *Life Sciences* 8(3):151-55.

Kuroda, K. and K. Takagi. 1968. "Physiologically active Substances in *Capsella bursa-pastoris*," *Nature* 220(5168):707-08.

Kuroda, K., and K. Takagi. 1969a. "Studies on *Capsella Bursa Pastoris*. I. General Pharmacology of Ethanol Extract of the Herb," *Archives Internationales de Pharmacodynamie et de Therapie* 178(2):382–91.

Kuroda, K. and K. Takagi. 1969b. "Studies on *Capsella bursa pastoris*. II. Diuretic, Anti-inflammatory and Anti-ulcer Action of Ethanol Extracts of the Herb." *Arch Int Pharmacodyn Ther* 178(2):392-99.

Kuusi et al. 1985. "The Bitterness Principle of Dandelion: II. Chemical Investigations." *Lebensm-Wiss Technol* 18:347-49.

Kyi, K.K. et al. 1977. "Hypotensive Property of *Plantago major* Linn.," *Union Burma J. Life Sci.* 4(1):167-71.

Lacey, Laurie. 1993. *Micmac Medicines: Remedies and Recollections*. Nimbus Publishing, 125pp.

Lambev, I. et al. 1981. "Study of the Anti-inflammatory and Capillary Restorative Activity of a Dispersed Substance from *Plantago major* L.," *Probl Vutr Med* 9:162-60.

Lamela, M. et al. 1985. "Effects of *Lythrum salicaria* in Normoglycemic Rats," *Journal of Ethnopharmacology* 14(1):83-91

Lamela, M. et al. 1986. "Effects of *Lythrum salicaria* Extracts on Hyperglycemic Rats and Mice," *Journal of Ethnopharmacology* 15(2):153-60.

Landys, Robin. 1997. *Herbal Defense*. New York: Warner Books, 562pp.

Lang, G. et al. 2001. "Non-Toxic Pyrrolizidine Alkaloids from *Eupatorium semialatum*," *Biochem Syst Ecol 2001* 29(2):143-47.

Lapinina, L. O., and T. F. Sisoeva. 1964. "Investigation of Some Plants to Determine their Sugar-lowering Action," *Farmatsevticcheski Zhurnal* 19(4):52-58.

Law, Donald. 1976. *Herbs and Herbal Remedies*. London: Foyles Hdbks., 63pp.

Lee, K., and JY Lin. 1988. "Antimutagenic Activity of Extracts from Anticancer Drugs in Chinese Medicine," *Mutation Research* 204(2):229-34.

Lee, K-H. et al. 1988. "The Cytotoxic Principles of *Prunella vulgaris, Psychotrial terpens, and Hyptis capitata*: Ursolic Acid and Related Derivatives," *Planta Medica* 54:308-11.

Leighton, Anna. 1985. *Wild Plant Use by the Woods Cree (Nihithawak) of East-Central Saskatchewan*. Mercury Series, Canadian Ethnology Service Paper No. 101. Ottawa: National Museums of Canada, 128pp.

REFERENCES

Lenfield, J. et al. 1986. "Anti-inflammatory Activity of Extracts from *Conyza canadensis*," *Pharmazie* 41(4):268-69.

Lepore, Donald. 1988. *The Ultimate Healing System: Breakthrough in Nutrition, Kinesiology and Holistic Healing Techniques—Course Manual*. Provo, UT: Woodland Bks., 402pp.

Leung, Albert Y. 1980. *Encyclopedia of Common Natural Ingredients Used in Food, Drugs, and Cosmetics*. New York: John Wiley & Sons.

Leung, Albert Y., and Steven Foster. 1996. *Encyclopedia of Common Natural Ingredients Used in Food, Drugs, and Cosmetics*, 2nd ed. New York: John Wiley & Sons, 688pp.

Lesuisse, D. et al. 1996. "Determination of Oenothein B as the Active 5-alpha-reductase-inhibiting Principles of the Folk Medicine Epilobium parvifloruam," *Journal of Natural Products* 59(5):490-92.

Leuschner, J. 1995. "Anti-inflammatory, Spasmolytic and Diuretic Effects of a Commercially-available *Solidago gigantea* Herb. Extract," *Arzneimforsch* 45(2):165-68.

Lewis, Walter H., and Memory P. F. Elvin-Lewis. 1977. *Medical Botany: Plants Affecting Man's Health*. New York: John Wiley Co., 515pp.

Lexa, A. et al. 1989. "Choleretic and Hepatoprotective Properties of *Eupatorium cannabinum* in the Rat," *Planta Medica* 55(2):127-32.

Lichius, J. J. and C. Muth. 1997. "The Inhibiting Effects of *Urtica dioica* Root Extracts on Experimentally induced Prostatic Hyperplasia in the Mouse," *Planta Medica* 63(4):307-10.

Lichius, J. J. et al. 1999. "Antiproliferative Effect of a Polysaccharide Fraction of a 20% Methanolic Extract of Stinging Nettle Root upon Epithelial Cells of the Human Prostate (LNCap), *Pharmazie* 54(10):768-71.

Lin, Y-C. et al. 1972. "Search for Biologically Active Substances in Taiwan Medicinal Plants. I. Screening for Anti-tumor and Anti-microbial Substances," *Chin J Microbiol* 5:76-78.

Lin, C. C. et al. 1996. "Anti-inflammatory and Radical-scavenge Effects of *Arctium lappa*," *American Journal of Chinese Medicine* 24(2):127-37.

Link, Mike. 1976. *Grazing: The Minnesota Wild Eater's Food Book*. Burnsville, MN: Voyageur Press, , 106pp

Linnaeus, C. 1829. *Vegetable Materia Medica*, in Whitlaw, C., *Whitlaw's New Medical Discoveries, with a Defence of the Linnaean Doctrine and A Translation of His Vegetable Materia Medica*. London: Privately Published.

Lipton, R.A. 1958. "The Use of *Impatiens biflora* (Jewelweed) in the Treatment of *Rhus* Dermatitis," *Ann. Allergy* 16:526-27.

List, P.H. and L. Horhammer, eds. 1969-79. *Hager's Handbuch der Pharmazeutischen Praxis*, 6 vols. Berlin: Springer-Verlag.

Lithander, A. et al. 1992. "Intracellular Fluid of Waybread (*Plantago major*) as a Prophylactic for Mammary Cancer in Mice," *Tumour Biol* 13(3):138-41.

Lloyd, John Uri, and G. G. Lloyd. 1930-31. *Drugs and Medicines of North America*, 2 vols., Cincinnati: J.U. & C.G. Lloyd, 1884-87; reprint ed (Lloyd Library Bulletin #29-31, Reproduction Series 9), Cincinnati:, Lloyd Library.

Lloyd, John Uri. 1929. *Origin and History of All the Pharmacopeial Vegetable Drugs, Chemicals, and Preparations,* 8th & 9th Decennial Revisions, Cincinnati: Caxton Press, 1921; reprint, Cincinnati: Lloyd Libary.

Locock, R.A. 1990. "Boneset (Eupatorium),"*Canadian Pharmaceutical Journal* 123(5):229-33.

Locock, R.A. 1965. *A Phytochemical Investigation of Eupatorium* Species. PhD Thesis. Ohio State University.

Long, D. et al. 1997. "Treatment of Poison Ivy/Oak Allergic Contact Dermatitis with an Extract of Jewelweed," *American Journal of Contact Dermatitis* 8(3):150-53.

Long, X., and J. Tian. 1990. "Antifatigue and Anoxia-tolerating Effects of the Root *of Tragopogon porrifolius* L.[E.T.]" *Chung Kuo Chung Yao Tsa Chih* 15(12):741-43, 765.

Lucas, Jannette May. 1945. *Indian Harvest: Wild Food Plants of America*. Philadelphia: J. B. Lippincott Co., 118pp.

Lust, John. 1974. *The Herb Book*. New York: Bantam Books, 659pp.

(MH 6:4:12) 1994-95. anon. "Essential Oil Research: Mint Gel," *Medical Herbalism* 6(4):12.

(MH 7:4:10-13) 1995-96. "Clinical Correspondence," *Medical Herbalism* 7(4):10-13.

(MH 9:2:12) Stansbury, Jill. 1997. "Botanical Therapies for Fibrocystic Breast Disease," *Medical Herbalism* 9(2):1, 8-13.

(MH 9:4:20) Daucus, Sasha. 1997. "'A Simple' Cure for Brown Recluse Bites," *Medical Herbalism* 9(4):20.

Mabey, Richard et al, eds. 1988. *The New Age Herbalist*. New York: Collier Bks., 288pp.

McClintock, Walter. 1923. *Old Indian Trails*. Boston: Houghton Mifflin Co., 336pp.

McCutcheon, A. R. et al. 1995. "Antiviral Screening of British Columbian Medicinal Plants*," Journal of Ethnopharmacology* 49(2):101-10.

Macfarlane, W. V. 1963. "The Stinging Properties of *Laportea*," *Economic Botany* 17(4):303.

McGuffin, Michael, et al. 1997. *American Herbal Products Association's Botanical Safety Handbook*. Boca Raton: CRC Press, 256pp.

MacLeod, Heather, and Barbara MacDonald. 1988. *Edible Wild Plants of Nova Scotia*. Halifax, NS: Nimbus, 135pp.

McQuade-Crawford, Amanda. 1996. *The Herbal Menopause Book*. Freedom, CA: Crossing Press, 218pp.

McQuade-Crawford, Amanda. 1997. *Herbal Remedies for Women*. Rocklin, CA: Prima Publishing, 291pp.

McIntyre, Anne. 1995. *The Complete Woman's Herbal*. New York: Henry Holt Co., 287pp.

REFERENCES

Maksyutina, N. P. et al. 1978. "Chemical Composition and Hypocholesterolemia Action of some Drugs from *Plantago major* Leaves. Part 1: Polyphenolic Compounds," *Farm Zh* (Kiev) 4:56-61.

Maksyutina, N.P. 1971. "Hydroxycinnamic Acids from *Plantago major* and *P. lanceolata,*" *Khim Prir Soedin* 7(6):795, 824-25.

Malten, K. E. 1983. "Chicory Dermatitis from September to April," *Contact Dermatitis* 9:232

Manning, David, and Dan and Nancy Jason. 1972. *Some Useful Wild Plants—For Nourishment and Healing.* Rev. and enlarged ed. Vancouver: Talonbooks, 174pp.

Manthorpe, R. et al. 1985. "Primary Sjogren's Syndrome Treated with Efamol/Efavit: A Double-blind Crossover Investigation," *Rheumatol Int* 4(4):165-67.

March, Kathryn G., and Andrew L.March. 1986. *The Wild Plant Companion: A Fresh Understanding of Herbal Food and Medicine.* N.p.: Meridian Hill Pubns., 166pp.

Marker, Russell E. 1947. "Analysis of Trilliums, Smilacina and Clintonia*," Journal of the American Chem. Society* 69:2242.

Marker, Russell E. et al. 1942. "Sterols. CXLVI. Sapogenins .LX. Some New Sources of Diosgenin," *Journal of the American Chem. Society* 64:1283-85.

Markova, H. et al. 1997. "*Prunella vulgaris*: A Rediscovered Medicinal Plant," *Ceska Slov. Farm.* 46(2):58-63.

Martin, Laura C. 1984. *Wild Flower Folklore.* Charlotte, NC: East Woods Press, 256pp.

Martinez, G. et al. 1993. "Effect of the Alkaloid Derivative Ukrain in AIDS Patients with Kaposi' Sarcoma," *Int Congress AIDS* 9(1):401.

Mascolo, N. et al. 1987. "Biological Screening of Italian Medicinal Plants for Anti-inflammatory Activity," *Phytotherapy Research* 1:28-31.

Matev, M. et al. 1982. "Clinical Trial of *Plantago major* Preparation in the Treatment of Chronic Bronchitis [E.T.]," *Vutreshni Bolesti* 21(2):133-37.

Mattson, Morris. 1841. The American Vegetable Practice, 2 vols in 1, Boston: Daniel L. Hale, 592pp.

Maussert, Otto. 1940. *Herbs for Health*, 3rd ed. San Francisco: by the author, 205pp.

Mayes, Vernon O., and Barbara Bayles Lacey. 1990. *Nanise'—A Navajo Herbal: One Hundred Plants from the Navajo.* Tsaile, AZ: Navajo Community College Press, 153pp.

Mechling, W. H. 1959. "The Malecite Indians, with Notes on the Micmacs," *Anthropologica* 8:239-63.

Medsger, Oliver Perry. 1972. *Edible Wild Plants.* New York: Collier Bks., 323pp.

Medve, Richard J. and Mary Lee. 1990. *Edible Wild Plants of Pennsylvania and Neighboring States.* University Park: Pennsylvania State University Press, 242pp.

Meeker, James E. et al. 1993. *Plants Used by the Great Lakes Ojibwe.* Odanah: Great Lakes Indian Fish and Wildlife Commission, 440pp.

Messenger, Will. 1992. "Common Smartweed: *Polygonum hydropiper*," *Coltsfoot* 13:2-5.

Metzner, J. et al. 1982. "Antiphlogistic and Analgesic Effect of Leiocarposide, a Phenolic Biglucoside of *Solidago virgaurea*," *Pharmazie* 39:869

Meurer-Grimes, Barbara et al. 1996. "Antimicrobial Activity in Medicinal Plants of the *Scrophulariaceae* and *Acanthaceae*," *International Journal of Pharmacognosy.*

Meyer, Clarence. 1973. *American Folk Medicine.* New York: Thomas Y. Crowell, 296pp.

Meyer, Joseph F. 1960. *The Herbalist.* Rev. ed. N.p: Clarence Meyer, 304pp.

Michael, Pamela. 1980. *All Good Things Around Us: A Cookbook and Guide to Wild Plants and Herbs.* New York: Holt, Rinehart and Winston, 240pp.

Middleton, Elliott, Jr. 1988. "Plant Flavonoid Effects on Mammalian Cell Systems," in Craker, Lyle E. and James E. Simon, eds. *Herbs, Spices, and Medicinal Plants: Recent Advances in Botany, Horticulture, and Pharmacology.* 4 vols. Phoenix: Oryx Press, 3:103–44.

Midge, Manik D. and A. V. Rama Rao. 1975. "Synthesis of Eupatilin & Eupafolin, the Two Cytotoxic Principles from the *Eupatorium* Species,"*Indian Journal of Chemistry* 13:541-42

Miller, Lucinda G. & Wallace J. Murray, eds. 1998. *Herbal Medicinals: A Clinician's Guide* (New York: Pharmaceutical Products Press.

Miller, F. M., and L. M. Chow. 1954. "Alkaloids of *Achillea millefolium* L. 1. Isolation and Characterization of Achilleine," *J Amer. Chem Soc.,* 76:1353.

Mills, Simon, and Kerry Bone. 2000. *Principles and Practice of Phytotherapy: Modern Herbal Medicine.* New York: Churchill Livingstone, 643pp.

Millspaugh, Charles F. 1974. *American Medicinal Plants*, Philadelphia, John C. Yorston, 1892; reprint, New York: Dover Publications, 806pp.

Min, B. S. et al. 1999. "Inhibitory Constituents Against HIV-1 Protease from Agastache rugosa," *Arch Pharm Res* 22(1):75-77.

Mittman, Paul, et al. 1990. "Randomized, Double-blind Study of Freeze-dried *Urtica dioica* in the Treatment of Allergic Rhinitis," *Planta Medica* 56(1):44–47.

Miyazawa, M. and H. Kameocka. 1983. "Constituents of the Essential Oil from *Rumex crispus*," *Yakagaku* 32:45-47.

Mockle, J. Auguste. 1955.*Contributions a l'etude des plantes medicinales du Canada*, Paris ed., Jouve. A doctoral thesis in the Faculty of Pharmacy at the Univ. of Paris, dealing with flora indigenous to the Quebec, Ontario area of Canada.

Moerman, Daniel E. 1977. *American Medical Ethnobotany: A Reference Dictionary.* New York: Garland Publ., 1977, 527pp.

Moerman, Daniel. 1996. "An Analysis of the Food Plants and Drug Plants of Native North America," *Journal of Ethnopharmacology* 52:1-22.

REFERENCES

Moerman, Daniel. 1986. *Medicinal Plants of Native America*. 2 vols. *Technical Reports*, No. 19. Ann Arbor: Univ. of Michigan Museum of Anthropology, 912pp.

Mohney, Russ. 1975. *Why Wild Edibles? The Joys of Finding, Fixing, and Tasting West of the Rockies*. Seattle: Pacific Search, 317pp.

Mooney, J. 1890. "Cherokee Theory and Practice of Medicine," *Journal of American Folklore* 3:44-50.

Moore, Michael. 1994a. *Herbal Repertory in Clinical Practice*. 3rd ed. Albuquerque: Southwest School of Botanical Medicine, unpaginated.

Moore, Michael. 1994b. *Herbal Materia Medica*. 4th ed. Albuquerque: Southwest School of Botanical Medicine, unpaginated.

Moore, Michael. 1990. *Los Remedios: Traditional Herbal Remedies of the Southwest*. Santa Fe: Red Crane Books, 108pp.

Moore, Michael. 1989. *Medicinal Plants of the Desert and Canyon West*. Santa Fe: Museum of New Mexico Press, 216pp.

Moore, Michael. 1979. *Medicinal Plants of the Mountain West*. Santa Fe: Museum of New Mexico Press, 200pp.

Moore, Michael. 1993. *Medicinal Plants of the Pacific West*. Santa Fe: Red Crane Books, 359pp.

Morita, K. et al. 1984. "A Desmutagenic Factor Isolated from Burdock (*Arctium lappa* Linne.)," *Mutation Research* 129(1):25-31.

Morita, K. et al. 1985. "Chemical Nature of A Desmutagenic Factor from Burdock (*A. lappa* L.)," *Agric. Biol. Chem.* 49:925-32.

Morley, Thomas. 1969. *Spring Flora of Minnesota, Including Common Cultivated Plants*. Minneapolis: Univ. Minnesota Press, 283pp.

Mornis, Risa. 1998. *An Herbal Feast: Herbalists Share their Favorite Recipes*. New Canaan, CT: Keats Publishing, 262pp.

Morton, Julia. 1974. *Folk Remedies of the Low Country*. Miami: E. A. Seamann Publ., 176pp.

Moskalenko, S. A. 1986. "Preliminary Screening of Far-Eastern Ethnomedicinal Plants for Antibacterial Activity," *Journal of Ethnopharmacology* 15:231-59.

Moss, Ralph W. 1998. *Herbs Against Cancer: History and Controversy*. Brooklyn: Equinox Press, 300pp.

Mowrey, Daniel B. 1986. *The Scientific Validation of Herbal Medicine*. New Canaan: Keats Publ. Co., 316pp.

Moyle, John B., and Evelyn W. Moyle. 1977. *Northland Wild Flowers: A Guide for the Minnesota Region*. Minneapolis: Univ. of Minnesota Press, 236pp.

Muller, et al. 2000. "Antiproliferative Effect on Human Prostate Cancer Cells by a Stinging Nettle Root (*Urtica dioica*) Extract," *Planta Medica* 66(1):44-47.

Munson, J. Patrick. 1981. "Contributions to Osage and Lakota Ethnobotany," *Plains Anthropology* 26:229-40.

Murai, M. et al. 1995. "Phenylethanoids in the Herb of *Plantago lanceolata* and Inhibitory Effect on Arachidonic Acid-Induced Mouse Ear Edema," *Planta Medica* 61:479-80.

Murray, Michael, and Joseph Pizzorno. 1998. *Encyclopedia of Natural Medicine*. Rocklin, CA: Prima Publishing, 946pp.

Naegele, Thomas. 1980. *Edible and Medicinal Plants of the Great Lakes*. Calumet: Survival Seminars, 428pp.

Naiman, Ingrid. 1999. *Cancer Salves: A Botanical Approach to Treatment*. Seventh Ray Press, 272pp.

Navarro, E. et al. 1994. "Diuretic Action of an Aqueous Extract *of Lepidium latifolium L.*" *Journal of Ethnopharmacology* 41(1-2):65-69.

Neef, Cilli, et al. 1996. "Platelet Anti-aggregating Activity of *Taraxacum officinale* Weber*," Phytotherapy Research* 10:S138-40.

Nene, Y. L., and K. Kumar. 1966. "Antifungal Properties of *Erigeron linifolius* Willd. Extracts," *Naturwissenschaften* 53:14, 363.

Newall, Carol et al. 1996. *Herbal Medicines: A Guide for Health-Care Professionals*. London: Pharmaceutical Press, 296pp.

Niering, William A., and Nancy C. Olmstead. 1979. *The Audubon Society Field Guide to North American Wildflowers: Eastern Region*. New York: Alfred A. Knopf, 887pp.

Niethammer, Carolyn. 1974. *American Indian Food and Lore*. New York: Collier Bks., 191pp.

Nishikawa, H. 1949. "Screening Tests for Antibiotic Action of Plant Extracts," *Japanese Journal of Experimental Medicine* 20(3):337-49.

Nocton, James J. et al. 1994. "Detection of *Borrelia burgdorferi* by Polymerase Chain Reaction in Synovial Fluid from Patients with Lyme Arthritis," *New England Journal of Medicine* 330(4):229–34.

North, Pamela. 1967. *Poisonous Plants & Fungi, in Color*. London: Pharmaceutical Society of Great Britain, 161pp.

Nuttall, Thomas. 1905. *A Journal of Travels in the Arkansas Territory During the Year 1819*, Vol. 12 in Reuben Gold Thwaites, ed. Early Western Travels. Cleveland: Arthur H. Clarke Co., 1905.

Obertreis, B. et al. 1996. "Antiphlogistic Effects of *Urtica dioica* folia Extract in Comparison to Caffeic Malic Acid," *Arzneim Forsch Drug Res* 46(1):52-56.

O'Céirín, Cyril, and Kit O'Céirín. 1978. *Wild and Free: Cooking from Nature*. Dublin, Ireland: O'Brien Press, 160pp.

Ody, Penelope. 1993. *The Complete Medicinal Herbal*. New York: Dorling Kindersley, 192pp.

REFERENCES

Ohigashi, H. et al. 1986. "Search for Possible Antitumor Properties by Inhibition of 12-0-tetradecanoyl phorbol-13-acetate-induced Epstein-Barr Virus Activation, Ursolic Acid, and Oleanolic Acid from an Anti-inflammatory Chinese Medicinal Plant, *Glechoma hereracea* L.," *Cancer Lett* 30(2):143-51.

Okpanyi, S. N. et al. 1989. "Anti-inflammatory, Analgesic, and Antipyretic Effect of Various Plant Extracts and their Combinations in an Animal Model [ET]," *Arzneimforsch* 39(6):698-703.

Okuda, T. et al. 1989. "Ellagitannins as Active Constituents of Medicinal Plants," *Planta Medica* 55:117-22.

Okuyama, E. et al. 1983. "Isolation and Indentification of Ursolic Acid-Related Compounds as the Principles of *Glechoma hederaceae* Having an Anti-ulcerogenic Activity," *Shoyakugaku Zasshi* 37(1):18-19.

Okwuasaba, F. et al. 1987, "Comparison of the Skeletal Muscle Relaxant Properties of *Portulacaoleracea* Extracts with Dantholene Sodium and Methoxyverampil," *Journal of Ethnopharmacology* 20:85-106

Okwuasaba, F. et al. 1987.. "Investigations into the Mechanism of Action of Extracts of *Portulaca oleracea*," 21:91-97.

Oliver, F. et al. 1991. "Contact Urticaria Due to the Common Stinging Nettle *(Urtica dioica)*—Histological, Ultrastructural and Pharmacological Studies," *Clinical and Experimental Dermatology* 16(1):1–7.

Oliver-Bever, Bep. 1986. *Medicinal Plants in Tropical West Africa*. Cambridge: Cambridge Univ. Press, 366pp.

Oliver-Bever, B. and G. R. Zahnd. 1979-80. "Plants with Oral Hypoglycaemic Action*," Quarterly Journal of Drug Research* 17:139-96.

Olsen, Cynthia. 1998. *Essiac: A Native Herbal Cancer Remedy*, 2d ed.. Pagosa Springs, CO: Kali Press, 129pp.

Olsen, Larry Dean. 1973. *Outdoor Survival Skills*. 4th ed. Provo, UT: Brigham Young Univ. Press, 188pp.

Orbell, A. 1967. *A Compendium of Botanical Remedies: Physiomedical Practice*. Exeter, UK: Ilford,

Oswalt, W. H. 1957. "A Western Eskimo Ethnobotany," *Anthropological Papers of the University of Alaska* 6:17-36.

Ownbey, Gerald B. 1971. *Common Wild Flowers of Minnesota*. Minneapolis: Univ. Minnesota Press, 331pp.

Ownby, Gerald B., and Thomas Morley. 1991. *Vascular Plants of Minnesota: A Checklist and Atlas*. Minneapolis: Univ. Minnesota, 306pp.

(PDRHrbs) 1998. *PDR for Herbal Medicines*. Montvale NJ: Medical Economics Co., 1244pp.

Pahlow, Manfred. 1993. *Healing Plants*. Hauppage: Barron's Educational Series, 224pp.

Palmer, 1878. "Plants Used by the Indians of the United States," *American Naturalist* 12:593-606, 646-55

Palmer, G. 1975. *Shuswap Indian Ethnobotany*. Las Vegas: Univ. Nevada Dept. Anthropology, 70.

Panciera, R. et al. 1990. "Acute Oxalate Poisoning Attributable to Ingestion of Curly Dock (*Rumex crispus*) in Sheep," *J Am Vet Med Assoc* 196:1981-90.

Parke-Davis. 1890. *Organic Materia Medica*. Detroit: Parke-Davis.

Pashby, N.L. et al. 1981. "A Clinical Trial of Evening Primrose Oil in Mastalgia," *British Journal of Surgery* 68:801.

Pedersen, Mark. 1998. *Nutritional Herbology*, rev. ed. Wendell W. Whitman Co., 336pp.

Peng, Y. et al. 1983. "65 Cases of Urinary-tract Infection Treated by Total Acid of *Achillea alpina*," *Journal of Traditional Chinese Medicine* 3(3):217-18.

Peng, Y. et al. 1983. *Bull. Chine. Mat. Med.* 8:41.

Pengelly, Andrew. 1997. *The Constituents of Medicinal Plants*. 2nd ed. Merriwa, Australia: Sunflower Herbals, 109pp.

Perry, F. 1952. "Ethno-botany of the Indians in the Interior of British Columbia," *Museum and Art Notes* 2(2):36-43.

Peterson, W. 1851. "On *Eupatorium perfoliatum*," *Amer. J. Pharm.* 17:206-10.

Phelan, J.T. and J. Juardo. 1963. "Chemosurgical Management of Carcinoma of the External Ear, *Surg Gynecol Obstet* 117:224-46.

Phelan, J.T. and J. Juardo. 1963. "Chemosurgical Management of Carcinoma of the Nose," *Surgery* 53(3):310-14.

Phelan, J.T. et al. 1962. "The Use of Mohs' Chemosurgery Technique in the Management of Superficial Cancers," *Surg Gynecol Obstet* 114:25-30.

Phillips, Charles D.F. 1897. *Materia Medica and Therapeutics—Vegetable Kingdom*. New York: William Wood & Co.

Phillips, Jan. 1979. *Wild Edibles of Missouri*. Jefferson City, MO: Missouri Dept. of Conservation, 248pp.

Phillips, Roger and Nicky Fox. 1990. *The Random House Book of Herbs*. New York: Random Hse, 192pp.

Phillips, Roger. *Wild Food*. 1986. Boston: Little, Brown, and Co., 192pp.

(PhytochemDB) Phytochemical Database. Web resource at: http://probe.nal.usda.gov:83

Pierce, Andrea. 1999. *The American Pharmaceutical Association Practical Guide to Natural Medicines*. NewYork: William Morrow & Co., 728pp.

Pond, Barbara. 1974. *A Sampler of Wayside Herbs: Rediscovering Old Uses for Familiar Wild Plants*. Riverside, CT: Chatham Press, 126pp.

Popowska, E. et al. 1975. *Acta pol Pharm* 32:491.

Porcher, Francis Peyre. 1863. *Resources of the Southern Fields and Forests, Medical, Economical, and Agricultural, Being Also Medical Botany of the Confederate States*. Richmond: Charleston, Evans and Cogswell, 601pp.

REFERENCES

Potterton, David, ed. 1983. *Culpeper's Color Herbal*. New York: Sterling Publ. Co., 224pp.

Prakash, A. O. 1984. "Biological Evaluation of Some Medicinal Plant Extracts for Contraceptive Efficacy,"*Contracep Deliv Syst* 5:9.

Priest, A.W. and L.R. Priest. 1982. *Herbal Medication: A Clinical and Dispensatory Handbook*. Esssex, UK: L.N. Fowler & Co.

Racz-Kotilla, E. et al. 1974. "The Action of *Taraxacum officinale* Extracts on the Body Weight and Diuresis of Laboratory Animals," *Planta Medica* 26(3):212-17.

Radin, Paul. 1923. *The Winnebago Tribe*. Bureau of American Ethnology Annual Report #37. Washington DC: Smithsonian Institution, 511pp.

Rafinesque, Constantine S., 1828-30. *Medical Flora, or Manual of Medical Botany of the United States*, 2 vols. Phila: Samuel C. Atkinson, , 268 + 276pp.

Ramm, S. and C. Hansen. 1995. "Brennessel-Extrakt bei Rheumatischen Beschwerden." *Dtsch Apoth Ztg* 135(Suppl.):3-8.

Ramm, S. and C. Hansen. 1996. "Brennesselblatter-extrakt bei Arthrose und Rheumatoider Arthritis," *Therapiewoche* 28:3-6.

Randall, C. et al. 1999. "Nettle Sting of *Urtica dioica* for Joint Pain: An Exploratory Study of this Complementary Therapy," *Complement Ther Med* 7(3):126-31.

Randall, C. et al. 2000. "Randomized, Controlled Trial of Nettle Sting for Treatment of Base-of-Thumb Pain," *Journal of the Royal Society of Medicine* 93(6):305-09.

Rao, K.V. and F.M. Alvarez. 1981. "Antibiotic Principle of *Eupatorium capillifolium*," *Journal of Natural Products* 44(3):252-56.

Rauha, J. P. et al. 2000. "Antimicrobial Effects of Finnish Plant Extracts Containing Flavonoids and Other Phenolic Compounds," *Int J Food Microbiol* 56(1):3-12.

Raymond, Marcel. 1945. "Notes ethnobotaniques sur les Tete de Boule de Manouan," in *Etudes ethnobotaniques quebecoises, Contributions de l'Institut botanique l'Universite de Montreal* 55:113-35.

Reagan, Albert B. 1928. "Plants Used by the Bois Forte Chippewa (Ojibwa) Indians of Minnesota," *Wisconsin Archaeologist* 7(4):230-48.

Regnier, F. E. et al. 1967. "Studies on the Composition of the Essential Oils of Three *Nepeta* Species," *Phytochemistry* 6(9):1281–89.

Reid, Daniel P. 1992. *Chinese Herbal Medicine*. Boston: Shambala, 174pp.

Reynolds, J. E. F., ed. 1982. *Martindale: The Extra Pharmacopoeia*, 28th ed. London: The Pharmaceutical Press.

Rice, K. C., and R. S. Wilson. 1976. "(-)-3-Isothujone, A Small Nonnitrogenous Molecule with Antinociceptive Activity in Mice," *J Med Chem* 19:1054-57.

Richardson, Mary Ann, and Tina Sanders. 2000. "Flor-Essence® Herbal Tonic Use in North America: A Profile of General Consumers and Cancer Patients," *HerbalGram* 50:40-46.

Richardson, Joan. 1981. *Wild Edible Plants of New England: A Field Guide, Including Poisonous Plants Often Encountered*, ed. Jane Crosen. Yarmouth, ME: DeLorme Publ. Co., 217pp.

Riddle, John M. 1985. *Dioscorides on Pharmacy and Medicine*. Austin: Univ. of Texas Press, 298pp.

Riehemann, et al. 1999. "Plant Extracts from *Urtica dioica*, an Antirheumatic Remedy, Inhibits Proinflammatory Transcription Factor NK-kappa," *FEBS Letters* 442(1):89-94.

Rinzler, Carol Ann. 1991. *The Complete Book of Herbs, Spices and Condiments: From Garden to Kitchen to Medicine Chest*. New York: Henry Holt and Co., 199pp.

Ringbom, T. et al. 1998. "Ursolic Acid from *Plantago major*, a Selective Inhibitor of Cyclooxygenase-2 Catalyzed Prostaglandin Biosynthesis," *Journal of Natural Products (Lloydia)* 61(10):1212-15.

Robbins, Wilfred et al. 1916. *Ethnobotany of the Tewa*, Bureau of American Ethnology, Bulletin 55, Washington DC: Smithsonian Institution, 124pp.

Roder, E. 1992. "Pyrrolizidenhaltige Arnzneipflanzen," *Deutsche Apotheker Zeitung* 132(45):2427-35.

Rodriguez, P. et al. 1995. "Allergic Contact Dermatitis Due to Burdock (*Arctium lappa*)," *Contact Dermatitis* 33:134-35.

Rogers, Carol. 1995. *The Woman's Guide to Herbal Medicine*, London: Hamish Hamilton, 217pp.

Rogers, Dilwyn J. 1980. *Lakota Names and Traditional Uses of Native Plants by Sicangu (Brule) People in the Rosebud Area, South Dakota*. St. Francis, SD: Rosebud Educational Soc., 112pp.

Roman, R. R. et al. 1992. *Archives of Medical Research* 23(1):59-64.

Romero, John Bruno. 1954. *The Botanical Lore of the California Indians*. New York: Vantage Press, 161pp.

Rousseau, J. J. 1947. "Ethnobotanique Abenakise," *Archives de Folklore* 11:145-82.

Rousseau, Jacques. 1945. "Le folklore botanique de Caughnawaga: Etudes Ethnobotaniques quebecoises," *Contributions de l'Institut Botanique de l'Universite Montreal* 55:7-74.

Rousseau, Jacques. 1945b. "Le folklore botanique de l'Ile aux Coudres," *Contributions de l'Institut Botanique de l'Universite Montreal* 55:75-111.

Rücker, G. et al. 1991. "Antimalarial Activity of Some Natural Peroxides," *Planta Medica* 57(3):295–96.

Rumesson, J. J. et al. 1990. "Fructans of Jerusalem Artichokes: Intestinal Transport, Absorption, Fermentation, and Influence on Blood Glucose, Insulin, and C-peptide Responses in Healthy Subjects," *American Journal of Clinical Nutrition* 52:675-81.

Russell, Helen Ross. 1975. *Foraging for Dinner: Collecting and Cooking Wild Foods*. Nashville: Thomas Nelson Inc., Pubs., 255pp.

St. Claire, Debra. 1997. *The Herbal Medicine Cabinet: Preparing Natural Remedies at Home*. Berkeley: Celestial Arts, 150pp.

REFERENCES

Sakai, Saburo. 1963. "Pharmacological Actions of *Verbena officinalis L.*," *Gifu Daigaku Igakubu Kiyo* 11(1):6-17.

Salvucci, M.E. et al. 1987. "Purification and Species Distribution of Rubisco activase," *Plant Physiol* 84:930-36.

Santillo, Humbart. 1984. *Natural Healing with Herbs*, ed. S. Dharmananda Prescott, AZ: Hohm Press, 370pp.

Sanyal, A., and K. C. Varma. 1969. "In Vitro Antibacterial and Antifungal Activity of *Mentha Arvensis* var. *Piperascens* Oil Obtained from Different Sources," *Indian Journal of Microbiology* 9(1):23–24.

Sarbhoy, A. K. et al. 1978. "Efficacy of Some Essential Oils and their Constituents on A Few Ubiquitous Molds," *Zentralbl Bakteriol* [Naturwiss] 133(7-8):723-25.

Saunders, Charles F. 1976. *Edible and Useful Wild plants of the United States and Canada*. New York: Dover Pubns., 275pp.

Saxena, P. R. et al. 1965. "Identification of Pharmacologically Active Substances in the Indian Stinging Nettle, *Urtica Parviflora* (Roxb.)," *Canadian Journal of Physiology and Pharmacology* 43:869–76.

Sayer, L.U. 1905. *A Manual of Organic Materia Medica and Pharmacognosy*. Philadelphia: P. Blakiston's Son & Co., 692pp.

Schauenberg, Paul, and Ferdinand Paris. *Guide to Medicinal Plants*. New Canaan, CT: Keats Publ., Co., 349pp.

Schneider, H. J. et al. 1995. "Treatment of Benign Prostatic Hyperplasia. Results of a Treatment Study with the Phytogenic Combination of Sabal extract WS 1473 und Urtica extract WS 1031 in Urologic Specialty Practices [ET]," *Fortschr Med* 113(3):37-40.

Schofield, Janice J. 1989. *Discovering Wild Plants: Alaska, Western Canada, the Northwest*. Anchorage: Alaska Northwest Bks., 354pp.

Schonefeld, G. et al. 1984. *Klinische und Experimentelle Urologie* 4:179.

Schottner, M. et al. 1997. "Lignans from the Roots of *Urtica dioica* and their Metabolites Bind to Human Sex Hormone Binding Globulin (SHBG)," *Planta Medica* 63(6):529-32.

Schulte, K. E. et al. 1967. "Polyacetylenes in Burdock Root," *Arzneimittel-Forschung* 17(7):829-33.

Scott, Julian. 1990. *Natural Medicine for Children*. New York: Avon Books, 191pp.

Scudder, John Milton. 1870. *Specific Medication & Specific Medicines*. Cincinnati: Wilstach, Baldwin, & Co., 253pp.

Scully, Virginia. 1970. *A Treasury of American Indian Herbs: Their Lore and their Use for Food, Drugs, and Medicine*. New York: Crown Books, 306pp.

Selway, J. W. T. 1986. "Antiviral Activity of Flavones and Flavanes," in V. Cody, et al, eds. *Plant Flavonoids in Biology and Medicine: Biochemical, Pharmacological, and Structure-Activity Relationships*. New York: Alan R. Liss, pp. 521–36.

Serkedjieva, J. et al. 1990. "Antiviral Activity of the infusion (SHS-174) from Flowrers of *Sambucus nigra* L., aerial parts of *Hypericum perforatum* L., and roots of *Saponaria officinalis* L. against Influenza and Herpes simplex Viruses*,*" *Phytotherapy Research* 4:97-100.

Serkedjieva, J. 2000. "Combined Antiinfluenza Virus Activity of *Flos Verbasci* Infusion and Amantadine Derivatives," *Phytotherapy Research* 14(7):571-74.

Serkedzhieva, I., and N. Mandova. 1986. "Antiviral Action of a Polyphenol Complex Isolated from the Medicinal Plant *Geranium sanguineum* L. II. Its Inactivating Action on the InfluenzaVirus," *Acta Microbiol Bulg* 18:78-82.

Serkedzhieva, I., and N. Manlova et al. 1987. "Antiviral Action of a Polyphenol Complex Isolated from the Medicinal Plant *Geranium sanguineum* L.. V. Mechanism of the Anti-Influenza Effect *in vitro*," *Acta Microbiol Bulg* 21:66-71.

Serkedzhieva, I., and Manlova et al. 1988. "Antiviral Action of a Polyphenol Complex Isolated from the Medicinal Plant *Geranium sanguineum* L. VI. Reproduction of the Influenza Virus Pretreated with the Polyphenol Complex," *Acta Microbiol Bulg* 22:16-21.

Shanberg, Karen, and Stan Tekiela. 1991. *Plantworks: A Wild Plant Cookbook, Field Guide and Activity Book*. Cambridge, MN: Adventure Pubns., 159pp.

Sharma, O.P. et al. 1998. "A Review of the Toxicosis and Biological Properties of the Genus *Eupatorium*," *Nat. Toxins* 6(1):1-14.

Shear, M. J. et al. 1960., in J. Hartwell, "Plant Remedies for Cancer," *Cancer Chemother Rep* 7:19-24

Shemluck, Melvin. 1982. "Medicinal and Other Uses of the *Compositae* by Indians in the United States and Canada," *Journal of Ethnopharmacology* 5:303-58.

Sherry, C.J., and J. P. Mitchell. 1982. "The Behavioral Effects of the 'Lactone-free' Hot Water Extract of Catnip *(Nepeta cataria)* on theYoung Chick," *Quart. J. Crude Drug Res.* 21:89-92.

Sherry, C.J., and P. S. Hunter. 1979. "The Effect of an Ethanolic Extract of Catnip *(Nepeta cataria)* on the Behavior of theYoung Chick," *Experientia* 35:237-38.

Sherry, C. J., and J. A. Koontz. 1979. "Pharmacologic Studies of 'Catnip Tea': The Hot Water Extract of *Nepeta cataria*," *Quarterly Journal of Crude Drug Research* 17(2):68–71.

Shipochliev, T. 1981. "Extracts from a Group of Medicinal Plants Enhancing the Uterine Tonus," *Vet Med Nauki* (Sofia) 18(4):94-98

Shipochliev, T. 1981b. "Anti-inflammatory Action of a Group of Plant Extracts," *Vet Med Nauki* (Sofia) 18(6):87-91.

Shipochliev, T. and G. Fournadjiev. 1984. "Spectrum of the Anti-inflammatory Effect *of Arctostaphylos uva-ursi* and *Achillea millefolium L.*," *Prob Vutr Med* 12:99-107.

Shook, Dr. Edward E. 1978. *Advanced Treatise in Herbology*. Beaumont, CA: Trinity Center Press, 360pp.

Siena, S. et al. 1989. "Activity of Monoclonal Antibody-Saporin-6 Conjugate Against B-Lymphoma Cells," *Cancer Research* 49:3328-32.

Silver, A. A., and J. C. Krantz. 1931. "The Effect of the Ingestion of Burdock Root on Normal and Diabetic Individuals: A Preliminary Report," *Annals of Internal Medicine* 5:274-84.

Silverman, Maida. 1990. *A City Herbal: A Guide to the Lore, Legend & Usefulness of 34 Plants that Grow Wild in Cities, Suburbs, & Country Places.* Boston: David R. Godine, 1990, 181pps.

Silverstein, Alvin et al. 1990. *Lyme Disease: The Great Imitator. How to Prevent and Cure It.* Lebanon, NJ: Avstar Pub., 104pp.

Sim, Stephen K. 1967. *Medicinal Plant Glycosides: An Introduction for Pharmacy Students.* Toronto: Univ. of Toronto Press, 76pp.

Simon, J. A. and E. S. Hudes. 1999. "Relationship of Ascorbic Acid to Blood Lead Levels," *JAMA* (23-30 June) 281(24):2289-93.

Slagowska, A. et al. 1987. "Inhibition of Herpes Simplex Virus Replication by *Flox verbasci* Infusion," *Pol J. Pharmacol Pharm* 39(1):55-61.

Smith, Ed. 1999. *Therapeutic Herb Manual.* Williams, OR: by the author, 131pp.

Smith, Harlan. I. 1929. "Materia Medica of the Bella Coola and Neighboring Tribes of British Columbia," *Bulletin of the National Museum of Canada:Annual Report 1927.* 56:47-68.

Smith, Huron H. 1933. "Ethnobotany of the Forest Potawatomi Indians." *Bulletin of the Public Museum of the City of Milwaukee* 7(1):1-230

Smith, Huron H. 1923. "Ethnobotany of the Menomini Indians," *Bulletin of the Public Museum of the City of Milwaukee* 4(1):1-174.

Smith, Huron H. 1928. "Ethnobotany of the Meskwaki Indians," *Bulletin of the Public Museum of the City of Milwaukee* 4(2):175-326.

Smith, Huron H. 1932. "Ethnobotany of the Ojibwe Indians,*" Bulletin of the Public Museum of Milwaukee* 4(3):327-525.

Smith, L.W. and C.C.J. Culvenor. 1981. "Plant Sources of Hepatoxic Pyrrolizidine Alkaloids,*" Journal of Natural Products (Lloydia)* 44(2):129-52.

Smith, Peter. 1901. *The Indian Doctor's Dispensatory, Being Father Smith's Advice Respecting Diseases and their Cure.* Cincinnati: 1813; reprint, Bulletin of the Lloyd Library 2, Reproduction Series 2, Cincinnati: Lloyd Library.

Smith, Warren G. 1973. "Arctic Pharmacognosia," reprinted from *Arctic* 26(4) (Dec.):324-33.

Smythe, B. B. 1901. "Preliminary List of Medicinal and Economic Kansas Plants," *Kansas Academy of Science Transactions* 18:191-209.

Sokeland, J., and J. Albrecht. 1997. "Combination of *Sabal* and *Urtica* Extract vs. Finasteride in Benign Prostatic Hyperplasia (Aiken stages 1 to II). Comparison of Therapeutic Effectiveness in a One-year, Double-blind Study [ET]," *Urologe A* 36(4):327-33.

Sonneland, J. 1972. "The 'Inoperable' Breast Carcinoma: A Successful Result using Zinc chloride Fixative," *American Journal of Surgery* 124(3):391-93.

Soule, Deb. 1995. *The Roots of Healing: A Woman's Book of Herbs*. New York: Citadel Press, 306pp.

Spalding, L. 1819. *History of the Introduction and Use of Scutellaria lateriflora, Scullcap, as A Remedy for Preventing and Curing Hydrophobia Occasioned by the Bite of Rabid Animals*. New York: Treadwell.

Speck, F. G. 1937. "Catawba Medicines and Curative Practices," *Publications of the Philadelphia Anthropological Society* 1:179-98.

Speck, Frank G. 1941. "A List of Plant Curatives Obtained from the Houma Indians of Louisiana," *Primitive Man* 14(4):49-75.

Speck, Frank G. 1917. "Medicine Practices of the Northeastern Algonquians," *Proceedings of the 19th International Congress of Americanists—1915*, 303-21.

Speck, F. G. 1915. *The Nanticoke Community of Delaware*. Museum of the American Indian, Heye Foundation, 88pp.

Speck, F. G. 1942. "Rappahannock Herbals, Folk-lore and Science of Cures," *Proceedings of the Delaware County Institute of Science* 10:7-55.

Spellenberg, Richard. 1979. *The Audubon Society Field Guide to North American Wildflowers: Western Region*. New York: Alfred A. Knopf, 862pp.

Spencer, Edwin Rollin. 1957. *Weeds*. Rev. ed. New York: Charles Scribner's Sons, 333pp.

Spoerke, David G. 1990. *Herbal Medications*. Santa Barbara: Woodbridge Press, 192pp.

Stanford, E. E. 1927. "*Polygonum hydropiper* in Europe and North America," *Rhodora* 29:77-87.

Stary, Frantisek. 1983. *A Concise Guide in Color: Herbs*. London.

Steedman, Elsie V. 1928. "Ethnobotany of the Thompson Indians of British Columbia, Based on Field Notes by James A. Teit," *Bureau of American Ethnology Annual Report* 45:441-522.

Steinmetz, E. F. 1957. *Codex Vegetabilis*. Amsterdam: by the author, unpaginated.

Steinmetz, E. F. 1959. *Materia Medica Vegetabilis*. Amsterdam: Keizerarauht, 714pp.

Stephens, Homer A. 1980. *Poisonous Plants of the Central United States*. Lawrence: Regents of Kansas, 166pp.

Stevenson, Matilda Coxe. 1915. "Ethnobotany of the Zuñi Indians" *Thirtieth Annual Report of the Bureau of American Ethnology, 1908-09*. Washington DC: Government Printing Office, 31-102.

Stewart, Anne Marie, and L. Knonoff. 1975. *Eating from the Wild*. New York: Ballantine Bks., 387pp.

Stewart, Hilary. 1981. *Wild Teas, Coffees, and Cordials*. Vancouver: Douglas and McIntyre, 127pp.

Stickl, O. 1929. "Chemotherapeutische Versuche gegen das Ubertraghore Mausecarcinom," *Virchow's Arch Pathol Anat* 270:801-67.

Stratton, Robert. 1943. *Edible Wild Greens and Salads of Oklahoma*. Stillwater, OK: Oklahoma A.M. College, 29pp.

Strike, Sandra S., and Emily D. Roeder. 1994. *Ethnobotany of the California Indians—Vol. 2: Aboriginal Uses of California's Indigenous Plants.* USA: Koeltz Scientific Books, 250pp.

Stuart, Malcolm, ed. 1982. *VNR Color Dictionary of Herbs and Herbalism.* New York: Van Nostrand Reinhold Co., 160pp.

Sturtevant, E. Lewis. 1972. *Sturtevant's Edible Plants of the World,* ed. U. P. Hedrick. 1919; reprint, New York: Dover Pubns., 686pp.

Sultana, S. et al. 1995. "Crude Extracts of Hepatoprotective Plants, *Solanum nigra and Cichorium intybus,* Inhibit Free Radical-Mediated DNA Damage," *Journal of Ethnopharmacology* 45:189-92.

Sund, J.M. and M.J. Wright. 1959. "Control Weeds to Prevent Lowland Abortion in Cattle," *Down to Earth* 15(1):10-13

Sund, J.M. and M.J. Wright. 1959. "Weeds Containing Nitrates Cause Abortion in Cattle," *Agronomy Journal* 49(5):278-79.

Susnik, F. 1982. "Present State of Knowledge of the Medicinal Plant *Taraxacum officinale* Weber," *Med Razgledi* 21:323-28.

Suter, C. M., ed. 1951. *Medicinal Chemistry,* 3 vols. London: John Wiley & Sons, 1200pp.

Swank, George R. 1932. *The Ethnobotany of the Acoma and Laguna Indians.* MA Thesis, University of New Mexico, 86pp.

Swanston-Flatt, S. K. et al. 1989. "Glycaemic Effects of Traditional European Plant Treatments for Diabetes: Studies in Normal and Streptozotocin Diabetic Mice," *Diabetes Research* 10(2):69-73.

Swanton, John R. 1928. "Religious Beliefs and Medical Practices of the Creek Indians," *Forty-second Annual Report of the Bureau of American Ethnology, 1924-25.* Washington DC: Government Printing Office, 472-672.

Sweet, Muriel. 1976. *Common Edible & Useful Plants of the West.* Happy Camp, CA: Naturegraph Publishers, 64pp.

Szczawinski, Adam F., and George A. Hardy. 1971. *Guide to Common Edible Plants of British Columbia (British Columbia Provincial Museum Handbooks, 20).* Victoria, B.C.: British Columbia Provincial Museum, 90pp.

Szczawinski, Adam F., and Nancy J. Turner. 1978. *Edible Garden Weeds of Canada (Edible Wild Plants of Canada, 1).* Ottawa: National Museum of Natural Sciences, National Museums of Canada, 184pp.

Szczawinski, Adam F., and Nancy J. Turner. 1980. *Edible Wild Greens of Canada (Edible Wild Plants of Canada, 4).* Ottawa: National Museum of Natural Sciences, National Museums of Canada, 179pp.

Tabba, H. D. et al. 1989. "Isolation, Purification, and Partial Characterization of Prunellin, an Anti-HIV Compound from the Aqueous Extracts of *Prunella vulgaris,*" *Antiviral Res* 11(5-6):263-73.

Takasaki, M. et al. 1999. "Anti-carcinogenic Activity of *Taraxacum* plant," *Biol Pharm Bull* 22(6):602-10.

Tantaquidgeon, Gladys. 1942. *A Study of Delaware Indian Medicine Practice and Folk Beliefs.* Harrisburg: Pennsylvania Historical Commission, 91pp.

Tantaquidgeon, Gladys. 1972. *Folk Medicine of the Delaware and Related Algonkian Indians.* Harrisburg: Pennsylvania Historical and Museum Commission, 145pp.

Tantaquidgeon, Gladys. 1928. "Mohegan Medicinal Practices, Weather-lore, and Superstition," *Forty-third Annual Report of the Bureau of American Ethnology 1925-1926,* Washington DC: Government Printing Office, 264-79.

Tatum, Billy Joe. 1976. *Billy Joe Tatum's Wild Foods Cookbook and Field Guide,* ed. Helen Witty. NY: Workman Publ. Co., 268pp.

Taylor, Lyda Averhill. 1940. *Plants Used as Curatives by Certain Southeastern Tribes.* Cambridge: Botanical Museum of Harvard University, 399pp.

Tecce, R. et al. 1991."Saporin 6 Conjugated to Monoclonal Antibody Selectively Kills Human Melanoma Cells," *Melanoma Res* 1(2):115-23.

Teit, James A. 1928. "The Salishan Tribes of the Western Plateau*",* ed. Franz Boas, *Forty-fifth Annual Report of the Bureau of American Ethnolog,y,* 23–396.

Thayer, Sam. 2001. "The Milkweed Phenomenon: You Most Certainly Cannot Believe Everything You Read," *The Forager* 1(2):2-4.

Thomas, Richard. 1993. *The Essiac Report: The True Story of a Canadian Herbal Cancer Remedy and of the Thousands of Lives it Continues to Save.* Los Angeles: ATIN, 95pp. + many appendices.

Thurston, E. Laurence. 1969. "The Morphology and Toxicology of Plant Stinging Hairs," *Botanical Review* 35(4):393–412.

Thurston, E. Laurence. 1974. "Morphology, Fine Structure, and Ontology of the Stinging Emergence of *Urtica Dioica,*" *American Journal of Botany* 61(8):809–17.

Tierra, Michael, ed. 1992. *American Herbalism: Essays on Herbs & Herbalism by Members of the American Herbalists Guild.* Freedom: Crossing Press, 321pp.

Tierra, Michael. 1992. *Planetary Herbology: An Integration of Western Herbs into the Traditional Chinese and Ayurvedic Systems.* Twin Lakes, WI: Lotus Press, 485pp.

Tierra, Michael. 1990. *The Way of Herbs.* New York: Pocket Books, 378pp.

Tilgner, Sharol. 1999. *Herbal Medicine from the Heart of the Earth.* Creswell, OR: Wise Acres Press, 384pp.

Tintera, John. 1955. "The Hypoadrenocortical State and its Management," *New York State Journal of Medicine* 55:1869-76.

Tintera, John. 1966. "Stabilizing Homeostasis in the Recovered Alcoholic through Endocrine Therapy: Evaluation of the Hypoglycemic Factor," *Journal of the American Geriatrics Society* 14(2):126-50.

Tintera, John. 1959. "What You Should Know About Your Glands and Allergies," *Woman's Day* 2:28-29, 92.

Tokuda, et al. 1986. "Inhibitory Effects of Ursolic Acid and Oleanolic Acid on Skin Tumor Promotion by 12-O-tetradecanolyphorbol-13-acetate," *Cancer Lett.* 33(3):279-85.

Tomikel, John. 1976. *Edible Wild Plants of Eastern United States and Canada.* Elgin, PA: Allegheny Press, 100pp.

Tomikel, John. 1986. *Edible Wild Plants and Useful Herbs.* Elgin, PA: Allegheny Press, 176pp.

Torres, I. C., and J. C. Suarez. 1980. "A Preliminary Study of Hypoglycemic Activity of *Lythrum salicaria*," *Journal of Natural Products* 43(5):559-63.

Tozyo, T. et al. 1994. "Novel Antitumor Sesquiterpenoids in *Achillea millefolium*," *Chem Pharm Bull* (Tokyo) 42:1096-1100.

Train, Percy et al. 1988. *Medicinal Uses of Plants by Indian Tribes of Nevada*, Beltsville, MD: U.S.D.A, 1957; reprint ed., rev., Lawrence, MA: Quarterman Pubns., 139pp.

Trease, George Edward, and William Charles Evans. 1989. *Pharmacognosy*, 13th ed. London: Baillière Tindall.

Trease, George Edward, and William Charles Evans. 1983. *Pharmacognosy*, 12th ed. London: Baillière Tindall, 812pp.

Trease, George Edward, and William Charles Evans. 1973. *Pharmacognosy*, 11th ed. London: Baillière Tindall.

Trease, George Edward, and William Charles Evans. 1972. *Pharmacognosy*, 10th ed. London: Baillière Tindall.

Trease, George Edward, and William Charles Evans. 1957. *Pharmacognosy*, 7th ed. London: Baillière Tindall.

Tsuda, Y. and Leo Marion. 1963. "The Alkaloids of *Eupatorium maculatum*," *Canadian Journal of Chemistry* 41:1919-23.

Tull, Delena. 1987. *A Practical Guide to Edible & Useful Plants, Including Recipes, Harmful Plants, Natural Dyes & Textile Fibers.* Austin: Texas Monthly Press, 400pp.

Tunon, H. et al. 1995. "Evaluation of Anti-inflammatory Activity of Some Swedish Medicinal Plants : Inhibition of Prostaglandin Synthesis and PAF-induced Exocytosis," *Journal of Ethnopharmacology* 48(2):61-76

Turner, N.C. and M. A. M. Bell. 1971. "The Ethnobotany of the Coast Salish Indians of Vancouver Island," *Economic Botany* 25:60-104.

Turner, Nancy J., and Adam F. Szczawinski. 1979. *Edible Wild Fruits and Nuts of Canada (Edible Wild Plants of Canada, 3).* Ottawa: National Museum of Natural Sciences, National Museums of Canada, 212pp.

Turner, Nancy J. 1973. "The Ethnobotany of the Bella Coola Indians of British Columbia," *Syesis* 6:193-230.

Turner, Nancy J. et al. 1980. *Ethnobotany of the Okanagan-Colville Indians of British Columbia and Washington* . Victoria: British Columbia Provincial Museum, 110pp.

Turner, Nancy J. 1973b. "The Ethnobotany of the Southern Kwakiutl Indians of British Columbia," *Economic Botany* 27:257-310.

Turner, Nancy J. 1975. *Food Plants of British Columbia Indians, Part 1: Coastal Peoples (British Columbia Provincial Handbooks, 34)*. Victoria: Brit. Columbia Provincial Museum, 264pp.

Turner, Nancy J. 1981. "A Gift for the Taking: The Untapped Potential of Some Food Plants of North American Native Peoples," *Canadian Journal of Botany* 59(11):2331–57.

Turner, Nancy J. et al. 1990. *Thompson Ethnobotany: Knowledge & Usage of Plants by the Thompson Indians of British Columbia*. Victoria: Royal British Columbia Museum, 335pp.

Turner, Nancy J., and Adam F. Szczawinski. 1978. *Wild Coffee and Tea Substitutes of Canada (Edible Wild Plants of Canada, 2)*. Ottawa: National Museum of Natural Sciences, National Museums of Canada, 111pp.

Tyler, Varro E. 1993. *The Honest Herbal: A Sensible Guide to the Use of Herbs and Related Remedies*. 3^d ed. New York: Pharmaceutical Products Press, 375pp.

Tyler, Varro E. 1985. *Hoosier Remedies* West LaFayette: Purdue Univ. Press, 212 pp.

Tyler, Varro E. et al. 1988. *Pharmacognosy*. 9th ed. Philadelphia: Lea & Febiger, 519pp.

Underhill, J. E. 1974. *Wild Berries of the Pacific Northwest on the Bush . . . On the Table, in the Glass*. Saanichton, B.C.: Hancock Hse., 128pp.

United States Dept. of Agriculture. 1971. *Common Weeds of the United States*. New York: Dover Pubns., 463pp.

Vance, F. R. et al. 1984. *Wildflowers of the Northern Great Plains*. Minneapolis: Univ. of Minnesota Press, 336pp.

Vermathen & Glasl. 1993. "Effect of the Herb Extract of *Capsella bursa-pastoris* on Blood Coagulation," *Planta Medica* 59 (Suppl.):A670.

Verzan-Petri, G. and Banh-Nhu. 1977. *Scienta Pharm* 45, c. 24.

Vestal, Paul A., and Richard Evans Schultes. 1939. *The Economic Botany of the Kiowa Indians*. Cambridge: Botanical Museum of Harvard University, 110pp.

Vestal, Paul A. 1952. The Ethnobotany of the Ramah Navaho. *Papers of the Peabody Museum of American Archaeology and Ethnology* 40(4):

Viereck, Eleanor G. 1987. *Alaska's Wilderness Medicines: Healthful Plants of the Far North*. Edmonds, WA: Alaska Northwest Publ. Co., 108pp.

Vincent, E. & G. Segonzac. 1948. "Higher Plants Having Antibiotic Properties," *Toulouse Medicale* 49:669.

Vogel, Virgil J. 1970. *American Indian Medicine*. Norman: Univ. of Oklahoma Press, 585pp.

Vollmar, Angelika et al. 1986. "Immunologically Active Polysaccharides *Eupatorium cannabinum* and *Eupatorium perfoliatum*," *Phytochemistry* 25(2):377-81.

Vontobel et al. 1985. "Results of a Double-blind Study on the Effectiveness of ERU (extractum radicis Urticae) Capsules in Conservative Treatment of Benign Prostatic Hyperplasia [ET]," *Urologe A* 24(1):49-51.

REFERENCES

Vuilleumier, B.S. 1973. "The Genera of *Lactucacaceae* (*Compositae*) in the Southeastern United States," *J. Arnold Arboreum* 54:42-93.

Wagner, H. et al. 1989. "Biologically Active Compounds from the Aqueous Extract of *Urtica Dioica*," *Planta Medica* 55(5):452–54.

Wagner, H. et al. 1972. "Flavonol-3-Glycosides in Eight Eupatorium Species,"*Phytochemistry* 11:1504-05.

Wagner, H. and K. Jurcik. 1991. "Immunologic Studies of Plant Combination Preparations: *In-vitro* and *in-vivo* Studies on the Stimulation of Phagocytosis[E.T.]," *Arzneimittelforschung* 41(10):1072-76.

Wagner, H. et al. 1985. "Immunostimulierend wirkende Polyscaccharide(Heteroglykane) aus hoheren Pflanzen,"*Arzneim.-Forsch./Drug Res.* 34(1):659-61 and 35(2):1069-75.

Wagner, H. & Proksch, A. 1985. "Immunostimulatory Drugs of Fungi & Higher Plants," in H. Wagner et al, eds., *Economic and Medicinal Plant Research*," 6 vols. New York: Academic Press: 1:113-53.

Wagner, H. et al. 1985. "*In vitro* Phagozytose-Stimulierung durch isolierte Pflanzenstoffe gemessen im Phagozytose-Chemolumineszenz(CL)-Modell," *Planta Medica* 51:139-44

Wagner, H. 1988. "Non-Steroid, Cardioactive Plant Constituents," in idem et al., eds. *Economic and Medicinal Plant Research*. 5 vols. San Diego: Academic Press, 2:17–38.

Walker, Marilyn. 1984. *Harvesting the Northern Wild: A Guide to Traditional and Contemporary Uses of Edible Forest Plants of the Northwest Territories*. Yellowknife, NW Territories, Canada: by the author, 224pp

Wallis, W.D. 1922. "Medicines Used by the Micmac Indians," *American Anthropologist* 24:24-30.

Wallis, T.E. 1967. *Textbook of Pharmacognosy*, 5th ed. London: Churchill Ltd.

Walters, Richard. 1993. *Options:The Alternative Cancer Therapy Book*. Garden City Park: Avery Publ. Group, 396pp.

Wang, H. K. et al. 1998. "Recent Advances in the Discovery and Development of Flavonoids and their Analogous Antitumor and Anti-HIV Agents," *Adv. Exp. Med. Biol.* 439:191-225.

Ward-Harris, J. 1983. *More than Meets the Eye: the Life and Lore of Western Wildflowers*. Toronto: Oxford Univ. Press, 242pp.

Washburn, Homer C. and Walter H. Blome. 1927. *Pharmacognosy and Materia Medica*. London: John Wiley & Sons, 586pp.

Weatherbee, Ellen Elliott, and James Garnett Bruce. 1979. *Edible Wild Plants of the Great Lakes Region*. Ann Arbor, MI: by the authors, 69pp.

Weed, Susun S. 1986. *Wise Woman Herbal for the Childbearing Year*. Woodstock, NY: Ash Tree Publ., 171pp.

Weiner, Michael. 1980. *Earth Medicine—Earth Food: Plant Remedies, Drugs, and Natural Foods of the North American Indians*. Rev. and expanded ed. New York: Collier Bks., 230pp.

Weiner, Michael A. 1994. *Herbs that Heal: Prescription for Herbal Healing.* Mill Valley, CA: Quantum Bks, 436pp.

Weiss, Rudolf Fritz. 1960. *Lehrbuch der Phytotherapie.* Stuttgart, Germany: Hippokrates-Verlag, 408pp.

Weiss, Rudolf Fritz. 1988. *Herbal Medicine,* Trans. from the 6th German Ed. by A. R. Meuss Beaconsfield, UK: Beaconsfield Publishers, 362pp.

Wernert, Susan J., ed. 1982. *Reader's Digest North American Wildlife.* Pleasantville, NY: Reader's Digest Assn., 559pp. + index.

Westrich, Lolo. 1989. *California Herbal Remedies.* Houston: Gulf Publishing Co., 180pp.

Wheelright, Edith Grey. 1974. *Medicinal Plants and their History.* Boston: Houghton Mifflin Co., 1935; reprint, New York: Dover Publications, 288pp.

Wherry, Edgar T. 1948. *Wild Flower Guide: Northeastern and Midland United States.* New York: Doubleday and Co., 202pp.

White, Leslie A. 1945. "Notes on the Ethnobotany of the Keres," *Papers of the Michigan Academy of Science, Arts, and Letters* 30:557-68.

Wilkinson, R. E., and H. E. Jaques. 1979. *How to Know the Weeds,* 3[d] ed. Dubuque, IA: Wm. C. Brown Co., 235pp.

Willard, Terry. 1994. *Herbology 1.* Calgary, Alb., Canada: Wild Rose College of Natural Healing.

Willard, Terry. 1992a. *Edible and Medicinal Plants of the Rocky Mountains and Neighbouring Territories.* Calgary, Alb., Canada: Wild Rose College of Natural Healing, 278pp.

Willard, Terry. 1992b. *Textbook of Advanced Herbology.* Calgary, Alb., Canada: Wild Rose College of Natural Healing, 436pp.

Willard, Terry. 1993. *Textbook of Modern Herbology.* Rev. 2nd ed. Calgary, Alb., Canada: Wild Rose College of Natural Healing, 389pp.

Willard, Terry. 1991. *Wild Rose Scientific Herbal.* Calgary, Alb., Canada: Wild Rose College of Natural Healing, 416pp.

Willer, F. and H. Wagner. 1990. "Immunologically Active Polysaccharides and Lectins from the Aqueous Extracts of *Urtica dioica,*[E.T.]" *Planta Medica* 56(6):669.

Williams, Kim. 1977. *Eating Wild Plants.* Missoula, MT: Mountain Press Publ. Co., 180pp.

Williams, C. A. et al. 1996. "Flavonoids, Cinnamic Acids, and Coumarins from the Different Tissues and Medicinal Preparations of *Taraxacum officinale,*" *Journal of Phytochemistry* 42(1):121-27.

Willigmann, I. et al. 1991. "Occurrence of Omega-Fatty Acids in *Portulaca oleracea,*" *Planta Medica* 57(Suppl. 2):A91.

Winder, W. 1846. "On Indian Diseases and Remedies," *Boston Medical and Surgical Journal* 34(1):10-13.

Winston, David. 1992. "Nvwote: Cherokee Medicine and Ethnobotany," in Tierra, 1992, op. cit., 86-99.

REFERENCES

Winston, David. 1999. *Herbal Therapeutics: Specific Indications for Herbs & Herbal Formulas.* 6th ed. [New Jersey:] Herbal Therapeutics Research Library, 55pp.

Winston, David. 1998. "Little-known, but Important, Herbal Medicines for the Pharmacy," discourse given at the "Medicines from the Earth," herbal medicine convention held at Black Mountain, NC.

Winterhoff, H. et al. 1994. "Endocrine Effects of *Lycopus europaeus* L. Following Oral Application," *Arzneimittelforschung* 44:41-45.

Winterhoff, H. et al. 1988. "On the Antigonadotropic Activity of *Lithospermum* and *Lycopus* species and some of their Phenolic Constituents," *Planta Medica* 54(2):101-06.

Woerdenbag, H.J. 1986. "*Eupatorium cannabinum* L.: A Review Emphasizing the Sesquiterpene Lactones and their Biological Activity," *Pharm Weekly Sci* 17(8):245-51.

Woerdenbag, H..J. 1993. "Eupatorium Species*,"* in *Adverse Effects of Herbal Drugs 2*, ed. P.A.G.M. Smet et al. Berlin: Springer Verlag, 171-94.

Wood, Edelene. 1990. *A Taste of the Wild.* Elgin, PA: Allegheny Press, 160pp.

Wood, Horatio et al. 1937. *The Dispensatory of the United States of America.* Centennial (22nd)Ed. Phila: J. B. Lippincott Co., 1894pp.

Wood, Horatio et al. 1926. *The Dispensatory of the United States of America.* 21st Ed. Phila: J. B. Lippincott Co., 1792pp.

Wood, G. P., and E. H. Ruddock. 1925. *Vitalogy, or Encyclopedia of Health and Home, Adapted for Family Use.* Chicago:

Wood, Matthew. 1997. *The Book of Herbal Wisdom.* Berkeley: North Atlantic Books, 580pp.

Woodward, Lucia. 1985. *Poisonous Plants: A Color Field Guide.* New York: Hippocrene Bks., 192pp.

Wren, R. C. 1988. *Potter's New Cyclopaedia of Botanical Drugs and Preparations*, rewritten E. M. Williamson & F. J. Evans. London: Saffron Walden/New York: C.W. Daniel Co., 400pp.

Wren, R. C. 1972. *Potter's New Cyclopaedia of Medicinal Herbs and Preparations*, re-edited and enlgd. R. W. Wren. New York: Harper/Colophon Bks, 400pp.

Wright, S. and J. F. Burton. 1982. "Evening Primrose Oil Improves Atopic Eczema," *Lancet* 2:1120-22.

Wunderlin, R. P. and R. F. Lockey. 1988. "Questions and Answers," *JAMA* 260:3064-65.

Wyman, Leland C., and Stuart K. Harris. 1951. *The Ethnobotany of the Kayenta Navaho.* Univ. of New Mexico Pubns. in Biology, No. 5. Albuquerque: Univ. New Mexico Press, 66pp.

Xia, Yanxing, X. 1983. "The Inhibitory Effect of Motherwort Extract on Pulsating Myocardial Cells in vitro," *Journal of Traditional Chinese Medicine,"* 3(3):185-88.

Yamasaki, K. et al. 1996. "Anti-HIV-1 Activity of *Labiatae* Plants, Especially Aromatic Plants," *Int Conf. AIDS* 11:65, Abstract Mo.A.1062.

Yamasaki, K. et al. 1993. "Screening Test of Crude Drug Extract on Anti-HIV Activity[E.T.]," *Yakugaku Zasshi* 113(11):818-24.

Yamashita, K. et al. 1984. "Effects of Fructooligosaccharides on Blood Glucose and Serum Lipids in Diabetic Subjects," *Nutrition Research* 4:491-96.

Yanchi, Liu. 1995. *The Essential Book of Traditional Chinese Medicine*, Volume 2: *Clinical Practice*. New York: Columbia Univ. Press, 1988, 479pp.

Yanovsky, E. 1936. *Food Plants of the North American Indians*, USDA Misc. Publ. 237. Washington DC: Government Printing Office, 1936; reprint, New York: Gordon Press, 84pp.

Yao, X. J. et al. 1992. "Mechanism of Inhibition of HIV-1 Infection *in vitro* by Purified Extract of *Prunella vulgaris*," *Virology* 187:56-62.

Yoshikawa, M. et al. 1996. "Medicinal Foodstuffs. II. On the Bioactive Constituents of the Tuber of *Sagittaria trifolia* L. (Kuwai, Alismataceae): Absolute Stereostructures of Trifoliones A, B, C, and D, Sagittariosides A and B, and Arabinothalictoside," *Chem Pharm Bull.*(Tokyo) 44:492-99.

Young, Kay. 1993. *Wild Seasons: Gathering and Cooking Wild Plants of the Great Plains.* Lincoln: Univ. Nebraska Press, 318pp.

Youngken, Heber W. 1924. "The Drugs of the North American Indian," *American Journal of Pharmacy* 96:485-502.

Youngken, Heber W. 1925. "The Drugs of the North American Indian—Part II," *American Journal of Pharmacy* 97:158-85, 257-71.

Youngken, Heber W. 1948. A *Text Book of Pharmacognosy*. 6th ed. New York: Blakiston Publ. Co.,

Zafar, M.M. and A. Hameed. 1992. *Journal of Ethnopharmacology* 30(2):223-26.

Zennie, Thomas M. and C. Dwayne Ogzewella. 1977. "Ascorbic Acid and Vitamin A Content of Edible Wild Plants of Ohio and Kentucky," *Economic Botany* 31(1):76-79.

Zevin, Igor Vilevich. 1997. *A Russian Herbal: Traditional Remedies for Health and Healing.* Rochester. Healing Arts Press, 250pp.

Zgorniak-Nowosielska, I. et al. 1991. "Antiviral Activity of *Flos verbasci* Infusion Against Influenza and *Herpes simplex* Viruses," *Arch Immunol Ther Exp* 39(1-2):103-08.

Zhang, C. F. 1984. "Progress in the Experimental Research on the Function of Huo Xue Hoa Yu of *Leonurus*," *Zheng Xi Yi Jie He Za Zhi* 4(10):638-40.

Zhang, S. Q. et al. 1988. *Journal of Traditional Chinese Medicine* 8(4):254-56.

Zhao, X.L. et ai. 1987. "A Comparative Study on the Pyrrolizidine Alkaloid Content and Pattern of Hepatic Pyrrolic Metabolite Accumulation in Mice Given Extracts of *Eupatorium* Plant Species, *Crotalaria assamica*, and Indian Herbal Mixture," *American Journal of Chinese Medicine* 15(1-2):59-67.

Zheng, M. 1990. "Experimental Study of 472 Herbs with Antiviral Action Against the Herpes Simplex Virus[E.T.]," *Chung His I Chieh Ho Tsa Chih* 10:39-41

Zigmond, Maurice. 1981. *Kawaiisu Ethnobotany.* Salt Lake City: Univ. Utah Press, 293pp.

REFERENCES

Ziyin, Shen, and Chen Zelin. 1996. *The Basis of Traditional Chinese Medicine*. Boston: Shambala, 244pp.

Zou, Qi-Zum, et al. 1989. "Effect of Motherwort on Blood Hyperviscosity," *American Journal of Chinese Medicine* 17(1-2):65-70.

INDEX

*

This comprehensive (indeed, nearly exhaustive!) index has been specially arranged for easy visibility and access to subjects, i.e., it has been designed as a "cued" index, visually representing various terms by categories, as follows:

⇒ ***DISEASES/CONDITIONS*** are in caps, bold, & italics.

⇒ ***Physiological functions*** of plants are in lower-case letters, bold, & italics.

⇒ *Latin names* of plants are in italics

⇒ Plant chemicals (including nutrients) are underlined.

About the Author

Matthew Alfs, M.H. is an herbalist who has been trained in a variety of healing traditions, including modern Western phytotherapy, Amerindian plant medicine, the herbal therapeutics of America's Eclectic physicians of the late 1800s and early 1900s, and Oriental medicine. A graduate of Wild Rose College of Natural Healing, he is a nationally recognized educator in the field of western herbology, being a frequently invited lecturer and teacher at colleges, universities, hospitals, and other institutions. In addition, his lively outdoor workshops on wild plants, now in their sixth year, are eagerly awaited by many Minnesotans and Wisconsinites each summer.

A seasoned herbal practitioner, Mr. Alfs is in great demand for his healing expertise, operating an office in the Twin Cities suburban area. He also serves as a natural-products coordinator for a major Minnesota university. Memberships include the American Herbalists Guild, the American Botanical Council, and the American Herb Association.

The author may be contacted for **LECTURE, WORKSHOP**, and **TEACHING** offers
at the following address:

Matthew Alfs
c/o Old Theology Book House
P O Box 120342
New Brighton MN 55112

Or you may inquire to him more speedily at his personal e-mail:
MHMinn@aol.com

If you would like to inquire into the possibility of having an
HERBAL CONSULTATION
with
Matthew Alfs, M.H.
for a concern that you have, you may contact him at either of the above addresses.

FURTHER COPIES OF THIS BOOK may be obtained through *your local bookstore* or by ordering from the publisher directly. Regarding the latter, please enclose the cover price + $4 for shipping & handling per book. (Checks, money orders, MasterCard and Visa are accepted. If paying with credit card, however, you need to *include your phone number* and *specify your name as it appears on the card)*. Send your order to:

Old Theology Book House
P O Box 120342
New Brighton MN 55112